ENZYMOLOGY AND MOLECULAR BIOLOGY OF CARBONYL METABOLISM 5

ADVANCES IN EXPERIMENTAL MEDICINE AND BIOLOGY

ENZYMOLOGY AND MOLECULAR BIOLOGY OF CARBONYL METABOLISM 5

Edited by

Henry Weiner
Purdue University
West Lafayette, Indiana

Roger S. Holmes
Griffith University
Brisbane, Queensland, Australia

and

Bendicht Wermuth
Inselspital
Berne, Switzerland

SPRINGER SCIENCE+BUSINESS MEDIA, LLC

Library of Congress Cataloging-in-Publication Data

Enzymology and molecular biology of carbonyl metabolism 5 / edited by
  Henry Weiner, Roger S. Holmes, and Bendicht Wermuth.
      p. ; . -- (Advances in experimental medicine and biology ; v.
  372)
      Proceedings of the Seventh International Workshop on Enzymology
  and Molecular Biology of Carbonyl Metabolism, held July 3-7, 1994,
  in Palmerston North, New Zealand--T.p. verso.
      Includes bibliographical references and index.
      ISBN 978-0-306-44989-5
      1. Aldehyde dehydrogenase--Congresses. 2. Aldose reductase-
  -Congresses. 3. Alcohol dehydrogenase--Congresses. 4. Carbonyl
  compounds--Metabolism--Congresses.   I. Weiner, Henry. II. Holmes,
  Roger S. III. Wermuth, Bendicht. IV. International Workshop on
  Enzymology and Molecular Biology of Carbonyl Metabolism (7th : 1994
  : Palmerston North, N.Z.) V. Series.
      [DNLM: 1. Aldehyde Dehydrogenase--physiology--congresses.
  2. Aldehyde Reductase--physiology--congresses. 3. Alcohol
  Dehydrogenase--physiology--congresses.   W1AD559 v.372 1995 / QU
  140E61 1995]
  QP603.A35E574   1995
  599'.019258--dc20
  DNLM/DLC
  for Library of Congress   95-7575
                                                                 CIP

Proceedings of the Seventh International Workshop on Enzymology and Molecular Biology of Carbonyl
Metabolism, held July 3–7, 1994, in Palmerston North, New Zealand

ISBN 978-0-306-44989-5   ISBN 978-1-4615-1965-2 (eBook)
DOI 10.1007/978-1-4615-1965-2

© 1995 Springer Science+Business Media New York
Originally published by Plenum Press, New York in 1995

10 9 8 7 6 5 4 3 2 1

PREFACE

Since the inception of these meetings in 1982, they have always been a satellite of the International Society for Biomedical Research on Alcoholism meeting. At our 1992 meeting in Dublin we learned that the next ISBRA meeting would be held in Brisbane, Australia. As the scientific organizer of all our previous meetings, I was very concerned about holding a meeting in the Southern Hemisphere for fear that many of our potential participants would not travel that far. I am pleased to say that I was proven to be incorrect. Nearly 90 scientists from a dozen countries participated at our seventh conference. At this meeting, like at all our previous ones, much new information about the three enzyme systems was presented. Of equal importance was, like at all our previous meetings, the extreme openness of the participants to discuss ideas, future directions and unpublished data.

On behalf of all the participants I wish to express our sincere thanks to our Massey University colleagues for the excellent organization of this Palmerston North, New Zealand meeting. These included Kathryn Kitson, Michael Hardman, Paul Buckley, Trevor Kitson and Len Blackwell. At this meeting a few new innovations were introduced. Though posters are common at many meetings, bush walks and visits to nature preserves to see kiwi birds are not. Our hosts were able to secure support from the International Union of Biochemistry and Molecular Biology, which allowed us to pay partial travel expenses for a few younger scientists, and from the Alcohol Advisory Council of New Zealand and from Glaxo New Zealand, Ltd., which allowed us to hold down the cost for all participants. I acknowledge their support and thank them, as well as the National Institute on Alcohol Abuse and Alcoholism, for their support. I also thank my two co-editors, Roger Holmes and Benz Wermuth, for their hard work in helping me review all the manuscripts. Most important, I wish to thank all the participants for making this another exciting experience for all of us attending the meeting.

Our eighth meeting will be held in South Dakota, June 29-July 4, 1996. I invite scientists from around the world to contact me if they are interesting in attending this meeting.

Henry Weiner
West Lafayette, Indiana
September, 1994

CONTENTS

ALDO-KETO REDUCTASE ALDEHYDE DEHYDROGENASE

1

SITE DIRECTED MUTAGENESIS TO PROBE FOR ACTIVE SITE COMPONENTS OF LIVER MITOCHONDRIAL ALDEHYDE DEHYDROGENASE

Henry Weiner, Jaume Farrés, Ujjwal J. Rout, Xinping Wang and Chao-Feng Zheng

Biochemistry Department
Purdue University
W. Lafayette, IN 47907-1153

Mammalian aldehyde dehydrogenase was first purified to homogeneity in our laboratory (Feldman and Weiner, 1972a). The enzyme, isolated from horse liver, was found to be an esterase as well as a non-specific NAD-dependent dehydrogenase. At that time we did not know that the enzyme we were studying (pI = 5) was from mitochondria. We found a second minor form of the enzyme with a pI = 6; later it was shown that this was the cytosolic form of the enzyme. During our initial attempts to purify the mitochondrial isoenzyme we found a third form with a pI value of 4.8. It had kinetic properties identical with those of the pI 5.0 enzyme form, but it was less stable when heated at 52°C. As we became more experienced with isolating the enzyme we failed to identify the low pI form. Though never proven, we felt that this was a degraded form of the mitochondrial isoenzyme.

In the middle 1970s, when automatic protein sequencing equipment became available, a sample of the enzyme was sent to a colleague. He informed us that it was not possible to obtain a sequence. Many years later when we were studying the beef mitochondrial enzyme we found again that the N-terminal residue was blocked when the enzyme was isolated from fresh tissue (Guan *et al.*, 1988). If, though, it was isolated from frozen tissue a ragged N-terminal sequence was obtained. We found that same phenomenon to occur if the enzyme was isolated from horse or from rat liver. If fresh tissue was used, the N-terminal residue of the protein was blocked, but, if the tissue was frozen prior to isolating the protein, an amino acid sequence could be obtained. We later found that the mitochondrial enzyme was N-terminally acetylated (Weiner *et al.*, 1991). Apparently during isolation from frozen tissue a protease can remove some of the N-terminal residues.

Our initial kinetic studies with the horse liver enzyme led us to conclude that the enzyme functioned with ordered binding and that NAD was the lead substrate. The rate limiting step for the enzyme was thought to be the deacylation step (Feldman and Weiner, 1972b). Other investigators studying enzyme from different species concluded that NADH dissociation could be rate limiting. In Figure 1 is presented a proposed reaction pathway.

Enzymology and Molecular Biology of Carbonyl Metabolism 5
Edited by H. Weiner *et al.*, Plenum Press, New York, 1995

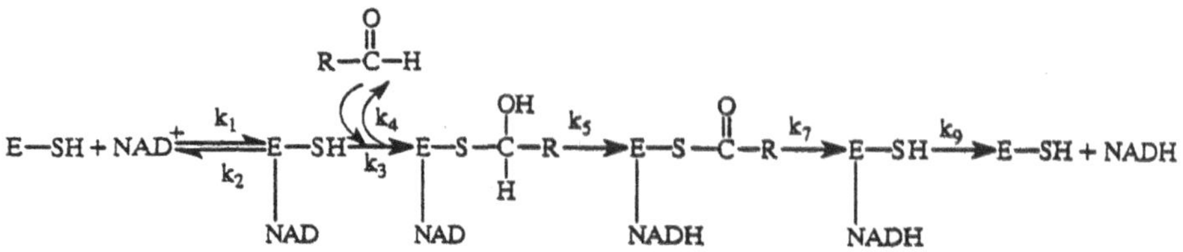

$$
\text{E—SH + NAD}^+ \underset{k_2}{\overset{k_1}{\rightleftharpoons}} \text{E—SH} \xrightarrow[k_3]{k_4} \text{E—S—}\overset{\overset{\text{OH}}{|}}{\underset{\underset{\text{H}}{|}}{\text{C}}}\text{—R} \xrightarrow{k_5} \text{E—S—}\overset{\overset{\text{O}}{\|}}{\text{C}}\text{—R} \xrightarrow{k_7} \text{E—SH} \xrightarrow{k_9} \text{E—SH + NADH}
$$

Figure 1. Model showing the reaction pathway for the aldehyde dehydrogenase catalyzed oxidation of an aldehyde.

No allosteric modulators of enzyme activity have been reported to exist. Coenzyme was found to activate the esterase reaction; it has been suggested that the presence of coenzyme increases the nucleophilicity of the active site (Takahashi and Weiner, 1981). We found that magnesium ions activated the dehydrogenase reaction. While studying the horse liver mitochondrial enzyme we observed that in the presence of Mg^{+2} ions the enzyme functioned with full-site reactivity, not half-of-the-site reactivity, as found in the absence of the ion. We went on to show that the presence of the ion caused the tetrameric enzyme to dissociate into a pair of dimers (Takahashi and Weiner, 1980). At least for the horse liver enzyme, the tetrameric enzyme seems to possess just two functioning subunits; when it was dissociated to the dimeric form, each subunit functioned. Thus far all mitochondrial class 2 ALDHs studied have been shown to be activated by the presence of Mg^{2+} ions. In contrast, cytosolic class 1 isozymes are inhibited in the presence of the ion (Weiner and Takahashi, 1983).

Unlike many other dehydrogenases, the detailed mechanism for the oxidation of substrate is not known. It was proposed that the enzyme functioned with covalent catalysis where a nucleophile attacked the carbonyl group of the aldehyde and this adduct was oxidized, as was illustrated in Figure 1. We suggested that the active site should possess a general base to help in the deacylation step, as illustrated in Figure 2.

Identification of the Components of the Active Site

Prior to the advent of molecular biology, active site components of enzymes were identified by chemical modification studies. Protein sequences were often compared but, until cDNA sequencing became common place, only limited sequence data existed. Thus, it was not possible to compare enzymes from many species to determine which residues were strictly conserved. Recently the amino acid sequences of all known ALDHs were compared and, surprisingly, few residues were conserved (Hempel *et al.*, 1993). Those that were, other than glycines, could be components of the active site.

Chemical modification studies were employed to determine the possible active site nucleophile of the enzyme. It is not possible to review those studies, but two different cysteine residues, 302 (Hempel and Pietruszko, 1981) and 49 (Tu and Weiner, 1988) and one

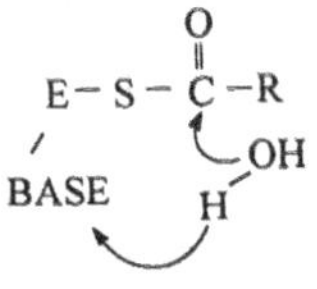

$$
\begin{array}{c}
\text{E}-\text{S}-\overset{\overset{\text{O}}{\|}}{\text{C}}-\text{R} \\
\text{BASE}\quad\text{H}
\end{array}
$$

Figure 2. General base facilitated deacylation of the acyl intermediate.

serine residue, 74 (Loomes *et al.*, 1990), were suggested to be possible candidates for the nucleophilic residue. Chemical modification also identified glutamate 268 (Abriola *et al.*, 1990) as being an essential component of the enzyme.

We undertook a series of studies employing site directed mutagenesis to try to determine the components of the active site. A summary of these finding will be presented in this report. Details of the experimental protocols were presented in our published papers.

The basic strategy for all our studies was to remove the potential nucleophilic side chain -SH or -OH by changing the residue to an alanine. In addition, the residue was converted into the other amino acids. The constructs were expressed in *E. coli* and the recombinantly expressed enzymes purified to homogeneity (Jeng and Weiner, 1991). The same general approach was used with the other mutations to be described; the substitutions depended upon what residues were being investigated.

NUCLEOPHILE

The proposed mechanism for the oxidation of aldehydes by ALDH requires that a nucleophilic residue attacks the carbonyl group and the adduct is then oxidized. This covalently bound, oxidized product would have to hydrolyzed, as was shown in Figure 1. Any one of three adducts could form. If it were with lysine, the oxidized bound product would be an amide. If it were with serine or cysteine, the covalent product would be an ester. No one has suggested that a lysine was involved in the initial step, but good evidence has been presented supporting the role of cysteine and possibly of serine.

Serine

Serine 74 of the rat liver mitochondrial enzyme was mutated to an alanine (Rout and Weiner, 1994). The mutant enzyme was found to have a Vmax of just 10% of the native enzyme. Totally unexpected was the observation that the Km for NAD increased 100 fold, from 20 to 2100 µM. The Km for aldehyde did not change. The fact that the enzyme was active showed that serine 74 was not the essential nucleophile in the active site. This residue is not highly conserved among non-mammalian forms of the enzyme supporting the conclusion that it might not be the active site residue.

The corresponding cysteine and threonine mutant enzymes were constructed. Both of these recombinantly expressed enzymes behaved like the alanine mutant. There seems to be a requirement for the presence of a serine at position 74 which can not be replaced by a similar amino acid residue. Its absence rendered the enzyme less catalytically active and caused NAD to bind poorly. How the serine -OH side chain is interfering with NAD binding is not known.

Cysteine

The two candidates for the active site nucleophile were C49 and C302. At a previous workshop we presented data to show that converting cysteine 49 to alanine did not affect the activity of the enzyme (Weiner *et al.*, 1991). In contrast, we reported that C302A was void of catalytic activity. Finding an inactive enzyme after performing a mutation does not prove that the residue was involved in the catalytic step. Inactivity could come from loss of coenzyme binding, subunit dissociation or a conformational change, to mention just a few possibilities. We since investigated in detail the reason why the mutation caused an inacti-vation of the enzyme. Though it is impossible to unequivocally prove a point, we were able to dismiss all possibilities except that the residue was involved in the catalytic process.

Table 1. Selective kinetic constants for recombinantly expressed rat liver
mitochondrial aldehyde dehydrogenase and selective mutations at cysteine 302.

Enzyme	kcat(M^{-1})[a]	Km(μM)[b]	Kd(μM)[c]
302C	110	0.15	2
302A	0	-	18
302S	0.8	800	nd[d]

[a]Vmax assay
[b]Km for propionaldehyde
[c]Kd for NADH
[d]not determined

The C302A mutant was still a tetramer and bound NADH with essentially the same K_d as did native enzyme. We were able to recover catalytic activity if a serine residue replaced the cysteine (Farrés *et al.*, 1995). Serine is known to function as a nucleophile in many enzyme, such as proteolytic enzymes, forming an acyl ester intermediate with substrate. Thus it was not unexpected that a C302S mutant would be active.

Though serine can act as a nucleophile in many enzymes, in general an hydroxyl group is less nucleophilic than is an -SH group. Though Km is not a true measure of binding, the data in Table 1 show that it was more difficult for the aldehyde to interact with the serine-mutant than with the native enzyme. This is consistent with the notion that it was more difficult to perform the initial nucleophilic attack on the substrate when a poor nucleophilic residue was at the active site. In fact, a detailed kinetic study, including pre-steady state analysis, showed that with the serine mutant the rate limiting step became the initial attack of the nucleophile at the carbonyl carbon (k_3 in Figure 1) and not deacylation (k_7). The fact that 302A mutant was catalytically inactive and the serine mutant had a very low specific activity, led us to suggest that indeed cysteine 302 is the active site nucleophilic residue of aldehyde dehydrogenase.

GENERAL BASE

Though not known to actually exist, in all likelihood it is necessary to invoke the action of a general base in the deacylation of the covalently bound, oxidized intermediate. In addition, it is very possible that the initial activation of the active site cysteine requires the involvement of a general base. Histidine often serves such a role. No direct evidence existed to suggest that a histidine in ALDH functioned in that capacity. We initially tried to chemically modify the residues with diethylpyrocarbonate but could not achieve complete inactivation. More recently, it was suggested that glutamate 268 could function as the

Table 2. Effect of histidine to alanine mutations on the activity of recombinantly expressed rat liver mitochondrial aldehyde dehydrogenases.

Enzyme	Vmax (%)
Native	100
29A	217
156A	101
235A	49
291A	88

[a]From Zheng and Weiner, 1993.

necessary general base (Abriola *et al.*, 1990). The importance of these residues was investigated.

Histidines

Though there are histidine residues conserved among all mammalian ALDHs, none are conserved among all the known enzymes. This is in contrast to, say, cysteine 302 or glutamate 268. H235 is the most conserved of the residues. Converting it to an alanine caused just a 50% loss in catalytic activity. Mutational analysis revealed that none of the conserved histidine residues were essential for activity (Zheng and Weiner, 1993). In Table 2 are summarized the pertinent findings from our published study.

Glutamate 268

The importance of the conserved glutamate 268 residue was noted from chemical modification studies prior to our knowledge that it was a residue conserved among all known aldehyde dehydrogenases. Conversion of the acidic residue to an amide (E268Q) caused a dramatic lose in catalytic activity of the human mitochondrial enzyme form to occur (Wang and Weiner, 1995). We found that the specific activity of the enzyme decreased to just 0.4% of that of the native enzyme. Even though the resulting enzyme had a low specific activity it was possible to determine the Km for aldehyde and NAD with the mutant enzyme. The values were essentially identical to what we found with the native enzyme. Thus glutamate 268 seems to perform an essential role in the catalytic process.

MECHANISM OF ACTIVATION OF THE NUCLEOPHILE BY THE GENERAL BASE

The best model for the activation of a nucleophilic amino acid by a general base is perhaps that of the proteolytic enzymes. It is known that serine 195 of chymotrypsin or trypsin is activated by the action of a histidine. Of equal importance is the fact that the histidine interacts with an aspartate to form a charge relay system. Many other enzymes have been shown to possess what appears to be a charge relay to facilitate the removal of a proton from an active site residue.

A model for the activation of the active site nucleophile is presented in Figure 2. An additional amino acid component would have to presented if the charge relay mechanism were functioning in ALDH, as illustrated in Figure 3. It is not possible to state with certainty what is the role of glutamate 268. It could be the general base (B_2) or it could be the third component (B_1) in the charged relay. One would have to know the spatial relationship between the active site cysteine and the glutamate before one could unequivocally know its precise role in the catalytic process.

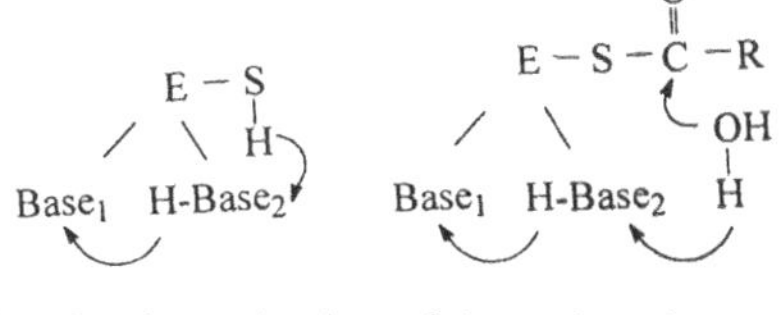

Figure 3. Charge relay mechanism for the activation of the active site cysteine or for the deacylation step. Here Base$_2$ is the general base and the proton is passed through the relay to Base$_1$.

SUMMARY

Mutational analysis allowed us to rule out an essential role for the histidine residues and for serine 74 in mammalian aldehyde dehydrogenase. The later though, was found to be important in coenzyme interaction. The function of the serine could not be replaced by threonine or by cysteine. The absolute requirement for cysteine 302 and for glutamate 268 was verified using mutational analysis. The fact that these two residues are completed conserved among all aldehyde dehydrogenases is consistent with their being essential in the catalytic process.

ACKNOWLEDGEMENT

This work was supported in part by Grant AA05812 from the National Institute on Alcohol Abuse and Alcoholism. H.W. was a recipient of a Senior Scientist Award (AA00038) from the same Institute.

REFERENCES

Abriola, D.P., MacKerell, A.D., and Pietruszko, R., 1990, Correlation of Loss of Activity of Human Aldehyde Dehydrogenase with Reaction of Bromoacetophenone with Glutamic Acid-268 and Cysteine-302 Residues. *Biochem. J.* 266:179-187.

Farrés, J., Wang, T.T.Y., Cunningham, S.J., and Weiner, H., 1995, Investigation of the Active Site Cysteine Residues of Rat Liver Mitochondrial Aldehyde Dehydrogenase by Site-Directed Mutagenesis, submitted.

Feldman, R.I. and Weiner, H., 1972a, Horse Liver Aldehyde Dehydrogenase. I. Purification and Characterization. *J. Biol. Chem.* 247:260-266.

Feldman, R.I. and Weiner, H., 1972b, Horse Liver Aldehyde Dehydrogenase. II. Kinetics and Mechanistic Implications of the Dehydrogenase and Esterase Activity. *J. Biol. Chem.* 247:267-272.

Guan, K.-L., Pak, Y.K., Tu, G.-C., Cao, Q.-N. and Weiner, H., 1988, Purification and Characterization of Beef and Pig Liver Aldehyde Dehydrogenase. *Alcoholism: Clin. Exp. Res.* 12:713-719.

Hempel, J., Nicholas, H. and Lindahl, R., 1993, Aldehyde dehydrogenase: Widespread Structural and Functional Diversity Within a Shared Framework. *Protein Sci.* 2:1890-1900.

Hempel, J.D. and Pietruszko, R., 1981, Selective Chemical Modifications of Human Liver Aldehyde Dehydrogenase E_1 and E_2 by Iodoactamide. *J. Biol. Chem.* 256: 10889-10896.

Jeng, J.-J. and Weiner, H., 1991, Purification and Characterization of Catalytically Active Precursor of Rat Liver Mitochondrial Aldehyde Dehydrogenase Expressed in *E. coli. Arch. Biochem. Biophys.* 289:214-222.

Loomes, K.M., Midwinter, G.G., Blackwell, L.F., and Buckley, P.D., 1990, Evidence for Reactivity of Serine-74 with Trans-4(N,N-dimethylamino) Cinnamaldehyde During Oxidation by the Cytoplasmic Aldehyde Dehydrogenase from Sheep Liver. *Biochemistry* 29:2070-2075.

Rout, U.K. and Weiner, H., 1994, Involvement of Serine 74 in the Enzyme-Coenzyme Interaction of Rat Liver Mitochondrial Aldehyde Dehydrogenase. *Biochemistry* 33:8955-8961.

Takahashi, K. and Weiner, H., 1980, Magnesium Stimulation of Catalytic Activity of Horse Liver Aldehyde Dehydrogenase. *J. Biol. Chem.* 255:8206-8209.

Takahashi, K. and Weiner, H., 1981, NAD-Activation of the Esterase Reaction of Horse Liver Aldehyde Dehydrogenase. *Biochemistry* 20:2720-2726.

Tu, G.-C. and Weiner, H., 1988, Identification of the Cysteine Residue in the Active Site of Horse Liver Mitochondrial. *J. Biol. Chem.* 263:1212-1217.

Weiner, H. and Takahashi, K., 1983, Effects of Magnesium and Calcium Ions on Mitochondrial and Cytosolic Liver Aldehyde Dehydrogenase. *Pharmac. Biochem. Behav.* 18:109-112.

Wang, X.-P. and Weiner, H., 1995, Investigation of the Role of Glutamate 268 in Human Liver Aldehyde Dehydrogenase, *Biochemistry*, in press.

Weiner, H., Farrés, J., Wang, T.T.Y., Cunningham, S.C., Zheng, C.-F. and Ghenbot, G., 1991, Probing the Active Site of aldehyde Dehydrogenase by Site Directed Mutagenesis. In *Enzymology and Molcular Biology of Carbonyl Metabolism III*. Eds. H. Weiner, B. Wermuth and D.W. Crabb. Plenum Press, New York.

Zheng, C.-F. and Weiner, H., 1993, Role of the Conserved Histidine Residues in Rat Liver Mitochondrial Aldehyde Dehydrogenase as Studied by Site Directed Mutagenesis. *Arch. Biochem. Biophys.* 305:460-466.

2

SUBSTRATE BINDING POCKET STRUCTURE OF HUMAN ALDEHYDE DEHYDROGENASES

A Substrate Specificity Approach

Shih-Jiun Yin, Ming-Fang Wang, Chih-Li Han, and Sung-Ling Wang

Department of Biochemistry
National Defense Medical Center
Taipei, Taiwan
Republic of China

INTRODUCTION

Human aldehyde dehydrogenases (AlDH), catalyzing irreversible oxidation of various aliphatic and aromatic aldehydes to the corresponding carboxylic acids, constitute a complex enzyme family (Pietruszko, 1983; Hempel and Jornvall, 1989; Hempel and Lindahl, 1989; Yoshida et al., 1991). There are at least five AlDHs that have been purified and characterized from human liver or stomach: AlDH1 and AlDH2 (E1 and E2; Greenfield and Pietruszko, 1977), AlDH3 (Wang et al., 1990), AlDH4 (E4, glutamic g-semialdehyde dehydrogenase; Forte-McRobbie and Pietruszko, 1986), and g-aminobutyraldehyde dehydrogenase (E3, Kurys et al., 1989). Kinetically, the AlDHs can be classified into two groups according to their Km values for acetaldehyde. The low Km (mM range) forms comprise AlDH1, AlDH2, and g-aminobutyraldehyde dehydrogenase (Greenfield and Pietruszko, 1977; Kurys et al., 1989). AlDH3 and AlDH4 belong to the high Km (mM range) forms (Yin et al., 1989; Forte-McRobbie and Pietruszko, 1986). Structurally, AlDH1 and AlDH2 show highest sequence homology (68%) among the above AlDHs (Hempel et al., 1984, 1985; Hsu et al., 1985). The degree of positional identity between AlDH1/2, AlDH3, and g-aminobutyraldehyde dehydrogenase is 30-40% (Hsu et al., 1985, 1992; Yin et al., 1991; Kurys et al., 1993), suggesting a distant relationship. AlDH4 has a larger subunit which shows no significant sequence homologies to all the known AlDHs (Forte-McRobbie and Pietruszko, 1986; Hempel et al., 1992), thus excluding it from the AlDH family. Recently, genes, i.e. *AlDH5* through *AlDH8*, encoding four additional human AlDHs have been cloned and characterized (Hsu et al., 1991, 1994). The kinetic properties of these new AlDH forms remain to be elucidated.

AlDH exhibits very broad substrate specificities (Pietruszko, 1983; Weiner, 1989). Many aldehyde metabolites in the metabolism of hydrocarbons, monoamines, diamines, polyamines, steroids and retinol (Ambroziak and Pietruszko, 1991; Monder et al., 1982; Dockham et al., 1992; Yoshida et al., 1992) as well as in lipid peroxidation (Algar and Holmes, 1989; Evces and Lindahl, 1989; Lindahl and Petersen, 1991) have been identified as active substrates. This suggests that AlDH may have a relatively wide substrate pocket. X-ray crystallographic structure of the enzyme is not yet available, though crystallization and preliminary analysis has appeared (Rose et al., 1990; Hurley et al., 1993). In this paper, we assessed and compared the topology of the substrate pockets in human AlDH1, AlDH2, and AlDH3 by using a substrate specificity approach. Our results indicate that the binding pockets are hydrophobic in nature. The data also suggest that the pocket in these three enzyme forms may be grossly in a barrel shape, which appear similar in length but varied in local topography such as the regions binding to the first 2 carbons of the substrate.

EXPERIMENTAL PROCEDURES

Autopsy livers were collected within 48 hours after death and stored at -70C. Livers with active AlDH2 or with prominent AlDH3 activity band on isoelectric focusing gels (Yin et al., 1988) were selected for purification. AlDH1 and AlDH2 were purified to apparent homogeneity according to Ikawa et al. (1983). The subunit molecular weights of AlDH1 and AlDH2 were determined by SDS-polyacrylamide gel electrophoresis to be 57,000 and 55,000 daltons, respectively. AlDH3 (AA form) (Yin et al., 1988) was purified to apparent homogeneity via CM-Sephadex, AMP-agarose, Sephacryl S-200 and DEAE-cellulose chromatographic steps. The enzyme purity was checked by SDS-polyacrylamide gel electrophoresis as a single protein staining band and the subunit molecular weight was 54,000 daltons. The protein concentrations of AlDH1/2 and AlDH3 were determined by Lowry (1951) and Bradford (1976) methods respectively, using bovine serum albumin as a standard.

Kinetic assays were performed with a Cary 2290 spectrophotometer in 50 mM HEPES, pH 7.5 and 25C, containing 0.5 mM NAD^+, 1 mM EDTA, 1.7% (v/v) acetonitrile (as an aldehyde carrier), and varied concentrations of substrate, unless otherwise noted. Milli-Q SP reagent water (Millipore Corporation, Bedford, Mass.) was further distilled and used for preparation of buffer solutions. The buffer was filtered through a Pyrex fritted funnel (fine porosity) to eliminate fine particles which interfered with kinetic assay at high sensitivity scale. All the aldehyde substrates were redistilled under normal or reduced pressure before use. Stoppered cuvettes were used for kinetic assay of acetaldehyde because of its low boiling point. The kinetic data were evaluated with HYPER or SUBIN statistical programs (Cleland, 1979). The coefficients of variation for Km, V, V/Km, and Ki were less than 15%. Hill coefficient and $S_{0.5}$ were analyzed according to Whitehead (1978).

RESULTS AND DISCUSSION

Substrate specificities of AlDH3 are shown in Table 1. Saturated straight chain aliphatic aldehydes (2-10 carbons) were all active substrates. The Km values decreased progressively without interruption as chain lengths increased. This suggests that AlDH3 has a hydrophobic substrate pocket, probably in a barrel shape which facilitates van der Waals contact interaction between the hydrocarbon chain of the substrate and the neutral side groups of the lining amino acid residues in the pocket. The maximal velocity (V) values remained relatively constant irrespective of chain length, suggesting that the rate-limiting step in catalysis may be NADH dissociation that is often seen in pyridine nucleotide linked dehydrogenases. The catalytic efficiency, V/Km value, for decaldehyde was about 40,000-fold greater than that for acetaldehyde, indicating long chain aldehydes may be of physiological relevance. This is compatible with the proposed physiological role of rat class 3 AlDHs in the oxidation of medium (6 to 9 carbon) chain length saturated and unsaturated aldehydes generated by the peroxidation of cellular lipids (Evces and Lindahl, 1989; Lindahl and Petersen, 1991).

Table 1. Substrate Specificity of AlDH3 with Straight Chain
Aliphatic Aldehydes

Substrate	K_m	V	V/K_m
	mM	µmol/min/mg	
Acetaldehyde	85	25	0.29
Propionaldehyde	15	28	1.9
Butyraldehyde	2.5	20	8.1
Valeraldehyde	0.41	19	46
Caproaldehyde	0.058	22	370
Heptaldehyde	0.013	25	1900
Octaldehyde	0.0042	27	6300
Nonaldehyde	0.0023	30	13000
Decaldehyde	0.0028	30	11000

Long chain aldehydes showed substrate inhibition. The inhibition constants (K_i) for octaldehyde, nonaldehyde, and decaldehyde for AlDH3 were determined to be 2.1, 0.16, and 0.30 mM, respectively. These K_i values may represent the dissociation constant of the substrate and the enzyme-NADH binary complex assuming formation of abortive ternary complex of enzyme-NADH-aldehyde at very high substrate concentrations. Since human AlDH3 exhibited similar low K_m values (Table 1) and also similar K_i values for nonaldehyde and decaldehyde, it would imply that from the bottom of active site to the outer surface of the substrate pocket may be 9-10 carbons in length. It has been reported that horse alcohol dehydrogenase (ADH) has a deep hydrophobic substrate pocket 5-10 A wide and 20 A long (i.e., about 10 carbon chain in length) from the protein surface to the active site zinc atom (Eklund and Branden, 1987). Our kinetic data suggest that the structure of substrate pocket in human AlDH3 might be grossly similar to that of horse ADH.

To study the substrate pocket topography in low K_m AlDHs, we applied a similar kinetic approach using various straight chain aliphatic aldehydes as topographic probes. AlDH1 exhibited progressively decreasing K_m values from acetaldehyde to valeraldehyde (Table 2). The K_m values for caproaldehyde to decaldehyde were too low (< 0.15 mM) to be precisely measured. However, the V values remained relatively constant for the 2-10 carbon substrates. This is consistent with the proposed mechanism that rate-limiting step of

Table 2. Substrate Specificity of AlDH1 with Straight Chain
Aliphatic Aldehydes

Substrate	K_m	V	V/K_m
	µM	µmol/min/mg	
Acetaldehyde	150	1.4	0.0097
Propionaldehyde	21[1]	1.1	0.13
Butyraldehyde	1.8	1.1	0.62
Valeraldehyde	0.55	1.6	3.0
Caproaldehyde	< 0.15	1.4	> 9.5
Heptaldehyde	< 0.15	1.7	> 11
Octaldehyde	< 0.15	1.7	> 12
Nonaldehyde	< 0.15	1.4	> 9.3
Decaldehyde	< 0.15	1.4	> 9.6

[1]$S_{0.5}$-value; Hill coefficient, 0.74.

Table 3. Substrate Specificity of AlDH2 with Straight Chain
Aliphatic Aldehydes

Substrate	Km	V	V/Km
	μM	μmol/min/mg	
Acetaldehyde	0.59	0.45	0.75
Propionaldehyde	0.34	0.38	1.1
Butyraldehyde	0.35	0.38	1.1
Valeraldehyde	0.27	0.31	1.2
Caproaldehyde	0.24	0.47	1.9
Heptaldehyde	< 0.15	0.34	> 2.3
Octaldehyde	< 0.15	0.31	> 2.1
Nonaldehyde	< 0.15	0.35	> 2.3
Decaldehyde	< 0.15	0.20	> 1.3

human cytosolic AlDH1 is NADH dissociation (Vallari and Pietruszko, 1981). All the substrates studied fitted to Michaelis-Menten kinetic equation except that propionaldehyde showed negative cooperativity with a Hill coefficient of 0.74 (Table 2). The length of the substrate pocket could not be estimated by Km changing due to the very low values. It, however, could be assessed by substrate inhibition studies. Ki values for heptaldehyde, octaldehyde, nonaldehyde , and decaldehyde were determined to be 650, 210, 110, and 45 mM, respectively. These dissociation constant data strongly suggest that the hydrophobic substrate pocket of AlDH1 may be at least 10 carbons in length, similar to that in AlDH3.

Kinetic constants for oxidation of straight chain aldehydes of mitochondrial AlDH2 are shown in Table 3. V values remained relative constant as those found for AlDH3 and AlDH1 (Tables 1 and 2). In contrast, the Km values were merely 2-fold difference between acetaldehyde and caproaldehyde. Actually, Km values for propionaldehyde up to caproaldehyde were similar. This would imply that there are no significant van der Waals interactions between the C4-C6 chain region of the substrate and the enzyme. This might be due to more wide of the pocket at this binding region. Km values for heptaldehyde (instead of caproaldehyde in the cytosolic AlDH1) to decaldehyde were very low (< 0.15 mM). This would suggest that the substrate pocket of AlDH2 is also generally hydrophobic in nature. Substrate inhibition could be detected for valeraldehyde and the longer aldehydes. The extents of the inhibition were, however, less than that of AlDH1. Consequently the substrate inhibition constants could not be measured (due to solubility limitation of the long chain aldehydes). This is consistent with the proposed mechanism that deacylation of the thioacyl intermediate is rate-limiting in mitochondral AlDH catalyzed reaction (Feldman and Weiner, 1972; Weiner et al., 1976).

To assess the topography of the first 2-carbon binding region in the substrate pocket of low Km AlDHs, branched chain aliphatic aldehydes were used as probes (Table 4). Km values for the substrate with a substituted *tert*-butyl group, i.e., 2,2-dimethylpropionaldehyde, for AlDH1 and AlDH2 were 3 and 4 fold greater than those with the 2,2-dimethylpropyl group (3,3-dimethylbutyraldehyde), respectively. This would imply that a steric hindrance may occur for binding of the bulky *tert*-butyl to the carbon-2 binding region in the substrate pocket of these two AlDHs. These structural implications are compatible with the observed 2-fold decrease in Km for the less bulky isoproyl group as compared to that for the *tert*-butyl group (Table 4). For AlDH1, Km value for the *tert*-butyl group substrate was 4-70 fold lower than those for the substrates with methyl or ethyl group. In contrast, Km value for the *tert*-butyl group substrate for AlDH2 was 1.5-3 fold higher than those with methyl or ethyl group. This increase in Km for bulky substrate, however, appeared not to occur for the 1-carbon longer substrate with a substituted 2,2-dimethylpropyl group (Table 4). Based on these kinetic data, we propose that in AlDH2 the pocket region binding

Table 4. Comparison of Km Values for Oxidation of Straight and
Branched Chain Aliphatic Aldehydes of AlDH1 and AlDH2

Substrate	Km	
R-CHO	AlDH1	AlDH2
	μM	μM
Methyl	150	0.59
Ethyl	8.2	0.34
Propyl	1.8	0.35
Butyl	0.55	0.27
Isopropyl	1.4	0.42
Isobutyl	1.2	0.44
tert-Butyl	2.3	0.91
2,2-Dimethylpropyl	0.7	0.22

to the first 2 carbons of the substrate appears narrower than that binding to carbon-3 and the following carbons. This is in agreement with the similar Km values for propionaldehyde to caproaldehyde for AlDH2 (Table 3). Since mitochondrial AlDH2 exhibited an extremely low Km value for acetaldehyde, the pocket region binding to the first 2 carbons of substrate in this enzyme might be narrower than that in cytosolic AlDH1.

X-ray crystal structure of horse ADH has indicated that at the bottom of the substrate binding pocket, nicotinamide ring of the coenzyme forms one side for binding ethanol via van der Waals contact interactions (Eklund and Branden, 1987). Side chains of serine 48 and phenylalanine 93 are also involved in the ethanol binding. By analogy to ADH, we propose that nicotinamide ring of NAD^+, neutral side chains of amino acid residues in the active site, and the mercaptomethyl group of the catalytic cysteine may contribute to binding acetaldehyde in AlDH1, AlDH2, and AlDH3. It should be pointed out that using comparisons of Km values of aldehydes to probe the topographic structure of substrate pocket may not be necessarily valid. Km value may be equal, larger or smaller as compared to the value of dissociation constant of the enzyme and the substrate depending on kinetic mechanisms (Fersht, 1985). Hence, substrate pocket structure of human AlDHs needs to be elucidated by future X-ray crystallographic experiments.

To compare the oxidation of acetaldehyde by AlDHs at a more physiological condition, kinetic constants were also measured in 0.1 M sodium phosphate buffer (no carrier acetonitrile added) (Table 5). At low acetaldehyde concentrations, catalytic efficiency, V/Km, of mitochondrial AlDH2 was found to be 120 and 9,200-fold greater than that of cytosolic AlDH1 and AlDH3, respectively. This indicates that AlDH2 can be the principal enzyme responsible for acetaldehyde oxidation *in vivo*, in agreement with the previous reports that mitochondria is the major site of the oxidation of this metabolite during ethanol metabolism (Parrilla et al., 1974; Weiner, 1987). Interestingly, about 50% of Orientals lack measurable mitochondrial AlDH2 activity (Harada et al., 1980). This is due to a single glutamate/lysine exchange at position 487 which causes a more than 250-fold increase in Km for NAD^+ and 10-fold decrease in V for the mutant enzyme (Farrés et al., 1994). This single amino acid mutation can cause acetaldehyde-induced sensitivity reaction during ethanol consumption (Mizoi et al., 1979; Enomoto et al., 1991), thereby affecting development of alcoholism and alcoholic cirrhosis (Thomasson et al., 1991; Chao et al., 1994). Since the high catalytic efficiency of AlDH2 is mainly attributed to its low Km for acetaldehyde (even more than 200-fold lower than that for NAD^+), it would be reasonably anticipated that

Table 5. Comparison of Kinetic Constants for Oxidation of
Acetaldehyde of AlDH1, AlDH2 and AlDH3[1]

	Km	V	V/Km
	µM	µmol/min/mg	
AlDH1	33	0.63	0.019
AlDH2	0.20	0.47	2.3
AlDH3	75000	19	0.00025

[1]Assayed in 0.1 M sodium phosphate, pH 7.5, at 25°C containing 0.5 mM
NAD$^+$,1 mM EDTA, and varied concentrations of acetaldehyde.

acetaldehyde should fit nicely into the substrate pocket through van der Waals force and other forces (if any) as suggested by the kinetic data shown in Tables 3-5.

CONCLUSIONS

Human aldehyde dehydrogenases can be categorized into two groups according to their Km values for acetaldehyde. Substrate pocket structures of the high-Km AlDH3 (75 mM) and the low-Km AlDH1 (33 mM) and AlDH2 (0.20 mM) have been assessed by a kinetic substrate specificity approach using various straight and branched chain aliphatic aldehydes as topographic probes.

Km values for straight chain aliphatic aldehydes (2-10 carbons) for AlDH3 decrease progressively as the chain lengths increase; the V values, however, remain relatively constant. The catalytic efficiency of AlDH3, V/Km, for decaldehyde is about 40,000-fold greater than that for acetaldehyde. AlDH3 exhibits substrate inhibition for octaldehyde or the longer aldehydes. The inhibition constants decrease as the chain lengths increase. These results suggest that the substrate pocket of AlDH3 is hydrophobic in nature, probably in a barrel shape with 10-carbon chain in length. Straight chain aliphatic aldehydes are also active substrates for AlDH1 and AlDH2 with longer one being a better substrate. Comparisons of kinetic constants for straight and branched chain aldehydes for AlDH2 suggest that the pocket region binding to the first two carbons of the substrate appears narrower, showing variations in the local topography of the substrate pocket.

Despite 30-70% sequence divergence between AlDH1, AlDH2, and AlDH3, our kinetic results suggest that they may exhibit grossly barrel-shaped substrate pocket, similar in length but varied in width, lined with neutral amino acid residues. Interestingly, X-ray crystallographic and computer graphic studies have indicated that human alcohol dehydrogenases have a deep hydrophobic substrate binding pocket. Similar structures of substrate pockets in ADH and AlDH imply that, in addition to ethanol/acetaldehyde pair, these two principal alcohol metabolizing enzymes may have other physiological, pharmacological or toxicological substrates similar in structure.

ACKNOWLEDGMENTS

This work was supported by grants from the National Science Council (NSC 81-0412-B016-31 and 82-0412-B016-33), the Department of Health, and Academia Sinica, Republic of China.

REFERENCES

Algar, E. M., and Holmes, R. S., 1989, Purification and properties of mouse stomach aldehyde dehydrogenase. Evidence for a role in the oxidation of peroxidic and aromatic aldehydes, *Biochim. Biophys. Acta*, 995:168.

Ambroziak, W., and Pietruszko, R., 1991, Human aldehyde dehydrogenase. Activity with aldehyde metabolites of monoamines, diamines, and polyamines, *J. Biol. Chem.*, 266:13011.

Bradford, M. M., 1976, A rapid and sensitive method for the quantitation of microgram quantities of protein utilizing the principle of protein-dye binding, *Anal. Biochem.*, 72:248.

Chao, Y.-C., Liou, S.-R., Chung, Y.-Y., Tang, H.-S., Hsu, C.-T., Li, T.-K., and Yin, S.-J., 1994, Polymorphism of alcohol and aldehyde dehydrogenase genes and alcoholic cirrhosis in Chinese patients, *Hepatology*, 19:360.

Cleland, W. W., 1979, Statistical analysis of enzyme kinetic data, *Methods Enzymol.*, 63A:103.

Dockham, P. A., Lee, M.-O., and Sladek, N. E., 1992, Identification of human liver aldehyde dehydrogenases that catalyze the oxidation of aldophosphamide and retinaldehyde, *Biochem. Pharmacol.*, 43:2453.

Eklund, H., and Branden, C.-I., 1987, Alcohol dehydrogenase, in "Biological Macromolecules and Assemblies," Vol 2, F. A. Jurnak and A. McPherson, eds., John Wiley & Sons, New York, p. 74.

Enomoto, N., Takase, S., Yasuhara, M., and Takada, A., 1991, Acetaldehyde metabolism in different aldehyde dehydrogenase-2 genotypes, *Alcohol. Clin. Exp. Res.*, 15:141.

Evces, S., and Lindahl, R., 1989, Characterization of rat cornea aldehyde dehydrogenase, *Arch. Biochem. Biophys.*, 274:518.

Farrés, J., Wang, X., Takahashi, K., Cunningham, S. J., Wang, T. T., and Weiner, H., 1994, Effects of changing glutamate 487 to lysine in rat and human liver mitochohdrial aldehyde dehydrogenase. A model to study human (Oriental type) class 2 aldehyde dehydrogenase, *J. Biol. Chem.*, 269:13854.

Feldman, R. I., and Weiner, H., 1972, Horse liver aldehyde dehydrogenase. II. Kinetics and mechanistic implications of the dehydrogenase and esterase activity, *J. Biol. Chem.*, 247:267.

Fersht, A., 1985, "Enzyme Structure and Mechanism," 2nd ed., W. H. Freeman, New York, p. 98.

Forte-McRobbie, C. M., and Pietruszko, R., 1986, Purification and characterization of human liver "high Km" aldehyde dehydrogenase and its identification as glutamic g-semialdehyde dehydrogenase, *J. Biol. Chem.*, 261:2154.

Greenfield, N. J., and Pietruszko, R., 1977, Two aldehyde dehydrogenases from human liver. Isolation via affinity chromatography and characterization of the isozymes, *Biochim. Biophys. Acta*, 483:35.

Harada, S., Misawa, S., Agarwal, D. P., and Goedde, H. W., 1980, Liver alcohol dehydrogenase and aldehyde dehydrogenase in the Japanese: Isoenzyme variation and its possible role in alcohol intoxication, *Am. J. Hum. Genet.*, 32:8.

Hempel, J., von Bahr-Lindstrom, H., and Jörnvall, H., 1984, Aldehyde dehydrogenase from human liver. Primary structure of the cytoplasmic isoenzyme, *Eur. J. Biochem.*, 141:21.

Hempel, J., Kaiser, R., and Jörnvall, H., 1985, Mitochondrial aldehyde dehydrogenase from human liver. Primary structure, difference in relation to the cytosolic enzyme, and functional correlations, *Eur. J. Biochem.*, 153:13.

Hempel, J., and Jörnvall, H., 1989, Aldehyde dehydrogenases - Structure, in "Human Metabolism of Alcohol," Vol II, K. E. Crow and R. D. Batt, eds., CRC, Boca Raton, p. 77.

Hempel, J., Lindahl, R., 1989, Class III aldehyde dehydrogenase from rat liver: Superfamily relationship to classes I and II and functional interpretations, in "Enzymology and Molecular Biology of Carbonyl Metabolism 2," H. Weiner and T. G. Flynn, eds., Alan R. Liss, New York, p. 3.

Hempel, J., Eckey, R., Berie, D., Romovacek, H., Agarwal, D. P., and Goedde, H. W., 1992, Human liver glutamic g-semialdehyde dehydrogenase: Structural relationship to the yeast enzyme, *Comp. Biochem. Physiol.*, 102B:791.

Hsu, L. C., Tani, K., Fujiyoshi, T., Kurachi, K., and Yoshida, A., 1985, Cloning of cDNAs for human aldehyde dehydrogenases 1 and 2, *Proc. Natl. Acad. Sci. USA*, 82:3771.

Hsu, L. C., and Chang, W.-C., 1991, Cloning and characterization of a new functional human aldehyde dehydrogenase gene, *J. Biol. Chem.*, 266:12257.

Hsu, L. C., Chang, W.-C., Shibuya, A., and Yoshida, A., 1992, Human stomach aldehyde dehydrogenase cDNA and genomic cloning, primary structure, and expression in Escherichia coli, *J. Biol. Chem.*, 267:3030.

Hsu, L. C., Chang, W. C., Lin, S. W., and Yoshida, A., 1994, Cloning and characterization of genes encoding four additional human aldehyde dehydrogenase isozymes, Seventh International Workshop on Enzymology and Molecular Biology of Carbonyl Metabolism, Palmerston North, New Zealand, p. 5.

Hurley, T. D., Yang, Z., Bosron, W. F., and Weiner, H., 1993, Crystallization and preliminary X-ray analysis of bovine mitochondrial aldehyde dehydrogenase and human glutathione-dependent formaldehyde dehydrogenase, in "Enzymology and Molecular Biology of Carbonyl Metabolism 4," H. Weiner, D. W. Crabb and T. G. Flynn, eds., Plenum, New York, p. 245.

Ikawa, M., Impraim, C. C., Wang, G., and Yoshida, A., 1983, Isolation and characterization of aldehyde dehydrogenase isozymes from usual and atypical human livers, *J. Biol. Chem.*, 258:6282.

Kurys, G., Ambroziak, W., and Pietruszko, R., 1989, Human aldehyde dehydrogenase. Purification and characterization of a third isozyme with low Km for g-aminobyturaldehyde, *J. Biol. Chem.*, 264:4715.

Kurys, G., Shah, P. C., Kikonyogo, A., Reed, D., Ambroziak, W., and Pietruszko, R., 1993, Human aldehyde dehydrogenase. cDNA cloning and primary structure of the enzyme that catalyzes dehydrogenation of 4-aminobutyraldehyde, *Eur. J. Biochem.* 218:311.

Lindahl, R., and Evces, S., 1984, Rat liver aldehyde dehydrogenase. I. Isolation and characterization of four high Km normal liver isozymes, *J. Biol. Chem.*, 259:11986.

Lindahl, R., and Petersen, D. R., 1991, Lipid aldehyde oxidation as a physiological role for class 3 aldehyde dehydrogenases, *Biochem. Pharmacol.*, 41:1583.

Lowry, O. H., Rosebrough, N. J., Farr, A. L., and Randall, R. J., 1951, Protein measurement with the Folin phenol reagent, *J. Biol. Chem.*, 193:265.

Mizoi, Y., Ijiri, I., Tatsuno, Y., Kijima, T., Fujiwara, S., and Adachi, J., 1979, Relationship between facial flushing and blood acetaldehyde levels after alcohol intake, *Pharmacol. Biochem. Behav.*, 10:303.

Monder, C., Purkaystha, A. R., and Pietruszko, R., 1982, Oxidation of the 17-aldol (20b hydroxy-21-aldehyde) intermediate of corticosteroid metabolism to hydroxy acids by homogeneous human liver aldehyde dehydrogenases, *J. Steroid Biochem.*, 17:41.

Parrilla, R., Ohkawa, K., Lindros, K. O., Zimmerman, U.-J. P., Kobayashi, K., and Williamson, J. R., 1974, Functional compartmentation of acetaldehyde oxidation in rat liver, *J. Biol. Chem.*, 249:4926.

Pietruszko, R., 1983, Aldehyde dehydrogenase isozymes, *Isozymes Curr. Top. Biol. Med. Res.*, 8:195.

Rose, J. P., Hempel, J., Kuo, I., Lindahl, R., and Wang, B.-C., 1990, Preliminary crystallographic analysis of class 3 rat liver aldehyde dehydrogenase, *Proteins Struct. Funct. Genet.*, 8:305.

Thomasson, H. R., Edengerg, H. J., Crabb, D. W., Mai, X.-L., Jerome, R. E., Li, T.-K., Wang, S.-P., Lin, Y.-T., Lu, R.-B., and Yin, S.-J., 1991, Alcohol and aldehyde dehydrogenase genotypes and alcoholism in Chinese men, *Am. J. Hum. Genet.*, 48:677.

Vallari, R. C., and Pietruszko, R., 1981, Kinetic mechanism of the human cytoplasmic aldehyde dehydrogenase E1, *Arch. Biochem. Biophys.*, 212:9.

Wang, S.-L., Wu, C.-W., Cheng, T.-C., and Yin, S.-J., 1990, Isolation of high-Km aldehyde dehydrogenase isozymes from human gastric mucosa, *Biochem. Int.*, 22:199.

Weiner, H., Hu, J. H. J., and Sanny, C. G., 1976, Rate-limiting steps for the esterase and dehydrogenase reaction catalyzed by horse liver aldehyde dehydrogenase, *J. Biol. Chem.*, 251:3853.

Weiner, H., 1987, Subcellular localization of acetaldehyde oxidation in liver, *Ann. N. Y. Acad. Sci.*, 492:25.

Weiner, H., 1989, Role of alcohol and aldehyde dehydrogenase in vivo: Speculations on their natural substrates, in "Human Metabolism of Alcohol," Vol II, K. E. Crow and R. D. Batt, eds., CRC, Boca Raton, p. 147.

Whitehead, E. P., 1978, Cooperativity and the method of plotting binding and steady-state kinetic data, *Biochem. J.*, 171:501.

Yin, S.-J., Chang, T.-C., Chang C.-P., Chen, Y.-J., Chao, Y.-C., Tang, H.-S., Chang, T.-M., and Wu, C.-W., 1988, Human stomach alcohol and aldehyde dehydrognase (ALDH): A genetic model proposed for ALDH III isoenzymes, *Biochem. Genet.*, 26:343.

Yin, S.-J., Liao, C.-S., Wang, S.-L., Chen, Y.-J., and Wu, C.-W., 1989, Kinetic evidence for human liver and stomach aldehyde dehydrogenase-3 representing a unique class of isozymes, *Biochem. Genet.*, 27:321.

Yin, S.-J., Vagelopoulos, N., Wang, S.-L., and Jörnvall, H., 1991, Structural features of stomach aldehyde dehydrogenase distinguish dimeric aldehyde dehydrogenase as a "variable" enzyme. "Variable" and "constant" enzymes within the alcohol and aldehyde dehydrogenase families, *FEBS Lett.*, 283:85.

Yoshida, A., Hsu, L. C., and Yasunami, M., 1991, Genetics of human alcohol-metabolizing enzymes, *Progr. Nucleic Acid Res. Mol. Biol.*, 40:255.

Yoshida, A., Hsu, L. C., and Dave, V., 1992, Retinal oxidation activity and biological role of human cytosolic aldehyde dehydrogenase, *Enzyme*, 46:239.

3

HUMAN CLASS 1 ALDEHYDE DEHYDROGENASE

Expression and Site-Directed Mutagenesis

Kerrie M. Jones, Trevor M. Kitson, Kathryn E. Kitson,
Michael J. Hardman and John W. Tweedie

Department of Chemistry and Biochemistry
Massey University
Palmerston North, New Zealand

INTRODUCTION

Human Class 1 and Class 2 aldehyde dehydrogenases have been sequenced at both the protein (Hempel et al., 1984, 1985) and DNA level (Hsu et al., 1988, 1989). Studies on the tertiary structure of aldehyde dehydrogenase are in progress (Baker et al., 1995, sheep Class 1; Hurley and Weiner, 1992, beef Class 2), but are not sufficiently advanced to suggest which amino acid residues are important in catalysis. Cys 302 is the only completely conserved cysteine in all known forms of the enzyme (Hempel et al., 1993), and labelling by various substrates and substrate analogues (von Bahr-Lindstrom et al., 1985; Kitson et al., 1991; Pietruszko et al., 1993) has implicated this residue as the probable active site nucleophile. This has been confirmed for the Class 2 enzyme by site-directed mutagenesis (Weiner et al., 1991). In order to establish whether Cys 302 is also the active site nucleophile for Class 1 aldehyde dehydrogenase we decided to carry out mutagenesis at Cys 302. A separate mutant was constructed in which Cys 301, the adjacent residue, was changed to alanine while Cys 302 was left unchanged.

The rationale for preparing the C301A form of cytosolic aldehyde dehydrogenase was to test the possible involvement of Cys 301 in the reaction of the enzyme with 2,2'-dithiodipyridine. Previous work (Kitson and Loomes, 1985; Loomes and Kitson, 1989) has led to the proposal that both disulfiram and 2,2'-dithiodipyridine oxidise a pair of cysteine residues (A and B) to form a cystine disulphide bridge, but that the order of reaction of the cysteines with the two modifiers is different. That is, cysteine A is modified rapidly by disulfiram leading to inactivation of the enzyme, and then the diethyldithiocarbamyl moiety is displaced from the enzyme by cysteine B. On the other hand, cysteine B is first modified by 2,2'-dithiodipyridine giving rise to an enzyme form which is approximately twice as active as the native form (when assayed with 1 mM acetaldehyde), but subsequently the 2-thiopyridyl label is slowly displaced by cysteine A with concomitant loss of enzyme activity. Experiments with methyl diethylthiocarbamyl disulphide and methyl 2-pyridyl disulphide (both of which donate a thiomethyl group to the enzyme, but which have different effects on enzyme activity) support the above proposal (Loomes and Kitson, 1989).

It is highly probable that Cys 302 is cysteine A in the scenario described above. A possible candidate for cysteine B is the adjacent Cys 301 which is present in many known aldehyde dehydrogenase sequences (including the sheep liver cytosolic enzyme with which the original 2,2'-dithiodipyridine work was done; Stayner and Tweedie, 1995). It is unusual for contiguous cysteine residues to exist in the form of cystine, but it is geometrically possible and there are some examples in the literature (Kao and Karlin, 1986). We thought it worthwhile, therefore, to change Cys 301 to alanine and examine the effect of 2,2'-dithiodipyridine on the mutant enzyme.

METHODS

Expression and Purification of Human Cytosolic Aldehyde Dehydrogenase

The cDNA for human cytosolic aldehyde dehydrogenase was obtained from Professor H. Weiner, Purdue University. The cDNA was cloned into the pT7-7 vector (Tabor & Richardson, 1985) under the control of the $\Phi10$ promotor which is specific for the RNA polymerase from bacteriophage T7. Human AlDH cDNA was expressed from this vector after transformation into *E. coli* SRP84/pGP1-2 {F⁻ *ilv his supo strA* (ts)*htpR*165::Tn10 *galOP*IS1 Δlon Δbio (lBam N⁺ *cI*857 H1[*cro⁻RAJ⁻bio*] *uvrB*)}. The resident plasmid (pGP1-2) carries the DNA for T7 RNA polymerase under the control of the strong λ promotor pL, together with the *cI*857 temperature-sensitive λ repressor expressed from the *lac* promoter. (Tabor & Richardson, 1985). Induction of AlDH expression was accomplished by heat shock to inactivate the heat-sensitive λ repressor. Cells were grown at 30 °C to an OD_{600} of ~1, transferred to 42 °C and maintained at this temperature for 5 minutes and then incubated at 30 °C overnight.

All buffers used in the purification contained 3 mM EDTA and 1 mM DTT and were degassed and bubbled with nitrogen before use. The harvested cells were suspended in 30 mM pH 6.0 sodium phosphate buffer, disrupted using a French press and treated with protamine sulphate. After centrifugation, the supernatant was passed through a CM-Sephadex column which had been equilibrated with the same buffer. AlDH was eluted immediately after the void volume and was dialysed against 30 mM pH 7.4 phosphate buffer, containing 50 mM NaCl, and then loaded onto a *p*-hydroxyacetophenone-Sepharose affinity column (Ghenbot & Weiner, 1992). AlDH was eluted using 10 mM *p*-hydroxyacetophenone (in the same buffer). The enzyme was concentrated and the buffer changed to 30 mM Bis-Tris pH 6.0 (containing 10% glycerol) using a centricon-30 concentrator (Amicon) and the purified enzyme was stored frozen at -70 °C. In the case of inactive enzyme prepared following site-directed mutagenesis, the enzyme protein was detected using polyclonal antibodies prepared in rabbits to sheep liver Class I aldehyde dehydrogenase, which shares 91% sequence identity with the human enzyme (Stayner and Tweedie, 1995).

Site-Directed Mutagenesis of Cytosolic Aldehyde Dehydrogenase

The uracil-selection method of Kunkel *et al.* (1987) was used for mutagenesis. The cDNA for AlDH was subcloned into M13 and single-stranded DNA was produced in the *dut⁻, ung⁻* strain CJ236. Redundant oligonucleotide primers were designed to introduce a number of mutations into the cDNA. The mutagenic oligonucleotides were annealed to the uracil-containing template and the complementary strand was synthesised using T4 DNA polymerase and T4 DNA ligase. The double-stranded DNA was introduced by transformation into the *dut⁺ ung⁺* strain TG1. Clones containing the desired mutations were selected by sequencing of the region near the site of mutation and each cDNA was completely sequenced to check that the desired mutation was the only change present. Mutant forms of AlDH were expressed and purified as described above.

RESULTS AND DISCUSSION

Expression of Aldehyde Dehydrogenase

Aldehyde dehydrogenase activity was present both in the soluble fraction and in inclusion bodies after lysis of the bacterial cells, as found for both the class 1 and class 2 enzymes by Zheng *et al.* (1993). Similar levels of expression were found after heat shock at 42 °C for either 2 or 10 mins and 5 mins was used for subsequent experiments. A growth temperature of 30 °C was used to reduce the formation of inclusion bodies. Since the level of expression was reasonably good, and about 50% of the activity was found in the soluble fraction, no further attempts were made to purify the enzyme that was present in the insoluble fraction.

Purification of Wild Type Aldehyde Dehydrogenase

The purification scheme outlined above allowed preparation of about 10 mg of wild type Class 1 aldehyde dehydrogenase from a 3.2 litre bacterial culture. The purified enzyme showed a single protein band with molecular weight about 53000 on SDS gels stained either with Coomassie blue or using a silver stain (Fig. 1). Isoelectric focussing on thin layer polyacrylamide gels gave a single protein band which showed aldehyde dehydrogenase activity with activity staining using acetaldehyde, NAD⁺, phenazine methosulphate and nitroblue tetrazolium. The isoelectric point of the enzyme was pH 5.2. The specific activity of the purified wild type enzyme was 0.34 μmol min⁻¹ mg⁻¹ (Table 1). This is slightly higher than the value of 0.22 μmol min⁻¹ mg⁻¹ reported by Yoshida *et al.*(1992). This is perhaps not surprising, since Yoshida *et al.* prepared the enzyme from autopsy livers and used a purification method involving five column chromatography steps, rather than two as in the method above. The use of nitrogen-saturated buffers was also found to be important in

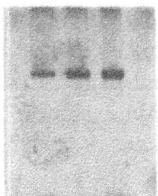

coomassie stained   **silver stained**

1µl   2µl   4µl   0.1µl 0.5µl 1µl

Figure 1. SDS polyacrylamide gel electrophoresis of purified recombinant wild type class 1 AlDH. A range of loadings was used, as indicated; the protein concentration was about 1 mg/ml.

Table 1. Comparison of the activity of the wild type human cytosolic aldehyde dehydrogenase with that of the C301A mutant under various conditions

	Specific activity (μmol min^{-1} mg^{-1})	
	wild type	C301A
specific activity as dehydrogenase	0.34	0.30
specific activity as esterase	0.12	0.12

	Activity in presence of modifiers (% of control)	
	wild type	C301A
disulfiram (1 μM)	3.5	3.7
diethylstilbestrol (10 μM)	289	313
2,2'-dithiodipyridine (1 μM)	196	216
2,2'-dithiodipyridine (1 μM) (added in absence of NAD$^+$)	162	178

Results are averages of 2 to 6 determinations. All assays were carried out at 25 °C in 50 mM phosphate buffer, pH 7.4. Dehydrogenase assays used 1 mM NAD$^+$ and 1 mM acetaldehyde; esterase assays used 50 μM p-nitrophenyl acetate. Wild type enzyme concentration was 94 nM; C301A concentration was 83 nM. Enzyme concentration was determined from A$_{280}$ assuming the same extinction coefficient as for sheep liver cytosolic aldehyde dehydrogenase (Dickinson *et al.*, 1981). Modifiers were added as 20 μL of a solution in ethanol; the same volume of ethanol was added to the control assays. Except where stated, modifiers were added to the enzyme in the presence of NAD$^+$, and acetaldehyde was added 1 min later to initiate the assay.

maintaining activity of the enzyme during and after purification. Zheng *et al.* (1993) reported a specific activity of 0.23 - 0.31 for the cloned and expressed enzyme assayed at pH 9.0 with propionaldehyde as substrate. Although they used a similar purification scheme, comparison of the final specific activities is not valid since different pH, buffers and substrates were used.

Purification and Characterisation of Mutant Forms of Aldehyde Dehydrogenase

Although the redundant oligonucleotide used for mutagenesis included other mutations (such as those for C302A), the only mutant forms of AlDH we have so far expressed and characterised are those with cysteine 302 changed to serine (C302S) and cysteine 301 changed to alanine (C301A). Both were purified to give a single major protein band on SDS gels. (Fig. 2). The C302S enzyme was essentially inactive when assayed under standard conditions, but the protein showed reaction with antibody for the sheep AlDH. (Fig. 2). These results confirm that cysteine 302 is likely to be the active site nucleophile, in agreement with previous results from chemical modification studies, and for site-directed mutagenesis of the mitochondrial enzyme (Weiner *et al.*, 1991).

The results in Table 1 show that C301A behaves in a similar manner to the wild type towards 2,2'-dithiodipyridine. Both forms of the enzyme are activated by reaction with this modifier either in the absence or presence of NAD$^+$. (A minor difference from previous work with sheep liver cytosolic aldehyde dehydrogenase is that the latter enzyme is not much affected by 2,2'-dithiodipyridine unless reaction occurs in the presence of NAD$^+$; Kitson 1982a). Clearly we can now say that the residue that first reacts with 2,2'-dithiodipyridine (referred to as cysteine B in the Introduction) is not Cys-301, since its absence does not preclude activation of the enzyme by the modifier. Neither is it Cys-302, of course, since that would lead to inactivation, not activation. Thus we conclude that a third, and currently

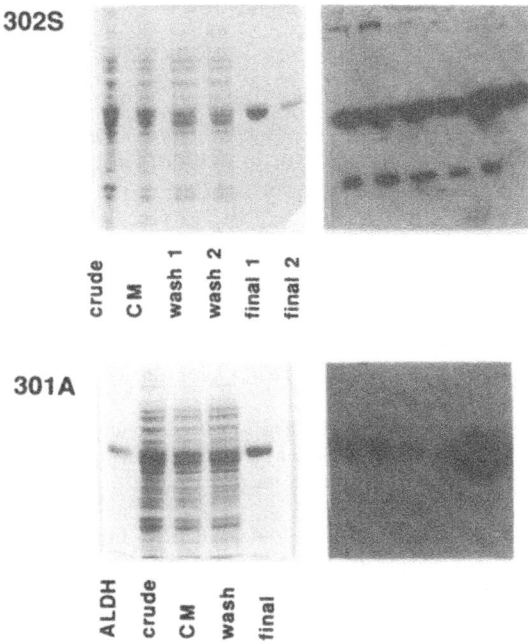

Figure 2. SDS polyacrylamide gel electrophoresis of the C302S and C301A mutants. The gels show impure fractions during the preparation of the mutant proteins, and the final purified enzymes. AlDH indicates sheep liver class 1 AlDH, as a marker. Gels on the left were stained with Coomassie blue, those on the right show reaction with antibody prepared against sheep class 1 AlDH. The lower antibody-reactive band in the C302S preparation is presumed to be an AlDH fragment of lower molecular weight. It was not present when unbound protein was re-run through the affinity column (final 2).

unidentified, cysteine residue is present in or near the active site, close enough to form a disulphide bridge to Cys-302.

Table 1 also shows that changing residue 301 from Cys to Ala has no significant effect on the specific activity of the enzyme as a dehydrogenase or as an esterase, on the activation of the dehydrogenase activity by diethylstilbestrol (Kitson, 1982b), or on the inactivation of the enzyme by disulfiram. Thus although this residue is adjacent to the enzyme's catalytic nucleophile, it appears to have no influence on the enzymic activity. This conclusion is consistent with the fact that Cys 301 is not conserved across the full range of aldehyde dehydrogenase sequences so far determined (Hempel *et al.*, 1993).

It is interesting that the expressed enzyme forms studied here have a small residual activity after disulfiram treatment. With Class 1 aldehyde dehydrogenase from sheep liver, interpretation of a similar observation was always complicated by the possibility of contamination of the enzyme by traces of the disulfiram-resistant mitochondrial (Class 2) aldehyde dehydrogenase (Dickinson *et al.*, 1981). This explanation for the residual activity is of course not tenable for the recombinant enzymes expressed in *E. coli* which have been used in this work. Thus the disulfiram-resistant residual activity is real and may be connected with the observed 'half-of-the-sites' reactivity of the enzyme (Pietruszko *et al.*, 1993). Although all four subunits of the enzyme have identical sequence, the quaternary structure may be such that two of them are disulfiram-sensitive and responsible for most of the enzymic activity, whereas two are disulfiram-resistant and have a very low activity.

ACKNOWLEDGEMENTS

We wish to thank Dr H. Weiner for donation of the cDNA for human cytosolic aldehyde dehydrogenase and for helpful advice, Massey University for a Postdoctoral Fellowship (to KMJ) and the Palmerston North Medical Research Foundation and Massey University Research Fund for financial support.

REFERENCES

Baker, H.M., Brown, R.L., Dobbs, A.J., Kitson, K.E., Kitson, T.M. and Baker, E.N., 1995, Crystallisation of sheep liver cytosolic aldehyde dehydrogenase in a form suitable for high resolution X-ray structural analysis, These Proceedings.

Dickinson, F.M., Hart, G.J. and Kitson, T.M., 1981, The use of pH-gradient ion-exchange chromatography to separate sheep liver cytoplasmic aldehyde dehydrogenase from mitochondrial enzyme contamination, and observations on the interaction between the pure cytoplasmic enzyme and disulfiram, Biochem. J., 199: 573-579.

Ghenbot, G. & Weiner, H. 1992, Purification of liver aldehyde dehydrogenase by p-hydroxyacetophenone-Sepharose affinity matrix and the coelution of chloramphenicol acetyl transferase from the same matrix with recombinantly expressed aldehyde dehydrogenase, Protein Expression Purif. 3: 470-478.

Hempel, J., von Bahr-Lindström, H. & Jörnvall, H., 1984, Aldehyde dehydrogenase from human liver: primary structure of the cytoplasmic isoenzyme, Eur. J. Biochem., 141: 21-35.

Hempel, J., Kaiser, R. & Jörnvall, H., 1985, Mitochondrial aldehyde dehydrogenase from human liver: primary structure, differences in relation to the cytosolic enzyme, and functional correlations, Eur. J. Biochem., 153: 13-28.

Hempel, J., Nicholas, H. & Lindahl, R., 1993, Aldehyde dehydrogenases: widespread structural and functional diversity within a shared framework, Protein Science, 2: 1890-1900.

Hsu, L.C., Bendel, R.E. & Yoshida, A., 1988, Genomic structure of the human mitochondrial aldehyde dehydrogenase gene, Genomics, 2: 57-65.

Hsu, L.C., Chang, W.-C. & Yoshida, A., 1989, Genomic structure of the human cytosolic aldehyde dehydrogenase gene, Genomics, 5: 857-865.

Hurley, T.D. & Weiner, H., 1992, Crystallisation and preliminary X-ray investigation of bovine liver mitochondrial aldehyde dehydrogenase, J. Mol. Biol., 227: 1255-1257.

Kao, P.N. and Karlin, A., 1986, Acetylcholine receptor binding site contains a disulfide cross-link between adjacent half-cystinyl residues, J. Biol. Chem., 261: 8085-8088.

Kitson, T.M., 1982a, Further studies of the action of disulfiram and 2,2'-dithiodipyridine on the dehydrogenase and esterase activities of sheep liver cytoplasmic aldehyde dehydrogenase, Biochem. J., 203: 743-754.

Kitson, T.M., 1982b, The activation of aldehyde dehydrogenase by diethylstilboestrol and 2,2'-dithiodipyridine, Biochem. J., 207: 81-89.

Kitson, T.M. & Loomes, K.M., 1985, Modification of thiol groups in cytoplasmic aldehyde dehydrogenase, Alcohol, 2: 97-101.

Kitson, T.M., Hill, J.P. & Midwinter, G.G., 1991, Identification of a catalytically essential nucleophilic residue in sheep liver cytoplasmic aldehyde dehydrogenase, Biochem. J., 275: 207-210.

Kunkel, T.A., Roberts, J.D. and Zakour, R.A., 1987, Rapid and efficient mutagenesis without phenotypic selection, Methods Enzymol., 154: 367-382.

Loomes, K.M. and Kitson, T.M., 1989, Studies of the action of symmetrical and mixed disulphide thiol-modifying reagents on the esterase activity of cytoplasmic aldehyde dehydrogenase, Biochim. Biophys. Acta, 998: 1-6.

Pietruszko, R., Abriola, D.P., Blatter, E.E. and Mukerjee, N., 1993, Aldehyde dehydrogenase: aldehyde dehydrogenation and ester hydrolysis, In Enzymology and Molecular Biology of Carbonyl Metabolism (Weiner, H., Crabb, D.W. and Flynn, T.G., eds), pp. 221-232, Plenum Press, New York and London.

Stayner, C.K. and Tweedie, J.W., 1995, Cloning and characterisation of the cDNA for sheep liver cytosolic aldehyde dehydrogenase, These Proceedings.

Tabor, S. & Richardson, C.C., 1985, A bacteriophage T7 RNA polymerase/promoter system for controlled exclusive expression of specific genes, Proc. Natl. Acad. Sci. U.S.A., 82: 1074-1078.

von Bahr-Lindström, H., Jeck, R., Woenckhaus, C., Sohn, S., Hempel, J. & Jörnvall, H., 1985, Characterization of the coenzyme binding site of liver aldehyde dehydrogenase: differential reactivity of coenzyme analogues, Biochemistry, 24: 5847-5851.

Weiner, H., Farres, J., Wang, T.T.Y., Cunningham, S.J., Zheng, C.-F. & Ghenbot, G., 1991, Probing the active site of aldehyde dehydrogenase by site directed mutagenesis, in Enzymology & Molecular Biology of Carbonyl Metabolism 3 (Weiner, H., Wermuth, B. & Crabb, D.W., Eds.), pp. 13-17, Plenum Press, New York.

Yoshida, A., Hsu, L.C. & Davé, V., 1992, Retinal oxidation activity and biological role of human cytosolic aldehyde dehydrogenase, Enzyme 46: 239-244.

Zheng, C.-F., Wang, T.T.Y. and Weiner, H., Cloning and expression of the full-length cDNAs encoding human liver class 1 and class 2 aldehyde dehydrogenase, Alcohol. Clin. Exp. Res., 17: 828-8. Clin. Exp. Res., 17: 828-831.

4

NITRATE ESTERS AS INHIBITORS AND SUBSTRATES OF ALDEHYDE DEHYDROGENASE

Regina Pietruszko, Neeta Mukerjee, Erich E. Blatter, and Teresa Lehmann

Center of Alcohol Studies
Rutgers University
Piscataway, New Jersey 08855-0969

Nitrate esters are used clinically as the smooth muscle relaxants (Needleman, 1975) in the treatment of angina pectoris. Guanine cyclase-mediated relaxation of aortic strips is dependent on the conversion of nitrate esters to nitric oxide by tissue sulfhydryl groups (Needleman and Johnson, 1973). The commonly accepted scheme for conversion of nitrate esters to nitric oxide (Ignarro et al., 1981) is shown below where C = glutathione or cysteine:

(a) $C\text{-}SH + RONO_2 \rightarrow C\text{-}S\text{-}NO_2 + ROH$

(b) $C\text{-}S\text{-}NO_2 + C\text{-}SH \rightarrow C\text{-}S\text{-}S\text{-}C + NO_2^-$

In the presence of H^+:

(c) $NO_2^- + H^+ \rightarrow HONO$

(d) $HONO \rightarrow NO$

(e) $C\text{-}SH + NO \rightarrow C\text{-}S\text{-}NO$

Isosorbide dinitrate (ISDN) and nitroglycerin are the nitrate esters used clinically in the USA. Ingestion of ISDN or nitroglycerin medication combined with ethanol consumption causes unpleasant symptoms (Dipalma, 1982) similar to those of the alcohol aversive drug, disulfiram (Kitson, 1977).

The structure of ISDN (1,4:3,6-dianhydrosorbitol-2,5-dinitrate) consists of two nearly planar cis-fused (in the form of "V") five-membered rings with nitrate ester groups situated outside (exo, position 2) or inside (endo, position 5) of the "V" shaped wedge (Figure 1,A). Thus, the two nitrate groups are not stereochemically equivalent. Hydrolysis of ISDN at position 5 yields isosorbide-2-mononitrate (IS2MN, Figure 1B), hydrolysis at position 2 yields isosorbide-5-mononitrate (IS5MN, Figure 1C) and complete hydrolysis yields isosorbide (Figure 1D).

Enzymology and Molecular Biology of Carbonyl Metabolism 5
Edited by H. Weiner *et al.*, Plenum Press, New York, 1995

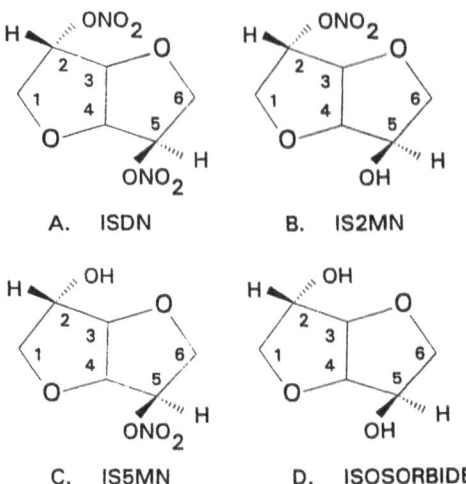

Figure 1. Chemical structures of isosorbide dinitrate (ISDN, A), isosorbide-2-mononitrate (IS2MN, B), isosorbide-5-mononitrate (IS5MN, C) and isosorbide (isosorbide, D).

It is demonstrated in this report that nitrate esters inactivate aldehyde dehydrogenase via the formation of a covalent bond. Here we report on aldehyde dehydrogenase-mediated formation of products from ISDN and on identification and quantitation of these products. The results suggest that the reactive species that inactivates aldehyde dehydrogenase is formed during the enzyme-catalyzed reaction and involves the covalent-enzyme-acyl intermediate.

EXPERIMENTAL

Materials

ISDN (containing 60% w/w lactose), methemoglobin, p-nitrophenyl acetate, 5,5-dithiobis nitrobenzoic acid (Ellman's reagent), bromoacetophenone and glutathione were from Sigma. IS2MN and IS5MN (without lactose) were a gift from Schwarz Pharma (Milwaukee, Wisconsin). Benzene, ethyl acetate, acetic acid, HPLC grade methanol, HPLC grade acetonitrile, chloral hydrate and diphenylamine were from Fisher Scientific. Propanal, 2-mercaptoethanol, N-(1-Naphthyl) ethylene-diamine dihydrochloride, sulfanilic acid, isosorbide and lactose were from Aldrich. NAD (grade 1) and nitrate reductase kit (#905658) were from Boehringer Mannheim. Sodium nitrate and sodium nitrite were from Mallinckrodt. [^{14}C(U)]-ISDN (specific radioactivity = 168.2 mCi/mmol) was from New England Nuclear. Stock solutions of IS2MN and IS5MN were made in water; ISDN was dissolved in 1:1 ethanol:water. All buffers were degassed and saturated with N_2 before use to prevent air oxidation of enzyme cysteines. Human liver aldehyde dehydrogenases were purified to homogeneity, stored and dialyzed before use as described previously (Blatter et al., 1990; Mukerjee and Pietruszko, 1992).

Determination of Enzyme Activity

Dehydrogenase and esterase activities and protein concentrations were determined as described previously (Mukerjee and Pietruszko, 1992). The concentration of active enzyme was calculated based upon the maximal specific activity of E2 = 1.6 μmol/min/mg, MW of E2 = 217,000, maximal specific activity of E1 = 0.6 μmol/min/mg, MW of E1 = 219,000 and two active sites per molecule (Ambroziak et al., 1989).

Enzyme Inactivation

Phosphate buffer, pH 7.0, 30mM, containing 1mM EDTA was used in all experiments. Inactivation was initiated by addition of the inhibitor solution to the enzyme solution in buffer. Samples were periodically withdrawn and assayed at 100-300 dilution. Following overnight incubation at 25°C the enzyme was precipitated by heating at 75-80°C for 5 min; following short incubations at 25°C and all incubations at 0°C the enzyme was precipitated by addition of trifluoroacetic acid (5% v/v final concentration) and the precipitated protein was removed by centrifugation.

Thin Layer Chromatography

Machery Nagel silica gel plates (4 x 8 cm) and a solvent system consisting of benzene: ethyl acetate: acetic acid (80:20:5 v/v) were used (Needleman and Hunter, 1965). IS2MN, IS5MN and ISDN were detected with 1% diphenylamine.

Quantitation of IS2MN, IS5MN, and ISDN Using HPLC

IS2MN, IS5MN and ISDN were resolved by reversed phase HPLC employing a Waters C_{18} μBondapak column and a linear gradient from 100 % water to 100 % methanol over 15 min at a flow rate of 1 ml/min and detected by absorbance at 230 nm.

Determination of Nitrite, Nitrate and Sulfhydryl Groups

Nitrite was determined by the diazotization and coupling procedure (Bell et al., 1963). Calibration curves were obtained by using sodium nitrite as a standard. Nitrate was determined by using nitrate reductase kit - Boehringer Mannheim. An extinction coefficient of 6.3 mM^{-1} cm^{-1} was used for NADPH. Sodium nitrate was used as control. The sulfhydryl groups were titrated by Ellman's reagent (Habeeb, 1972).

RESULTS AND DISCUSSION

Both E1 and E2 isozymes of human aldehyde dehydrogenase were inhibited by ISDN, IS2MN and IS5MN. The E2 isozyme was also effectively inhibited by nitroglycerin. Thus, previously reported work on inhibiton of human erythrocyte aldehyde dehydrogenase by ISDN and nitroglycerin (Towell et al., 1985) was confirmed by these experiments. The surprising finding was that the inhibition of both isozymes was stable during exhaustive dialysis, suggesting that it must have occurred via formation of a covalent bond. This allowed use of saturation kinetics (Kitz and Wilson, 1962) to obtain more information. The results obtained from saturation kinetics for the cytoplasmic E1 isozyme are shown in Figure 2 (A,B,C) for IS2MN, IS5MN and ISDN, respectively. With all three compounds Y-axis intercepts were obtained. The results of saturation kinetics for the E2 isozyme are shown in Figure 3 (D,E,F). Pronounced Y-axis intercepts were obtained with all three compounds demonstrating that all three inhibitors inactivated both isozymes of aldehyde dehydrogenase in a specific manner. The first order rate constants k_3 of the covalent bond formation (shown in Table 1) are the reciprocals of the Y-axis intercepts. Since saturation occurred in all cases,

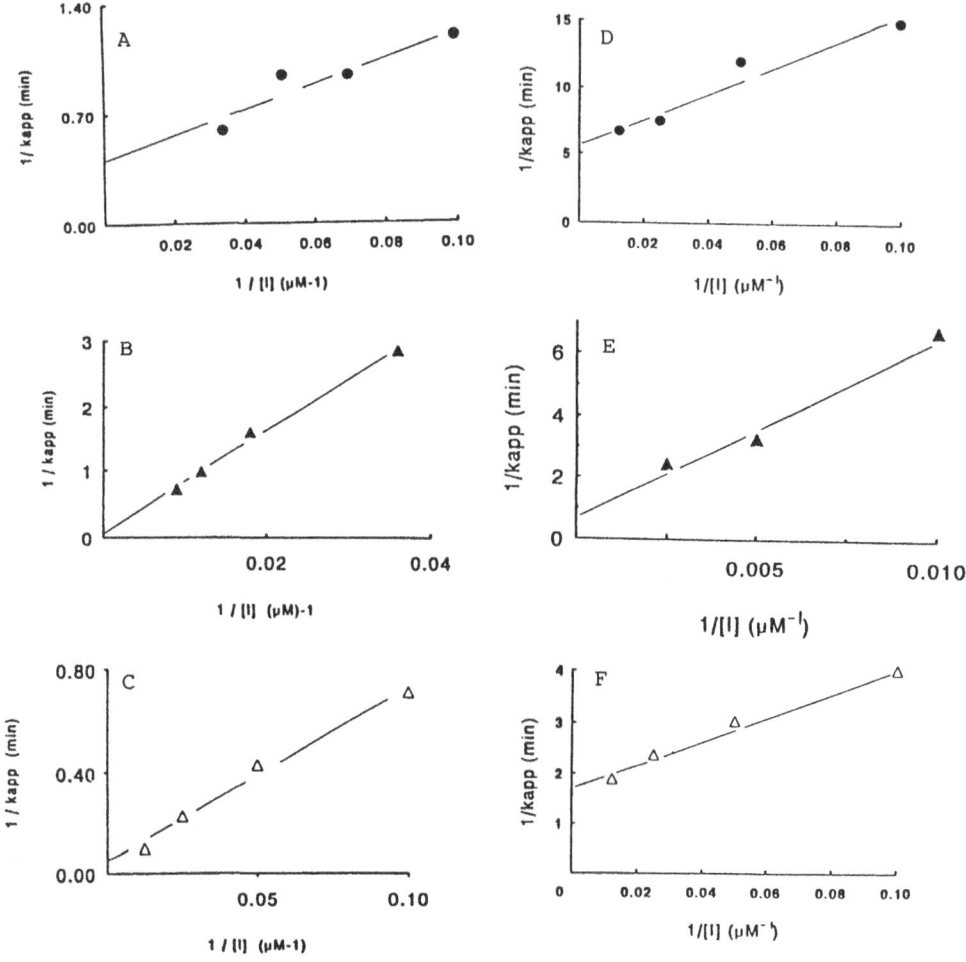

Figure 2. Secondary plots from saturation kinetics of IS2MN, IS5MN and ISDN with E1 and E2 isozymes of human aldehyde dehydrogenase. A,B,C = E1 isozyme; D,E,F = E2 isozyme. A,D = IS2MN; B,E = IS5MN; C,F = ISDN.

these data suggested that all three inhibitors reacted at the active sites of both E1 and E2 isozymes.

In addition to saturation kinetics, other evidence indicated that the reaction that resulted in inhibition of aldehyde dehydrogenase occurred at the active site. The inactivation of both isozymes by isosorbide esters was promoted by NAD and virtually abolished by chloral (an aldehyde-substrate-competitive inhibitor). K_i values determined from saturation kinetics (in the absence of NAD) for both the E1 and E2 isozymes are shown in Table 1. The K_i values for IS5MN for both isozymes are large relative to those of IS2MN and ISDN demonstrating that IS5MN was not as well recognized by aldehyde dehydrogenase as IS2MN and ISDN. The K_i values for IS2MN were the lowest demonstrating that IS2MN was best recognized of all three nitrate esters. ISDN, which contains both structures, resembled IS2MN more than IS5MN in its K_i values. IS5MN,however, can form covalent bonds much faster than IS2MN (see k_3 values in Table 1). The rate constants of covalent bond formation

Table 1. Binding Constants and Rate Constants for Covalent Bond Formation with IS2MN, IS5MN, ISDN, and Nitroglycerin

Enzyme	Compound	K_i (μM)	k_3 (min^{-1})	k_3/K_i $(min^{-1}\mu M^{-1})$
E1	IS2MN	20.5	2..5	0.12
"	IS5MN	1408	18	0.01
"	ISDN	75	16	0.2
E2	IS2MN	14	0.17	0.01
"	IS5MN	474	0.9	0.002
"	ISDN	12	0.5	0.04
"	NITROGLYCERIN	44	6	0.14

Data for E1 from reference [19]; data for E2 were calculated by averaging results from several different experiments. All experiments (except nitroglycerin) were at 25°C; nitroglycerin data were obtained at 0°C.

for ISDN resemble those of IS5MN. Thus it appears that ISDN is recognized by the enzyme via its nitrate at position 2, while its reaction occurs at nitrate at position 5.

Binding Constants and Rate Constants for Covalent Bond Formation with IS2MN, IS5MN, ISDN and Niroglycerin.

An interesting point emerging from this comparison was the fact that K_i values were all lower for the E2 isozyme than the E1 isozyme showing that all three isosorbide esters bound better to E2 isozyme. Despite better binding to the E2 isozyme the inhibition of the E1 isozyme occurred faster than that of the E2 isozyme. The reason for greater susceptibility of the E1 isozyme to nitrate ester inhibition lies in the velocities of covalent bond formation (see k_3 values in Table 1) which are much greater for the E1 than the E2 isozyme. The k_3/K_i ratios (Table 1) also demonstrate that all three isosorbide esters are more potent inhibitors of the E1 isozyme than of the E2 isozyme; the larger the k_3/K_i value, the better the inhibitor. This agrees well with our previous work with irreversible inhibitors of aldehyde dehydro-genase where E1 isozyme was always more susceptible to inhibiton than the E2 isozyme (Hempel and Pietruszko, 1981; Vallari and Pietruszko, 1982; MacKerell et al., 1986).

Nitroglycerin was tested only with the E2 isozyme. The inactivation was so rapid at 25°C that the experiments had to be performed at 0°C. As with isosorbide esters, the E2 isozyme showed saturation kinetics with nitroglycerin. The Ki value (Table 1) for nitroglyc-erin resembles those for IS2MN and ISDN while the rate constant of the covalent bond formation is much larger, providing an explanation for the greater reactivity of nitroglycerin.

The above results confirmed that inhibition of aldehyde dehydrogenase by nitrate esters occurred in an irreversible manner. The chemical structure of nitroglycerin is that of a trinitrate ester, ISDN (Figure 1) is a dinitrate ester of isosorbide while IS2MN and IS5MN are mononitrate esters of isosorbide; all have no reactive groups that form covalent bonds with protein nucleophiles. Aldehyde dehydrogenase, however, catalyzes ester hydrolysis. Incubation of ISDN with aldehyde dehydrogenase resulted in formation of mononitrate products. These products (IS2MN and IS5MN) were identified by both HPLC (Figure 3) and TLC and were quantitated via HPLC. In the absence of enzyme ISDN did not form any products. The same products, however, could be readily demonstrated following incubation

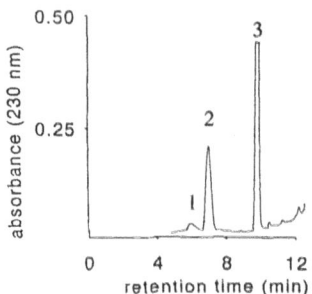

Figure 3. Products separated by HPLC following incubation of the E1 isozyme with ISDN. Peak 1 = IS2MN; Peak 2 = IS5MN; Peak 3 = ISDN.

of ISDN with 2-mercaptoethanol at concentrations comparable to the total concentration of the enzyme SH groups (of which there are 44 per E1 tetramer and 36 per E2 tetramer). Enzyme concentrations employed in these experiments were high (10 - 90 μM).

The above experiments suggested that the observed product formation from ISDN could have occurred by the non-enzymatic reaction with the enzyme sulfhydryl groups. In order to determine if product formation occurred enzymatically it was necessary to employ inactive enzyme. Iodoacetamide (Hempel and Pietruszko, 1981) and disulfiram (Vallari and Pietruszko, 1982), both known to react at the active site of aldehyde dehydrogenase, were initially employed for the E1 isozyme inactivation (Table 2). In both cases, following incubation with ISDN, there was a considerable decrease in the amount of the product formed, relative to that by the unmodified control, indicating that a large part of the product must have been formed by the residues with which iodoacetamide and disulfiram reacted. In both cases, however, the amount of product formed by the inactivated enzyme was not insignificant (Table 2). Since, the E1 isozyme modified by either iodoacetamide or disulfiram still retained some catalytic activity (Table 2) it was impossible to decide whether product formed by the modified enzyme was a result of non-enzymic process. By the use of bromoacetophenone (MacKerell et al., 1986; Abriola et al., 1987; Abriola et al., 1990), which totally abolished activity (Table 2), the enzymic and non enzymic components of product formation could be finally identified and separated. When bromoacetophenone-inactivated

Table 2. Effect of Active Site Modification of the E1 Isozyme with Irreversible Inhibitors on Product Formation from ISDN

Inhibitor	Activity Remaining After Modification (% Control)	ISDN Incubation at 25°C	Products Formed from ISDN % Control	
			IS2MN	IS5MN
Iodoacetamide	7	16 hrs	65	39
Disulfiram	13	48 hrs	41	8
Bromoacetophenone	0	16 hrs	12	27
Bromoacetophenone	0	10 min	0	10 -25

Modification conditions are described in detail [14,15,16].

Table 3. Enzyme-Catalyzed Product Formation from ISDN and Activity
Loss

Time (min)	Active Enzyme (μN)	Inactivated Enzyme (μN)	Products (μM)		Sum of Products (μM)	Ratio of Sum of Products to Inactivated Enzyme
			IS2MN	IS5MN		
0	56	0	0	0		
0.3	48	8	3.5	2	5.5	0.7
1.0	38	19	13	7	20	1.1
4.0	19	37	31	15	46	1.2
8.0	22	44	30	19	49	1.1

E1 (8.9 mg/ml, SA = 0.4 μmol/min/mg) was incubated with 1mM ISDN at 0°C. Enzyme normality (μN), calculated from activity loss, was based on two active sites per molecule of 219,000 daltons. Bromoacetophenone-modified control incubated with ISDN in the same conditions formed no mononitrate products.

E1 isozyme was incubated with ISDN for short time periods (10 min) there was virtually no product formation as compared to longer incubation times (16 hr) and to unmodified enzyme. These experiments indicated that products were produced by the non-enzymic process at a considerably slower rate than the enzyme-catalyzed reaction. Following this, the enzyme-catalyzed product formation could be easily separated from the non-enzymic reaction by lowering the incubation temperature with ISDN to 0°C (Table 3).

The products of enzymic degradation of ISDN were identified as IS2MN (major product) and IS5MN (minor product) (see Table 3). The results presented in Table 3 also demonstrate that inactivation of the E1 isozyme occurs simultaneously with metabolism. In fact, only a single catalytic turnover is required to inactivate the enzyme. Thus, an enzyme inhibitor that irreversibly inactivates aldehyde dehydrogenase, must be generated during metabolism.

Metabolism of ISDN could be monitored by HPLC where mononitrate products were identified and determined quantitatively. With the mononitrates, because isosorbide does not absorb, metabolism was studied via disappearance of substrates. At low (40 μM) substrate concentration at 0°C in the presence of E1 isozyme, IS2MN was found to disappear much faster than either IS5MN or ISDN (Mukerjee and Pietruszko, 1994). Thus, IS2MN which had the lowest K_i value as an inhibitor (see Table 1), was also the best substrate, confirming that nitrate at position 2 was best recognized by aldehyde dehydrogenase active site. The results obtained via metabolism, like those from saturation kinetics also suggest that ISDN is bound to aldehyde dehydrogenase via nitrate at position 2, while major metabolism occurs at position 5.

To obtain more information about the reaction mechanism with nitrate esters, inactivation of the enzyme was carried out in the presence of ^{14}C-labeled ISDN. No incorporation of ^{14}C-labeled ISDN into the enzyme was observed. This suggested that inactivation might be occurring via nitration of an active site residue. Nitrate or nitrite, however, were never detected among reaction products at 0°C where only enzyme-catalyzed product formation occurred. It is possible that at 0°C the acyl intermediate is stable. Nitrite (but not nitrate) was detected following incubation with the enzyme at 25°C, but in less than

Table 4. Products Formed by Alkaline Hydrolysis of 1mM ISDN

Conditions of Hydrolysis	Products Following Neutralization				Nitrite Direct
	IS2MN	IS5MN	Nitrite	Nitrate	
50mM phosphate pH 12.5 with NaOH, 16hr,25°C	132	186	65	0	-
1N NaOH, 16hr, 37°C	201	362	15	0	1300
5N KOH, 10 min, 100°C	-	-	-	-	1476

Neutralization was done by addition of HCl to pH 7.0. Direct determination of nitrate was not possible because the procedure was enzymatic.

stoichiometric amounts relative to alcohol products formed (Mukerjee and Pietruszko, 1994).

Hydrolysis of ISDN with alkali followed by identification and quantitation of products was, therefore, attempted. It can be seen (Table 4) that hydrolysis of ISDN by NaOH results in formation of nitrite and not nitrate, as anticipated. Exact quantitation of the nitrite product was also difficult. If the pH of the reaction mixture was adjusted to pH 7.0 prior to assay the amount of nitrite was low and considerably less than stoichiometric. There was no nitrate detected.

If pH was rapidly adjusted to pH 1 by addition of sulfuric acid, the amount of nitrite was almost stoichiometric. Other hydrolysis products were identified as IS2MN, IS5MN and isosorbide. Hydrolysis of organic nitrate esters to nitrite products was previously described (Ignarro, 1990; Boschan et al., 1955). In the cases where nitrite was the product the alcohol was converted to an unsaturated or to a carbonyl compound by the reactions shown below:

(a) Nucleophilic substitution:

$$HO^- + RONO_2 \rightarrow ROH + NO_3^-$$

(b) Elimination of ß-hydrogen:

$$HO^- + RCH_2CH_2ONO_2 \rightarrow RCH = CH_2 + H_2O + NO_2^-$$

(c) Elimination of α-hydrogen:

$$HO^- + RCH_2ONO_2 \rightarrow RCH = O + H_2O + NO_2^-$$

For this reason TLC plates, used for identification of hydrolysis products, were sprayed with 2,4-dinitrophenyl hydrazine, or with alkaline permanganate, but no carbonyl or unsaturated compound was detected. In partial sodium hydroxide hydrolysis, as in other non-enzymic reactions, IS5MN was the major product.

Since alkaline hydrolysis of ISDN yielded nitrite and not nitrate, possibility of a nitrite thioester enzyme intermediate was also considered. Nitrite thioesters (Oae and Shinhama, 1993) are usually red or green colored compounds with absorbance between 250-360 nm; however no change color or of absorbance was ever observed upon reaction of aldehyde dehydrogenase with isosorbide dinitrate. Nitrite can be also displaced from nitrite thioesters by mercuric salts (Saville, 1958) and subsequently assayed. When this was attempted, following inhibiton of aldehyde dehydrogenase with ISDN, no nitrite was detected, suggesting that nitrite thioester was not an intermediate.

During this investigation it has been demonstrated that the number of enzyme sulfhydryl groups decreased during enzyme inactivation (2SH groups were lost for 80% activity loss of the E1 isozyme). Also, the activity of the inactivated enzyme could be fully restored by treatment with 2-mercaptoethanol, suggesting involvement of enzyme sulfhydryl groups in the process. Thus it appears that the enzyme-intermediate in nitrate ester inactivation would most likely involve a sulfhydryl group or a pair of sulfhydryl groups. Both E1 and E2 isozymes contain a cysteine as the catalytic residue (cysteine 302; Kitson et al., 1991; Blatter et al., 1992), both also contain an adjacent cysteine in position 301. The E2 isozyme also contains an adjacent cysteine in position 303. The enzyme intermediate in both metabolism and inactivation could be a nitrate thioester involving residue 302. Reaction of nitrate thioesters with sulfhydryl groups leads to reduction of nitrate to nitrite and the concomitant formation of a disulfide.

Inactivation of aldehyde dehydrogenase by nitrate esters could occur by the following mechanism. Nitrate ester (eg. ISDN) is recognized and bound by the enzyme as a substrate. ISMN is released as a product with a simultaneous formation of a nitrate thioester with the catalytic sulfhydryl residue, cysteine 302. The nitrate thioester is then attacked by the adjacent sulfhydryl (cysteine 301) or another cysteine adjacent in tertiary structure, reducing the nitrate to nitrite and forming a disulfide between cysteine 302 and this other cysteine. This would result in an inactive enzyme; reduction of this disulfide, involving the catalytic residue, with 2-mercaptoethanol would fully restore catalytic activity.

Thus, it appears that inactivation of aldehyde dehydrogenase by nitrate esters is mechanism-based and occurs via esterase function of aldehyde dehydrogenase. Esterase reaction of aldehyde dehydrogenase has been generally considered to be unphysiological and a simple outcome of a catalytic mechanism involving a covalent intermediate. The above results suggest that esterase reaction of aldehyde dehydrogenase may be physiologically important. A disulfiram-like reaction to nitroglycerin and ISDN is well recognized *in vivo* in man following antianginal medication and consumption of alcoholic beverage. This aversive reaction most likely occurs as a result of aldehyde dehydrogenase inactivation by a mechanism analogous to that discussed in this paper.

ACKNOWLEDGEMENTS

Supported by R. Brinkley Smithers Institute For Alcoholism Prevention, Research Scientist Award K05 AA00046 and Charles and Johanna Busch Memorial Fund. We wish to express our gratitude to Schwarz Pharma, Milwaukee, Wisconsin for providing us with gift samples of isosorbide-2-mononitrate, isosorbide-5-mononitrate and isosorbide.

REFERENCES

Abriola, D.P., Fields, R., Stein, S., MacKerell, A.D., Jr. and Pietruszko, R. (1987) Active site of aldehyde dehydrogenase. Biochemistry 26, 5679-5684.

Abriola, D.P., MacKerell, A.D. Jr., and Pietruszko, R (1990) Correlation of loss of activity of human aldehyde dehydrogenase with reaction of bromoacetophenone with glutamic acid 268 and cysteine 302 residues. Partial sites reactivity of aldehyde dehydrogenase. Biochem. J. 266, 179-187.

Ambroziak, A., Kosley, L. L., and Pietruszko, R. (1989) Human aldehyde dehydrogenase: coenzyme-binding studies. Biochemistry, 28, 5367-5373.

Bell, F.K., O'Neil J.J and Burgison R. M. (1963) Determination of the oil/water distribution coefficients of glyceryl trinitrate and two similar nitrate esters. J. Pharm Sci, 52, 637-639.

Blatter, E. E., Abriola, D. P., and Pietruszko, R. (1992) Aldehyde dehydrogenase: covalent intermediate in aldehyde dehydrogenation and ester hydrolysis. Biochem. J. 282, 353-360.

Blatter, E.E., Tasayco, M.L., Prestwich, G., and Pietruszko, R. (1990) Chemical modification of aldehyde dehydrogenase by a vinyl ketone analog of an insect pheromone. Biochem. J. 272, 351-358.

Boschan, R., Merrow, R.T., and Van Dolah, R.W. (1955) The chemistry of nitrate esters. Chem. Rev. 55, 485-510.

Dipalma, J. F., (1982) The nitrites and nitrates. Am. Fam. Physician, 25, 216-218.

Habeeb, A.F.S.A., (1972) Reaction of protein sulfhydryl groups with Ellman's Reagent. Methods in Enzymology, 25, 457-464.

Hempel, J.D., and Pietruszko, R. (1981) Selective chemical modification of human liver aldehyde dehydrogenases E1 and E2 with iodoacetamide. J. Biol. Chem. 256, 10889-10896.

Ignarro, L. J. (1990) Biosynthesis and metabolism of endothelium-derived nitric oxide. Ann. Rev. Pharmacol. Toxicol. 30, 535-560.

Ignarro, L.J., Lippton, H., Edwards, J.C., Baricos, W.H., Hyman, A.L., Kadowitz, P.J., and Gruetter, C.A. (1981) Mechanism of vascular smooth muscle relaxation by organic nitrates, nitrites, nitroprusside and nitric oxide. Evidence for the involvement of S-nitrosothiols as active intermediates. J. Pharmacol. Expl. Ther. 218, 739-749.

Kitson, T. M. (1977) A disulfiram-ethanol reaction. J. Stud. Alc, 38, 96-113.

Kitson, T.M., Hill, J.P., and Midwinter, G.G. (1991) Identification of a catalytically essential nucleophile residue in sheep liver cytoplasmic aldehyde dehydrogenase. Biochem. J. 275, 207-210.

Kitz, R., and Wilson, I. B. (1962) Esters of methanesulfonic acid as irreversible inhibitors of acetylcholinesterase. J. Biol. Chem, 237, 3245-3249.

MacKerell, A.D., Jr., MacWright, R.S., and Pietruszko, R. (1986) Bromoacetophenone as an affinity reagent for human liver aldehyde dehydrogenase. Biochemistry, 25, 5182-5189.

Mukerjee, N., and Pietruszko, R. (1992) Human mitochondrial aldehyde dehydrogenase substrate specificity: comparison of dehydrogenase with esterase reaction. Arch. Biochem. Biophys. 299, 23-29.

Mukerjee, N., and Pietruszko, R. (1994) Inactivation of human aldehyde dehydrogenase by isosorbide dinitrate. J. Biol. Chem. 269, 21664-21669.

Needleman, P. (1975) Pharmacological and biochemical interactions of organic nirates with sulfhydryls: possible correlations with the mechanism for tolerance development, vasodilation, and mitochondrial and enzyme reactions. in Organic Nitrates, pp 98-114, New York, Springer-Verlag, (Needleman, P ed.).

Needleman, P., and Hunter, F. E. (1965) The transformation of glyceryl trinitrate and other nitrates by glutathione-organic nitrate reductase. Mol. Pharmacol, 1, 77-86.

Needleman, P., and Johnson, E.M. (1973) Mechanism of tolerance development to organic nitrates. J. Pharmacol. Expl.Ther. 184, 709-715.

Oae, S., and Shinhama, K. (1993) Organic thionitrites and related substances. A review. Organic Prep. Proc. Intl. 15, 167-198.

Saville, B. (1958) A scheme for the colorimetric determination of microgram amounts of thiols. Analyst (London), 83, 670-672.

Towell, J., Garthwaite, T., and Wang, R. (1985) Erythrocyte aldehyde dehydrogenase and disulfiram-like side effects of hypoglycemics and antianginals. Alcoholism: Clin. Exp. Res, 9, 438-442.

Vallari, R.C., and Pietruszko, R. (1982) Human aldehyde dehydrogenase: mechanism of inhibition by disulfiram. Science, 216, 637-639.

5

USE OF A CHROMOPHORIC REPORTER GROUP TO PROBE THE ACTIVE SITE OF CYTOSOLIC ALDEHYDE DEHYDROGENASE

Trevor M. Kitson and Kathryn E. Kitson

Department of Chemistry and Biochemistry
Massey University
Palmerston North
New Zealand

INTRODUCTION

The concept of using a covalently-linked chromophoric group to 'report' on the environment of an enzyme's active site was first expounded by Burr and Koshland (1964). In the work reported here, we have used 3,4-dihydro-3-methyl-6-nitro-2*H*-1,3-benzoxazin-2-one (DMNB) to provide a reporter group for examining the nature of the active site in cytosolic aldehyde dehydrogenase from sheep liver.

Aldehyde dehydrogenase catalyses the dehydrogenation of aldehydes by NAD^+ and also the hydrolysis of esters such as *p*-nitrophenyl acetate. Previous work has established that Cys-302 (the only completely conserved cysteine residue in all known aldehyde dehydrogenase sequences; Hempel *et al.*, 1991, 1993) constitutes the nucleophile that attacks both the aldehyde substrate 4-*trans*-(N,N-dimethylamino)cinnamaldehyde and the ester substrate analogue 4-*trans*-(N,N-dimethylamino)cinnammoylimidazole (Blatter *et al.*, 1992), and is modified by several other reagents (von Bahr-Lindström *et al.*, 1985; Abriola *et al.*, 1990; Blatter *et al.*, 1990). We have shown that *p*-nitrophenyl [^{14}C]dimethylcarbamate (an analogue of the typical activated ester substrates that the enzyme hydrolyses) also specifically labels Cys-302 (Kitson *et al.*, 1991). We reasoned that the cyclic compound DMNB (which is a close structural relation of *p*-nitrophenyl dimethylcarbamate) would react similarly with the enzyme, placing at the active site a covalently-linked *p*-nitrophenol moiety as shown here:

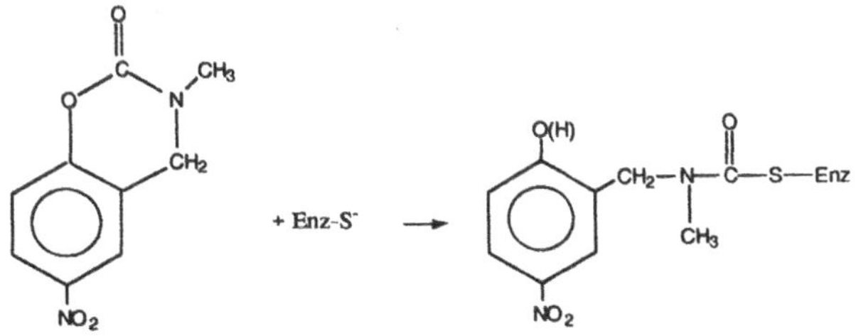

Enzymology and Molecular Biology of Carbonyl Metabolism 5
Edited by H. Weiner *et al.*, Plenum Press, New York, 1995

The synthesis and properties of DMNB were reported and its action on chymotrypsin was examined (Kitson and Freeman, 1993), followed by preliminary studies with aldehyde dehydrogenase (Kitson and Kitson, 1994). The compound has proved to be an excellent reporter-group reagent for the following reasons. It has the potential to modify any enzyme with p-nitrophenyl esterase activity. It is only minimally larger than the common substrate p-nitrophenyl acetate so steric hindrance should not be a problem in its use. It is unlikely to react with an enzyme elsewhere than with the catalytically essential nucleophile. The nitrogen atom adjacent to the carbonyl group enormously reduces the latter's electrophilicity resulting in an acylation step that is very much slower than with an ester substrate, but which does nevertheless occur. However, the rate of the subsequent deacylation step (which would follow with a normal substrate) is reduced to zero, providing an inert, stable acyl-enzyme for examination at leisure. This acyl-enzyme carries a p-nitrophenol moiety that, when ionised, absorbs strongly in a region of the visible spectrum well away from any complicating absorbance arising from protein or NADH, for example. Furthermore, the p-nitrophenol group's normal pK_a value (~7) is conveniently right in the middle of the pH range of physiological significance, and so studying perturbations in this pK_a caused by changes in the reporter group's environment is likely to provide valuable and meaningful information. We report below on the current state of our work with DMNB and sheep liver cytoplasmic aldehyde dehydrogenase.

EXPERIMENTAL

Cytosolic aldehyde dehydrogenase was purified from sheep liver and modified by DMNB as reported previously (Kitson and Kitson, 1994). DMNB was synthesised as described by Kitson and Freeman (1993).

Determination of the pK_a of the P-Nitrophenol Group in DMNB-Labelled Aldehyde Dehydrogenase

Samples of labelled enzyme solution (0.5 mL) were mixed with 1.5 mL of various buffers (McKenzie, 1969) and the absorbance spectrum from 300-550 nm was recorded at 20'C with a Varian Cary 1 spectrophotometer. In measuring the amplitude of the p-nitrophenoxide peak, the A_{550} was taken as zero. In some cases after scanning, diethylstilboestrol in ethanol (15 μL) was added to give a concentration of 50 μM and then NAD$^+$ in 10 mM phosphate buffer, pH 7.4 (25 μL), was added to give a concentration of 2 mM, and the absorbance spectrum was re-scanned after each of the additions. The data were plotted by using Enzfitter (Leatherbarrow, 1987) to compute the best theoretical titration curve.

Titration of DMNB-Labelled Aldehyde Dehydrogenase with NAD$^+$

DMNB-labelled aldehyde dehydrogenase (2.5 mL) was mixed with 0.5 mL of 0.2 M phosphate buffer, pH 7.4. The small value of A_{440} of this solution was subtracted from subsequent measurements. Several sequential additions of NAD$^+$ in 10 mM phosphate buffer, pH 7.4, were then made at 25'C, each aliquot being 5 μL and giving an increase in NAD$^+$ concentration of 1 μM. The A_{440} was measured after each addition. This was followed by several more 5 μL additions each giving an increase in NAD$^+$ concentration of 10 μM. The total increase in volume of the solution was 5%. Finally, solid NAD$^+$ (~1 mg) was added in order to determine the limiting A_{440} in the presence of excess NAD$^+$. In a similar procedure, sequential additions of NADP$^+$ (10 μL, 30 μM) were made.

Stopped-Flow Investigation of Binding of NAD⁺ to DMNB-Modified Aldehyde Dehydrogenase

Modified enzyme from one syringe was mixed with NAD^+ solution from the other syringe of a stopped-flow spectrophotometer (Hi-Tech Scientific) at 25'C. Solutions contained 70 mM phosphate buffer, pH 7.4, enzyme (4.7 µM after mixing), and various concentrations of NAD^+. The increase in A_{440} was shown to conform closely to a single exponential curve.

RESULTS AND DISCUSSION

The Nature of the Substrate Binding Site in Aldehyde Dehydrogenase

The pK_a of the enzyme-linked reporter group derived from DMNB is 7.21 after the enzyme has been denatured with perchloric acid and redissolved in urea solution (Kitson and Kitson, 1994). This is very similar to the pK_a of free p-nitrophenol in aqueous solution, i.e. 7.15 (Kitson and Freeman, 1993), and this is to be expected, since under these conditions any peculiar enzymic microenvironment will have been totally disrupted. However, on the undenatured enzyme, the reporter group shows a considerably higher pK_a value of 9.75-10.1 (Kitson and Kitson, 1994) and 9.92 (this work, see Fig. 1). This of course means that the environment of the site in which the reporter group finds itself strongly disfavours the ionisation of the p-nitrophenol group. Two explanations for this spring to mind. The first is that the active site of aldehyde dehydrogenase contains a negatively charged group that would interact unfavourably with the negative p-nitrophenoxide moiety. The second is that the active site is of much less polar nature than water, such that the neutral p-nitrophenol group is preferred over its corresponding ionised form.

At present we cannot unequivocally decide which of the two possible explanations described above is correct. Conceivably of course, both factors may contribute to some extent. Some previous observations (such that iodoacetamide but not iodoacetate inactivates aldehyde dehydrogenase, or that the enzyme is inactive towards glyoxylate as a substrate; Hempel and Pietruszko, 1981; Deady et al., 1985) can likewise be explained either in terms of a negatively charged active site or simply by a very non-polar one. However, the observation that a positively-charged analogue of disulfiram is as effective an inactivator of cytosolic aldehyde dehydrogenase in vitro as disulfiram itself, whilst a negatively-charged

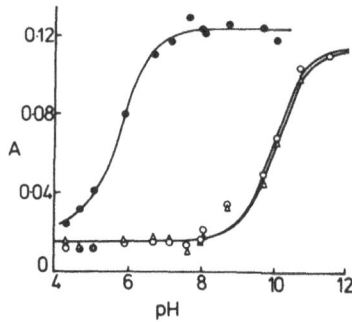

Figure 1. Ionisation profile of the DMNB reporter group bound to cytosolic aldehyde dehydrogenase. ○, in absence of diethylstilboestrol and NAD^+; Δ, in presence of diethylstilboestrol (50 µM);●, in presence of diethylstilboestrol (50 µM) and NAD^+ (2 mM).

analogue is much less potent (Kitson, 1976), supports the idea that the active site is negative in character better than the idea of its merely having a preference for non-polar molecules. Pietruszko *et al.* (1993) have argued, based on its nucleophilicity towards bromoacetophenone, that Glu-268 is present at or near the active site as a 'naked anion', and it may be this residue that inhibits the ionisation of the DMNB reporter group. On the other hand, there is much circumstantial evidence pointing to a hydrophobic character for aldehyde dehydrogenase's active site. For instance, the enzyme is active towards long-chain aldehydes, aromatic aldehydes and steroidal aldehydes. Even negatively charged aromatic aldehydes (2- and 4-carboxybenzaldehyde) are substrates (Deady *et al.*, 1985). Retinal is a particularly good substrate for the cytosolic isoenzyme (Yoshida *et al.*, 1993). Aldehyde dehydrogenase binds also to non-aldehyde steroids and analogues (such as progesterone and diethylstilboestrol; Kitson, 1982), to an acetophenone-linked affinity matrix (Ghenbot and Weiner, 1992), and of course to *p*-nitrophenyl esters and DMNB itself.

It is intriguing that the pK_a values of the DMNB reporter group when bound to chymotrypsin (9.03 and 6.94 before and after denaturation, respectively; Kitson and Freeman, 1993) are fairly similar to the values observed when bound to aldehyde dehydrogenase, and yet the λ_{max} values are quite different (395 and 433 nm, respectively). At this stage, however, we have insufficient information to conclude that this means that the binding site of aldehyde dehydrogenase is necessarily negatively charged. (That of chymotrypsin is known of course to be non-polar.) It would be informative to probe other well-studied enzymes with DMNB in order to build up a fuller picture of the relationship between λ_{max} of the reporter group and the character of the binding site.

Binding of NAD^+ to DMNB-labelled aldehyde dehydrogenase reduces the pK_a of the reporter group dramatically, from ~10 to 5.35 (Kitson and Kitson, 1994) or 5.78 (this work, in the presence also of diethylstilboestrol; see Fig. 1). In other words, the acidity of the *p*-nitrophenol group is tens of thousands of times greater in the presence of NAD^+ than in its absence. The only obvious explanation for this is that the ionised *p*-nitrophenoxide form of the reporter group is stabilised by interaction with a positive charge. Either NAD^+ causes a conformational change in the enzyme, bringing a positively charged side-chain into close proximity to the reporter group, or the positively charged nitrogen in the nicotinamide ring of NAD^+ itself provides the stabilising influence. The latter explanation is supported by the fact that, unlike NAD^+, NADH has very little effect on the pK_a of the reporter group (Kitson and Kitson, 1994). If the nicotinamide ring of NAD^+ and the DMNB-derived reporter group are indeed effectively in contact, then this tends to support the view that dehydrogenase and esterase sites of aldehyde dehydrogenase are identical, a belief which is not universally held (see Blackwell *et al.*, 1989, and references therein).

Diethylstilboestrol has been shown to be a competitive inhibitor of the hydrolysis of *p*-nitrophenyl pivalate catalysed by cytosolic aldehyde dehydrogenase, with a K_i of 8.3 μM (Kitson, 1989). It was reasoned that this hydrophobic, phenolic compound might also compete for the enzyme's binding site against the DMNB-derived reporter group if the latter is capable of pivoting into and out of this site. Experiment shows that this does not happen; a high concentration of diethylstilboestrol (50 μM) has little or no effect on the pK_a of the reporter group either in the presence or absence of NAD^+ (Fig. 1 and Kitson and Kitson, 1994).

The addition of $NADP^+$ to DMNB-modified aldehyde dehydrogenase at pH 7.4 results in no development of absorbance in the visible region, unlike what happens with NAD^+ (see Fig. 2). Moreover, the presence of $NADP^+$ does not interfere in any way with the ionisation of the reporter group that is induced by adding NAD^+. Of the several classes of aldehyde dehydrogenase, only the class 3 enzymes can utilise $NADP^+$ as coenzyme (Hempel *et al.*, 1993). The present work clearly shows that $NADP^+$ does not bind to the cytosolic (class 1) enzyme at all.

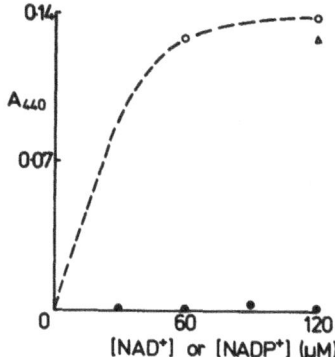

Figure 2. Comparison of the effects of NAD$^+$ and NADP$^+$ on the ionisation of the DMNB reporter group bound to cytosolic aldehyde dehydrogenase. ○, NAD$^+$; ●, NADP$^+$; △, NADP$^+$ (120 μM) followed by NAD$^+$ (60 μM).

The Binding of NAD$^+$ to DMNB-Modified Aldehyde Dehydrogenase

The wide difference in pK$_a$ values of the reporter group in the presence and absence of NAD$^+$ provides a sensitive probe for investigating the strength of binding and rate of binding of NAD$^+$ to the modified enzyme at neutral pH. Fig. 3 shows the increase in A$_{440}$ as small increments of NAD$^+$ concentration are made. The titration is biphasic. There is an initial marked increase in absorbance followed by a second phase in which the increase is less steep and apparently linear. Each phase is associated with approximately half the total maximal increase in absorbance produced by adding excess NAD$^+$. From the magnitude of

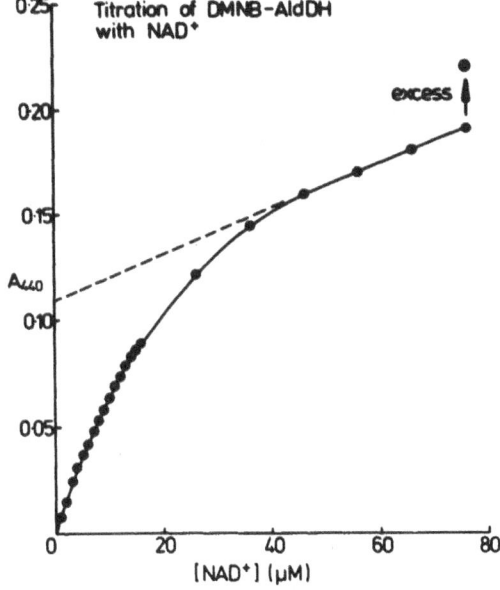

Figure 3. Titration of DMNB-modified cytosolic aldehyde dehydrogenase with NAD$^+$. The figure shows the increase in A$_{440}$ as small increments of NAD$^+$ concentration are made to DMNB-labelled enzyme (8.1 μM) at pH 7.4 and 25′C.

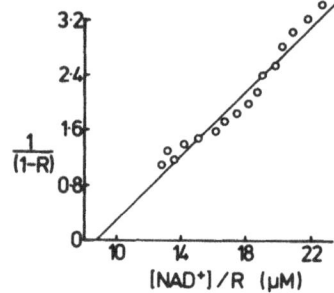

Figure 4. Scatchard plot constructed from the data shown in Fig. 3. R is the fractional saturation of the tightly binding NAD^+ site. Error is estimated to be ± 5%.

this limiting absorbance increase and the enzyme concentration, it can be calculated that there are approximately 1.65 reporter groups bound per enzyme tetramer (making the assumption that the extinction coefficient of the fully ionised reporter group is the same as that of free p-nitrophenoxide). This result is in line with previous studies that have suggested that aldehyde dehydrogenase has 'half-of-the-sites' reactivity (Ambroziak *et al.*, 1989; Weiner, 1982).

By back-projecting the apparently linear part of the titration curve shown in Fig. 3, the earlier data could be analysed as a Scatchard plot (Gutfreund, 1972); the result is shown in Fig. 4. This procedure, perhaps as to be expected, does not give a very good straight line, but it allows an approximate estimate for the dissociation constant to be calculated, and this is 3-4 μM. Apparently, therefore, DMNB-modified aldehyde dehydrogenase has two binding sites for NAD^+, one with K = 3-4 μM and one with a lower affinity for NAD^+. The dissociation constant for the native enzyme-NAD^+ complex has been reported to be 3.2 μM (from titrations in which the quenching of protein fluorescence was monitored) and 8 μM (from steady-state kinetic data) (Blackwell *et al.*, 1989). Thus it appears from these results that the presence of the reporter group does little if anything to strengthen the binding of NAD^+, even though we have suggested above that reporter group and NAD^+ may have a direct ionic interaction, and even though it has been shown previously that NAD^+ is not easily removed from DMNB-modified enzyme by either dialysis or gel filtration (Kitson and Kitson, 1994).

The yellow colour that develops rapidly on mixing NAD^+ with DMNB-modified enzyme can be readily monitored by stopped-flow spectrophotometry. A typical single-ex-

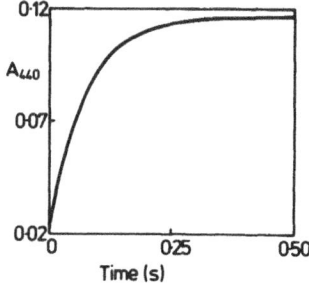

Figure 5. Stopped-flow spectrophotometer trace obtained at 440 nm when DMNB-modified cytosolic aldehyde dehydrogenase is mixed with NAD^+ at 25'C and pH 7.4. Final concentrations: enzyme, 4.7 μM; NAD^+, 100 μM.

ponential increase in absorbance is shown in Fig. 5. The magnitude of the absorbance increase and of the first-order rate constant as a function of NAD^+ concentration are shown in Fig. 6. Interestingly, the rate of colour development continues to increase with NAD^+ concentration beyond the point at which there is little further increase in the absorbance amplitude. An explanation of these results is complicated by the fact that several processes are in operation here. First, NAD^+ binds to the enzyme; secondly, there is probably an NAD^+-induced change in enzyme conformation (Blackwell *et al.*, 1989); and thirdly, at some stage the reporter group ionises. The relative rates of these processes are not known. In particular, the rate of ionisation of the reporter group may be limiting in some circumstances. For example, we have shown that mixing DMNB-labelled enzyme with a high concentration of pH 11.5 buffer in the stopped-flow spectrophotometer results in a value of 29 s^{-1} for the first-order rate constant for ionisation of the reporter group. Presumably the actual change in pH of the medium is virtually instantaneous; ionisation of the reporter group must await the diffusion into the active site of a water molecule to act as a proton acceptor. That this process is relatively slow supports the argument that the active site environment is hydrophobic.

Future work will attempt to clarify the results obtained above with NAD^+, and will also examine the competition between NAD^+ and NADH for the modified enzyme's binding site, since at neutral pH the enzyme-NAD^+ complex is yellow and the enzyme-NADH complex is colourless (Kitson and Kitson, 1994).

Crystallisation and X-Ray Studies of DMNB-Labelled Aldehyde Dehydrogenase

The cytosolic aldehyde dehydrogenase from sheep liver will crystallise readily under some conditions, but the crystals are poorly formed and diffract X-rays only weakly. However, after modification with DMNB in the manner used for the reporter group work reported above, the enzyme was found to give good diffraction-quality crystals, perhaps because having an occupied active site results in a more rigid enzyme conformation. The details of the crystallisation and X-ray work are reported elsewhere in this volume.

It is of interest to note that, if an analogue of DMNB can be synthesised with an iodo substituent in one of the vacant positions of the benzene ring, then this may provide an ideal vehicle for the incorporation of heavy atoms into aldehyde dehydrogenase. When the detailed tertiary structure of the enzyme becomes known, then it will of course be possible

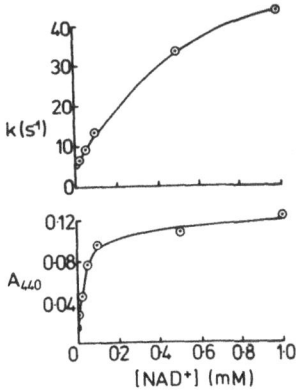

Figure 6. Plot of first-order rate constant and amplitude of absorbance change (from stopped-flow traces such as that shown in Figure 5) as a function of NAD^+ concentration.

to answer some of the questions posed above, such as: Is the active site of predominantly hydrophobic character or does it contain a negatively charged group, and if so is this Glu-268? Does the pyridinium nitrogen atom of NAD$^+$ interact directly with the p-nitrophenoxide moiety of the reporter group? Thus it may well turn out that reaction of the enzyme with DMNB leads to a solution of the tertiary structure that in turn leads neatly back to an explanation of what precisely happens in the reaction of the enzyme with DMNB.

ACKNOWLEDGMENTS

We are grateful to Mr Graham H. Freeman for assistance in the synthesis of DMNB, to Dr Paul D. Buckley for advice on operation of the stopped-flow spectrophotometer, and to the New Zealand Lottery Grants Board for funding assistance.

REFERENCES

Abriola, D.P., MacKerell, A.D. and Pietruszko, R., 1990, Correlation of loss of activity of human aldehyde dehydrogenase with reaction of bromoacetophenone with glutamic acid-268 and cysteine-302 residues, *Biochem. J.*, 266: 179-187.

Ambroziak, W., Kosley, L.L. and Pietruszko, R., 1989, Human aldehyde dehydrogenase: coenzyme binding studies, *Biochemistry*, 28: 5367-5373.

Blackwell, L.F., MacGibbon, A.K.H. and Buckley, P.D., 1989, Aldehyde dehydrogenases - kinetic characterisation, In Human Metabolism of Alcohol, vol. II (Crow, K.E. and Batt, R.D., eds), pp. 89-104, CRC Press, Boca Raton, Florida.

Blatter, E.E., Tasayco, M.L., Prestwich, G. and Pietruszko, R., 1990, Chemical modification of aldehyde dehydrogenase by a vinyl ketone analogue of an insect pheromone, *Biochem. J.*, 272: 351-358.

Blatter, E.E., Abriola, D.P. and Pietruszko, R., 1992, Aldehyde dehydrogenase: covalent intermediate in aldehyde dehydrogenation and ester hydrolysis, *Biochem. J.*, 282: 353-360.

Burr, M. and Koshland, D.E., 1964, Use of 'reporter groups' in structure-function studies of proteins, *Proc. Natl. Acad. Sci. USA*, 52: 1017-1024.

Deady, L.W., Buckley, P.D., Bennett, A.F. and Blackwell, L.F., 1985, Kinetics of inhibition and hysteresis of sheep liver cytoplasmic aldehyde dehydrogenase with glyoxylic acid: further evidence relating to the two-site model for aldehyde oxidation, *Arch. Biochem. Biophys.*, 243: 586-597.

Ghenbot, G. and Weiner, H., 1992, Purification of liver aldehyde dehydrogenase by p-hydroxyacetophenone-Sepharose affinity matrix and the coelution of chloramphenicol acetyl transferase from the same matrix with recombinantly expressed aldehyde dehydrogenase, *Protein Expression Purif.*, 3: 470-478.

Gutfreund, H., 1972, Enzymes: Physical Principles, pp. 68-72, Wiley, London.

Hempel, J.D. and Pietruszko, R., 1981, Selective chemical modification of human liver aldehyde dehydrogenases E$_1$ and E$_2$ by iodoacetamide, *J. Biol. Chem.*, 256: 10889-10896.

Hempel, J.D., Nicholas, H. and Jörnvall, H., 1991, Thiol proteases and aldehyde dehydrogenases: evolution from a common thiolesterase precursor? *Proteins: Struct., Funct. Genet.*, 11: 176-183.

Hempel, J.D., Nicholas, H. and Lindahl, R., 1993, Aldehyde dehydrogenases: widespread structural and functional diversity within a shared framework, *Protein Science*, 2: 1890-1900.

Kitson, T.M., 1976, The effect of some analogues of disulfiram on the aldehyde dehydrogenases of sheep liver, *Biochem. J.*, 155: 445-448.

Kitson, T.M., 1982, The activation of aldehyde dehydrogenase by diethylstilboestrol and 2,2'-dithiodipyridine, *Biochem. J.*, 207: 81-89.

Kitson, T.M., 1989, Kinetics of p-nitrophenyl pivalate hydrolysis catalysed by cytoplasmic aldehyde dehydrogenase, *Biochem. J.*, 257: 573-578.

Kitson, T.M. and Freeman, G.H., 1993, 3,4-Dihydro-3-methyl-6-nitro-2H-1,3-benzoxazin-2-one, a reagent for labelling p-nitrophenyl esterases with a chromophoric reporter group - synthesis and reaction with chymotrypsin, *Bioorg. Chem.*, 21: 354-365.

Kitson, T.M. and Kitson, K.E., 1994, Probing the active site of cytoplasmic aldehyde dehydrogenase with a chromophoric reporter group, *Biochem. J.*, 300: 25-30.

Kitson, T.M., Hill, J.P. and Midwinter, G.G., 1991, Identification of a catalytically essential nucleophilic residue in sheep liver cytoplasmic aldehyde dehydrogenase, *Biochem. J.*, 275: 207-210.

Leatherbarrow, R.J., 1987, Enzfitter Manual, Elsevier Science Publishers, Amsterdam.

McKenzie, H.A., 1969, pH and Buffers, In Data for Biochemical Research (Dawson, R.M.C., Elliott, D.C., Elliott, W.H. and Jones, K.M., eds), pp. 475-508, Oxford University Press, London.

Pietruszko, R., Abriola, D.P., Blatter, E.E. and Mukerjee, N., 1993, Aldehyde dehydrogenase: aldehyde dehydrogenation and ester hydrolysis, In Enzymology and Molecular Biology of Carbonyl Metabolism 4 (Weiner, H., Crabb, D.W. and Flynn, T.G., eds), pp. 221-232, Plenum Press, New York and London.

von Bahr-Lindström, H., Jeck, R., Woenckhaus, C., Sohn, S., Hempel, J. and Jörnvall, H., 1985, Characterisation of the coenzyme binding site of liver aldehyde dehydrogenase: differential reactivity of coenzyme analogues, *Biochemistry*, 24: 5847-5851.

Weiner, H., 1982, Aldehyde dehydrogenase, In Enzymology of Carbonyl Metabolism: Aldehyde Dehydrogenase and Aldo/Keto Reductase (Weiner, H. and Wermuth, B., eds), pp. 1-9, Alan R. Liss, New York.

Yoshida, A., Hsu, L.C. and Yanagawa, Y., 1993, Biological role of human cytosolic aldehyde dehydrogenase 1: hormonal response, retinal oxidation and implication in testicular feminisation, In Enzymology and Molecular Biology of Carbonyl Metabolism 4 (Weiner, H., Crabb, D.W. and Flynn, T.G., eds), pp. 37-44, Plenum Press, New York and London.

6

STUDIES OF THE ESTERASE ACTIVITY OF CYTOSOLIC ALDEHYDE DEHYDROGENASE USING STERICALLY HINDERED AND CYCLIC SUBSTRATES

Kathryn E. Kitson, Treena J. Blythe and Trevor M. Kitson

Department of Chemistry and Biochemistry
Massey University
Palmerston North
New Zealand

INTRODUCTION

Aldehyde dehydrogenase has the ability to catalyse the hydrolysis of p-nitrophenyl esters as well as the dehydrogenation of aldehydes by NAD^+. Some work has been interpreted in terms of two different types of active site (Blackwell $et\ al.$, 1983), but much other evidence points to the identity of the dehydrogenase and esterase sites (Loomes and Kitson, 1986; Kitson $et\ al.$, 1991). In particular, Cys-302 has been identified as the nucleophile that becomes acylated by the aldehyde substrate $trans$-4-(N,N-dimethylamino)cinnamaldehyde (Pietruszko $et\ al.$, 1993) and by the ester substrate analogue p-nitrophenyl dimethylcarbamate (Kitson $et\ al.$, 1991). Thus studying the esterase activity of aldehyde dehydrogenase should be capable of giving information about the active site of relevance to the dehydrogenase activity.

When the enzyme acts as a dehydrogenase, there is of course a requirement for NAD^+. However, the esterase activity can occur without NAD^+, and thus inhibition of the esterase activity by a particular substance will show how well that substance binds to the active site in the absence of NAD^+. The kinetics of inhibition of hydrolysis of the usual esterase substrate (p-nitrophenyl acetate) can be rather complicated (Kitson, 1989a), but simple competitive inhibition is observed with the substrate p-nitrophenyl pivalate (trimethylacetate) and inhibitors such as diethylstilboestrol ($K_i = 8.3\ \mu M$) and p-nitrobenzaldehyde ($K_i = 50\ \mu M$). This is because, with the sterically hindered pivalate substrate, the acylation step is rate-determining and so any subsequent binding of inhibitor to acyl-enzyme is not reflected in the inhibition pattern.

In this work, we have investigated the inhibition of hydrolysis of p-nitrophenyl pivalate by the following compounds. (1) p-Hydroxyacetophenone. This is used (in the absence of NAD^+) to elute aldehyde dehydrogenase from an acetophenone-linked affinity matrix (Ghenbot and Weiner, 1992). (2) Citral. This is a mixture of cis and $trans$ forms of

$Enzymology\ and\ Molecular\ Biology\ of\ Carbonyl\ Metabolism\ 5$
Edited by H. Weiner $et\ al.$, Plenum Press, New York, 1995

45

3,7-dimethyl-2,6-octadienal. The *trans* form resembles to some extent part of the common all-*trans* form of retinal. (3) Retinal. The cytoplasmic form of aldehyde dehydrogenase has an extremely high affinity for retinal; the K_m is reported to be 0.06 µM for the human enzyme (Yoshida *et al.*, 1993). We were interested in determining if this large hydrophobic aldehyde can bind to the enzyme in the absence of NAD⁺. With smaller substrates such as acetaldehyde and propanal (with which most kinetic studies have been performed), the aldehyde only binds to the enzyme after a conformational change induced by the prior binding of NAD⁺ (Blackwell *et al.*, 1989).

In the second part of the work reported here, we have examined the enzyme-catalysed hydrolysis of the lactone 6-nitrodihydrocoumarin, a cyclic analogue of the substrate *p*-nitrophenyl propanoate. Simple *p*-nitrophenyl esters (if not sterically hindered) show a rapid burst of *p*-nitrophenoxide production when acted on by cytosolic aldehyde dehydrogenase (Kitson, 1989a). We reasoned that the acyl-enzyme involved in the hydrolysis of the lactone would closely resemble the stable acyl-enzyme produced by modification of the enzyme with 3,4-dihydro-3-methyl-6-nitro-2*H*-1,3-benzoxazin-2-one (DMNB). (See elsewhere in this volume and Kitson and Kitson, 1994.) Therefore we predicted, from our previous work with DMNB, that the *p*-nitrophenol group temporarily covalently attached to the enzyme during the hydrolysis of 6-nitrodihydrocoumarin would be unionised in the absence of NAD⁺ and ionised in its presence. Thus a burst of absorbance in the visible spectrum would not be seen in the absence of NAD⁺ but would be observed in its presence. We report below that the first of these predictions was shown to be true. However, in the presence of NAD⁺, the stopped-flow spectrophotometer trace gave a totally unexpected result.

DMNB   6-nitrodihydrocoumarin

EXPERIMENTAL

Cytosolic aldehyde dehydrogenase was purified from sheep liver as reported previously (Kitson and Kitson, 1994). 6-Nitrodihydrocoumarin was prepared by nitration of dihydrocoumarin (Aldrich) using concentrated nitric and sulphuric acids in cold acetic acid (m.p. 134-135°C; Tobias *et al.*, 1969, give 130-130.7°C). Retinal (all-*trans*) was purchased from Sigma. It was dissolved in dimethylformamide to give a 10 mM solution and stored at -20°C in the dark. The solution was diluted to 1 mM with 70% aqueous methanol and suitable aliquots were taken for the inhibition studies. Equal amounts of retinal were added to the reference cell when performing spectrophotometric enzyme assays in the presence of retinal. The hydrolysis of *p*-nitrophenyl pivalate was studied in 50 mM phosphate buffer, pH 7.4, at 25°C, using a Varian Cary 1 spectrophotometer. The substrate was added in a small volume of ethanol. The formation of *p*-nitrophenoxide was followed at 400 nm. Best-fit straight lines were calculated for Lineweaver-Burk plots using Enzfitter (Leatherbarrow, 1987).

Stopped-flow studies were performed using a Hi-Tech Scientific instrument at 25°C. One syringe contained enzyme in 70 mM phosphate buffer, pH 7.4, with or without NAD⁺ (100 µM); the other syringe contained a mixture of 4.5 mL of 70 mM phosphate buffer, pH 7.4, and 0.5 mL of substrate in acetone, again with or without NAD⁺ (100 µM). Enzyme and substrate concentrations (after mixing) were 7.7 µM and 250 µM respectively. Non-cyclic ester hydrolysis was monitored at 400 nm; lactone hydrolysis was monitored at 440 nm, as this is the wavelength where our experience with DMNB led us to believe the acyl-enzyme would absorb (Kitson and Kitson, 1994).

RESULTS AND DISCUSSION

Inhibition Studies with a Sterically Hindered Ester Substrate

Fig. 1 shows that p-hydroxyacetophenone is a competitive inhibitor of the hydrolysis of p-nitrophenyl pivalate catalysed by cytosolic aldehyde dehydrogenase with a K_i of 0.31 mM. It has previously been reported that p-hydroxyacetophenone is a more potent inhibitor of the dehydrogenation reaction with the beef liver mitochondrial enzyme (Ghenbot and Weiner, 1992), with $K_i = 50$ µM. Whether this merely involves non-covalent binding of the inhibitor in the active site or whether reversible thiohemiketal formation occurs with the proposed catalytic nucleophile (Cys-302) has not been firmly established. We thought it relevant to examine the binding of p-hydroxyacetophenone in the absence of NAD^+, since the compound is used under these conditions to elute aldehyde dehydrogenase from an acetophenone-bearing affinity matrix (Ghenbot and Weiner, 1992). The fact that the inhibition of ester hydrolysis is relatively weak in the absence of NAD^+ (Fig. 1) agrees with the fact that a high concentration of p-hydroxyacetophenone (10 mM) is necessary to elute the enzyme from the affinity matrix. Although the published affinity chromatography method works well, making it extremely useful for isolation of pure aldehyde dehydrogenase, it would be interesting to determine if there is a specific requirement for p-hydroxyacetophenone or if any water-soluble aromatic compound would suffice for successful elution.

Evidently the presence of NAD^+ significantly tightens the binding of p-hydroxyacetophenone to aldehyde dehydrogenase. This is not unexpected from the fact that the enzyme follows an ordered pathway in which NAD^+ has to bind first before aldehyde will bind (Blackwell *et al.*, 1989). (We have also observed a similar effect with retinal; see below.) However, it is interesting that diethylstilboestrol is a much more potent inhibitor of the esterase activity in the absence of NAD^+ than is p-hydroxyacetophenone (Kitson, 1989a); K_i values are 8.3 µM and 0.31 mM, respectively. Diethylstilboestrol has a structural similarity to p-hydroxyacetophenone but is twice as large. Perhaps one of its aromatic rings occupies the substrate site (where p-nitrophenyl esters, aldehydes and p-hydroxyacetophenone bind) and the other aromatic ring occupies an adjacent position which (in the dehydro-

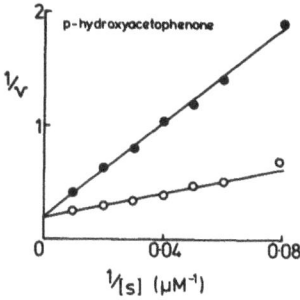

Figure 1. Lineweaver-Burk plot for the inhibition by p-hydroxyacetophenone of the hydrolysis of p-nitrophenyl pivalate catalysed by cytosolic aldehyde dehydrogenase at pH 7.4 and 25°C. inhibitor concentration was zero (o) or 1 mM (●). $^1/v$ is in arbitrary units.

genation reaction) would be the binding site for the nicotinamide ring of NAD^+. The fact that diethylstilboestrol (which lacks a carbonyl group) is a much better inhibitor than *p*-hydroxyacetophenone shows that reversible covalent interaction with Cys-302 is not necessary for effective inhibition, and therefore may not occur with *p*-hydroxyacetophenone (or the aldehydes discussed below).

p-hydroxyacetophenone   diethylstilboestrol

Fig. 2 shows that citral is a competitive inhibitor of the hydrolysis of *p*-nitrophenyl pivalate with a K_i of ~21 μM. Citral is a mixture of *cis* and *trans* isomers. The *trans* form is an analogue of part of the structure of all-*trans* retinal, as shown here:

retinal   citral

Citral evidently binds moderately well to aldehyde dehydrogenase in the absence of NAD^+, much more tightly than does *p*-hydroxyacetophenone, for instance. Perhaps therefore, aldehyde dehydrogenase 'recognises' aldehydes to some extent even without undergoing the postulated NAD^+-induced conformational change (Blackwell *et al.*, 1989). At present it is impossible to say whether citral binds non-covalently only, or whether it enters into reversible thiohemiacetal formation with Cys-302. Blatter *et al.* (1990) found that a vinyl ketone, $R-CO-CH=CH_2$ (where R is an unsaturated C_{13} chain), rapidly and irreversibly inactivates aldehyde dehydrogenase, presumably by Michael addition of Cys-302 to give $R-CO-CH_2CH_2-S$-Enzyme. With citral, our finding that it is a competitive inhibitor and not an irreversible stoicheiometric inactivator shows that no such reaction occurs as that observed with the vinyl ketone. Presumably citral binds so that the -SH of Cys-302 is not particularly close to carbon-3 and thus Michael addition to the α,β-unsaturated aldehyde group does not take place.

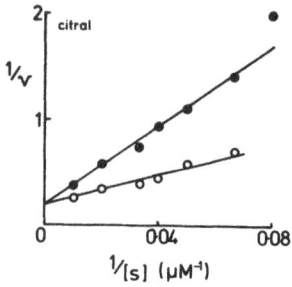

Figure 2. Lineweaver-Burk plot for the inhibition by citral of the hydrolysis of *p*-nitrophenyl pivalate catalysed by cytosolic aldehyde dehydrogenase at pH 7.4 and 25°C. Inhibitor concentration was zero (o) or 53 μM (●) $^1/v$ is in arbitrary units.

We worked with citral before investigating retinal, in anticipation that the latter would be problematic. Retinal is expensive, unstable, light-sensitive and has an appreciable absorption at 400 nm, the wavelength at which *p*-nitrophenoxide is monitored. However, it still proved possible to follow the enzyme-catalysed hydrolysis of *p*-nitrophenyl pivalate in the presence of a low concentration of retinal and we consistently found it to be an inhibitor, although reproducibility of individual experiments was not as good as with *p*-hydroxyacetophenone or citral. We conclude tentatively that retinal is a competitive inhibitor with a K_i of approximately 11 μM. Our best Lineweaver-Burk plot is shown in Fig. 3.

It has been reported that retinal is an extremely good substrate for the cytosolic isoenzyme of human aldehyde dehydrogenase, with $K_m = 0.06$ μM, leading to the proposal that its involvement in the metabolism of retinal may be one of the main raisons d'être for aldehyde dehydrogenase (Yoshida *et al.*, 1993). (The product, retinoic acid, binds to several receptors, and the complexes thus produced act as potent transcriptional regulators for the genes involved in cell differentiation and development of the embryo; Ragsdale and Brockes, 1991.) If the active site of the cytosolic enzyme is indeed 'designed' to accommodate retinal, then we considered it possible that retinal might bind tightly even in the absence of NAD⁺, and under these conditions it would then be a very potent inhibitor of ester hydrolysis. The fact that the K_i is not exceptionally low negates this hypothesis. It seems that in the absence of NAD⁺, retinal has little more affinity for the enzyme than citral, certainly not the extremely high affinity that it shows when actually acting as a substrate. Thus for retinal, as for simpler substrates (Blackwell *et al.*, 1989), the specific, tight, aldehyde-binding site must be created by the binding of NAD⁺.

Studies with a Cyclic Ester Substrate

In our work with DMNB (see Kitson and Kitson, 1994, and elsewhere in this volume) we have observed that a *p*-nitrophenol-containing reporter group covalently bound in the active site of cytosolic aldehyde dehydrogenase is ionised in the presence of NAD⁺ at pH 7.4, but unionised in its absence. We therefore predicted that the enzyme-catalysed hydrolysis of 6-nitrodihydrocoumarin (which is structurally closely related to DMNB; see Introduction) would show a burst of colour development in the presence of NAD⁺ but not in its absence.

Fig. 4A shows the typical burst of *p*-nitrophenoxide production observed in the stopped-flow spectrophotometer when enzyme is mixed with the simple ester substrate, *p*-nitrophenyl propanoate in the absence of NAD⁺. Also shown (Fig. 4B) is the absence of any equivalent burst with the lactone, 6-nitrodihydrocoumarin. There is no reason to expect that the

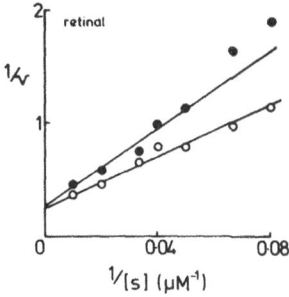

Figure 3. Lineweaver-Burk plot for the inhibition by retinal of the hydrolysis of *p*-nitrophenyl pivalate catalysed by cytosolic aldehyde dehydrogenase at pH 7.4 and 25°C. Inhibitor concentration was zero (o) or 6.7 μM (●). $^1/v$ is in arbitrary units.

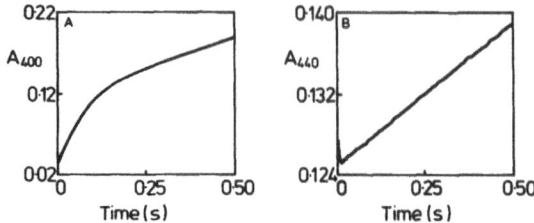

Figure 4. Stopped-flow spectrophotometer trace showing the early course of ester hydrolysis catalysed by cytosolic aldehyde dehydrogenase at pH 7.4 and 25°C in the absence of NAD^+. Substrate was (A) p-nitrophenyl propanoate and (B) 6-nitrodihydrocoumarin.

acylation step with the lactone should be rate-limiting; indeed the carbonyl group of the lactone is probably more open to nucleophilic attack than that of the equivalent non-cyclic ester as reflected in the considerably faster spontaneous hydrolysis of the former in the absence of enzyme. Only in the case of grossly hindered substrates such as p-nitrophenyl pivalate (Kitson, 1989a) does the acylation of the enzymic nucleophile become rate-limiting. Thus the lack of an observable burst with 6-nitrodihydrocoumarin is consistent with the p-nitrophenol group of the acyl-enzyme being unionised as predicted. In the case of DMNB we have concluded that ionisation of the reporter group is disfavoured either by a hydrophobic binding site or one that contains negative charge (Kitson and Kitson, 1994). With 6-nitrodihydrocoumarin the situation is evidently similar, the difference being that in this case the acyl-enzyme (lacking the modifying influence of the nitrogen atom in the equivalent species formed from DMNB) is not inert and can only be investigated on the time-scale of the stopped-flow spectrophotometer.

The presumed mechanism of the reaction of enzyme with 6-nitrodihydrocoumarin involves nucleophilic attack of the thiolate anion of Cys-302 on the carbonyl group, followed by collapse of the resulting tetrahedral intermediate with the expulsion of the p-nitrophenoxide 'leaving group' and formation of a thioester (to which the p-nitrophenoxide group is still covalently attached). The fact that no burst of p-nitrophenoxide formation is seen must mean that the anion becomes protonated faster than it is formed, either by a water molecule or an enzymic group. Pietruszko *et al.* (1993) have suggested that the carboxylate anion of the conserved Glu-268 (Hempel *et al.*, 1993) participates in the enzyme-catalysed reaction by removing the proton from the -SH group of Cys-302. We could speculate that it then rapidly donates this proton to the incipient p-nitrophenoxide leaving group as it is formed in the hydrolysis of 6-nitrodihydrocoumarin.

In the presence of 100 μM NAD^+, p-nitrophenyl propanoate does not exhibit a real burst of p-nitrophenoxide production, merely a slight irregularity before the linear steady

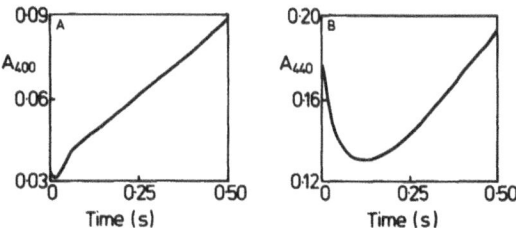

Figure 5. Stopped-flow spectrophotometer trace showing the early course of ester hydrolysis catalysed by cytosolic aldehyde dehydrogenase at pH 7.4 and 25°C in the presence of NAD^+ (100 μM). Substrate was (A) p-nitrophenyl propanoate and (B) 6-nitrodihydrocoumarin.

state rate is established (Fig. 5A). This steady state rate is 85% of the rate in the absence of NAD^+ under otherwise identical conditions. With DMNB or *p*-nitrophenyl dimethylcarbamate the presence of NAD^+ slows the rate of the very slow 'burst' seen with these substrate analogues (Kitson, 1989b; Kitson and Kitson, 1994). The hydrolysis of *p*-nitrophenyl pivalate, for which acylation is the rate-determining step, is also inhibited by the presence of NAD^+ (Kitson, 1989a). If NAD^+ likewise slows the acylation rate with *p*-nitrophenyl propanoate until this becomes almost rate-limiting, then this would explain why the burst has practically disappeared.

Fig. 5B shows that with 6-nitrodihydrocoumarin in the presence of 100 µM NAD^+ a most unexpected stopped-flow trace was observed. There is initially a large exponential decay in absorbance followed by a rise to a fast linear steady state rate (which soon starts to tail off due to substrate depletion; data not shown). The steady state rate is 8.1 times as fast as the equivalent rate in the absence of NAD^+. We had anticipated there might be a burst of colour development in this experiment since we know that the equivalent acyl-enzyme produced by DMNB is coloured in the presence of NAD^+. However, we are led instead to propose the following to explain the experimental results. It appears that the initial enzyme/substrate/NAD^+ complex is coloured, but not the acyl-enzyme/NAD^+ complex, accounting for the decline in absorbance as the enzymic nucleophile becomes acylated. (There is no sign of any very early build-up of absorbance; this must happen faster than detectable under the conditions used in the stopped-flow experiment. This would be consistent with rapid binding of substrate to enzyme/NAD^+, but it does not seem plausible that formation of the acyl-enzyme could be instantaneous, especially as we have noted above NAD^+ appears to slow the acylation rate. Even if it were the acyl-enzyme in the ionised *p*-nitrophenoxide form that is responsible for the absorbance at time zero in Fig. 5B, there is then no obvious mechanism for this to decline, before rising again as the product is released.) We speculate that the excited electronic state of the substrate is stabilised, perhaps by polar interactions with the positive charge of NAD^+ and the negative charge of Glu-268 as shown here:

Alternatively there may be the possibility of charge-transfer complex formation between the aromatic rings of the substrate and of NAD^+. If there is any truth in these suggestions, it has to account for a huge shift in λ_{max} of 6-nitrodihydrocoumarin from 285 nm (in free aqueous solution) to the region of 440 nm as observed in the stopped-flow experiments. Following attack of the anion of Cys-302 on the substrate, we propose that the incipient *p*-nitrophenoxide leaving group is protonated (again possibly involving Glu-268); this must be invoked to explain why the acyl-enzyme is apparently not coloured at this stage. It is possible that if this acyl-enzyme were not labile, it would then go on to adopt a conformation in which the ionised form is stabilised by NAD^+ as we have proposed for DMNB (Kitson and Kitson, 1994). However, the acyl-enzyme from 6-nitrodihydrocoumarin suffers rapid hydrolysis of course, and the observed absorbance rise must be mainly due to the liberation of product.

We intend to do much further work with cyclic substrates in an attempt to clarify the origin of the intriguing 'negative burst' seen in Fig. 5. Whatever explanation is adopted it must account for the facts that this phenomenon is seen with the lactone in the presence of NAD$^+$, but not in its absence, and not in the presence of NADH (unpublished results), and not under any circumstances that we have seen with non-cyclic p-nitrophenyl esters. As with many other problems with aldehyde dehydrogenase, a solution would probably be greatly helped by detailed knowledge of the three-dimensional structure of the active site.

ACKNOWLEDGMENTS

We are indebted to Mr Graham H. Freeman for synthesis of 6-nitrodihydrocoumarin, and to the New Zealand Lottery Grants Board for funding assistance towards purchase of a spectrophotometer. We also gratefully acknowledge financial support from the Neurological Foundation of New Zealand for our work with retinal.

REFERENCES

Blackwell, L.F., Bennett, A.F. and Buckley, P.D., 1983, Relationship between the mechanisms of the esterase and dehydrogenase activities of the cytoplasmic aldehyde dehydrogenase from sheep liver. An alternative view, *Biochemistry*, 22: 3784-3791.

Blackwell, L.F., MacGibbon, A.K.H. and Buckley, P.D., 1989, Aldehyde dehydrogenases - kinetic characterisation, In Human Metabolism of Alcohol, vol. II (Crow, K.E. and Batt, R.D., eds), pp. 89-104, CRC Press, Boca Raton, Florida.

Blatter, E.E., Tasayco, M.L., Prestwich, G. and Pietruszko, R., 1990, Chemical modification of aldehyde dehydrogenase by a vinyl ketone analogue of an insect pheromone, *Biochem. J.*, 272: 351-358.

Ghenbot, G. and Weiner, H., 1992, Purification of liver aldehyde dehydrogenase by p-hydroxyacetophenone-Sepharose affinity matrix and the coelution of chloramphenicol acetyl transferase from the same matrix with recombinantly expressed aldehyde dehydrogenase, *Protein Expression Purif.*, 3: 470-478.

Hempel, J.D., Nicholas, H. and Lindahl, R., 1993, Aldehyde dehydrogenases: widespread structural and functional diversity within a shared framework, *Protein Science*, 2: 1890-1900.

Kitson, T.M., 1989a, Kinetics of p-nitrophenyl pivalate hydrolysis catalysed by cytoplasmic aldehyde dehydrogenase, *Biochem. J.*, 257: 573-578.

Kitson, T.M., 1989b, The action of cytoplasmic aldehyde dehydrogenase on methyl p-nitrophenyl carbonate and p-nitrophenyl dimethylcarbamate, *Biochem. J.*, 257: 579-584.

Kitson, T.M. and Kitson, K.E., 1994, Probing the active site of cytoplasmic aldehyde dehydrogenase with a chromophoric reporter group, *Biochem. J.*, 300: 25-30.

Kitson, T.M., Hill, J.P. and Midwinter, G.G., 1991, Identification of a catalytically essential nucleophilic residue in sheep liver cytoplasmic aldehyde dehydrogenase, *Biochem. J.*, 275: 207-210.

Leatherbarrow, R.J., 1987, Enzfitter Manual, Elsevier Science Publishers, Amsterdam.

Loomes, K.M. and Kitson, T.M., 1986, Aldehyde dehydrogenase catalyses acetaldehyde formation from 4-nitrophenyl acetate and NADH, *Biochem. J.*, 238: 617-619.

Pietruszko, R., Abriola, D.P., Blatter, E.E. and Mukerjee, N., 1993, Aldehyde dehydrogenase: aldehyde dehydrogenation and ester hydrolysis, In Enzymology and Molecular Biology of Carbonyl Metabolism 4, (Weiner, H., Crabb, D.W. and Flynn, T.G., eds), pp. 221-232, Plenum Press, New York and London.

Ragsdale, C.W. and Brockes, J.P., 1991, Retinoids and their targets in vertebrate development, *Current Opinion in Cell Biol.*, 3: 928-934.

Tobias, P., Heidema, J.H., Lo, K.W., Kaiser, E.T. and Kézdy, F.J., 1969, The α–chymotrypsin-catalysed hydrolysis of lactones, *J. Amer. Chem. Soc.*, 91: 202-203.

Yoshida, A., Hsu, L.C. and Yanagawa, Y., 1993, Biological role of human cytosolic aldehyde dehydrogenase 1: hormonal response, retinal oxidation and implication in testicular feminisation, In Enzymology and Molecular Biology of Carbonyl Metabolism 4, (Weiner, H., Crabb, D.W. and Flynn, T.G., eds), pp. 37-44, Plenum Press, New York and London.

7

THE REDUCTION OF PROPIONIC ANHYDRIDE BY ALDEHYDE DEHYDROGENASE-NADH MIXTURES AT pH 7

Rosemary L. Motion, Jeremy P. Hill, Kimmo Wiltshire, Paul D. Buckley, and Leonard F. Blackwell

[1] Department of Chemistry and Biochemistry
Massey University
[2] Palmerston North, New Zealand.
Dairy Research Institute
Palmerston North, New Zealand

INTRODUCTION

Although the oxidation of aldehydes by aldehyde dehydrogenase is usually considered to be irreversible, Hart & Dickinson (1978) have shown that anhydrides of carboxylic acids could be reduced to aldehydes by enzyme.NADH mixtures. In a later study Motion et al (1988) demonstrated that acylation of aldehyde dehydrogenase.NADH complexes by acetic anhydride led to the production of acetaldehyde and NAD^+ in stoichiometric amounts. It is clear from these results, and the fact that 80% of the initial NADH concentration can be oxidised by anhydrides at an almost constant rate, that there is no intrinsic thermodynamic barrier to the reverse reaction once the appropriate acyl enzyme species is formed. The mechanism for reduction of aldehydes is presumably the reverse of that for aldehyde oxidation (Scheme I).

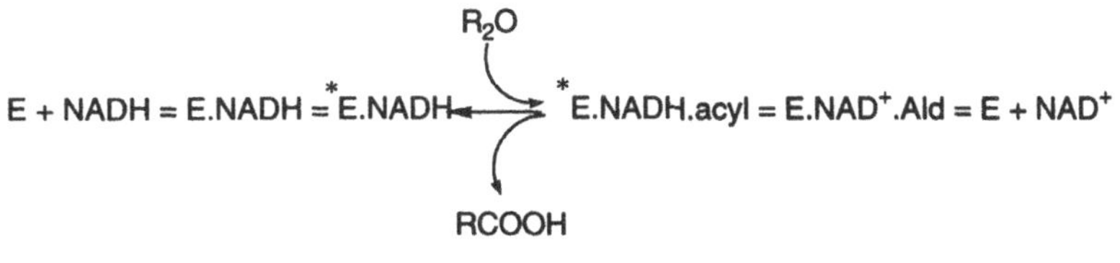

$$E + NADH = E.NADH = {}^*E.NADH \rightleftharpoons {}^*E.NADH.acyl = E.NAD^+.Ald = E + NAD^+$$

R_2O

RCOOH

$R_2O + H_2O = 2RCOOH$

Scheme I

In scheme I *E.NADH represents a conformationally rearranged form of E.NADH and it is the slow isomerisation of the two binary enzyme.NADH complexes (Blackwell *et al.*, 1987) which partly controls the steady state rate of aldehyde oxidation. In scheme I it is assumed that on the reverse reaction pathway acylation of the *E.NADH complex will occur to produce the same acylated complex as observed during enzyme catalysed oxidation of aldehydes.

Although some preliminary kinetic studies have been carried out on the reverse reaction with acetic anhydride (Hart & Dickinson, 1978; Motion *et al*, 1988), the precise details of the reduction of anhydrides by aldehyde dehydrogenase have not been established. We have studied the kinetics of the reverse reaction with propionic anhydride as acylating agent, with the aim of determining to what extent the reaction pathway for the reduction of propionic anhydride is indeed the microscopic reverse of the normal pathway for oxidising aldehydes.

EXPERIMENTAL PROCEDURES.

Cytoplasmic aldehyde dehydrogenase (ALDH) was prepared essentially as described by Hill *et al.*, (1992). NAD$^+$(grade III) and NADH (grade III) were from Sigma Chemical Co. (St. Louis, MO, U.S.A.), propionic anhydride was from Koch-Light Laboratories (Colnbrook, Bucks, U.K.).

METHODS

The active site concentration was determined from V_{max} measured by using 20 mM propionaldehyde and 1 mM NAD$^+$ at pH 7.6 and 25° as described previously Blackwell *et al.*, (1987) . Studies of the reduction of propionic anhydride were carried out by mixing enzyme (0.1 - 8.0 μM) and NADH (0.2 - 400 μM) in a mixture of 0.1 M Na$_2$SO$_4$ and 0.1 M NaNO$_3$ adjusted to pH 7.0 in a 3 cm^3 cuvette. Enzyme stock solutions contained 25 mM pH 7.3 phosphate buffer but the final buffer concentration in each experiment did not exceed 5 mM. The changes in absorbance at 340 nm were monitored using an Aminco DW 2a uv-visible spectrophotometer with an amount of NADH (in 25 mM phosphate buffer pH 7.3) in the reference cuvette equal to the initial free NADH concentration in the sample cuvette. A typical absorbance versus time trace is shown in Fig.1. All experiments were carried out at 25 °C.

Propionic anhydride (0.01 - 0.1 cm^3) was added last to the cuvette as a solution in acetonitrile such that the final volume of acetonitrile in reaction solutions did not exceed 3% of the total assay volume. MacGibbon *et al.*, (1978) demonstrated that this amount of acetonitrile has no effect on the enzyme activity. NADD was prepared as described by Hart and Dickinson (1978).

Pre-Steady-State and Stopped-Flow Studies

These were carried out on a Durrum-Gibson D110 stopped-flow spectrophotometer as described by Bennett *et al.*, (1982) in either the protein or nucleotide fluorescence modes. The data were analysed on a Cromemco Z2D microcomputer as described by Blackwell *et al.*, (1987). All solutions contained 0.1 M sodium sulphate and 0.1 M sodium nitrate. As for steady state experiments the final buffer concentration never exceeded 5 mM. For nucleotide fluorescence bursts a solution containing enzyme (2-10 μM) premixed with NADH (5-100 μM) was rapidly mixed against a solution containing NADH (at a concentration equivalent

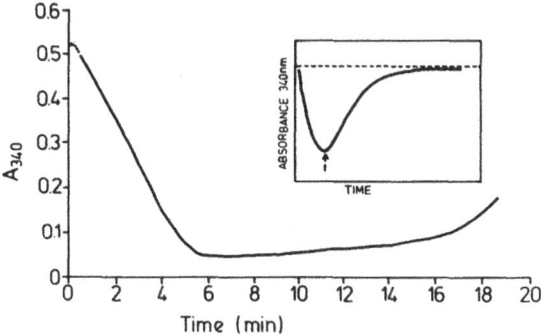

Figure 1. The solution initially contained enzyme (2.5 μM), NADH (2.5 μM) and propionic anhydride (2.6 mM). The insert shows a trace for which the turnover from the reverse reaction to the normal aldehyde oxidation pathway was immediate.

to the free NADH concentration in the first solution) and anhydride (0.1-8.0 mM) in acetonitrile. The anhydride solution was added directly to the pre-prepared NADH solution immediately prior to its introduction into the drive syringes of the stopped-flow spectrophotometer and used immediately. Fresh anhydride solutions were prepared frequently to minimise the effects of decreasing anhydride concentration as a result of hydrolysis.

RESULTS AND DISCUSSION

Utilisation of NADH During the Reverse Reaction

Hydrolysis of all the anhydrides used in this study occurred spontaneously in water alone, but constitutents of the reaction mixture also catalysed the hydrolysis. The most important catalyst appeared to be the phosphate buffer itself. Since the rate of hydrolysis of propionic anhydride was less in salt solutions and the rate of the reverse reaction was linear until NADH oxidation ceased even though the pH changed from 7 to 5.3 (Fig. 1) all subsequent work was carried out in this medium.

The utilisation of NADH through the reverse reaction pathway depended on the continued availability of sufficient anhydride as expected for scheme I. Under the most favourable conditions a maximum amount of 91 per cent of the initial NADH concentration was oxidised irrespective of the starting value for the NADH concentration. For example in salt solutions at pH 7.0, when the initial concentration of propionic anhydride was 1.3 mM (a large excess), the decrease in absorbance at 340 nm during the reverse reaction corresponded to a loss of 91 per cent of the initial NADH concentrations (as in Fig. 1) at all NADH concentrations up to 75 μM After the initial absorbance decrease, a pseudo-equilibrium condition was established for several minutes, during which there was no net change in absorbance. The NADH:NAD$^+$ ratio during the pseudo-equilibrium period was about 1:9 but the reverse reaction was still favoured, presumably because of the much tighter binding of NADH to the enzyme compared to NAD$^+$ (MacGibbon *et al*, 1977). Finally, when sufficient anhydride had been removed by hydrolysis, the absorbance increased due to the re-oxidation of the aldehyde on the normal dehydrogenase pathway.

At concentrations where the propionic anhydride was hydrolysed before the maximum amount of NADH (91%) had been oxidised, as expected a pseudo-equilibrium was not observed and instead there was a sharp turnover from NADH oxidation to NADH production

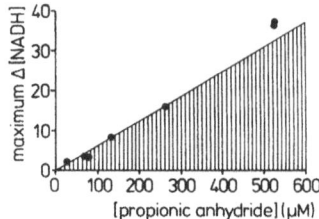

Figure 2. Optimum utilisation of NADH as a function of the Propionic Anhydride Concentration. The plot shows the highest NADH concentration at which the maximum proportion (91%) of the NADH was oxidised during the reaction. For concentrations outside the shaded area a smaller proportion of the NADH was oxidised as the propionic anhydride hydrolysed too rapidly for complete utilisation. The enzyme concentration was 2-3 μM in all cases.

(see insert to Fig.1). There was a linear relationship between the initial propionic anhydride concentration and the largest NADH concentration which could still be 91 per cent utilised before spontaneous hydrolysis of the anhydride allowed aldehyde oxidation to dominate (Fig. 2).

Dependence of Initial Rate on NADH and Propionic Anhydride Concentrations

The effect of the NADH concentration on the reverse reaction was determined by adding various amounts of NADH to assays containing enzyme (2.4 μM) and a high concentration of propionic anhydride (1.3 mM) in electrolyte solutions (see methods). The reaction rate increased in a hyperbolic fashion until a maximum rate was achieved at an NADH concentration of 30 μM (Fig. 3). A Lineweaver-Burk plot of the initial rate data at concentrations of NADH below 30 μM was linear giving a K_m value of 4.0 μM (Fig. 3, insert). This K_m value is similar to that for the dehydrogenase pathway (MacGibbon et al., 1977a), suggesting that nucleotide is binding in the normal dehydrogenase coenzyme binding site as shown in scheme 1. The anhydride concentration was then varied, however at very low concentrations of propionic anhydride, initial rates of reaction were difficult to obtain because of the rapid hydrolysis of the anhydride during the mixing time (5-7 seconds). Nevertheless, saturation of the steady-state rate was clearly evident as the concentration of propionic anhydride was increased to around 0.5 μM, suggesting a K_m value of less than 0.1 μM. Clearly the reverse reaction can continue even at very low propionic anhydride concentrations.

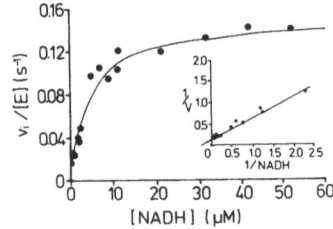

Figure 3. Effect of the NADH Concentration on the Rate of the Reverse Reaction. The NADH concentration was varied in solutions containng enzyme (2.4 μM) and propionic anhydride (1.3 mM) in electrolyte solution. The insert shows the Lineweaver-Burk plot of the data up to 21 M.

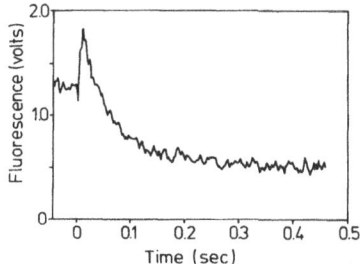

Figure 4. Pre-Steady-State Decrease in Nucleotide Fluorescence observed in Reverse Reactions. A solution of enzyme (10 M) and NADH (20 M) in electrolyte solution was rapidly mixed with a solution containing propionic anhydride (600 M) and NADH (12 M) in the stopped flow spectrophotometer. For the amplitude of this process 1 V corresponds to an enzyme concentration of 4.3 M.

PreSteady State Results

When enzyme premixed with NADH was rapidly mixed with a solution containing NADH and propionic anhydride there was an exponential decrease in nucleotide fluorescence (Fig. 4), followed by the linear steady state phase. At saturating concentrations of NADH and propionic anhydride, the rate constant for this burst process was 30 ± 10 s^{-1}, and the relative amplitude of the process was 0.5 ± 0.1 of the formal enzyme concentration. The fact that only half the enzyme concentration was involved in the burst process is consistent with either a rapid acylation of the *E.NADH complex causing quenching of the nucleotide fluorescence or with a rapid loss of NADH during the hydride transfer step. A transient decrease in protein fluorescence was also observed when a solution containing enzyme (13 μM) and NADH (60 μM) was pushed against a solution of propionic anhydride (3.9 mM) and NADH (51 μM). A rate constant of 25 ± 10 s^{-1} was obtained for this process, which was similar to that obtained in nucleotide fluorescence studies, suggesting that both fluorescence changes arise as a result of the same process. It seems unlikely that acylation of the *E.NADH complex could cause quenching of the protein fluorescence but it is known (MacGibbon *et al.*, 1977b) that NAD$^+$ causes more quenching of the protein fluorescence than does NADH. Hence transfer of a hydride ion to form NAD$^+$ would be associated with a loss in fluorescence amplitude as observed. The burst process therefore most likely represents the hydride transfer step.

Interestingly enough a value of 30 s^{-1} for the rate constant of the reverse hydride transfer step is required to explain the hyperbolic dependence of the nucleotide fluorescence burst observed on the normal pathway for aldehyde oxidation (MacGibbon *et al.*, 1977b).

Table 1. Isotope Effects in Reverse Steady State Reaction with Propionic Anhydride

[NADH] μM	$k_{cat}(H)$ s^{-1}	$k_{cat}(D)$ s^{-1}	k_H/k_D
10	0.88	0.33	2.67
20	0.81	0.28	2.89
50	0.60	0.17	3.53

[AlDH] = 1 μM   [Propionic Anhydride] = 833 μM

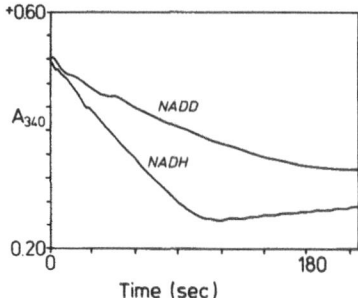

Figure 5. Isotope Effect on the Steady State Reverse Reaction. Enzyme (1 M), NADH, or NADD (50 M) and propionic anhydride (833 M) were mixed in electrolyte solutions at 25°C.

However as first reported by Dickinson (1978) and as observed again in the present study (Table 1), there was a kinetic isotope effect of about 3 on the steady-state phase of the reverse reaction, which requires that hydride transfer should be rate limiting in the steady-state phase of the reaction. Clearly, hydride transfer cannot be rate limiting in both the pre-steady state and the steady state phases of the reaction. There was also an isotope effect of about 3 on the turnover points in the reverse reaction (Fig. 5), which also cannot be explained on the basis of scheme I.

Time Course of the Reverse Reaction under Different Mixing Conditions.

In figure 6 a comparison is made between the changes in fluorescence observed in the stopped-flow spectrofluorimeter experiments when NADH is and is not premixed with the enzyme before mixing with propionic anhydride. When NADH was not pre-mixed with the enzyme a rapid increase in fluorescence was observed (Fig. 5, trace B) corresponding to about 40% of the control. Thereafter, as the reverse occurred, there was a steady decrease in fluorescence (as in Fig. 1) which continued until the propionic anhydride concentration was exhausted by hydrolysis. At this point the fluorescence began to increase again as NADH was reformed via the normal aldehyde oxidation pathway, ultimately reaching the control

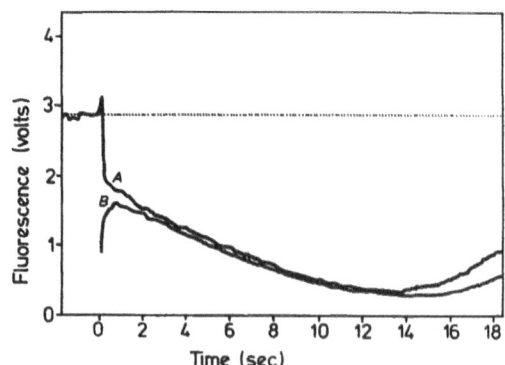

Figure 6. Comparison of Amplitudes of Changes in Nucleotide Fluorescence during NADH Binding and Pre-Steady-State Reverse Processes. Trace A: Enzyme (10 µM) premixed with NADH (20 M) was pushed against propionic anhydride (520 M). Trace B: Enzyme (10 M) was pushed against NADH (20 M) and propionic anhydride (520 M).

value for binding of NADH to the free enzyme (5.6 times the fluorescence of free NADH). These reactions were carried out as soon as possible after preparation of the propionic anhydride solution to minimise the extent of hydrolysis so that the propionic anhydride concentrations were comparable in each case.

For the alternative mixing condition the fluorescence of the initial mixture of enzyme and NADH rapidly dropped (as in Fig. 4) to about 40% of the control value, followed by a steady state fluorescence time course which was identical with that shown in figure 5B for the first mixing condition.

Mechanistic Significance of Kinetic Data

All the kinetic results can be understood in terms of the two kinetically significant enzyme.NADH complexes shown in scheme I. One of these, the *E.NADH complex, after fast acylation is presumed to undergo rapid hydride transfer resulting in loss of nucleotide fluorescence. It should be noted that the product of the acylation of this complex, the *E.NADH.acyl species, is also formed after hydride transfer on the normal aldehyde oxidation pathway.

The other complex, namely E.NADH, is not believed to exist in an acylated form on the aldehyde oxidation pathway. However, the pathway for the reverse reaction is not the microsopic reverse of the enzyme catalysed oxidation of aldehydes. Instead of acylation by a carboxylic acid (the product of aldehyde oxidation), acylation is by the much more reactive anhydride. In order to explain the kinetic data it is proposed that the anhydride can acylate the E.NADH complex. Then when NADH and propionic anhydride are rapidly mixed with free enzyme, the E.NADH complex which is formed in the first fast binding step, is acylated before the isomerisation to the *E.NADH complex can occur (see scheme II). From the results shown in figure 6, the fluorescence enhancement of the E.NADH.acyl form (equal to the total enzyme active site concentration) which results is significantly less than for the combined mixture of *E.NADH and E.NADH (approximately 40% less). In the E.NADH.acyl species hydride transfer must take place in a less favourable enzyme conformation and is therefore slow, and rate limiting, during the steady-state phase of the reverse reaction.

From figure 6 it is clear that, regardless of the mixing conditions, after the initial fast processes are complete the fluorescence enhancement is the same for both, and therefore the same enzyme.NADH species must be present.

How does the enzyme all get converted to this form under the other mixing condition? When NADH is premixed with the enzyme an equilibrium mixture of the two E.NADH complexes is initially present. On mixing with anhydride both the E.NADH and the *E.NADH forms are acylated. Although the *E.NADH.acyl species can undergo rapid hydride transfer, when this enzyme recycles on the reaction pathway it is the E.NADH complex which must be first formed. Immediate acylation of this E.NADH complex results

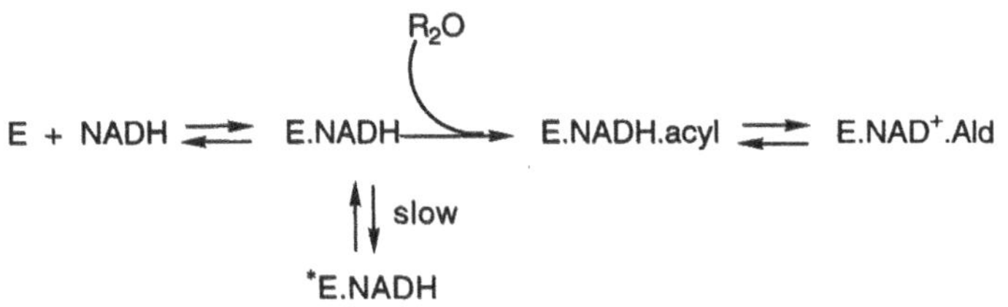

Scheme II

in the initial equilibrium mixture of E.NADH and *E.NADH being converted rapidly to E.NADH.acyl, the same species that accumulates under the other mixing condition.

The proposal that it is hydride transfer in the E.NADH.acyl complex that is rate limiting in the steady-state phase of the reaction, explains the isotope effect observed for this stage of the reaction. It should be noted that the slow isomerisation of the enzyme.NADH complexes which limits the steady-state rate on the normal aldehyde oxidation pathway, is too slow to account for the reverse steady-state rate of 0.8 s^{-1}. The pathway for the reverse reaction shown in scheme II allows this isomerisation step to be circumvented during the steady-state phase of the reverse reaction

The results of this study of the reverse reaction provide further evidence of the importance of the presence of two isomeric enzyme.NADH complexes which exist in two different conformations and which interconvert at a rate which limits the enzyme catalysed oxidation of aldehydes by sheep liver cytosolic aldehyde dehydrogenase. When enzyme and NADH are premixed, acylation with propionic anhydride results in about 50 per cent of the enzyme active sites undergoing a single turnover through the aldehyde oxidation pathway in the reverse direction. Subsequent steady-state reaction occurs via the second NADH complex. These complexes must not be ignored when experiments involving steady-state inhibitors are carried out under different mixing conditions.

ACKNOWLEDGMENTS

We acknowledge the award of a Massey University post-doctoral fellowship to Dr J.P.Hill.

REFERENCES

Bennett, A.F., Buckley, P.D. and Blackwell, L.F., 1982, Proton release during the pre-steady-state oxidation of aldehydes by aldehyde dehydrogenase. Evidence for a rate limiting conformation change, Biochemistry, 21, 4404-4413.

Blackwell, L.F., Motion, R.L., MacGibbon, A.K.H., Hardman, M.J. and Buckley, P.D., 1987, Evidence that the slow conformational change controlling NADH release from the enzyme is rate limiting during the oxidation of propionaldehyde by aldehyde dehydrogenase, Biochem. J., 242, 803-808.

Hart, G.J. and Dickinson, F.M., 1978, Partial reversal of the acetaldehyde and butyraldehyde oxidation reactions catalysed by aldehyde dehydrogenase from sheep liver, Biochem. J., 175, 753-756.

MacGibbon, A.K.H., Blackwell, L.F.and Buckley, P.D., 1977a, Kinetics of sheep liver aldehyde dehydrogenase, Eur. J. Biochem., 77, 93-100.

MacGibbon, A.K.H., Blackwell, L.F.and Buckley, P.D., 1977b, Pre-steady-state kinetic studies on cytoplasmic sheep liver aldehyde dehydrogenase, Biochem. J., 167, 469-477.

MacGibbon, A.K.H., Haylock, S.J., Buckley, P.D., and Blackwell, L.F.,1978, Kinetic studies on the esterase activity of cytoplasmic sheep liver aldehyde dehydrogenase, Biochem. J., 171, 533-538.

Motion, R.L., Buckley, P.D., Bennett, A.F. and Blackwell, L.F., 1988, Evidence that the cytoplasmic aldehyde dehydrogenase-catalysed oxidation of aldehydes involves a different active-site group from that which catalyses the hydrolysis of 4-nitrophenyl acetate, Biochem. J., 254, 903-906.

8

CLONING AND CHARACTERISATION OF THE cDNA FOR SHEEP LIVER CYTOSOLIC ALDEHYDE DEHYDROGENASE

Cherie K. Stayner and John W. Tweedie

Department of Chemistry and Biochemistry
Massey University
Palmerston North, New Zealand

INTRODUCTION

Sheep liver cytosolic aldehyde dehydrogenase (AlDH) has been studied extensively by others from this department (reviewed in Blackwell *et al.*, 1989). In order to extend these studies using site-directed mutagenesis it was necessary first to isolate and sequence the cDNA for this form of the enzyme. The cDNA sequences for the human liver cytosolic (Hsu *et al.*, 1989) and mitochondrial iso-forms of AlDH (Hsu *et al.*, 1988) were used to design a probe which would allow specific identification of cytosolic AlDH. We report here the isolation and sequencing of a full-length clone for sheep liver cytosolic AlDH.

METHODS

Isolation of RNA

Liver was obtained from a freshly-slaughtered Perendale ewe and the tissue was immediately frozen as small fragments in liquid nitrogen. Frozen tissue was stored at -70° until required. Frozen tissue ($\approx$ 1 g) was pulverised, while still frozen in liquid nitrogen, and the powdered tissue thawed in a solution containing guanidinium thiocyanate (4M), sodium citrate (25mM), 2-mercaptoethanol (10mM) and sodium *N*-lauryl sarcosine (0.5%). The thawed tissue was homogenised for 30 seconds in an Ultraturrax tissue homogeniser. Subsequent procedures for the isolation of total RNA were essentially as described by Chirgwin (Chirgwin *et al.*, 1979) with the addition of a second caesium chloride density gradient centrifugation step. The pellet of total RNA was resuspended in HEPES/HCl (10mM, pH 7.6), EDTA (1mM) and stored at -20°. Poly-A$^+$ RNA was isolated by absorption to oligo-dT cellulose (Aviv & Leder, 1972).

Preparation of sheep liver cDNA library

Single-strand cDNA was prepared from sheep liver poly-A$^+$ RNA using M-MLV RNaseH$^-$ reverse transcriptase (Superscript II™, BRL) to extend a primer of oligo-dT. The complementary cDNA strand was synthesised using DNA polymerase I. EcoR1 adaptors were ligated to the ends of the cDNA which was then ligated to phosphatase treated, EcoR1 cut, λ-ZAP™ vector arms (Stratagene). The ligated cDNA library was packaged (λ packaging extract from Stratagene) and used to infect E. coli (Strain XL1-Blue™, Stratagene). The library (0.5 ml) had a titre of 1.6 x10^6 pfu/μg of vector arms prior to amplification (0.8 x 10^6 pfu/ml) . The library was amplified prior to screening and the amplified library had a titre of 7 x 10^9 pfu/ml.

Screening the cDNA library

The amplified cDNA library was screened from plaque lifts to nitrocellulose filters. A total of 4 x 10^5 plaques were screened by hybridisation (65°, 0.1 x SSC) to a probe (described below) labelled by random priming with ^{32}P-dCTP. Selected positive clones identified at the first round were purified to single clones by two further rounds of screening. All λ clones were converted to the corresponding pBlueScript™ plasmids by the protocols supplied with the cloning kit (Stratagene).

Characterisation of cDNA clones

The cDNA clones were characterised by restriction mapping and by sequencing using a combination of single-strand and double-strand sequencing (Sequenase™, USB). Both strands of the clones were sequenced several times using specific primers about 300-400 nucleotides apart which were designed from sequence which had been already determined.

RESULTS AND DISCUSSION

Design of DNA probe specific for cytosolic AlDH

The nucleotide sequences of human liver cytosolic (Hsu et al., 1989) and mitochondrial (Hsu et al., 1988) AlDH cDNA's were aligned and inspected to identify regions where the sequence divergence between the two forms was greatest. An 193 bp Rsa1 fragment between nucleotides 236 and 429 in the human cytosolic cDNA was gel purified and used to screen the cDNA library.

Identification of cDNA clones containing sheep cytosolic AlDH sequences

The first round of screening yielded 21 positive clones. Ten of these were selected for further screening and four remained positive after the second round. A third round screen was carried out which yielded 100% positive clones for all four isolates. After conversion to plasmids these clones were shown to contain cDNA inserts of ≈2.1kb in length and two of the clones subsequently proved to be identical. The variation between clones was due to differences in the lengths of the 5'-untranslated region of the cDNA. The cDNA from all three clones was sequenced and the sequence of the longest cDNA is shown in Table 1.

The cDNA clone is 2107 nucleotides in length, including 41 nucleotides of 5'-untranslated sequence before the first ATG codon. The coding sequence is 1503 nucleotides, coding for a protein of 501 amino acids. There is a 700 nucleotide 3'-untranslated region,

Table 1. Nucleotide sequence of the cDNA for sheep liver cytosolic aldehyde dehydrogenase. The 5'-untranslated region (1-41) and the 3'-untranslated region (1548-2107) are shown in lower case. The open reading frame coding for the AlDH protein is shown in upper case, starting at the ATG codon (42-44) and ending at the TAA termination signal (1545-1547). The putative polyadenylation signal (2059-2064) is shown underlined. The 3'-terminal polyadenylate sequence starts at nucleotide 2079.

```
   1    atcgctgagc ctgtcacctg tgttcaggag cagaaccaac aATGTCGTCC
  51    TCAGCCATGC CAGACGTACC TGCCCCACTC ACCAATTTGC AGTTTAAATA
 101    TACTAAGATC TTCATAAACA ATGAATGGCA TAGTTCAGTG AGTGGTAAGA
 151    AATTTCCAGT CTTTAATCCC GCAACCGAGG AGAAACTCTG TGAGGTGGAA
 201    GAAGGAGATA AGGAGGATGT TGACAAAGCA GTGAAGGCTG CAAGACAAGC
 251    TTTTCAGATT GGCTCTCCAT GGCGTACTAT GGATGCTTCA GAGAGAGGAC
 301    GCTTGTTAAA CAAGTTGGCT GACTTAATTG AAAGAGATCG TCTGCTCCTG
 351    GCGACAATGG AGGCTATGAA TGGTGGAAAA CTATTTTCCA ATGCATATCT
 401    GATGGATTTA GGAGGCTGCA TAAAAACACT ACGCTACTGT GCAGGCTGGG
 451    CTGACAAGAT CCAGGGCCGT ACAATACCCA TGGATGGGAA CTTTTTTACA
 501    TATACAAGAA GTGAGCCTGT TGGTGTGTGT GGCCAAATCA TTCCTTGGAA
 551    TTTCCCGTTG CTCATGTTCC TCTGGAAGAT AGGGCCTGCC CTTAGCTGCG
 601    GAAACACAGT GGTTGTCAAA CCAGCAGAGC AAACCCCTCT GACTGCTCTT
 651    CACATGGGAT CTTTAATAAA AGAGGCAGGG TTTCCTCCTG GAGTAGTGAA
 701    TATTGTCCCT GGTTATGGGC CTACTGCAGG GGCAGCCATT TCTTCTCACA
 751    TGGATGTAGA CAAAGTGGCC TTCACAGGAT CGACAGAGGT TGGCAAACTG
 801    ATCAAAGAAG CTGCTGGGAA AAGCAATCTG AAAAGGGTTT CCCTGGAACT
 851    CGGGGGAAAG AGTCCTTGCA TTGTGTTTGC TGATGCCGAC TTGGACAATG
 901    CTGTTGAATT TGCACACCAA GGAGTATTCT ATCACCAGGG CCAATGTTGT
 951    ATAGCTGCAT CCCGTCTCTT TGTAGAAGAA TCAATTTATG ATGAGTTTGT
1001    TCGAAGGAGT GTTGAGCGGG CGAAAAAGTA TGTTCTTGGA AATCCTCTGA
1051    CCCCAGGAGT CAGTCAAGGC CCTCAGATTG ATAAAGAACA ATATGAAAAA
1101    ATACTTGACC TCATTGAAAG TGGGAAGAAG GAGGGGGCCA AGCTGGAATG
1151    TGGCGGAGGC CCTTGGGGGA ATAAAGGCTA CTTTATCCAA CCCACAGTGT
1201    TCTCTGATGT TACTGATGAT ATGCGCATTG CCAAAGAGGA GATATTTGGA
1251    CCTGTGCAGC AAATCATGAA GTTTAAGTCT TTAGATGATG TAATCAAGAG
1301    AGCAAACAAT ACTTTCTATG GGTTATCAGC AGGAATTTTT ACCAATGATA
1351    TTGATAAAGC CATCACAGTC TCCTCTGCTT TGCAGTCTGG AACCGTGTGG
1401    GTGAACTGCT ATAGTGTGGT ATCTGCCCAG TGCCCCTTTG GTGGATTCAA
1451    GATGTCTGGA AATGGACGAG AACTCGGAGA ATATGGTTTC CATGAATACA
1501    CAGAAGTCAA GACAGTCACA ATCAAAATTT CTCAGAAGAA CTCATAAact
1551    atgagaggac agagatgact cctcaatggc taagcatctc ctgcagtggc
1601    taatgtatct tagtggtttt aaatgcaaaa ttgttctttc cttgattctt
1651    taaacataag ctaatcatat tagcattaat actacacata gacaacttga
1701    cttttatgtt attctgaaag aatcatcagc cttctgctat gacacccaag
1751    ccctctctga aatgaaaatg atggacttgg atgcaatctc tctagctctg
1801    taatagccat gtgcttctct ctgtagttac ttgcctagga taatcacttt
1851    atagaagagg acaagttgtc atttagcatc tttctcttga tgacctcttg
1901    aagtacttac tatacctgtt aatttcagac taggtatgtt ctgttctccg
1951    tagtggttca gtccttggaa tttgttgaaa tgtttcctag aatgtcatgt
2001    ctgcttgtca aatgaagtaa tgcctgaaat acttagatgt aacctaaaca
2051    tacactgtaa taaaaacaac cttgcatgaa aaaaaaaaa aaaaaaaaaa
2101    aaaaaaa
```

Table 2. Alignment of the amino acid sequences of liver aldehyde dehydrogenases from human cytosol (top lines) (Hsu *et al.*, 1989), ovine cytosol (middle lines) (this work) and human mitochondria (bottom lines) (Hsu *et al.*, 1988). The two human sequences were separately aligned with the ovine cytosolic AlDH. The human mitochondrial AlDH sequence is shown from the serine residue at the N-terminus of the mature protein which is produced during translocation into the mitochondrial matrix.

```
        MSSSGTPDLPVLLTDLKIQYTKTFINNEWHDSVSGKKFPVFNPATEEELC
        ||||: ||:|. ||:|.:.|||.||||||.|||||||||||||||.||
   1    MSSSAMPDVPAPLTNLQFKYTKIFINNEWHSSVSGKKFPVFNPATEEKLC
        |..| ..||||   . :. :..|||||||||..|| |.||..||.|:| :|
        SSAAATQAVPAPNQQPEVFCNQIFINNEWHDAVSRKTFPTVNPSTGEVIC

        QVEEGDKEDVDKAVKAARQAFQIGSPWRTMDASERGRLLYKLADLIERDR
        :|||||||||||||||||||||||||||||||||||||||| |||||||||
   51   EVEEGDKEDVDKAVKAARQAFQIGSPWRTMDASERGRLLNKLADLIERDR
        :|.|||||||||||||||:||||| ||||.|||||:|||||||
        QVAEGDKEDVDKAVKAARAAFQLGSPWRRMDASHRGRLLNRLADLIERDR

        LLLATMESMNGGKLYSNAYLNDLAGCIKTLRYCAGWADKIQGRTIPIDGN
        |||||||.|||||:|||||| ||:||||||||||||||||||||:|||
   101  LLLATMEAMNGGKLFSNAYLMDLGGCIKTLRYCAGWADKIQGRTIPMDGN
        .||..::.||.| :|:||: .:|.|||||||||||||.::|:|||.||:
        TYLAALETLDNGKPYVISYLVDLDMVLKCLRYYAGWADKYHGKTIPIDGD

        FFTYTRHEPIGVCGQIIPWNFPLVMLIWKIGPALSCGNTVVVKPAEQTPL
        |||||| ||:||||||||||||||:|::|||||||||||||||||||
   151  FFTYTRSEPVGVCGQIIPWNFPLLMFLWKIGPALSCGNTVVVKPAEQTPL
        ||..||| |||||||||||| ||:||||.||.||.||:|.||||||
        FFSYTRHEPVGVCGQIIPWNFPLLMQAWKLGPALATGNVVVMKVAEQTPL

        TALHVASLIKEAGFPPGVVNIVPGYGPTAGAAISSHMDIDKVAFTGSTEV
        ||||::|||||||||||||||||||||||||||||||||||||||||
   201  TALHMGSLIKEAGFPPGVVNIVPGYGPTAGAAISSHMDVDKVAFTGSTEV
        |||.::.||||||||||||||||||:||||||||.|| |||||||||||:
        TALYVANLIKEAGFPPGVVNIVPGFGPTAGAAIASHEDVDKVAFTGSTEI

        GKLIKEAAGKSNLKRVTLELGGKSPCIVLADADLDNAVEFAHHGVFYHQG
        |||||||||||||||||.||||||||||:||||||||||||||:|||||||
   251  GKLIKEAAGKSNLKRVSLELGGKSPCIVFADADLDNAVEFAHQGVFYHQG
        |::|. |||.|||||.||||||||.||||:| ||| || ::|:::||
        GRVIQVAAGSSNLKRVTLELGGKSPNIIMSDADMDWAVEQAHFALFFNQG

        QCCIAASRIFVEESIYDEFVRRSVERAKKYILGNPLTPGVTQGPQIDKEQ
        |||||||||||||||||||||||||||||:|||||||||||.||||||||
   301  QCCIAASRLFVEESIYDEFVRRSVERAKKYVLGNPLTPGVSQGPQIDKEQ
        |||.|:|| ||:|.||||| |||.|||. |:|||:.. ..|||:|..|
        QCCCAGSRTFVQEDIYDEFVVRSVARAKSRVVGNPFDSKTEQGPQVDETQ

        YDKILDLIESGKKEGAKLECGGGPWGNKGYFVQPTVFSNVTDEMRIAKEE
        |:||||||||||||||||||||||||||||||||||:||||||||||
   351  YEKILDLIESGKKEGAKLECGGGPWGNKGYFIQPTVFSDVTDDMRIAKEE
        :.|||:.|:.||.|||||| |||| :::|||||||||:|| |:| |||||
        FKKILGYINTGKQEGAKLLCGGGIAADRGYFIQPTVFGDVQDGMTIAKEE

        IFGPVQQIMKFKSLDDVIKRANNTFYGLSAGVFTKDIDKAITISSALQAG
        ||||||||||||||||||||||||||||||||:||.|||||||:|||||.|
   401  IFGPVQQIMKFKSLDDVIKRANNTFYGLSAGIFTNDIDKAITVSSALQSG
        |||||  ||:|||.:::|: |||. |||.|::||.|:||| :| |||.|
        IFGPVMQILKFKTIEEVVGRANNSTYGLAAAVFTKDLDKANYLSQALQAG

        TVWVNCYGVVSAQCPFGGFKMSGNGRELGEYGFHEYTEVKTVTVKISQKN
        ||||||:|||||||||||||||||||||||||||||||||||||:|||||||
   451  TVWVNCYSVVSAQCPFGGFKMSGNGRELGEYGFHEYTEVKTVTIKISQKN
        ||||||.|.:||||||||.||||.||||||||||::.|||||||:|:.|||
        TVWVNCYDVFGAQSPFGGYKMSGSGRELGEYGLQAYTEVKTVTVKVPQKN

        S
        |
   501  S
        |
        S
```

including a poly-A tail of 29 residues. The 3'-UTR contains a poly-A recognition sequence (2059-2064) but does not appear to contain any U-rich mRNA de-stabilising regions.

Sequence comparisons of cDNA and conceptual translation

In order to confirm that we had in fact cloned the cytosolic iso-form of the sheep liver AlDH we compared the sequence of the cDNA shown in Table 1 with the human liver cytosolic and mitochondrial AlDH sequences. The cDNA sequence showed 90% identity with the human cytosolic AlDH and 65% identity with the mitochondrial iso-form. These differences in sequence were reflected in the sequence comparison at the amino acid sequence level which is shown in Table 2.

The amino acid sequence identity (similarity) between the human liver cytosolic iso-form and the putative sheep liver iso-form is 91% (97%) while the comparison with the human mitochondrial iso-form shows 68% identity (80%). The close match between the sheep cDNA and the human liver cytosolic AlDH at both the cDNA and amino acid sequence levels makes us confident that we have in fact isolated the cDNA for the sheep liver cytosolic iso-form. The amino acid sequences for residues 68-77 and 298-308 correspond exactly with the sequences of two peptides from sheep liver cytosolic AlDH which have been published previously (Loomes *at al.* 1990, Kitson *et al.* 1991).

CONCLUSIONS

The cDNA for sheep liver cytosolic aldehyde dehydrogenase has been cloned and sequenced. The predicted amino acid sequence of the protein coded for by the cDNA corresponds closely with that of the corresponding human protein. The cDNA has now been transferred to a plasmid vector (pT7-7) preparatory to expression of the protein in *E. coli*. The cDNA sequence described herein has been deposited at GenBank. (Accession No. U12761.)

ACKNOWLEDGEMENT

We are indebted to Professor Henry Weiner for providing us with the cDNA for human cytosolic and mitochondrial aldehyde dehydrogenases.

REFERENCES

Aviv, H. and Leder, P., 1972, Purification of biologically active globin messenger RNA by chromatography on oligo-thymidilic acid-cellulose., Proc. Natl. Acad. Sci. USA, 69:1408-1412.

Blackwell, L.F., Buckley, P.D. and MacGibbon, A.K.H., 1989, Aldehyde dehydrogenases - Kinetic characterisation., in "Human Metabolism of Alcohol" (Crow, K.E. and Batt, R.D., Eds), Vol. II pp 89-104., CRC Press, Boca Raton, Florida.

Chirgwin, J.M., Przybyla, A.F., MacDonald, R.J. and Rutter, W.J., 1979, Isolation of biologically active RNA from sources enriched in ribonuclease., Biochemistry, 18:5294-5299.

Hsu, L.C., Bendel, R.E. and Yoshida, A., 1988, Genomic structure of the human mitochondrial aldehyde dehydrogenase gene., Genomics, 2:57-65.

Hsu, L.C., Chang, W.C. and Yoshida, A., 1989, Genomic structure of the human cytosolic aldehyde dehydrogenase gene., Genomics, 5:857-865.

Kitson T.M., Hill, J.P. and Midwinter, G.G., 1991, Identification of a catalytically essential nucleophilic residue in sheep liver cytoplasmic aldehyde dehydrogenase., Biochem. J., 275:207-210.

Loomes K.M., Midwinter, G.M., Blackwell, L.F. and Buckley, P.D., 1990, Evidence for reactivity of serine-74 with *trans*-4-(*N,N*-dimethylamino)cinnamaldehyde during oxidation by the cytoplasmic aldehyde dehydrogenase from sheep liver., Biochemistry, 29:2070-2075.

9

CRYSTALLIZATION OF SHEEP LIVER CYTOSOLIC ALDEHYDE DEHYDROGENASE IN A FORM SUITABLE FOR HIGH RESOLUTION X-RAY STRUCTURAL ANALYSIS

Heather M. Baker, Rosemary L. Brown, Aaron J. Dobbs,
Kathryn E. Kitson, Trevor M. Kitson and Edward N. Baker

Department of Chemistry and Biochemistry
Massey University
Palmerston North
New Zealand

INTRODUCTION

Aldehyde dehydrogenase (AlDH) is an NAD^+-dependent enzyme that is widely distributed through many different species, including both eukaryotes and prokaryotes. In animals, aldehyde dehydrogenases are found in a number of different bodily locations where they exist as several distinct isozymes. These enzymes share the general role of detoxification of aldehydes (Jakoby and Ziegler, 1990) but the individual isozymes show differences in specificity and reactivity that suggest they may have additional, more specialised, roles.

Three main isozymes of AlDH have been identified on the basis of their amino acid sequences (Lindahl and Hempel, 1991). Class 1, which is typified by the mammalian cytosolic AlDH, and Class 2, represented by the mitochondrial enzyme, share ~70% sequence identity and are homotetramers of 4 x 55 kDa. The Class 3 isozymes, however, are homodimers of 2 x 50 kDa (Jones *et al*, 1988); these include a tumor-specific form of AlDH that is induced by various carcinogens (Campbell *et al*, 1989) and a corneal AlDH that is present in very high concentrations in mammalian cornea (Abedinia *et al*, 1990).

Although aldehyde dehydrogenases have been extensively characterised by kinetic and reactivity studies, and numerous amino acid sequences have been determined (Hempel *et al*, 1993) no three-dimensional structure is yet available for any AlDH. This presents a major impediment to a proper understanding of the functional properties of the enzyme. Moreover, although the 3D structures of all three classes of AlDH are expected to be broadly similar, despite the lower sequence identity (<30%) of the Class 3 enzyme (Rose

Enzymology and Molecular Biology of Carbonyl Metabolism 5
Edited by H. Weiner *et al.*, Plenum Press, New York, 1995

et al, 1990), it is unlikely that a single structure will answer some of the structural questions that we wish to address. Thus the structural basis for the high affinity of Class 1 AlDH for retinal as a substrate, compared with the inactivity of the Class 2 isozyme (Yoshida *et al*, 1992), and for the different reactivities with the drug disulfiram (Kitson, 1989), may depend on quite subtle effects. Ultimately, detailed 3D structures for all three isozymes will be needed if effects like these are to be understood, and if new structure-based drugs are to be designed.

Crystals of the mitochondrial (Class 2) AlDH from bovine liver (Hurley and Weiner, 1992) and of the Class 3 enzyme from rat liver (Rose *et al*, 1990) have been reported already. Here we present details of the first crystallization of a Class 1 AlDH, isolated from the cytosolic fraction of sheep liver. A feature of the work is the use of an active site reporter group, 3,4-dihydro-3-methyl-6-nitro-2H-1,3-benzoxazin-2-one (DMNB) (Kitson and Kitson, 1994), as a covalent modifier of the enzyme; this appears to have been a key factor in obtaining diffraction-quality crystals, after many unsuccessful earlier attempts.

MATERIALS AND METHODS

Cytosolic aldehyde dehydrogenase was purified from sheep liver as previously described (Kitson and Kitson, 1994). Crystallization experiments used either the free enzyme or the covalently-modified DMNB derivative; in the latter case the enzyme was reacted with DMNB and separated from excess modifier by passage down a column of Bio Gel P-6. All crystallization experiments were based on the vapour diffusion method, using either hanging drops or sitting drops (McPherson, 1982). The search for crystallization conditions requires a systematic variation of factors such as precipitant type and concentration, buffer and pH. Conditions under which the free enzyme could be crystallized were initially found using the screening procedures of Jancarik and Kim (1991). For later crystallization experiments, however, we used a more systematic approach, developed in our laboratory. This is a statistically-based procedure employing orthogonal arrays (Kingston *et al*, 1994), and has the advantage that the coverage of the various possible sets of conditions is thorough and analysable, but does not require a prohibitive number of experiments. Three separate crystallization searches were carried out, each using a different type of precipitant, ie. salts, non-ionic polymers (based on polyethylene glycols) and alcohols. All experiments were done at 4'C. Once initial crystallization conditions were found, these were refined by slight modifications, including changes in protein concentration and drop size and the use of additives such as β-octylglucoside.

RESULTS AND DISCUSSION

No crystals of AlDH were obtained using either salts or alcohols as precipitants. On the other hand both the free enzyme and the DMNB-modified enzyme were readily crystallized using polyethylene glycols. The optimal conditions for the crystallization of the free enzyme used a protein solution comprising AlDH at a concentration of 2.5 mg/ml, 1 mM NAD^+ and 5 mM dithiothreitol in 0.01 M Bis Tris Propane/HCl, pH 6.5, and a reservoir solution comprising 0.1 M sodium acetate buffer, pH 5.0, 0.17 M $MgCl_2$ and 12% polyethylene glycol 6000. Equal volumes (2-5 μl) of the two solutions were mixed and the drops were allowed to equilibrate with the reservoir solution. The crystals were, however, small and poorly-formed, and diffracted X-rays very weakly. The use of β-octylglucoside as an additive increased the size slightly but the crystals were still of poor quality.

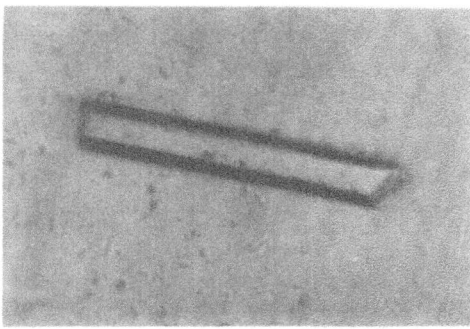

Figure 1. Crystals of DMNB-modified sheep liver cytosolic AlDH.

Greatly improved crystals were obtained using the DMNB-modified AlDH. The conditions differed slightly from those found for the free enzyme. The best crystals were obtained using 4 μl drops of protein solution (6 mg/ml AlDH, 1 mM NAD⁺, 5 mM dithiothreitol, 0.01 M HEPES, pH 7.0) which were mixed with equal volumes of a reservoir solution comprising 0.2 M Bis Tris Propane/HCl, pH 6.5, 0.17 M MgCl$_2$ and 6.5% methoxylpolyethylene glycol 5000. Rod-shaped crystals (Figure 1) appeared after 10-14 days at 4'C and reached dimensions typically 0.05 x 0.05 x 0.2 mm. Although small, these crystals diffracted well, giving measurable data to 2.8 Å resolution on a rotating anode generator. The improved quality of these crystals may imply that the free enzyme is able to exist in several conformations in solution, but that reaction with DMNB, which is presumed to bind to the conserved Cys 302 in the active site, converts all of the molecules to the same, more rigid, form. This would have evident advantages for crystallization.

The crystals are orthorhombic, with P222 symmetry and unit cell dimensions a = 80.7, b = 92.5, c = 151.6 Å. Based on systematically absent reflections the probable space-group is P2$_1$2$_1$2. The assumption of a dimer (2 x 55 kDa) in the asymmetric unit gives a value of V$_M$ of 2.8 Å^3/Da, and a solvent content of 56%, which is well within the normal range found for protein crystals (Matthews, 1968). The complete AlDH tetramer is presumably generated by a crystallographic two-fold axis. A partial native data set to 2.8 Å resolution has been collected on a Rigaku R-Axis IIc image plate detector on an RU200 rotating anode generator, and a full X-ray structural analysis is planned. The presence of a dimer in the asymmetric unit presents a favourable case for structure determination, as the unit cell size is manageable for data collection on a laboratory source, but it also gives the possibility of exploiting the two-fold redundancy by two-fold non-crystallographic symmetry averaging; this is a powerful means of improving initial phase information.

ACKNOWLEDGMENTS

We wish to thank Drs L.F. Blackwell, P.D. Buckley, M.J. Hardman and J.P. Hill for advice, encouragement and help with protein purification. We gratefully acknowledge research support from the Palmerston North Medical Research Foundation and from Massey University, including the award of a Graduate Assistantship to AJD. ENB also receives research support as an International Research Scholar of the Howard Hughes Medical Institute.

REFERENCES

Abedinia, M., Pain, T., Algar, E.M. and Holmes, R.S., 1990, Bovine corneal aldehyde dehydrogenase: the major soluble corneal protein with a possible dual protective role for the eye, *Exp. Eye Res.*, 51: 419-426.

Campbell, P., Irving, C.C. and Lindahl, R., 1989, Changes in aldehyde dehydrogenase during rat urinary bladder carcinogenesis, *Carcinogenesis*, 10: 2081-2087.

Hempel, J., Nicholas, H. and Lindahl, R., 1993, Aldehyde dehydrogenases: widespread structural and functional diversity within a shared framework, *Protein Science*, 2: 1890-1900.

Hurley, T.T. and Weiner, H., 1992, Crystallization and preliminary X-ray investigation of bovine liver mitochondrial aldehyde dehydrogenase, *J. Mol. Biol.*, 277: 1255-1257.

Jakoby, W.B. and Ziegler, D.M., 1990, The enzymes of detoxification, *J. Biol. Chem.*, 265: 20715-20718.

Jancarik, J. and Kim, S.-H., 1991, Sparse matrix sampling: a screening method for crystallization of proteins, *J. Appl. Cryst.*, 24: 409-411.

Jones, D.G., Brennan, M.D., Hempel, J. and Lindahl, R., 1988, Cloning and complete nucleotide sequence of a full-length cDNA encoding a catalytically functional tumour-associated aldehyde dehydrogenase, *Proc. Natl. Acad. Sci. USA*, 85: 1782-1786.

Kingston, R.L., Baker, H.M. and Baker, E.N., 1994, Search designs for protein crystallization based on orthogonal arrays, *Acta Cryst.*, D50: in press.

Kitson, T.M., 1989, Reactions of aldehyde dehydrogenase with disulfiram and related compounds, In Human Metabolism of Alcohol, (Crow, K.E. and Batt, R.D., eds), pp. 117-132, CRC Press, Boca Raton, Florida.

Kitson, T.M. and Kitson, K.E., 1994, Probing the active site of cytoplasmic aldehyde dehydrogenase with a chromophoric reporter group, *Biochem. J.*, 300: 25-30.

Lindahl, R. and Hempel, J., 1991, Aldehyde dehydrogenases: what can be learned from a baker's dozen sequences, *Adv. Exp. Med. Biol.*, 284: 1-8.

Matthews, B.W., 1968, Solvent content of protein crystals, *J. Mol. Biol.*, 33: 491-497.

McPherson, A., 1982, The Preparation and Analysis of Protein Crystals, Wiley, New York.

Rose, J., Hempel, J., Kuo, I., Lindahl, R. and Wang, B.-C., 1990, Preliminary crystallographic analysis of class 3 rat liver aldehyde dehydrogenase, *PROTEINS: Struct. Func. Genet.*, 8: 305-308.

Yoshida, A., Hsu, L. and Davé, V., 1992, Retinal oxidation activity and biological role of human cytosolic aldehyde dehydrogenase, *Enzyme*, 46: 239-244.

10

PROGRESS TOWARD THE TERTIARY STRUCTURE OF (CLASS 3) ALDEHYDE DEHYDROGENASE

Julie Sun[1], John Hempel[2], Ronald Lindahl[3], John Perozich[2], John Rose[1] and Bi-Cheng Wang[1]

Departments of Crystallography[1] and Molecular Genetics/ Biochemistry[2]
University of Pittsburgh
Pittsburgh, PA 15261
Department of Biochemistry and Molecular Biology
University of South Dakota[3]
Vermillion SD 57069

INTRODUCTION

We chose to pursue crystallography of the class 3 aldehyde dehydrogenase since, with subunits some 10% shorter than classes 1 and 2, and a dimeric rather than tetrameric quaternary structure, it presented a simpler molecule to solve. Further, despite a primary structure only some 30% identical to class 1 or 2 structures, secondary structural predictions suggested generally comparable tertiary structures (Hempel et al., 1989). The pTALDH expression system (Harper et al., 1988) provided a ready source of material. Subsequently, the observed low solubility of the enzyme (ca. 1mg/ml in the presence of NAD) was probably a further advantage aiding crystallization.

The crystallization and preliminary X-ray diffraction data of the rat class 3 AlDH (expressed in E. coli from pTALDH) have been reported previously (Rose et al., 1990). Large 0.8x0.3x0.2mm crystals can be obtained from 10% Polyethylene glycol (PEG 3350-3750) in 20mM PIPES, pH 6.2. The crystals belong to the monoclinic space group $P2_1$ with a=64.93Å, b=170.93Å, c=47.16Å and $\beta=110.25°$. Self-rotation function studies (Rossmann & Blow, 1962) indicate a non-crystallographic 2-fold axis at $\kappa=180°$, $\psi=90°$, and $\phi=20°/110°$ which is consistent with one AlDH dimer per assymetric unit and gives a solvent content of 50% (Vm=2.45Å^3/dalton, Matthews, 1968) which is in the normal range for protein crystals.

We here add to the details given in our preliminary report on the native crystals. Numerous compounds have been pursued in an effort to produce suitable heavy atom derivatives. Only p-chloromercuribenzoate (pCMB) yielded interpretable Pattersons with a suitable level of incorporation and R_{iso} values of 10-20%.

Table 1. A summary of native and pCMB derivative data of AlDH

	Native	HG-10	HG-20	HG-30
Resolution	2.5 Å	2.9 Å	2.7 Å	2.65 Å
No. of reflections	25,656	15,271	24,332	21,511
Rsym*	5.02 %	9.50 %	9.51 %	8.75 %
Initial heavy atom sites from Patterson map		4 sites	5 sites	2 sites
Final total heavy atom sites		6 sites	6 sites	6 sites

* $Rsym = \Sigma_{hkl} \Sigma_n |(I_{hkl})_n - <I_{hkl}>| / \Sigma_{hkl} \Sigma_n(I_{hkl})_n$

RESULTS AND DISCUSSION

After an intensive derivative search in which over 30 different heavy atom compounds were screened under a variety of soaking times and conditions an isomorphous p-chloromercuribenzoate (pCMB) derivative was obtained. The pCMB-soaked crystals were found to be sensitive to X-ray exposure with a useful crystal lifetime of less than 2 days. Refinement of the initial crystal soaking conditions resulted in improved crystal lifetime and provided a means of adjusting the occupancy of the heavy atom sites. Cross Fourier analysis

Table 2. The refined heavy atom parameters for the three Hg-derivative data sets

HG-10 (R_{cen}^*: 57.6%/5Å)

	X	Y	Z	Occupancy	B (Å^2)
HG1	0.432	0.000	0.824	1.327	25
HG2	0.268	0.688	0.099	1.627	25
HG3	0.123	0.576	0.579	0.986	25
HG4	0.071	0.900	0.870	1.251	25
HG5	0.010	0.582	0.745	1.879	25
HG6	0.424	0.020	0.112	0.526	25

HG-20 (R_{cen}^*: 56.1%/3.5Å)

	X	Y	Z	Occupancy	B (Å^2)
HG1	0.413	0.000	0.809	1.302	25
HG2	0.258	0.688	0.095	1.262	25
HG3	0.111	0.576	0.608	1.196	25
HG4	0.064	0.900	0.880	1.220	25
HG5	0.007	0.582	0.763	1.030	25
HG6	0.432	0.020	0.138	0.896	25

HG-30 (R_{cen}^*: 53.9%/3.5Å)

	X	Y	Z	Occupancy	B (Å^2)
HG1	0.412	0.000	0.807	0.746	25
HG2	0.266	0.688	0.100	1.919	25
HG3	0.122	0.576	0.605	0.910	25
HG4	0.065	0.900	0.873	1.351	25
HG5	0.005	0.582	0.761	0.675	25
HG6	0.427	0.020	0.118	0.540	25

* $R_{cen} = \Sigma [| |F_{PH}-F_P| - |F_{H(cal)}| |] / \Sigma |F_{PH}-F_P|$ for the centric reflections, $F_{H(cal)}$ is the calculated heavy atom scattering factor.

Table 3. Phasing statistics

	SIRAS phases (after solvent flattening)	Averaged phases (non-crystallographic symmetry averaging)
R-factor	0.248	0.294
Correlation Coefficient	0.967	0.945
Figure of Merit	0.770	0.715
Resolution (Å)	3.5 phase extend to 3.0Å	3.0
# of Reflections	15926	18514

showed six consistent heavy atom sites, three per monomer, with varying occupancies depending om the crystal soaking conditions used. A summary of native and pCMB data collected and heavy atom parameters is presented in Tables 1 and 2, respectively.

The heavy atom phase probabilities from the three data sets were merged and protein phases were calculated using the Iterative Single Isomorphous Replacement with Anomalous Scattering method (ISIRAS, Wang, 1985). Three filters each run with eight cycles of iteration were used to resolve the phase ambiguity to 3.5Å resolution. In the following two runs, the first four cycles of iteration were used to resolve the phase ambiguity to 3.5Å resolution and the remaining four cycles used to extend the phases to 3.0Å resolution. The final cycle had an average figure of merit of 0.77 for a total of 15926 reflections to 3Å resolution, a map inversion R-factor of 0.248 and a correlation coeficient of 0.967.

Non-crystallographic symmetry averaging (PHASES, Furey & Swaminathan, 1992) was then used to further improve the protein phases prior to visual inspection of the initial electron density maps. First the ISIRAS phases were imported into the PHASES package and the non-crystallographic 2-fold axis was refined (LSQROT). The refinement converged to $\kappa=180°$ $\psi=87.58°$, and $\phi=51.72°$ (the large numerical difference in ϕ from the native data is a convention-based difference in the two programs), $X_o=33.42$Å, $Y_o=49.87$Å, and $Z_o=50.16$Å. Next, non-crystallographic symmetry averaging (option 1, 50% solvent content) with solvent flattening and phase combination were carried out for 16 cycles. A comparison of the phasing statistics obtained from the ISIRAS and non-crystallographic symmetry averaging is presented in Table 3. Figures 1a and 1b show the skewed ISIRAS map before and after non-crystallographic symmetry averaging.

The non-crystallographic symmetry-averaged ISIRAS-phased electron density map $(2F_o-F_c)$ was displayed onto transparencies at a scale of 0.25cm/Å using a 1Å grid and stacked for visual inspection. Inspection of the mini-map clearly showed the dimer boundary with rough Cα locations of 324 amino acid residues (~70% total length). These Cα positions were adjusted to a neighbor-neighbor distance of ~3.8Å, Provisional polyalanine fitting was made using the updated "O" program (Jones et al., 1991).

The resulting polyalanine model for the class 3 AlDH monomer is shown in Figure 2a. The molecule is a prolate ellipsoid of dimensions 50Åx50Åx85Å. The molecule is composed of two subdomains connected by a peptide hinge with the N-terminal domain being somewhat larger than the C-terminal domain. A total of five right-handed α helices and several β strands have been identified thus far from the initial mini-map, and two α-helical segments designated CA176-CA186 and CA236-CA247 are shown in Figures 3a and 3b, respectively ("CA" emphasizes the fact that these numbers do not correspond to position numbers from the primary structure). The two monomers associate in the dimer in a head-to-tail manner, consistent with the non-crystallographic 2-fold axis of the dimer. A Cα trace for the dimer is shown in Figure 2b. Interactions are observed between the N and C termini of each monomer across the dimer interface with the C and N terminal residues of the other monomer, respectively.

As yet, it has not been possible to specifically identify any side-chains in the model. Toward that goal, however, we noted that the hinge region in the mini-map, where rapid

Y = -0.03

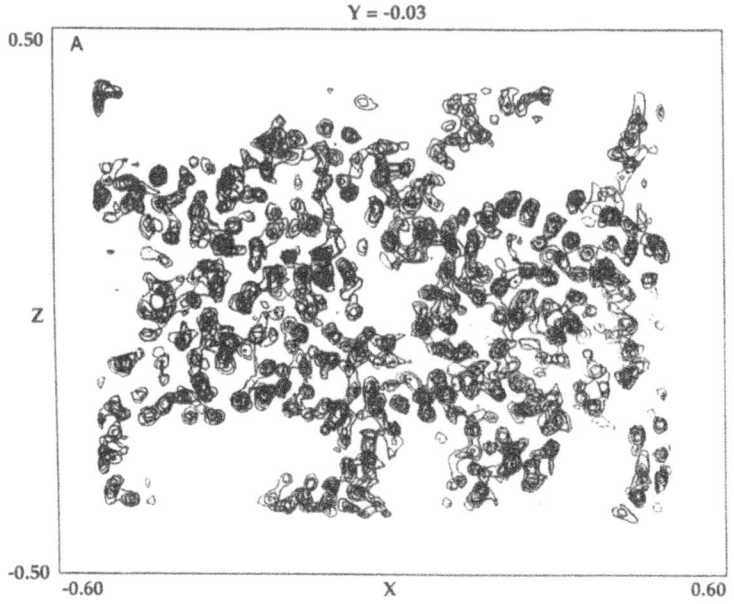

Y = -0.03

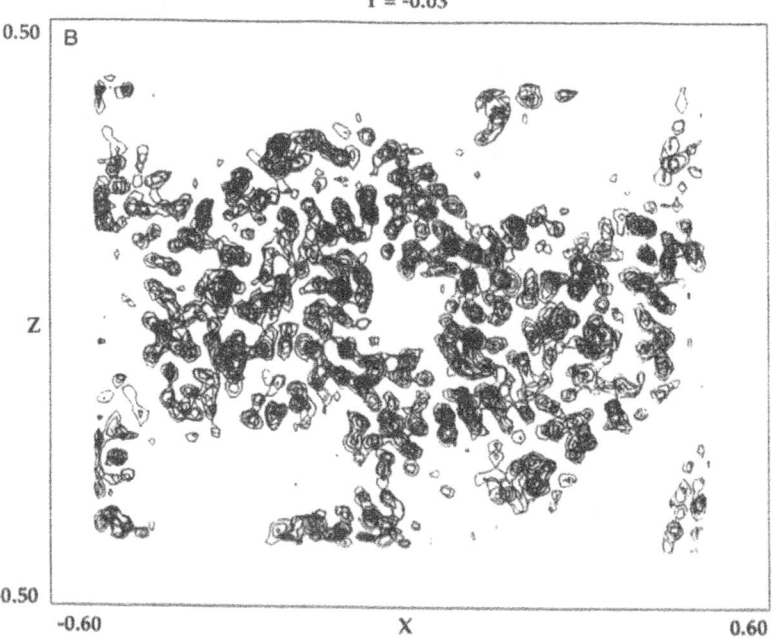

Figure 1. Projection of a 10Å thick slabs, from the solvent flattened ISIRAS-skewed map, looking down the non-crystallographic symmetry axis (A), and from averaging a skewed map of the dimer at 3Å resolution (B).

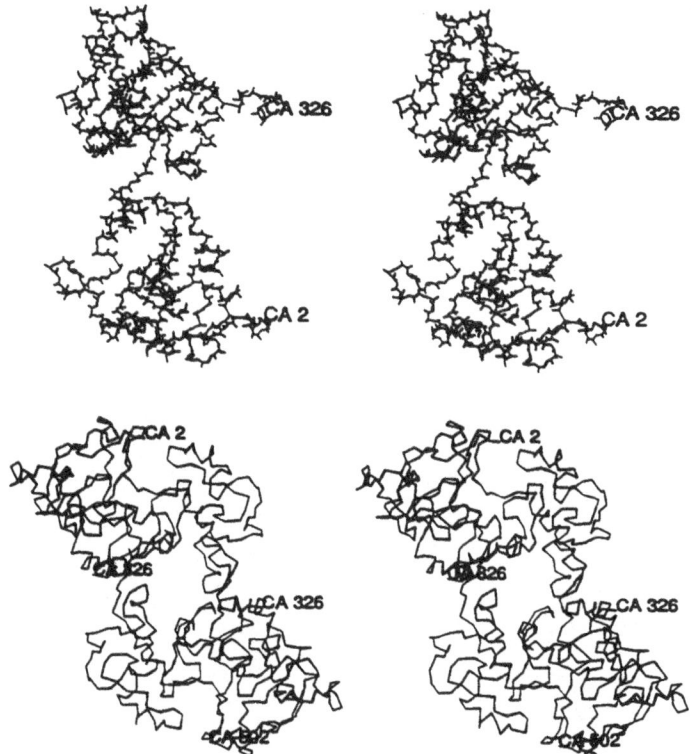

Figure 2. Top: Stereo-drawing of a polyalanine model of the AlDH **monomer**. Bottom: Stereo-drawing of an AlDH **dimer** (Cα model).

proteolysis would be expected, was at the approximate middle of the sequence, and that Loomes and Jörnvall (1991) had obtained limited proteolytic cleavages of classes 1 and 2 AlDHs at the approximate middle of those chains. Proteolysis of native class 3 AlDH with trypsin proceeded rapidly, yet produced major fragments of distinctly unequal size, a doublet around 40kDa and a diffuse band around 10kDa (data not shown). Trans-blotting to ImmobilonP (Millipore) membranes and Edman degradation of the excised smaller Coomassie blue-stained band yielded two derivatives at each cycle. However, interpretation of these data was consistent with major and minor fragments starting at positions 358 (EKPLALY...) and 376 (MIAETSS....), respectively. These positions are clearly not at the middle of the 452-residue structure, yet they may be relevant to one of the loops apparent in the C-terminal domain from the map (Figure 4).

ACKNOWLEDGEMENT

Supported by AA06985 (to JH, RL and B-CW).

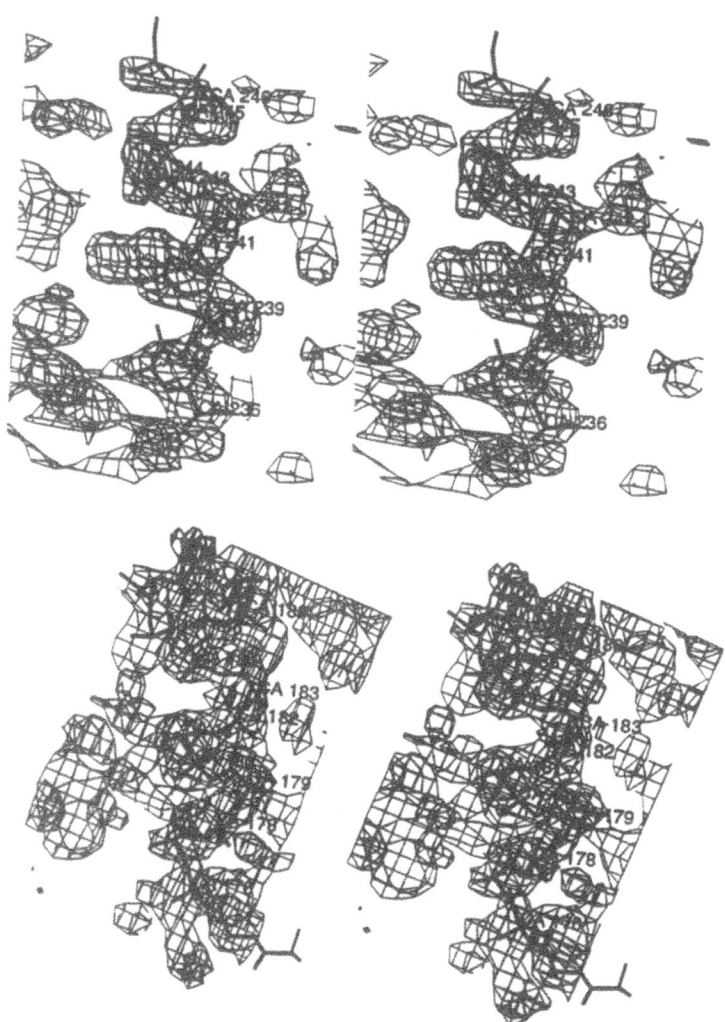

Figure 3. Representative electron density maps of AlDH. Top: the segment tentatively identified as residues CA-176-CA186. Bottom: the segment tentatively identified as residues CA236-CA247.

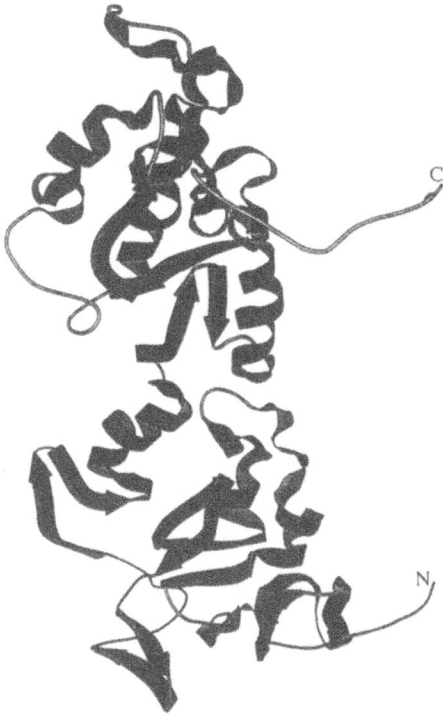

Figure 4. Ribbon diagram of the AlDH subunit model.

REFERENCES

Furey WF, Swaminathan S. (1990) PHASES-a program package for the processing and analysis of diffraction data from macromolecules. American Crystallographic Association, Fortieth Anniversary Meeting, New Orleans, abst PA33, 73.

Harper K, Jones D, Brennan MD, Lindahl R. (1988) Characterization of a functional recombinant rat liver aldehyde dehydrogenase. Biochem Biophys Res Comm 152:940-947.

Hempel J, Harper K, Lindahl, R. (1989) Inducible (class 3) aldehyde dehydrogenase from rat hepatocellular carcinoma: distant relationship to the class 1 and 2 enzymes from liver cytosol/mitochondria. Biochemistry 28:1160-1167.

Jones TA, Zou J-Y, Cowan SW. (1991) Improved methods for building protein models in electron density maps and the location of errors in these models. Acta Cryst A47:110-119.

Loomes, K, Jörnvall, H. (1991) Structural organization of aldehyde dehydrogenases probed by limited proteolysis. Biochemistry 30:8865-8870.

Matthews B. (1968) Solvent content of protein crystals. J Mol Biol 33:491-497.

Rose JP, Hempel, J, Kuo, I, Lindahl R, Wang B-C. (1990) Preliminary crystallographic analysis of rat class 3 aldehyde dehydrogenase. Proteins 8:305-308.

Rossmann M, Blow DM. (1962) Detection of sub-units within the crystallographic assymetric unit. Acta Crystal 15:24-30.

Wang B-C. (1985) Resolution of phase ambiguity in macromolecular crystallography. Meth Enzymol 115:90-112.

11

UDP-GLUCOSE DEHYDROGENASE
STRUTURAL CHARACTERISTICS

John Perozich, Amy Leksana, and John Hempel

Department of Molecular Genetics and Biochemistry
University of Pittsburgh
Pittsburgh, PA 15261

INTRODUCTION

UDP-glucose dehydrogenase (UDPGDH; EC 1.1.1.22) belongs to a small family of NAD$^+$-linked oxidoreductases which transfer four electrons per catalytic cycle. Other examples of four-electron-transferring enzymes include histidinol dehydrogenase, β-hydroxy-β-methylglutaryl-CoA reductase, and other nucleotide sugar dehydrogenases (Feingold and Franzen, 1981). The bovine liver enzyme is an apparent homohexamer of 52kDa subunits which appears to function as a trimer of dimers due to its half-sites reactivity to iodoacetate and iodoacetamide (Franzen et al., 1980). We have recently completed the first primary structure of UDPGDH from a mammalian source (bovine liver) (Hempel et al., 1994). Here we present additional details of the analysis and results of a search for potentially homologous proteins using profile analysis.

REACTION MECHANISM OF UDPGDH

UDPGDH performs two separate, but linked oxidations of UDP-D-glucose (UDP-Glc) to yield UDP-D-glucuronic acid (UDP-GlcA). According to the reaction mechanism proposed by Ordman and Kirkwood (1977) (Figure 1), the first half-reaction is initiated by the formation of a Schiff's base between the carbon-6 (C6) of UDP-Glc and the transfer of the *pro*-R hydrogen of C6 to the first equivalent of NAD$^+$ to produce NADH. Multiple sequence alignment of bovine liver UDPGDH with 4 other related sequences, including other four-electron-transferring, nucleotide sugar dehydrogenases, showed that two lysine residues are strictly conserved, Lys-219 and Lys-338. Thus, it is quite likely that one of these two lysines is the catalytic residue involved in the first half-reaction (Hempel et al., 1994).

The second half-reaction is catalyzed by a cysteine residue which attacks the Schiff's base to form a thiohemiacetal intermediate. Since the substrate remains covalently-bound to the enzyme, no UDP-6-aldehydo-D-glucose (UDP-AlGlc) is ever produced by the enzyme. However, if UDP-AlGlc is added as a substrate it can be either oxidized to UDP-GlcA or reduced to UDP-Glc (Nelsestuen and Kirkwood, 1971). The only strictly conserved cysteine

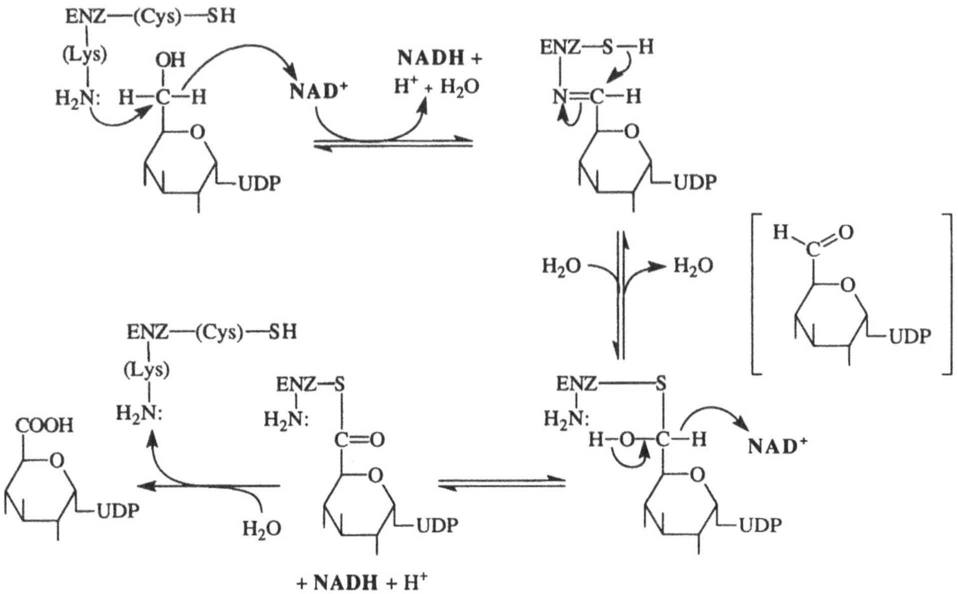

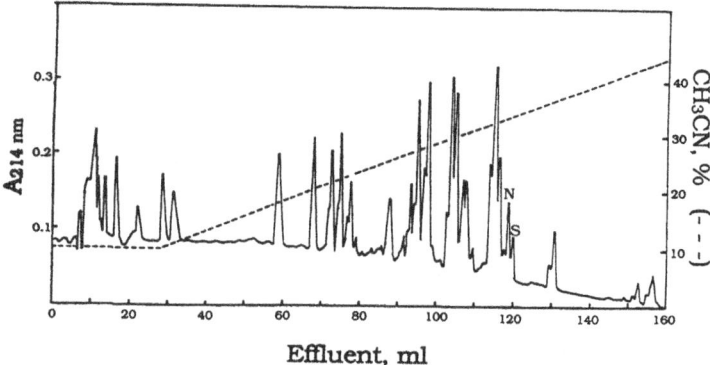

Figure 1. Reaction mechanism of bovine liver UDP-glucose dehydrogenase, based on that proposed by Ordman and Kirkwood (1977). The first half-reaction is the formation of a Schiff's base between a catalytic lysine and carbon-6 of UDP-glucose. The second half-reaction is catalyzed by a catalytic cysteine which attacks the Schiff's base to form a thiohemiacetal intermediate. The total reaction consumes 2 moles of NAD^+ per mole of UDP-glucose, removing four electrons per catalytic cycle in the form of 2 NADH molecules.

identified in the multiple alignment is Cys-275, identifying it as the probable catalytic residue for this second half-reaction. A tryptic peptide containing the active-site cysteine, which also corresponds to Cys-275, had earlier been identified (Franzen et al., 1981). The thiohemiacetal is then oxidized via another equivalent of NAD^+ to generate a thiolester, which

Effluent, ml

Figure 2. C_{18} reversed-phase HPLC separation of a peptide pool from a tryptic digest of NIPCAM-modified UDP-glucose dehydrogenase. The S-alkylated enzyme preparation was digested with TPCK-trypsin. The cleaved enzyme mixture was first pre-fractionated into pools via ion-exchange HPLC. These pools were applied to C_{18} reversed-phase HPLC in 0.1% aqueous trifluoroacetic acid and eluted with gradients of acetonitrile. Peaks containing peptides with either an Asn or a Ser at the residue corresponding to position 74 of UDPGDH are marked by a N or S, respectively.

is later hydrolyzed to yield UDP-GlcA. This final step is irreversible and appears to be rate-limiting.

Evidence has also been presented that an arginine residue may serve an unknown function in the enzyme mechanism (Feingold and Franzen, 1981). Positively-charged arginyl residues can often serve as recognition sites for negatively-charged groups of the substrate or cofactor molecule in the active site (Riordan et al., 1977). The only strictly conserved arginine residue is Arg-345 (Hempel et al., 1994), indicating this residue may possibly play that functional role.

The ability to oxidize both alcohols and aldehydes invokes the hypothesis that at some point in evolution, these four-electron-transferring dehydrogenases may have resulted from a gene fusion between an alcohol dehydrogenase (ADH) and an aldehyde dehydrogenase (AlDH). ADHs can themselves oxidize both alcohols and aldehydes given the proper conditions, as discussed elsewhere in this volume by N. Oppenheimer. However, ADHs do not contain a catalytic thiol residue and cannot form a covalently-bound intermediate as seen with AlDHs. Although a few similarities can be noted between UDPGDH and these two

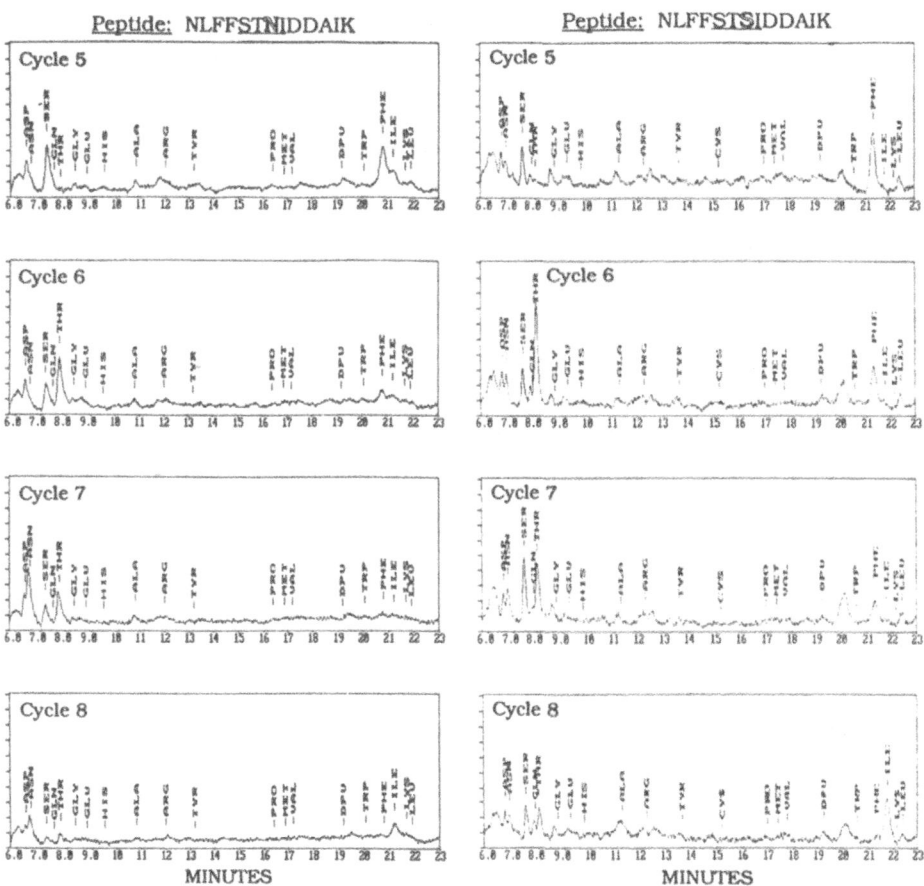

Figure 3. HPLC separations of PTH-derivatized amino acids from automated Edman degradation of peptides containing the Asn/Ser microheterogenity. Peptides were prepared and sequenced as described previously (Hempel et al., 1994). HPLC traces of derivatives from cycles 5-8 are shown, corresponding to the segment ST#1 of the peptides with the sequence NLFFST#IDDAIK, where # indicates either an Asn (left column) or Ser (right).

classes of enzymes, no sequence homology was found to either class, possibly indicating an example of convergent evolution (Hempel et al, 1994).

MICROHETEROGENEITY

Upon sequencing bovine liver UDPGDH, an Asn/Ser microheterogeneity at position 74 was clearly evident (Hempel et al., 1994). A tryptic digest of UDPGDH from an enzyme preparation from a single liver yielded two adjacent peaks on C_{18} reversed-phase HPLC (Figure 2). The first of the two peaks contained a peptide with the sequence: NLFFSTNID-DAIK (Figure 3). The second peak had a peptide with the sequence: NLFFSTSIDDAIK. Two peaks were similarly identified from a staphylococcal V8 (Glu-specific) protease digest of UDPGDH from an enzyme preparation from a different liver than that used for the tryptic digest (data not shown), clearly supporting a microheterogeneity.

The reasons for this microheterogeneity are not yet clear. One hypothesis is that these liver samples came from heterozygotes expressing both "Asn-74" and "Ser-74" alleles. Another possibility is that a recent gene duplication event has occurred, with the mutation at the codon for residue 74 being the only mutational event. Although thought to be homohexameric, functional UDPGDH could possibly require both Asn-74 and Ser-74 subunits. However, it is unlikely that this microheterogeneity could account for the half-sites reactivity displayed by bovine liver UDPGDH. Although a residue substitution between Asn and Ser would minimally require only one base change (AAU & AAC, Asn; AGU & AGC, Ser), both residues have similar secondary structural potential, preferring reverse turns, and Asn and Ser are frequently found to substitute for one another. One possible way that this microheterogeneity could affect enzyme function is if this residue is located within the active site pocket of the enzyme. A substitution of these two residues for one another may alter hydrogen bonding of UDPGDH to either the substrate or coenzyme molecules.

PROFILE ANALYSIS

As noted above, UDPGDH was aligned with 3 related sequences of other four-elec-tron-transferring dehydrogenase and a hypothetical sequence from *Salmonella typhimurium*. These sequences had approximately 30% identity between any pair, indicating that they are quite divergent from one another (Hempel et al., 1994). Another point of interest is that the number of instances where the same residue is shared by just 4 of the 5 sequences (24) is fewer than the number of totally conserved residues (27). This again further suggests that the sequences have evolved from a common ancestor for such a long period of time that the consensus residues reflect only those residues which are absolutely necessary for enzyme structure and/or function.

A similarity profile was generated from this alignment and used to search for homologous sequences (Gribskov et al., 1990). The search led to the identification (with marginal Z scores of at least 4.60) of two lactate dehydrogenases (LDHs) from *B. stearother-mophilus* and *B. megaterium*. These enzymes were probably identified mainly on the basis of the pattern of residues characteristic of a Rossman fold located near the amino-termini of both the LDHs and the nucleotide sugar dehydrogenases. These proteins also have a hydrophobic residue (Val in LDHs; Ile in UDPGDH) located four residues amino-terminal to and an Asp 20 residues downstream from the GXGXXG region. A profile was then generated from three bacterial LDHs from *B. stearothermophilus*, *B. megaterium*, and *Lactobacillus caseii*. When this profile was used to search the database, it revealed no nucleotide sugar dehydrogenases. One short chain ADH, 3-β-hydroxysteroid dehydrogenase

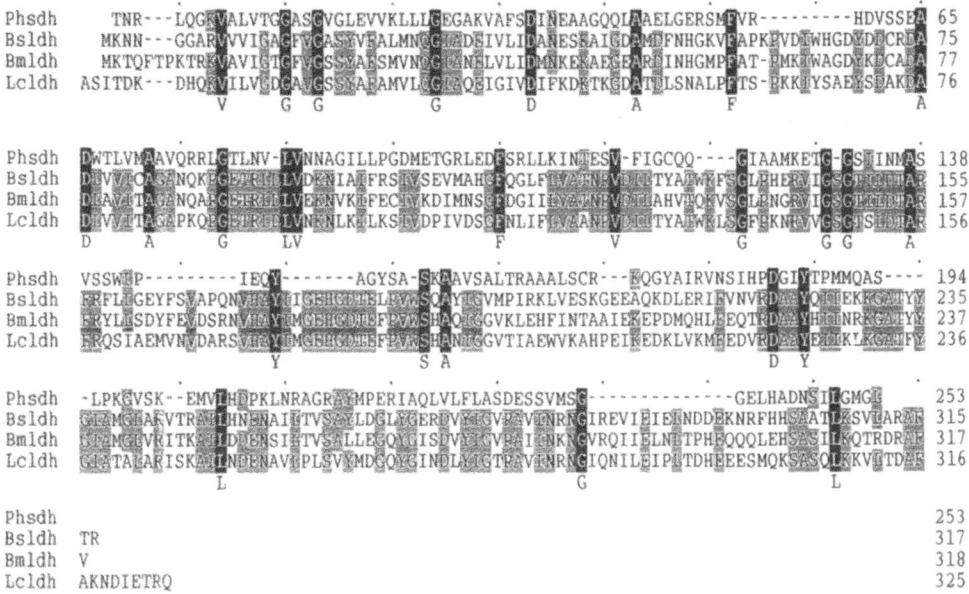

```
Phsdh    TNR---LQGKVALVTGGASCVGLEVVKLLIGEGAKVAFSDINEAAGQQLAAELGERSMEVR--------HDVSSEA  65
Bsldh    MKNN---GGARVVVIGAGFVGASYVEALMNGGHADEIVLIDANESHAIGDAMDFNHGKVEAPKEVDIWHGDYDICRDA  75
Bmldh    MKTQFTPKTRKVAVIGTGFVGSSYAESMVNGGCANLVLIDMNKEKAEGEARDINHGMPEAT-EMKIWAGDYKDCALA  77
Lcldh    ASITDK---DHQKVILVGDGAVGSSYAFAMVLQGIAQSIGIVDIFKDKTKGDATDLSNALPFTS-EKKIYSAEYSDAKDA  76
              V    G     G        G          D      A      F           A
```

```
Phsdh    DWTLVMAAVQRRLGTLNV-ILVNNAGILLPGDMETGRLEDESRLLKINTESV-FIGCQQ----GIAAMKETG-GSTINMAS  138
Bsldh    DVVVICAGANQKPGETRLDLVDKNIAIFRSIVSEVMAHGFQGLFLVATNPVDILTYAAWKFSGLRHERVIGSGTHGLMIAR  155
Bmldh    DLAVITAGANQAGESRLDLVEENVKIFECIVKDIMNSGFDGIILVVTNPVDILTYAHVEOKVSGLRNGRVIGSGTILDIAR  157
Lcldh    DIVVITAGAPKQEGETRLDLVNKNLKILKSIVDPIVDSCENLIFMVANPVDILTYAIWKILSGFEKNRVVGSGTSLDIAR  156
            D     A    G    LV        F       V          G        GG       A
```

```
Phsdh    VSSWNP--------IEQY------AGYSA-SKAAVSALTRAAALSCR---KQGYAIRVNSIHPDGIYTPMMQAS----  194
Bsldh    FRFLIGEYFSVAPQNVHAYIIGEHGDTELPVWSQAYIGVMPIRKLVESKGEEAQKDLERIFVNVRDAAYQIIEKKFGATYY  235
Bmldh    ERYLLSDYFEVDSRNVHAYIIMGEHGDTEFPVWSHAQIGGVKLEHFINTAAIEKEPDMQHLEEQTRDAAYEIINRKKGATYY  237
Lcldh    FRQSIAEMVNVDARSVHAYIIMGEHGDTEFPVWSHANIGGVTIAEWVKAHPEIKEDKLVKMHEDVRDAAYEIIKLKGATFY  236
            Y           S A                  D Y
```

```
Phsdh    -LPKGVSK--EMVLHDPKLNRAGRAYMPERIAQLVLFLASDESSVMSG------------GELHADNSILGMGL  253
Bsldh    GIAMGLARVTRMILHNENAILTVSAYLDGLYGERDVVETGVPAVINRNGIREVIEIELNDDEKNRFHHSAATLKSVLARAF  315
Bmldh    STIMGVRITKGIDDENSIHTVSALLEGOYGISDVVFGVPAIELNKNGVRQIIELNTPHPQQQLEHSASILLQTRDRAF  317
Lcldh    GIATALARISKEILNDENAVFPLSVYMDGOYGINDLYIGTPAVINRNGIQNILEIPLTDHEEESMQKSASOLRKVETDAF  316
            L                          G                     L
```

```
Phsdh                                                                   253
Bsldh    TR                                                              317
Bmldh    V                                                               318
Lcldh    AKNDIETRQ                                                       325
```

Figure 4. Multiple sequence alignment of three lactate dehydrogenases from *B. stearothermophilus* (Bmldh), *B. megaterium* (Bsldh), and *Lactobacillus caseii* (Lcldh) with *Pseudomonas testosteroni* 3-β-hydroxysteroid dehydrogenase (Phsdh).

from *Pseudomonas testosteroni*, was identified with a low Z score of 4.86, again presumably due in some part to the amino-terminally-located Rossman fold. An alignment of these four proteins (Figure 4) shows a limited degree of similarity. However, neither a profile generated from three related short chain ADHs, which included 3-β-hydroxysteroid dehydrogenase, 7-α-hydroxysteroid dehydrogenase from *E. coli*, and 20-β-hydroxysteroid dehydrogenase from *Streptomyces exfoliatus*, nor the LDH profile revealed any comparable similarities to the nucleotide sugar dehydrogenases. Also, no sequence homologies between bovine liver UDPGDH and any alcohol or aldehyde dehydrogenase, or to the two other known four-electron-transferring oxidoreductases, histidinol dehydrogenase and β-hydroxy-β-methylglutaryl CoA reductase, were identified by pairwise searches using FASTA (Pearson and Lipman, 1988) or profile analysis. Thus, it appears at present that the four-electron-transferring, nucleotide sugar dehydrogenases have evolved from a different ancestor than the other enzymes discussed above.

ACKNOWLEDGMENTS

Summer support through the MRAP program for Keisha Hortman, who performed many of the HPLC separations, is acknowledged.

REFERENCES

Feingold, D. S., and Franzen, J. S., 1981, Pyridine nucleotide-linked four-electron transfer dehydrogenases, *Trends Biochem. Sci.* **6**:103-106.

Franzen, B., Carruba, C., Feingold, D. S., Ashcom, J., and Franzen, J. S., 1981, Amino acid sequence of the tryptic peptide containing the catalytic-site thiol group of bovine liver uridine diphosphate glucose dehydrogenase, *Biochem. J.* **199**:599-602.

Franzen, J. S., Ashcom, J., Marchetti, P., Cardamone, J. J., and Feingold, D. S., 1980, Induced vs. preexisting asymmetry models for half-of-the-sites reactivity in bovine liver uridine diphosphoglucose dehydrogenase, *Biochim. Biophys. Acta* **614**:242-255.

Gribskov, M., Lüthy, R., and Eisenberg, D., 1990, Profile analysis, *Methods Enzymol.* **183**:146-159.

Hempel, J., Perozich, J., Romovacek, H., Hinich, A., Kuo, I., and Feingold, D. S., 1994, UDP-glucose dehydrogenase from bovine liver: Primary structure and relationship to other dehydrogenases, *Protein Sci.* **3**:1074-1080.

Nelsestuen, G. L., and Kirkwood, S., 1971, The mechanism of action of uridine diphosphoglucose dehydrogenase: Uridine diphosphodialdoses as intermediates, *J. Biol. Chem.* **246**:3828-3834.

Ordman, A. B., and Kirkwood, S., 1977, Mechanism of action of uridine diphosphoglucose dehydrogenase: Evidence for an essential lysine residue at the active site, *J. Biol. Chem.* **252**:1320-1326.

Pearson, W. R., and Lipman, D. J., 1988, Improved tools for biological sequence comparison, *Proc. Natl. Acad. Sci. USA* **85**:2444-2448.

Riordan, J. F., McElvany, K. D., and Borders, C. L., 1977, Arginyl residues: Anion recognition sites in enzymes, *Science* **195**:884-886.

12

KINETIC STUDIES ON CLASS 3 ALDEHYDE DEHYDROGENASE FROM BOVINE CORNEA

Ian K. Riley, Christopher A. Burrows, Michael J. Hardman and
Paul D. Buckley

Department of Chemistry and Biochemistry
Massey University
Palmerston North, New Zealand

INTRODUCTION

In 1973 Holt and Kinoshita first demonstrated the occurence of a bovine corneal protein with a mass of 54 kDa (BCP 54). This protein has subsequently been shown to be a class 3 aldehyde dehydrogenase which is 84% identical to rat tumor-associated AlDH at the amino acid level (Cooper *et al.*, 1991). Abedinia *et al.* (1990) purified the bovine corneal AlDH and demonstrated that it was the major soluble protein in bovine cornea, representing 20 to 40% of the soluble protein.

Bovine corneal aldehyde dehydrogenase exhibits activity with both NAD^+ and $NADP^+$ (Abedinia *et al.* 1990). Acetaldehyde, although a substrate, has a very high K_m (180 mM) (Abedinia *et al.* 1990); based on k_{cat}/K_m values, hexanal, Δ-2-hexenal and 4-hydroxynonenal are the best substrates for the enzyme (Abedinia *et al.* 1990). The enzyme probably plays a role in protecting the cornea from the damage which could be caused by aldehydes generated by peroxidation of membrane lipids.

Konishi and Mimura (1992) investigated activators and inhibitors of the bovine corneal AlDH. Disulfiram did not significantly affect the activity of the enzyme. *p*-Hydroxymercuribenzoate did not inhibit the NAD^+-dependent propionaldehyde dehydrogenase activity but completely inhibited the $NADP^+$-dependent activity at a concentration of 50 μM.

The aim of the present paper is to further investigate the kinetic properties of this enzyme, and in particular to study the pre-steady state phase of the reaction in order to learn which steps limit the rate during this phase of the reaction.

EXPERIMENTAL

Aldehyde dehydrogenase was isolated from fresh or frozen bovine cornea using the purification procedure described by Abedinia *et al.* (1990) but on a larger scale, beginning with 36 grams of corneal tissue, rather than 3 grams. Hexanal and benzaldehyde were

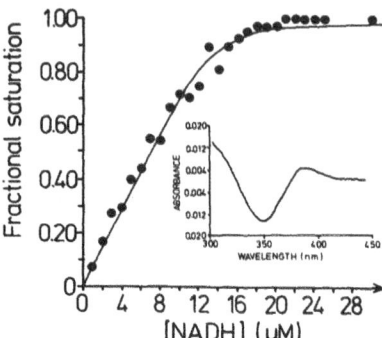

Figure 1. Titration of NADH binding sites of bovine corneal AlDH. Aliquots of NADH were added to both enzyme and reference solutions. Fractional saturation values from 3 titrations were calculated from changes in A_{348} and combined. The fitted curve corresponds to a K_d of 0.3 µM and an NADH binding site concentration of 13 µM. Inset: Difference spectrum between enzyme.NADH complex and free NADH.

purified by distillation before use. NAD^+ and NADH were from Sigma and were used without further purification.

The steady-state assay of Abedinia *et al.* (1990) was modified by the omission of BSA and pyrazole, which did not effect the assay results, and replacement of Tris-HCl buffer by MOPS buffer at pH 7.5. Stopped-flow studies were carried out on a Hi Tech Stopped Flow Spectrofluorimeter essentially as described by Hill *et al.* (1992). The difference spectrum between enzyme-bound and free NADH was determined on a Hewlett-Packard 8452 diode array spectrophotometer.

RESULTS AND DISCUSSION

The absorbance spectrum of enzyme-bound NADH differed from that of free NADH (Fig. 1 inset). The absorbance differences were small, about 0.02 absorbance units at 348 nm, but sufficient for use in an enzyme-NADH titration. The combined results of three titrations (shown in Fig. 1) gave a dissociation constant for NADH of less than 1 µM. Since

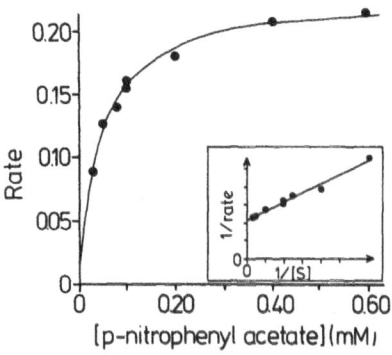

Figure 2. Esterase activity of bovine corneal AlDH, with *p*-nitrophenyl acetate as substrate. The line represents the computer-calculated fit to the Michaelis-Menten equation, with K_m = 46 ± 3 µM and Vmax = 0.229 ± .005 abs units min^{-1}.

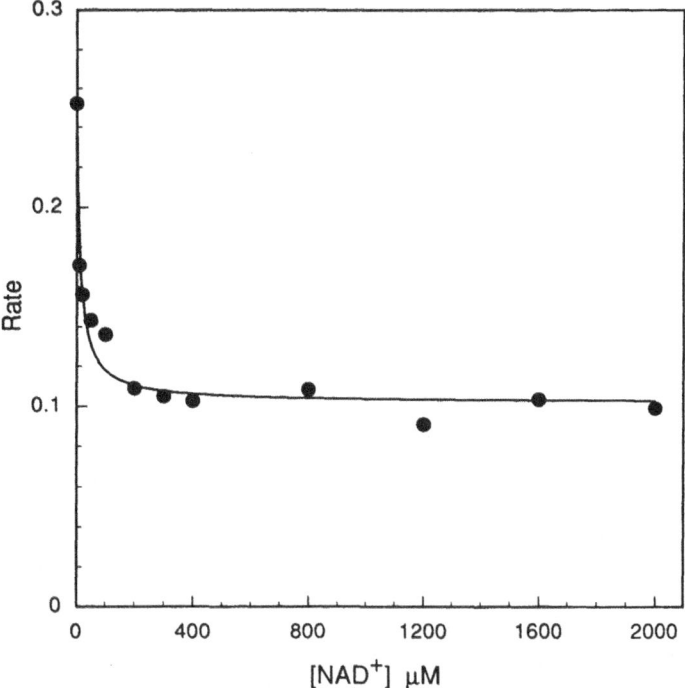

Figure 3. Effect of NAD$^+$ concentration on hydrolysis of *p*-nitrophenyl acetate catalysed by bovine corneal AlDH. The line represents the computer-calculated fit to a saturation equation with $K_i = 12 \pm 3$ µM and rates at zero and infinite NAD$^+$ of 0.249 and 0.103 respectively.

this sample of enzyme may have contained tightly-bound aldehyde, as discussed later, this may represent dissociation of NADH from the enzyme.NADH.aldehyde ternary complex.

Class 1 and class 2 aldehyde dehydrogenases display esterase activity with *p*-nitrophenyl acetate and related esters (MacGibbon et al., 1978; Takahashi K. and Weiner H, 1981). The corneal class 3 aldehyde dehydrogenase also showed esterase activity and the dependence on [ester] (Fig. 2) gives a K_m of 46 ± 3 µM. A significant difference from the Class 1 enzyme was in the effect of NAD$^+$ and NADH on the esterase activity. When either

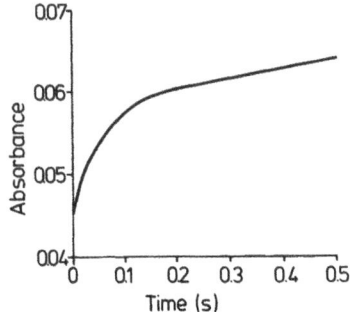

Figure 4. Time course of the absorbance change at 340 nm on mixing AlDH with NAD$^+$ and hexanal at pH 7.5. [NAD$^+$] = 2.0 mM, [hexanal] = 1.0 mM (after mixing). The burst rate constant for this run is 22.4 ± 0.2 s^{-1}.

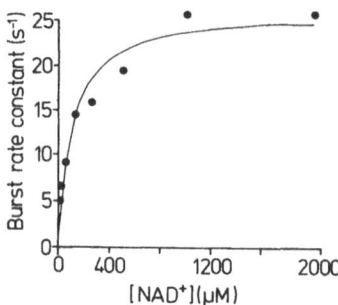

Figure 5. Dependence on [NAD$^+$] of the rate constant of the burst on mixing AlDH with NAD$^+$ and hexanal at pH 7.5.

NAD$^+$ (Fig. 3) or NADH was added to the esterase assay, inhibition was observed, with K_i values of about 10 μM in each case. Neither NAD$^+$ nor NADH completely inhibited the esterase activity. No evidence was obtained of the activation observed at the lowest NAD$^+$ and NADH concentrations with the class 1 enzyme (MacGibbon et al., 1978). It seems that for the class 3 aldehyde dehydrogenase, which is dimeric, there is only one type of coenzyme binding site.

Hexanal, which had the largest kcat/Km of all the substrates studied by Abedinia et al. (1990), was chosen as substrate for the majority of the stopped flow studies. Since the assays for dehydrogenase activity did not go to completion in Tris-HCl buffer, all subsequent work was done in MOPS or phosphate buffer.

In the first experiments when enzyme was mixed with NAD+ and hexanal, a burst in absorbance was observed at 340 nm, which could be fitted to a single exponential (Fig. 4). The burst rate constant showed typical saturation behaviour when the concentration of NAD$^+$ was varied (Fig. 5), and the amplitude of the burst decreased with enzyme concentration as expected. However, the burst rate constant and amplitude were independent of the hexanal concentration, even at an added aldehyde concentration of 5 μM. Finally a burst with the same amplitude and rate constant was observed when enzyme was mixed with NAD$^+$ in the absence of any added aldehyde.

There seemed to be two explanations of this unexpected result. Either some aldehyde had remained tightly bound to the enzyme throughout the purification procedure, which had not included a dialysis step, and gave rise to a burst in NADH production on mixing with

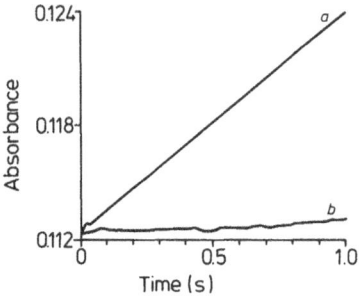

Figure 6. Time course of the absorbance change at 340 nm on mixing dialysed AlDH at pH 7.5 with NAD$^+$ and hexanal (trace a) or NAD$^+$ alone (trace b). [NAD$^+$] = 1.0 mM, [hexanal] (trace b) = 1.0 mM; concentrations are after mixing.

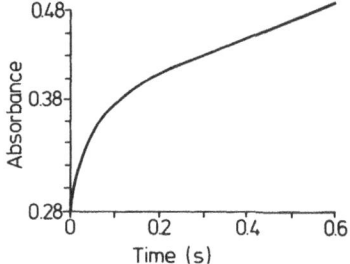

Figure 7. Time course of the absorbance change at 340 nm on mixing NAD$^+$ with dialysed AlDH which had been premixed with hexanal. [NAD$^+$] = 1.0 mM, [hexanal] = 1 mM (after mixing). The burst rate constant is 24.7 ± 0.3 s^{-1}.

NAD+. Or there was an absorbance change around 340 nm when NAD$^+$ bound to the free enzyme, and the stopped-flow experiments were simply measuring the binding of NAD$^+$ to the enzyme.

To check the first possibility the enzyme was dialysed to remove any tightly bound aldehyde. After dialysis against two changes of buffer over a 24 hour period the amplitude of the burst did decrease. More extensive dialysis over three days with more buffer changes gave enzyme which did not give a burst at 340 nm when mixed with NAD$^+$, although there was still a very slow increase in absorbance at this wavelength (see trace b in Fig. 6).

When this extensively dialysed enzyme was rapidly mixed in the stopped-slow spectrophotometer with NAD+ and hexanal no burst was observed, only the steady-state phase of the reaction (Fig. 6 trace a), a rather disappointing result since our aim was to study the fast steps before the steps on the pathway which limit steady-state turnover.

However, when enzyme and hexanal were premixed and then mixed with NAD$^+$ in the stopped-flow instrument we did observe a burst in absorbance (Fig. 7). A typical saturation curve was obtained when the NAD$^+$ concentration was varied (Fig. 8). Both the amplitude of the burst and the steady state rate decreased as the enzyme concentration was decreased. A few experiments were also carried out in which the concentration of hexanal, premixed with the enzyme before mixing with NAD$^+$, was varied. The burst amplitude was lower at lower aldehyde concentrations and no burst was observed at the lowest concentrations (5 to 20 μM). The burst rate constant at saturating [aldehyde] was the same as that observed with the unknown tightly bound aldehyde in the previous experiments, showing

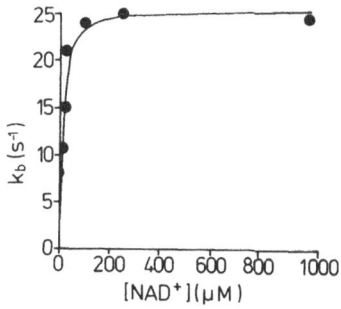

Figure 8. Dependence on [NAD$^+$] of the rate constant of the burst on mixing NAD$^+$ with dialysed AlDH which had been premixed with hexanal.

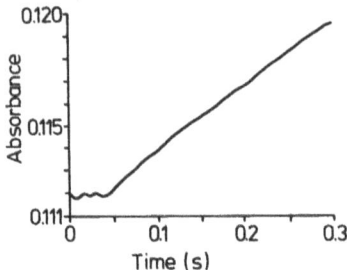

Figure 9. Time course of the absorbance change at 340 nm on mixing hexanal with dialysed AIDH which had been premixed with NAD$^+$. [NAD$^+$] = 1.0 mM, [hexanal] = 0.5 mM (after mixing).

that either the unknown aldehyde was similar in structure to hexanal, or the step controlling the burst was not hydride transfer, but some step which was insensitive to aldehyde structure.

One other possible mixing condition remained. When NAD$^+$ was premixed with enzyme before mixing with hexanal (at saturating levels of both substrate and cofactor) not only was no burst observed but there appeared to be a lag before the steady-state phase was established (Fig. 9). At a lower hexanal concentration (100 μM) a small rapid decrease in absorbance was observed before the expected steady-state increase. Clearly the pre-steady-state processes are more complex than a simple lag phase. Given that even with the extensively dialysed enzyme a slow change in absorbance was observed when NAD$^+$ was mixed with enzyme alone (Fig. 6) it is perhaps not surprising that complex behaviour may occur under these mixing conditions.

A burst in the production of NADH was not observed with benzaldehyde (5 mM) as substrate, whether the enzyme was premixed with benzaldehyde or with NAD$^+$. Similarly, when the disappearance of the chromophoric substrate trans-4-N,N-dimethylaminocin-namaldehyde (DACA) was monitored at 400 nm only the steady-state phase of the reaction was observed, regardless of the mixing conditions.

The stopped-flow results with hexanal can be summarised in the following scheme:

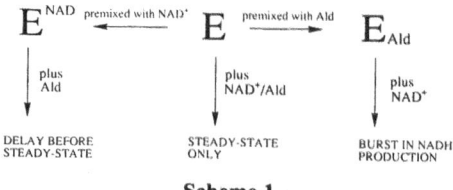

Scheme 1.

It is extremely unlikely that the extensively dialysed enzyme used in the burst experiments described above contained significant amounts of either NAD$^+$ or NADH. The NADH titration shows that even before dialysis NADH bound readily to the enzyme, a result which would have been impossible if the enzyme was saturated with NADH. If the undialysed enzyme contained bound NAD$^+$, a burst would not have been observed on rapidly mixing NAD$^+$ with it.

The appearance of a burst in the production of NADH only when the enzyme is premixed with hexanal requires that aldehyde must bind to the free enzyme. The binding results in a form of the enzyme which allows the observation of the rapid hydride transfer, but which is only accessible relatively slowly; otherwise a burst would have been observed when NAD$^+$ and hexanal were both rapidly mixed with the enzyme.

Similarly pre-mixing with NAD$^+$ must allow the enzyme to slowly adopt a form which is catalytically unfavourable and lies off the normal pathway, The process must be slow, since premixing of enzyme and NAD$^+$ is required in order to see the delays before the steady-state phase of the reaction.

One possible representation of a kinetic model which rationalises the experimental observations is shown in Scheme 2.

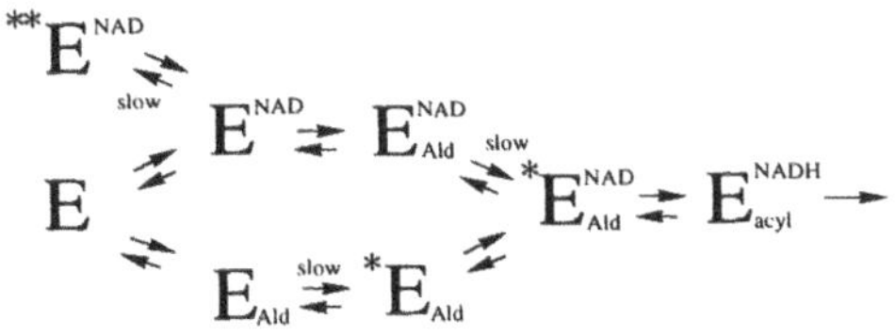

Scheme 2.

In this model, when enzyme and hexanal are premixed the enzyme can slowly adopt the $*E_{Ald}$ form. When the enzyme-hexanal equilibrium mixture is rapidly mixed with NAD$^+$, the fast hydride transfer step can be observed and there is a burst in the production of NADH. However, when the enzyme recycles through the pathway the steady-state is limited by the slow step (shown in this model as an isomerisation), whether NAD$^+$ or hexanal binds preferentially to the free enzyme. Similarly when NAD$^+$ and hexanal are mixed with the enzyme at the same time the slow isomerisation is rate limiting (whether NAD$^+$ or hexanal binds first to the free enzyme) and only the steady-state phase of the reaction is observed. Pre-mixing with NAD$^+$ results in a slow rearrangement to a form shown as $**E^{NAD}$. Only after the enzyme escapes from this dead-end form can the normal steady-state phase of the reaction be observed.

Although the details of the exact enzyme forms involved in the slow steps shown in the Scheme 2 can be changed, any kinetic model suggested must contain all the features described above.

CONCLUSIONS

Hexanal, and presumably structurally similar aldehydes, can bind to the class 3 bovine corneal aldehyde dehydrogenase in the absence of NAD$^+$. Other structurally different aldehydes such as benzaldehyde either cannot bind to the free enzyme, or if they do, cannot in the absence of NAD$^+$ cause the enzyme to adopt a form which allows the rapid hydride transfer to occur. For all of these substrates a step before hydride transfer is rate limiting during the steady-state phase of the reaction, perhaps a conformational change which is triggered by the binding of the aldehyde.

When mice were exposed to ultraviolet light, it was found that the decrease in the corneal aldehyde dehydrogenase levels parallels the appearance of corneal clouding (Downes et al., 1993). The presence of the enzyme is clearly essential for protection of the cornea against UV-B induced tissue damage.

The ability of aldehydes, such as hexanal, to bind to free enzyme may provide a mechanism for the protection of the cornea (regardless of the levels of NAD$^+$) from the rapid build up of aldehydes produced by sudden exposure of the eye to ultraviolet light. It has been suggested that the reason for the large amounts of this enzyme present in the cornea is to allow it to play a major role in filtering out UV-B ultraviolet light and protecting the retina from damage (Abedinia et al. 1990). From our results it seems that the corneal aldehyde dehydrogenase may also be present in large amounts in order to function as a binding protein to tie up reactive aldehydes before damage can occur which may lead to the clouding of the cornea.

REFERENCES

Abedinia, M., Pain, T. Algar, E.M. and Holmes, R., 1990, Bovine corneal aldehyde dehydrogenase: the major soluble corneal protein with a possible protective role for the eye, Exp. Eye Res., 51: 419-426.

Cooper, D.L., Baptist, E.W., Enghild, J.J., Isola, N.R. and Klintworth, G.K., 1991, Bovine corneal protein 54K (BCP 54) is a homologue of the tumor-associated (class 3) rat aldehyde dehydrogenase (RATALD), Gene, 98: 201-207.

Downes, J.E., Swann, P.G. and Holmes, R.S., 1993, Ultraviolet light-induced pathology in the eye: associated changes in ocular aldehyde dehydrogenase and alcohol dehydrogenase activities, Cornea, 12: 241-248.

Holt, W.S. and Kinoshita, J.H., 1973, The soluble proteins of the bovine cornea, Invest. Ophthalmol., 12: 114-126.

Hill, J.P., Buckley, P.D., Blackwell, L.F., Sime, R.M. and Kingston, R.L., 1992, Activation of aldehyde dehydrogenase at physiological temperatures, Biochem. Pharmocology, 44: 2425-2426.

Konishi, Y and Mimura, Y., 1992, Kinetic properties of the bovine corneal aldehyde dehydrogenase (BCP 54), Exp. Eye Res., 55: 569-578.

MacGibbon, A.K.H., Haylock, S.J., Buckley, P.D. and Blackwell, L.F., 1978, Kinetic studies on the esterase activity of cytoplasmic sheep liver aldehyde dehydrogenase, Biochem. J., 171: 533-538.

Takahashi, K. and Weiner, H., 1981, Nicotinamide adenine dinucleotide activation of the esterase reaction of horse liver aldehyde dehydrogenase, Biochemistry, 20: 2720-2726.

13

COVALENT MODIFICATION OF CLASS 2 AND CLASS 3 ALDEHYDE DEHYDROGENASE BY 4-HYDROXYNONENAL

Dylan P. Hartley, Ronald Lindahl[1] and Dennis R. Petersen

School of Pharmacy and Hepatobiliary Research Center
University of Colorado Health Sciences Center
Denver, CO
[1]Department of Biochemistry and Molecular Biology
University of South Dakota School of Medicine
Vermillion, SD

INTRODUCTION

The peroxidation of cellular membrane lipids gives rise to a number of chemically diverse aldehydes. Whereas a variety of different aldehydes are produced during this process, malondialdehyde and 4-hydroxy-2,3-*trans* nonenal (4-HNE) are the most abundant aldehydic products produced during lipid peroxidation. These two products of lipid peroxidation are ,-unsaturated aldehydes and, as a result, are relatively strong electrophiles which are thought to be an important factor in their cellular toxicities. For instance, the extreme electrophilic potential of 4-hydroxynonenal directs its interactions with nucleophiles such as the thiol, histidyl or amine groups of proteins. The potential of 4-HNE to alkylate specific functional groups of cellular proteins is proposed to be responsible for its general cytotoxic effects on hepatocytes and fibroblasts, disruption of cellular calcium homeostasis, inhibition of protein kinase C, and stimulation of neutrophil chemotaxis (see Esterbauer *et al.* 1991 for a comprehensive review concerning the biological effects of 4-HNE). Because of this extreme chemical reactivity toward a wide range of nucleophiles, it is difficult to identify, with a high degree of accuracy, specific cellular proteins that might be predisposed targets for interaction with 4-hydroxynonenal.

Given the cytotoxic nature of 4-HNE, most mammalian cells have efficient enzymatic pathways which maintain sub-micromolar concentrations of this aldehyde during mild or even severe oxidative stress. A recent and comprehensive review of the literature (Schaur *et al.* 1992) cites studies using subcellular fractions, purified enzymes or isolated hepatocytes demonstrating that 4-HNE can be oxidized via aldehyde dehydrogenases (ALDH 1.2.1.3), reduced by alcohol dehydrogenase (ADH 1.1.1.1) or conjugated with glutathione by glutathione *S*-transferase (GST 2.5.1.18). The fact that 4-HNE is a substrate for these diversified enzymes suggests that these proteins may have a greater potential to be inactivated by this aldehydic product of lipid peroxidation relative to other cellular proteins. Interestingly, these enzymes appear to biotransform 4-HNE to less reactive intermediates without major effects on their catalytic potentials. However, there is evidence that while the high-affinity form of ALDH present in rat liver mitochondria (Class 2 ALDH) is catalytically the most active in oxidizing 4-HNE (Mitchell and Petersen, 1987), which also functions as a potent competitive or mixed-type (irreversible) inhibitor of acetaldehyde oxidation by this enzyme (Mitchell and Petersen, 1991). In this instance, the inhibitory potential of 4-HNE would be consistent with numerous reports in the literature that a nucleophilic cysteine residue occupies the active site of ALDHs (Pietruszko *et al.* 1992; Tu and Weiner, 1988).

The multiple forms of ALDH present in cytosolic, mitochondrial and microsomal fractions of liver display a wide range of affinities for 4-HNE and thus can serve as models to probe the molecular mechanisms that predispose these and other enzymes to inactivation by this ,-unsaturated aldehyde. For instance, the high-affinity form of Class 2 ALDH from rat liver displays a K_m value of 18 M for 4-HNE while the tumor-associated Class 3 ALDH oxidizes 4-HNE with a K_m value of 1.3 mM (Mitchell and Petersen, 1987; Lindahl and Petersen, 1991). Likewise, the V/K value for the former enzyme is approximately 200-fold higher than the latter. We have therefore used these two forms of ALDH as models to assess the role of 4-HNE covalent interactions in differential enzymatic inactivation of these specific isoforms of ALDH.

METHODS AND PROCEDURES

Chemicals and Solutions

All analytical chemicals and reagents were purchased from Sigma Chemical Co. (St. Louis, MO). (*E*)-4-hydroxy-2-nonenal was liberated from the diacetyl form as described previously (Mitchell and Petersen, 1991) and diluted in deionized and distilled water.

Purification of Recombinant Class 3 and Rat Liver Class 2 ALDH

Recombinant Class 3 ALDH was expressed in DH5 *E. coli* and purified by the modification of the procedures outlined by Harper *et al.,* (1988). Whereas, rat liver Class 2 ALDH was isolated, from sonicated rat liver mitochondria through ammonium sulfate precipitation and DEAE-anion exchange chromatography as reported by Mitchell and Petersen (1987) and further purified using 5'-AMP sepharose affinity chromatography as outlined previously (Harper *et al.,* 1988). Briefly, bacterial crude extract containing recombinant Class 3 ALDH or DEAE eluent containing rat liver Class 2 ALDH were applied to a buffer exchange column prepared from Bio-Gel P-6DG polyacrylamide (Bio-Rad Laboratories, Richmond, CA) equilibrated with 100 mM potassium phosphate buffer system containing 1.0 mM 2-mercaptoethanol and 1.0 mM EDTA. For both recombinant Class 3 and rat liver Class 2 ALDH enzymes, enzymatically active fractions were collected from the buffer exchange column, pooled and applied to a 5'-AMP sepharose 4B affinity column equilibrated with the 100 mM potassium phosphate buffer system. After protein application, the column was washed successively with the 100 mM potassium phosphate system, 450 mM potassium phosphate system, and 25 mM sodium phosphate, whereafter recombinant Class 3 or rat liver Class 2 enzymes were eluted with 25 mM sodium phosphate containing 0.5 mg/ml NAD$^+$. Active fractions were pooled, placed in dialysis tubing (MW cutoff = 3,000 kDa), and concentrated against sucrose. In all instances, appropriate column eluent fractions were assessed for ALDH activity. All buffers were maintained at pH 7.4 and all purification steps were done at 4 C. Protein concentrations were determined by the method of Lowry *et al*., (1951).

Assay of Enzymatic Activity

Determination of enzymatic activity was measured spectrophotometrically at 340 nm by monitoring the production of NADPH or NADH associated with the oxidation of benzaldehyde or acetaldehyde by recombinant Class 3 or rat liver Class 2 ALDH, respectively. Enzymatic activity was measured routinely throughout the purification sequences, or after purified enzymes had been exposed to 4-HNE for a period of 2 hours (see below).

Covalent Modification of Recombinant Class 3 and Rat Liver Class 2 ALDH by 4-HNE

Aliquots of purified recombinant Class 3 or rat liver Class 2 ALDH (20 µg) in 25 mM sodium phosphate, pH 7.4, were incubated with various concentrations of 4-HNE (0-4 mM) for 2 hours at 37 C; the total incubation volume was 200 µl. At 2 hours, enzymes exposed to 4-HNE were derivatized with 2,4-dinitrophenylhydrazine (DNPH) as described by Benedetti *et al*., (1982) or 5,5'-dithio-*bis*-2-nitrobenzoic acid (DTNB) as outlined by DiMonte *et al*., (1984) for detection of protein hydrazones or protein sulfhydryls, respectively.

Detection of 4-HNE Altered ALDH Monomers by Sodium Dodecyl Sulfate-Polyacrylamide Gel Electrophoresis

Enzymes (20 µg of enzyme protein/incubate) were incubated for two hours with 4-HNE in 25 mM sodium phosphate, pH 7.4, were rapidly frozen with liquid nitrogen, lyophilized to dryness, taken up in 20 µl (1 µg/µl) sample buffer, and subjected to SDS-PAGE (20 µl/lane). Polyacrylamide gels for SDS-PAGE were prepared as 7.5% polyacrylamide protein resolution gels layered with a 4.0% polyacrylamide protein stacking gel. Proteins were detected with Brilliant Blue R-250 stain or with silver stain.

RESULTS

The effect of various concentrations of 4-hydroxynonenal on enzyme activity, the formation of protein-bound carbonyls and the disappearance of sulfhydryl groups in Class 3 ALDH is presented in Figure 1. These data demonstrate that incubating Class 3 ALDH with concentrations of up to 500 M 4-HNE for two hours had little effect on enzyme activity whereas exposure to 1.0 mM 4-HNE for this same duration of time destroyed approximately 50% of the initial enzyme activity. However, following incubation with 2.0 mM 4-HNE, only 10% of the initial Class 3 ALDH activity could be detected. The graphical data presented in the insert of Figure 1 show that exposure of the Class 3 enzyme to increasing concentrations of 4-HNE resulted in a proportional increase of enzyme-bound carbonyls, measured by hydrazone content, and a corresponding disappearance sulfhydryl group content of the enzyme. Quantitatively, these data indicate that when treated with 4-HNE, a 50% decrease in Class 3 ALDH activity corresponds to the incorporation of approximately 2 moles of 4-HNE per mole of enzyme subunit and the removal of nearly 50% of the titratable sulfhydryl groups.

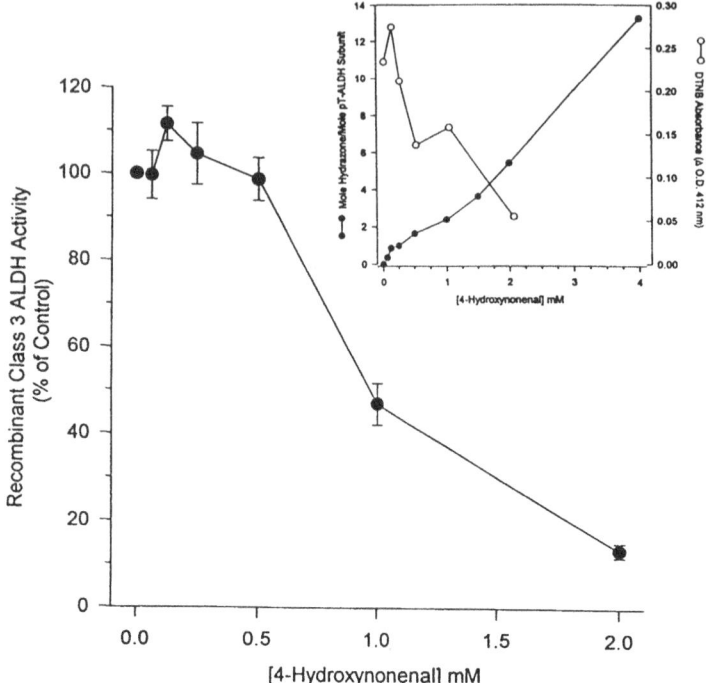

Figure 1. Concentration-dependent inactivation of the tumor-associated, recombinant Class 3 ALDH by 4-hydroxynonenal. The inset plot illustrates the 4-HNE concentration dependent formation of protein-mixed carbonyls (hydrazones) and decrease of sulfhydryl (thiol) content measured by DTNB absorbance.

The data presented in Figure 2 indicates that the high-affinity form Class 2 ALDH from rat liver is very sensitive to inactivation by 4-HNE as compared to the Class 3 form. At a concentration of 30 M 4-HNE, Class 2 ALDH activity decreased sharply by approximately 20% and the degree of Class 2 ALDH inactivation that occurred between 30 M and 125 M 4-HNE is extensive. Calculations from extrapolation indicate that approximately, 50% of the Class 2 ALDH activity is abolished at 4-HNE concentrations of 100 M. Data presented in the insert of Figure 2 indicate that a loss of 50% of Class 2 ALDH activity occurs upon incorporation of 2 moles of protein-bound carbonyls per subunit of enzyme and a loss of approximately 20% of the titratable sulfhydryl groups present in the enzyme.

It is apparent from Figures 3 and 4 that incubation of purified Class 2 and Class 3 ALDH with 4-HNE resulted in molecular modifications of these enzymes that significantly alters their electrophoretic mobility in SDS polyacrylamide gels. Figure 3 presents the SDS-PAGE electrophoretic profiles of Class 3 ALDH incubated with concentrations of 4-HNE ranging from 250 M to 4.0 mM. When stained with Coomassie Blue, the purified Class 3 ALDH was visualized as a major band at 51 kDA which corresponds to the monomeric molecular weight of this enzyme. An additional minor band of protein having a molecular weight slightly below 51 kDA was also visualized and most likely represents a modified recombinant Class 3 ALDH produced through proteolysis of the full-length protein when expressed in *E. coli*. The most apparent changes in electrophoretic patterns of the Class 3 monomer become visible at 1.5 mM 4-HNE where two bands of protein at approximately 100 kDa and 205 kDa become visible. The fact that these molecular weights are two or four times the monomeric subunit molecular weight indicate that these new bands represent

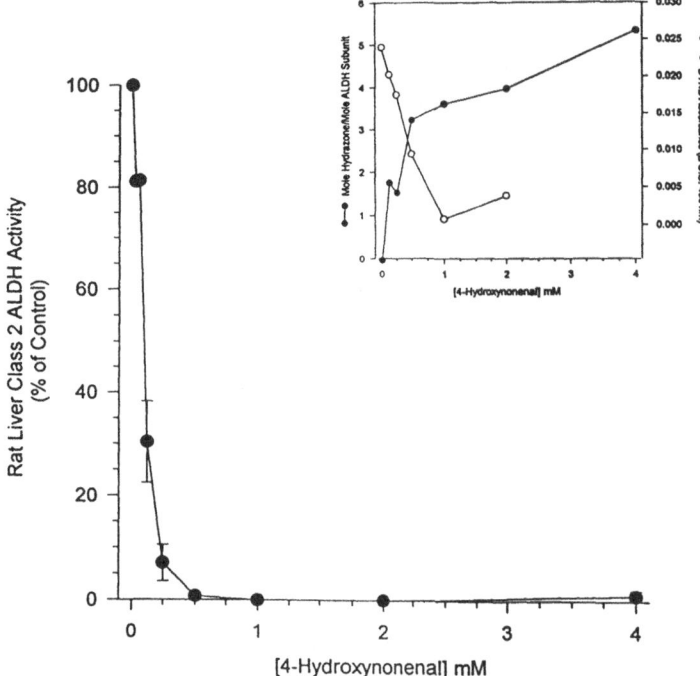

Figure 2. Inactivation of Class 2 ALDH by 4-hdroxynonenal. The inset presents the corresponding 4-HNE-mediated changes in protein-bound carbonyls and the related decrease in sulfhydryl residues measured by changes in DTNB absorbance.

dimers or tetramers stabilized through cross-linking by 4-HNE. It is also apparent from Figure 3 that when the Class 3 ALDH was incubated with 2.0 or 3.0 mM 4-HNE, the intensity of the 205 kDA band increased and protein complexes with even larger molecular weights appeared suggesting that the interactions of high concentrations of 4-HNE with Class 3 ALDH results in formation of multi-monomeric complexes.

The effect of 4-HNE concentrations ranging from 130 M to 4.0 mM on the electrophoretic mobility of Class 2 ALDH is presented in Figure 4. For this isoform of ALDH the

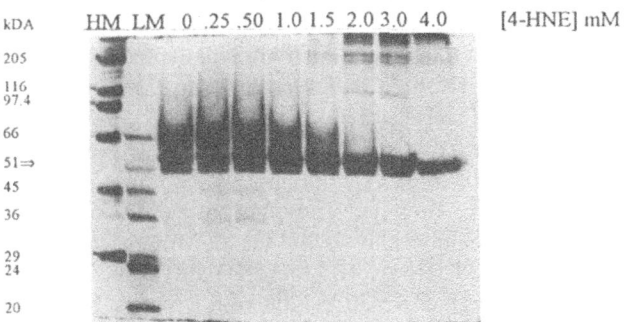

Figure 3. Effect of 4-hydroxynonenal on the electrophoretic mobility of Class 3 ALDH in 7.5% SDS gels. Lanes HM and LM represent high and low molecular weight standards respectively. Protein bands were stained with Coomassie Brilliant Blue R-250.

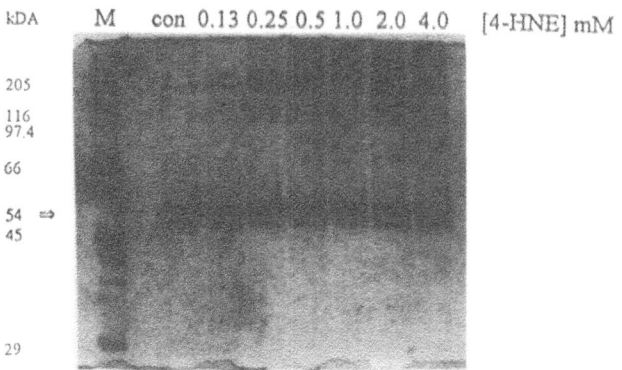

Figure 4. Modification of Class 2 ALDH electrophoretic mobility in SDS gels by 4-hydroxynonenal. Lane M presents the electrophoretic mobility of molecular weight standards. Protein bands were visualized by silver staining.

protein bands were visualized by a silver staining procedure which revealed what appeared to be a single and minor contaminant detected at approximately 200 kDA, in the purified preparation of Class 2 ALDH. This SDS-PAGE gel displayed the appearance of a novel band of 100 kDA and an increase in the intensity of the 205 kDA band after incubation of the Class 2 enzyme with 250 M 4-HNE. The intensity of both the 100 kDA and 205 kDA bands increased proportional to 4-HNE concentrations up to 1.0 mM, after which point the staining intensity of protein complexes with molecular weights greater than 205 kDA increased dramatically. Consistent with the inactivation data presented in Figure 2, the results in Figure 4 indicate that Class 2 ALDH is also more liable than the Class 3 enzyme to molecular interactions with 4-HNE which result in formation of 4-HNE multi-monomeric complexes.

DISCUSSION

The data presented here demonstrate that Class 2 and Class 3 forms of ALDH readily form adducts with 4-HNE which is one of the predominant products of lipid peroxidation. Based on the loss of enzyme activity, these results also reveal that of the two ALDH isoforms, Class 2 ALDH is by far the most sensitive to alkylation by this electrophilic ,-unsaturated aldehyde. The susceptibility of Class 2 and Class 3 ALDH to alkylation by 4-HNE is consistent with a previous report that glyceraldehyde 3-phosphate dehydrogenase (GAPDH) is also inactivated by 4-HNE (Uchida and Stadtman *et al.*, 1993). These investigators attributed the sensitivity of GAPDH inactivation to 4-HNE alkylation of cysteine 149 which occupies the active site of this enzyme. It is interesting to note that the active site of Class 2 and Class 3 ALDH, is thought to be cysteine 302 which could also be a susceptible site for adduct formation by 4-HNE. However, previous studies (Mitchell and Petersen, 1987 and Lindahl and Petersen, 1991) have shown that the K_m values for the oxidation of 4-HNE by Class 2 ALDH and Class 3 ALDH are 18 M and 1.3 mM respectively, resulting in a V/K parameter for the Class 2 enzyme which is 200-fold greater. In the present study, the observation that Class 2 ALDH is the most sensitive to inactivation by this aldehydic product of lipid peroxidation suggests that because of this catalytic specificity, this particular isoform of ALDH may be predisposed adduct formation. The potential effect of this adduct formation on catalytic activity is evident from our previous observation (Mitchell and Petersen, 1991) that 4-HNE functions as a competitive or mixed-type inhibitor of acetaldehyde oxidation by the Class 2 ALDH with an apparent K_i value of 0.58 M.

The results presented in Figures 2 and 3 are suggestive of some interesting differences in the effect of protein-bound carbonyls on the activity of Class 2 and Class 3 ALDH. Protein bound carbonyls originate from Michael addition reactions of the ,-unsaturated double bond of 4-HNE with nucleophilic thiol, histidyl or amine groups of cysteine, histidine or lysine, respectively. Comparatively, the data in Figure 1 indicate that Class 3 ALDH has more nucleophilic sites per subunit for formation of protein-bound carbonyls in that incubation with 4.0 mM 4-HNE resulted in adduction of approximately 13 bound carbonyls per subunit compared to only 5 per subunit for the Class 2 enzyme. These data also reveal that there are substantial differences between the Class 2 and Class 3 ALDHs in the number of protein-bound carbonyls formed per subunit and the extent of enzyme inhibition. For instance, the Class 3 ALDH still retains 20% of its activity with 14 protein-bound carbonyls per subunit while the Class 2 enzyme is completely inhibited with only 2 to 3 protein-bound carbonyls per subunit. A similar pattern is observed with respect to the number of titratable thiol groups in the Class 2 and Class 3 enzyme. The data in Figure 1 indicate that the Class 3 enzyme retained 50% of its initial activity when approximately 50% of accessible thiol groups were adducted with 4-HNE where as the Class 2 ALDH lost all activity when 50% of the thiol groups were presumably involved in formation of protein-bound carbonyls. Collectively, these data suggest that 4-HNE adduct formation with cysteine residues plays a relatively important role in the inactivation of these ALDH isoforms and that the critical thiol group(s), possibly cysteine 302, is more protected in the Class 3 ALDH. This notion of active-site adduction by this aldehydic product of lipid peroxidation would also be consistent with the relative affinities of these two ALDH isoforms for 4-HNE.

The SDS-PAGE gels presented in Figures 3 and 4 demonstrate that the chemical interactions of 4-HNE with Class 2 and Class 3 ALDH are complex in that they involve inter-and intrasubunit cross-linking. This is visualized by formation of 4-HNE-induced complexes of both Class 2 and Class 3 monomers which, based on molecular weight estimates, most likely represents dimeric and tetrameric structures of the respective subunits. While the physiological significance of the 4-HNE mediated cross-linking remains to be established, the mechanisms proposed to be responsible for formation of these inter- and intrasubunit complexes are presented in Scheme 1.

The ability of 4-HNE to form intra- and intersubunit cross-links is based on its propensity to participate in Michael addition reactions with thiol, histidyl and amine groups of proteins. These stable, covalent bonds would be responsible for forming protein-bound carbonyls that would retain a free carbonyl (aldehyde) group that could then form a Schiff base linkage with a primary amine present in the same or different ALDH subunit. However, our data indicate that formation of these secondary, intersubunit cross-links do not occur at lower concentrations of 4-HNE. Thus, the enzymatic inactivation of Class 2 and Class 3 ALDH is most likely due to Michael addition reactions of 4-HNE with nucleophilic residues at and/or adjacent to the active site or possibly coenzyme binding site of these enzymes.

The results of the present study raise some interesting questions concerning the inactivation of the high-affinity form of Class 2 ALDH during hepatocellular oxidative stress and lipid peroxidation. In this context, the results of the present study are consistent with earlier studies from our laboratory (Hjelle and Petersen, 1983) demonstrating a rapid loss of Class 2 ALDH activity *in vivo* following administration of the prooxidant CCl_4 to rats. The ability of chronic ethanol ingestion to initiate excessive hepatic lipid peroxidation would be expected to have a similar effect on Class 2 ALDH which is consistent with decreased activities of this enzyme in the livers of alcoholic subjects (Jenkins and Peters, 1980; Jenkins *et al.* 1984). It is also noteworthy that the inducible Class 3 ALDH is more resistant to alkylation by 4-HNE. The fact that this tumor-associated enzyme is inducible in several

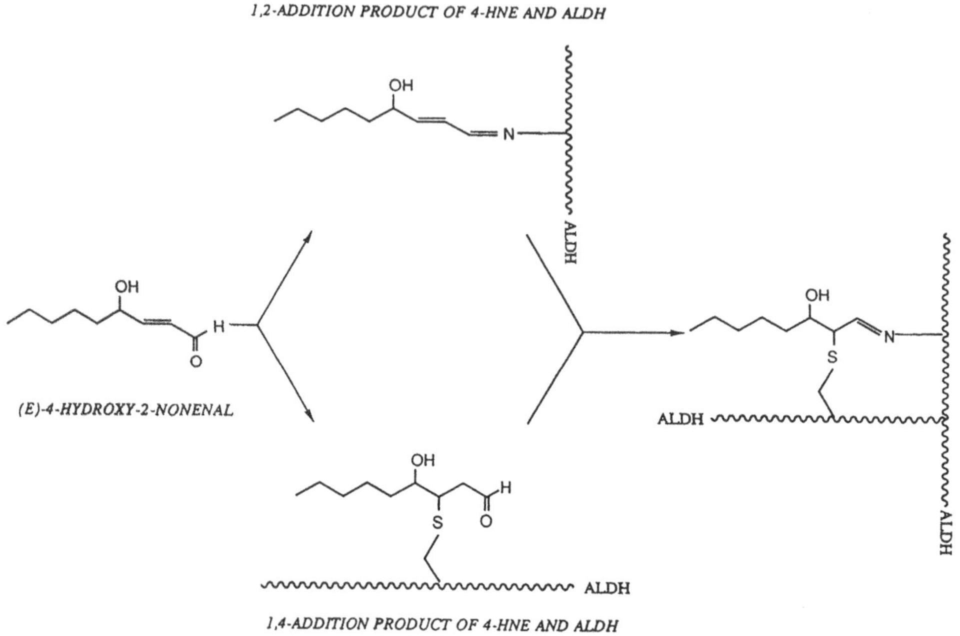

1,2-ADDITION PRODUCT OF 4-HNE AND ALDH

Scheme 1. Hypothetical scheme illustrating 1,2- and 1,4-addition reactions of 4-hydroxynonenal with alde-hyde dehydrogenases. As illustrated, the addition reactions could occur within or between ALDH subunits.

tissues following xenobiotic exposure and tumorigenesis may be associated with the ability of certain cells to withstand oxidative damage as well as exposure to chemicals such as cyclphosphamide that have potent protein alkylating potential.

ACKNOWLEDGEMENTS

This work was supported by NIH/AA 05370 Predoctoral Fellowship to D.P.H., NIH/AA 09300 to D.R.P. and CA21103 to R.L.

REFERENCES

Benedetti, A., Esterbauer, H., Ferrali, M., Fulceri, R., and Comporti, M. (1982) Evidence for aldehydes bound to liver microsomal protein following CCl_4 or $BrCCl_3$ poisoning. *Biochim. Biophys. Act.* 711:345-356.

Di Monte, D., Ross, D., Bellomo, G., Eklow, L., Orrenius, S. (1984) Alterations in intracellular thiol homeostasis during the metabolism of menadione by isolated rat hepatocytes. *Arch. Biochem. Biophys.* 235:334-342.

Esterbauer, H., Schaur, R.J. and Zollner, H. (1991) Chemistry and biochemistry of 4-hydroxynonenal malon-dialdehyde and related aldehydes. *Free Rad. Biol. & Med.* 11:81-128.

Harper, K., Jones, D.E., Brennan, M.D., and Lindahl, R. (1988) Characterization of a functional recombinant rat liver aldehyde dehydrogenase: expression as a non-fusion protein in *E. coli. Biochem. Biophysic. Res. Commun.* 152:940-947.

Hjelle, J.J. and Petersen, D.R. (1983) Time course of carbon tetrachloride-induced decrease in mitochondrial aldehyde dehydrogenase activity. *Toxicol. Appl. Pharmcolog.* 67:159-165.

Jenkins, W.J., and Peters, T.J. (1980) Selectively reduced hepatic aldehyde dehydrogenase in alcoholics. *Lancet* 2:628-629.

Jenkins, W.J., Cakebread, K. and Palmer, K.R. (1984) Effect of alcohol consumption on hepatic aldehyde dehydrogenase activity in alcoholic patients. *Lancet* 1:1084-1050.

Lindahl, R. and Petersen, D.R. (1991) Lipid aldehyde oxidation as a physiological role for class 3 aldehyde dehydrogenase. *Biochem. Pharmacol.* 41:1583-1587.

Lowry, O.H., Rosebrough, N.J., Farr, A.L., and Randall, R.J. (1951) Protein measurement with the Folin phenol reagent. *J. Biol. Chem.* 193:265-275.

Mitchell, D.Y., and Petersen, D.R. (1987) The oxidation of ,-unsaturated aldehydic products of lipid peroxidation by rat liver aldehyde dehydrogenase. *Toxicol. Appl. Pharmacol.* 87:403-410.

Mitchell, D.Y. and Petersen, D.R. (1991) Inhibition of rat hepatic mitochondrial aldehyde dehydrogenases-mediated acetaldehyde oxidation by trans-4-hydroxynonenal. *Hepatology* 13:728-734.

Pietruszko, R., Blatter, E., Abriola, D.F. and Prestwich, G. (1990) Localization of Cysteine 302 at the active site of aldehyde dehydrogenase. In "Enzymology and Molecular Biology of Carbonyl Metabolism, III" p19-30. Ed by H. Weiner, B. Wermuth and D. Crabb, Plenum Press, NY.

Schaur, R.J., Zollner, H. and Esterbauer, H. (1990) Biological effects of aldehydes with particular attention to 4-hydroxynonenal and malondialdehyde. p141-163, Vol III ed. C. Vigo-Pelfrey, Boca Raton, FL, CRC Press Inc.

Tu, G-C and Weiner, H. (1988) Identification of the cysteine residue at the active site of horse liver mitochondrial aldehyde dehydrogenase. *J. Biol. Chem.* 263:1212-1217.

Uchida, K. and Stadtman, E.R. (1993) Covalent attachment of 4-hydroxynonenal to glyceraldehyde-3-phosphate dehydrogenase. *J. Biol. Chem.* 268:6388-6393.

14

CONSTITUTIVE AND OVEREXPRESSED HUMAN CYTOSOLIC CLASS-3 ALDEHYDE DEHYDROGENASES IN NORMAL AND NEOPLASTIC CELLS/SECRETIONS

Norman E. Sladek, Lakshmaiah Sreerama and Ganaganur K. Rekha

Department of Pharmacology
University of Minnesota Medical School
3-249 Millard Hall, 435 Delaware Street, S.E.
Minneapolis, MN 55455

INTRODUCTION

Cytosolic class-3 aldehyde dehydrogenases have been the subject of extensive investigation in recent years because they are demonstrated determinants of cellular sensitivity to certain widely used anticancer drugs, namely, cyclophosphamide, ifosfamide, 4-hydroperoxycyclophosphamide and mafosfamide (cellular sensitivity to these drugs decreases as the cellular concentration of cytosolic class-3 aldehyde dehydrogenase increases) [Bunting and Townsend, 1993; Sreerama and Sladek, 1993a,b, 1994a; Rekha et al., 1994; Sreerama et al., 1994a], and because they may aid in protecting cells by catalyzing the detoxification of otherwise toxic aldehydes arising from lipid peroxidation therein [reviewed in Lindahl, 1992]. Moreover, it has been proposed that corneal cytosolic class-3 aldehyde dehydrogenase protects the eye against UV-B light-induced damage, in part, by directly absorbing it [Downes et al., 1994 and references cited therein]. Of particular interest, for obvious reasons, is the regulation of their expression, particularly as such regulation is influenced by environmental, dietary and therapeutic agents.

Recent investigations in our laboratory reveal that overexpression of cytosolic class-3 aldehyde dehydrogenases can be effected by at least three different mechanisms. Our findings in that regard and our experience with potential inhibitors of these enzymes, also of substantial interest for obvious reasons, are summarized in the paragraphs that follow. Also presented is evidence suggesting that the cytosolic class-3 aldehyde dehydrogenase present in the neoplastic cells is a slight variant of that present in normal cells.

Enzymology and Molecular Biology of Carbonyl Metabolism 5
Edited by H. Weiner *et al.*, Plenum Press, New York, 1995

CONSTITUTIVE OVEREXPRESSION OF CYTOSOLIC CLASS-3 ALDEHYDE DEHYDROGENASE BROUGHT ABOUT BY OXAZAPHOSPHORINES

High levels of a truncated, but enzymatically fully active, version of the cytosolic class-3 aldehyde dehydrogenase present at very low levels in cultured breast adenocarcinoma MCF-7/0 cells are found in such cells made resistant to cyclophosphamide and related compounds by growing them in the presence of gradually increasing concentrations of 4-hydroperoxycyclophosphamide for several months; the subline thus derived is termed MCF-7/OAP, Tables 1 and 2. Glutathione S-transferase and DT-diaphorase, but not cytochrome P450 IA1, are also overexpressed in this subline, Table 1. Enzyme overexpression is stable, i.e., it is retained, seemingly indefinitely, when the inducing agent is removed from the culture medium.

The molecular alterations underlying the constitutive expression of high levels of the truncated cytosolic class-3 aldehyde dehydrogenase are unknown. One possibility is that two mutations occurred, one accounting for the expression of the truncated enzyme in place of, or in addition to, the usual enzyme, and a second allowing the constitutive, relatively elevated, expression of the truncated class-3 aldehyde dehydrogenase (and of the usual enzyme if the truncated enzyme is expressed in addition to the usual enzyme as seems to be the case, see below), as well as that of glutathione S-transferase and DT-diaphorase, because of, most probably, some change in the ARE regulatory pathway, see below. 4-Hydroperoxy-cyclophosphamide is a demonstrated mutagen [Hales, 1982].

Table 1. Enzyme activities and glutathione levels in Lubrol-treated whole homogenates prepared from methylcholanthrene- and catechol-treated human breast adenocarcinoma MCF-7/0 and MCF-7/OAP cells.

Cell Line/Treatment	mIU/10^7 Cells				nmol GSH/10^7 Cells
	ALDH-3	GST	DT-D	P450 IA1	
MCF-7/0	2	25	82	0.03	185
MCF-7/0/MC	310	150	495	0.48	190
MCF-7/0/CAT	768	250	6,395	0.03	182
MCF-7/OAP	254	157	340	0.03	199
MCF-7/OAP/MC	3,534	192	15,300	0.67	180
MCF-7/OAP/CAT	1,593	227	12,400	0.03	185

[a]Composite of information heretofore unpublished and data published in Sreerama and Sladek [1993a,b, 1994a] and Sreerama et al. [1994a]. Treatment with 3 µM methylcholanthrene (MCF-7/0/MC, MCF-7/OAP/MC) or 30 µM catechol (MCF-7/0/CAT, MCF-7/OAP/CAT) was for 5 days as described previously [Sreerama and Sladek, 1993b, 1994a; Sreerama et al., 1994a]. Quantification of glutathione (GSH) levels and of cytosolic class-3 NADP-linked aldehyde dehydrogenase (ALDH-3), glutathione S-transferase (GST), NADH-linked DT-diaphorase (DT-D) and cytochrome P450 IA1 (P450 IA1) activities was as described previously [Sreerama and Sladek, 1993a, 1994a; Sreerama et al., 1994a]. Substrates were 4 mM benzaldehyde, 1 mM 1-chloro-2,4-dinitrobenzene, 0.04 mM 2,6-dichlorophenol-indophenol and 0.005 mM 7-ethoxyresorufin, respectively.

Table 2. Physical and catalytic properties of human class-3 aldehyde dehydrogenases obtained from various sources: differences.

Property	Source of Cytosolic Class-3 Aldehyde Dehydrogenase		
	Stomach Mucosa Saliva Breast	MCF-7/0 MCF-7/0/MC MCF-7/0/CAT Colon C Warthin Tumor Mucoepidermoid Carcinoma	MCF-7/OAP
NAD Km (µM)	~45	~45	550
Esterolytic activity (mIU/mg)	~8,700	~8,700	3,350
pI Range	5.75 - 6.35	5.75 - 6.35	6.0 - 6.45
Subunit M_r (kDa)	54.5	54.5	40
Recognition of subunit by anti-stomach mucosa cytosolic class-3 aldehyde dehydrogenase IgY	Yes	Yes	No
NADP-dependent catalysis of acetaldehyde oxidation	++	++	-

[a]Composite of information heretofore unpublished and data published in Sreerama and Sladek [1993a,b, 1994a,b], Rekha et al. [1994] and Sreerama et al. [1994a,b].

TRANSIENT OVEREXPRESSION OF CYTOSOLIC CLASS-3 ALDEHYDE DEHYDROGENASE BROUGHT ABOUT BY BIFUNCTIONAL INDUCERS

Overexpression of a cytosolic class-3 aldehyde dehydrogenase, as well as of glutathione S-transferase, DT-diaphorase and cytochrome P450 IA1, can also be induced in MCF-7/0, as well as in MCF-7/OAP, cells by growing them in the presence of Ah receptor ligands, e.g., methylcholanthrene, for a few days, Table 1. However, in this case, overexpression of these enzymes is transient in that enzyme activities return to basal levels in a few days when the inducing agent is removed from the culture medium [Sreerama and Sladek, 1994a]. The class-3 aldehyde dehydrogenase overexpressed in methylcholanthrene-treated MCF-7/0 cells exhibits physical and catalytic properties identical to those exhibited by the class-3 aldehyde dehydrogenase expressed at very low levels in untreated MCF-7/0 cells, Table 2. However, while the class-3 aldehyde dehydrogenase overexpressed in methylcholanthrene-treated MCF-7/OAP cells is yet to be fully characterized, preliminary studies (isoelectric focusing, subunit molecular weight, antibody recognition) indicate an increased expression of both the enzyme present at very low levels in untreated MCF-7/0 cells and the truncated version of this enzyme known to be constitutively expressed at relatively high levels in untreated MCF-7/OAP cells (data not shown) suggesting that the presence of a relatively low level of the MCF-7/0-type class-3 aldehyde dehydrogenase in MCF-7/OAP cells has gone undetected thus far, most probably because of the relatively large amounts of the truncated version of the enzyme that is present in these cells.

MCF-7/0 cells are known to be estrogen receptor-positive. Interestingly, expression of class-3 aldehyde dehydrogenase and other enzymes could be induced in all of the cell lines known to express estrogen receptors and thus far tested, but not in any of those known not to and thus far tested, Table 3. Consistent with our observations, others have shown that overexpression of cytochrome P450 IA1 can be induced by another Ah receptor ligand, viz., 2,3,7,8-tetrachlorodibenzo-*p*-dioxin (TCDD), in estrogen receptor-positive, but not in estrogen receptor-negative, cultured human breast (adeno)carcinoma cell lines [Vickers et al., 1989; Arellano et al., 1993].

Transcriptional activation of the relevant genes by Ah receptor ligands, termed bifunctional inducers because they induce certain "phase I", e.g., cytochrome P450 IA1, as well as certain "phase II", e.g., class-3 aldehyde dehydrogenase, drug metabolizing enzymes (see legend to Figure 1 for further discussion), is not fully understood, but appears to be via xenobiotic responsive elements (XRE) present in the 5'-flanking region of these genes, as

---

Figure 1. Schematic summarizing the present understanding of the molecular mechanisms by which bifunctional inducers, e.g., dioxin (TCDD) and polycyclic aromatic hydrocarbons (PAH) such as methylcholanthrene, and monofunctional inducers, e.g., phenolic antioxidants such as catechol, may effect the increased expression of phase I enzymes, viz., cytochrome P450 IA1 (CYP IA1) and cytochrome P450 IA2 (CYP IA2), and/or phase II enzymes, viz., cytosolic class-3 aldehyde dehydrogenase (ALDH-3), glutathione S-transferase (GST), DT-diaphorase (DT-D), UDP-glucuronosyl transferase (UDP-GT), epoxide hydrolase (EH) and carbonyl reductase (CBR) [modified from Prochaska and Talalay, 1988; Nebert et al., 1990; Belinsky and Jaiswal, 1993; Jaiswal, 1994]. As originally defined by Williams [1967] and still defined in current pharmacology textbooks, e.g., Benet and Sheiner [1990], phase I enzymes are those that catalyze asynthetic reactions, e.g., oxidations, reductions and hydrolyses, and phase II enzymes are those that catalyze synthetic reactions, e.g., conjugations. Albeit conveniently when discussing monofunctional and bifunctional inducers of enzyme activity, DT-diaphorase and, more recently, cytosolic class-3 aldehyde dehydrogenase have thus been incorrectly referred to as phase II enzymes by several investigators, e.g., Talalay et al. [1987], Nebert et al. [1990] and Lindahl [1992]. Certain isozymes of the glutathione S-transferase, UDP-glucuronosyl transferase and DT-diaphorase families of enzymes, and cytosolic class-3 aldehyde dehydrogenase, do have in common the fact that they are each induced by so-called bifunctional inducers (defined as agents that induce these enzymes as well as cytochrome P450s IA1 and IA2; monofunctional inducers are defined as agents that induce certain isozymes of the glutathione S-transferase, UDP-glucuronosyl transferase and DT-diaphorase families of enzymes, and cytosolic class-3 aldehyde dehydrogenase, but not cytochrome P450s IA1 and IA2 [reviewed in Talalay et al., 1987; Belinsky and Jaiswal, 1993]). Additionally, the so-called antioxidant responsive element (ARE) is present in the 5'-upstream region of the genes coding for each of these enzymes [Rushmore and Pickett, 1990; Rushmore et al., 1990, 1991; Favreau and Pickett, 1991; Jaiswal, 1991; Li and Jaiswal 1992; Asman et al., 1993; Nebert 1994]. With the caveat that the upstream region of the cytochrome P450 IA2 gene has yet to be completely ascertained and published, this element is (apparently) not present in the 5'-upstream region of the genes coding for cytochrome P450s IA1 and IA2 [Jaiswal et al., 1985; Neuhold et al., 1986; Prochaska and Talalay, 1988]. On the other hand, xenobiotic responsive elements (XRE) are found in the 5'-upstream region of each of the genes coding for the six enzymes under consideration [Nebert and Jones, 1989; Rushmore and Pickett, 1990; Rushmore et al., 1990, 1991; Favreau and Pickett, 1991; Jaiswal, 1991; Li and Jaiswal 1992; Asman et al., 1993; Nebert 1994]. Thus, when discussing monofunctional and bifunctional inducers of enzyme activity, it would be perhaps more accurate and meaningful to classify the enzymes currently referred to as phase I enzymes in this context as XRE-regulated enzymes, and those currently referred to as phase II enzymes in this context as XRE/ARE-regulated enzymes. Monofunctional inducers would be those that induce only the XRE/ARE-regulated family of enzymes and bifunctional inducers would be those that induce both the XRE/ARE- and XRE-regulated families of enzymes. Epoxide hydrolase also appears to belong to the XRE/ARE-regulated family of enzymes although the definitive data (5'-upstream sequence) is yet to be obtained [Meijer and DePierre, 1987; Talalay et al., 1987; Talalay, 1994]. Further, we are tentatively adding carbonyl reductase to the XRE/ARE-regulated family of enzymes because it is induced by both monofunctional and bifunctional inducers [Forrest et al., 1990]; again, it remains to be ascertained whether there are XRE and ARE in the 5'-upstream region of the gene coding for this enzyme. The reader is advised that XRE are also referred to as DRE (dioxin responsive elements) or AhRE (aryl hydrocarbon responsive elements), and that ARE are also referred to as EpRE (electrophile responsive elements) [Friling et al., 1990; Nebert et al., 1990; Nebert, 1994]. Details of the mechanism by which enzyme expression is induced via ARE are not known; intriguing is the possibility that Jun and Fos proteins may be involved [Jaiswal, 1994]. AhR, aromatic hydrocarbon receptor; hsp90, heat shock protein 90 kDa; ARNT, aromatic hydrocarbon receptor nuclear translocator; CR, coding region.

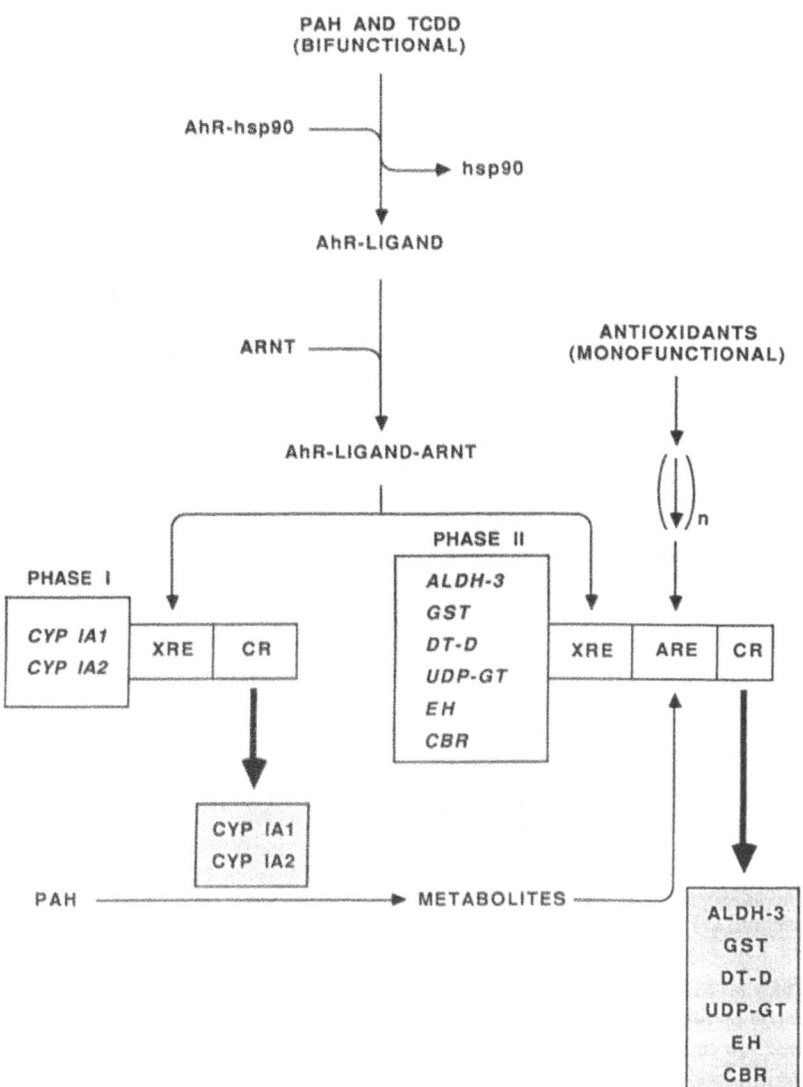

Table 3. Receptor status and the ability of methylcholanthrene and catechol to induce cytosolic class-3 aldehyde dehydrogenase activity in cultured cells.

Malignancy	Cell Line	Constitutive ALDH-3	Receptor Status[b]		ALDH-3 Induced By	
			Ah	Estrogen	MC	CAT
Breast (Adeno)carcinomas	MCF-7/0	+	+	+	Yes	Yes
	MCF-7/OAP	+++	?	?	Yes	Yes
	T-47D	+/-	+	+	Yes	Yes
	ZR-75-1	+/-	+	+	Yes	Yes
	MDA-MB-231	+/-	+	-	No	Yes
	SK-BR-3	+/-	?	-	No	Yes
Colon Carcinomas	RCA	+/-	?	?	No	?
	HCT-116b	+/-	?	?	No	Yes
	Colon C	++++	?	?	No	Yes
Ovarian Carcinomas	OVCAR-3	+/-	?	?	No	?
	OVCAR-432	+/-	?	?	No	?
Myeloma	8226/S	+/-	?	?	No	?

[a]Composite of information heretofore unpublished and data published in Sreerama and Sladek [1993a,b, 1994a,b], Rekha et al. [1994] and Sreerama et al. [1994a]. Cells were grown (monolayer) in the presence of 3 μM methylcholanthrene (MC) or 30 μM catechol (CAT) for 5 days and harvested. Cytosolic class-3 aldehyde dehydrogenase (ALDH-3) activity (NADP-linked enzyme-catalyzed oxidation of benzaldehyde by Lubrol-treated whole homogenates) was quantified as previously described [Sreerama and Sladek, 1993a].

[b]According to literature reports [Engel and Young, 1978; Vickers et al., 1989; Safe et al., 1991; Taylor-Papadimitriou et al., 1993].

well as by the antioxidant responsive elements (ARE) present in the same region of these genes except in the case of cytochrome P450s IA1 and, putatively, IA2, Figure 1.

TRANSIENT OVEREXPRESSION OF CYTOSOLIC CLASS-3 ALDEHYDE DEHYDROGENASE BROUGHT ABOUT BY MONOFUNCTIONAL INDUCERS

Overexpression of cytosolic class-3 aldehyde dehydrogenase, as well as of glutathione S-transferase and DT-diaphorase, but not of cytochrome P450 IA1, can be induced by phenolic antioxidants, e.g., catechol, in cultured human breast (adeno)carcinoma, and colon carcinoma, cell lines, regardless of whether or not they express estrogen receptors, Tables 1 and 3. Again, overexpression of these enzymes is transient in that enzyme activities return to basal levels in a few days when the inducing agent is removed from the culture medium [Sreerama et al., 1994a]. Again also, the class-3 aldehyde dehydrogenase overexpressed in catechol-treated MCF-7/0 cells exhibits physical and catalytic properties identical to those exhibited by the class-3 aldehyde dehydrogenase expressed at very low levels in untreated MCF-7/0 cells, Table 2, and preliminary studies (isoelectric focusing, subunit molecular weight, antibody recognition) indicate an increased expression of both the enzyme present at very low levels in untreated MCF-7/0 cells and the truncated version of this enzyme known to be constitutively expressed at relatively high levels in untreated MCF-7/OAP cells (data not shown).

Transcriptional activation of the relevant genes by these agents, termed monofunctional inducers because they induce only certain "phase II" enzymes, appears to be via ARE present in the 5'-flanking region of these genes, Figure 1.

Table 4. Catalysis of aldophosphamide and benzaldehyde oxidation by (semi)purified NAD-dependent class-3 aldehyde dehydrogenases: relative rates.

Source		(nmol Aldophosphamide oxidized/min/mg) (1000) (nmol Benzaldehyde oxidized/min/mg)
Normal	Stomach mucosa	0.29
	Saliva	0.32
Neoplasm	Colon C	2.85
	MCF-7/0	2.20
	MCF-7/0/MC	3.63
	MCF-7/0/CAT	3.32
	Warthin Tumor	3.03
	MCF-7/OAP	8.08

[a]Composite of information heretofore unpublished and data published in Sreerama and Sladek [1993a, 1994a,b], Rekha et al. [1994] and Sreerama et al. [1994a]. MCF-7/0 cells were treated with methylcholanthrene (MC) and catechol (CAT), enzymes were purified, and enzyme-catalyzed oxidation of aldophosphamide (160 μM) and benzaldehyde (4 mM) was quantified, as described previously [Sreerama and Sladek, 1994b]. The cofactor was NAD (1 mM). Additional experimental details can be found in Sreerama and Sladek [1993a, 1994a].

PHYSICAL AND KINETIC PROPERTIES OF CYTOSOLIC CLASS-3 ALDEHYDE DEHYDROGENASES FOUND IN HUMAN NORMAL AND MALIGNANT CELLS/SECRETIONS

Physical and kinetic characterization of human cytosolic class-3 aldehyde dehydrogenases purified from normal (stomach mucosa, saliva, breast) and neoplastic (cultured breast adenocarcinoma MCF-7, colon carcinoma C, Warthin tumor and mucoepidermoid carcinoma) cells/secretions suggests that, although seemingly identical in every other way, Table 2, the enzyme present in neoplastic cells differs from that present in normal cells/secretions in that the former exhibits much greater ability to catalyze the oxidative detoxification of cyclophosphamide and related anticancer drugs, Table 4. Given that, thus far, the putative variant has been found only in neoplastic cells, an attractive, albeit unlikely, possibility is that it is tumor-specific and thus of potential diagnostic value.

INHIBITORS OF CYTOSOLIC CLASS-3 ALDEHYDE DEHYDROGENASE-MEDIATED CATALYSIS

Inhibitors of cytosolic class-3 aldehyde dehydrogenase would be of obvious experimental value and could be of therapeutic value, e.g., they could be used to sensitize otherwise insensitive tumor cells to cyclophosphamide when the basis for the insensitivity is relatively elevated levels of cytosolic class-3 aldehyde dehydrogenase. Unfortunately, excluding competing alternative substrates, there are no known inhibitors of cytosolic class-3 aldehyde dehydrogenase. Thus, known inhibitors of class-1 aldehyde dehydrogenases, viz., disulfiram

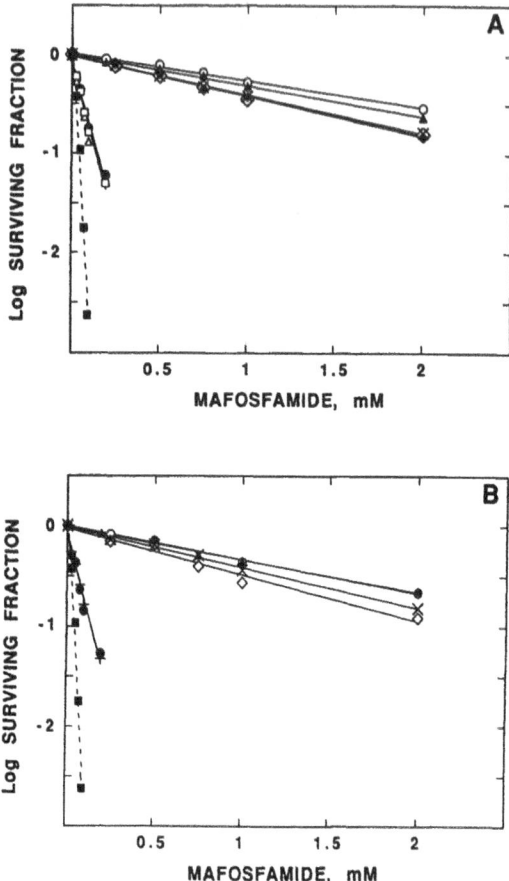

Figure 2. Sensitivity of cultured MCF-7/OAP (**A**) and catechol-treated MCF-7/0 (**B**) cells to mafosfamide in the absence and presence of various aldehydes or known inhibitors of class-1 aldehyde dehydrogenases. Exponentially growing MCF-7/OAP cells, and MCF-7/0 cells that had been grown in the presence of catechol (30 μM) for 5 days, were harvested and preincubated at 37°C with 0.1 mM cyanamide (✕), 0.05 mM disulfiram (♦) or 1 mM chloral hydrate (◊) for 60 min; 5 mM acetaldehyde (▲), 5 mM benzaldehyde (●), 1 mM octanal (△), 5 mM 4-pyridinecarboxaldehyde (□) or 0.2 mM 4-(diethylamino)benzaldehyde (+) for 5 min; or vehicle (○) for 5 or 60 min. Mafosfamide or vehicle was then added, and the incubation was continued for an additional 30 min after which time the cells were reharvested, resuspended in drug-free growth medium and cultured. A colony-forming assay was used to determine surviving fractions. Each point is the mean of measurements on triplicate cultures. Sensitivity of untreated MCF-7/0 cells to mafosfamide (■) is shown for comparison. Aldehyde dehydrogenase activities (4 mM NADP; 4 mM benzaldehyde) in Lubrol-treated whole homogenates of MCF-7/OAP, MCF-7/0 and catechol treated MCF-7/0 cells were 249, 1.9 and 759 mIU/10^7 cells, respectively. Additional experimental details can be found in Sreerama and Sladek [1993a, 1994a].

and chloral hydrate, do not, at pharmacologically relevant concentrations, inhibit cytosolic class-3 aldehyde dehydrogenases in vitro [Sreerama and Sladek, 1993a, 1994a] and, predictably, do not sensitize cultured MCF-7/OAP, MCF-7/0/MC or MCF-7/0/CAT cells to mafosfamide, Figure 2 and L. Sreerama and N. E. Sladek, unpublished observations. Further, cyanamide, another known inhibitor of class-1 aldehyde dehydrogenases, did not restore the sensitivity of cultured MCF-7/OAP, MCF-7/0/MC or MCF-7/0/CAT cells to mafosfamide either, Figure 2 and L. Sreerama and N. E. Sladek, unpublished observations. However, cyanamide, per se, does not inhibit class-1 aldehyde dehydrogenases, i.e., it is a proinhibitor.

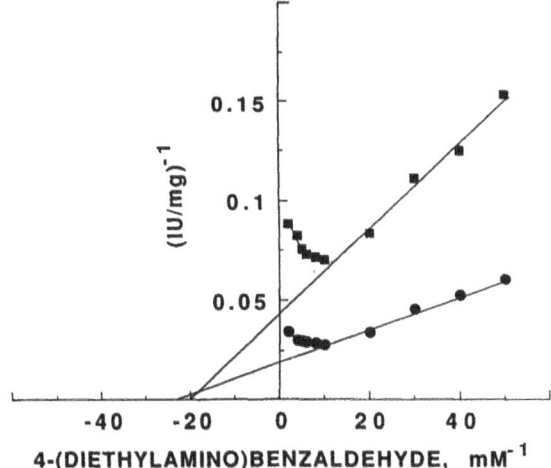

Figure 3. Catalysis of 4-(diethylamino)benzaldehyde oxidation by colon C class-3 aldehyde dehydrogenase: Lineweaver-Burk analysis. Colon C cytosolic class-3 aldehyde dehydrogenase was purified, and initial rates of enzyme catalyzed 4-(diethylamino)benzaldehyde oxidation were measured, as described previously [Rekha et al., 1994]. Cofactor was 1 mM NAD (■) or 4 mM NADP (●). Data points represent the means of duplicate determinations.

Bioactivation is catalase-mediated [DeMaster et al., 1984]. Although catalase is virtually ubiquitously distributed, the possibility that MCF-7/OAP, MCF-7/0/MC and MCF-7/0/CAT cells lack this enzyme cannot be dismissed and thus it cannot be concluded that "activated" cyanamide does not inhibit class-3 aldehyde dehydrogenase. Perhaps relevant to the foregoing is a report suggesting that cyanamide inhibits rat cytosolic class-3 aldehyde dehydrogenase in intact animals [Vasiliou and Marselos, 1989].

According to Russo et al. [1989], 4-(diethylamino)benzaldehyde is a potent inhibitor of mouse class-1 aldehyde dehydrogenase. Although an aldehyde, it is purportedly not a substrate for this enzyme nor for mouse class-2 aldehyde dehydrogenase which it also inhibits, albeit much less well. This agent sensitized cultured human tumor cells expressing relatively large amounts of cytosolic class-3 aldehyde dehydrogenase, viz., MCF-7/0/CAT and colon C, to mafosfamide, Figure 2B and Rekha et al. [1994]. However, subsequent experiments revealed that it was yet another, in fact, quite good (Km ~50 μM), substrate for human cytosolic class-3 aldehyde dehydrogenase, Figure 3, and thus that its ability to restore sensitivity of cultured MCF-7/0/CAT and colon C cells to mafosfamide was because it served as an alternative substrate for the detoxifying enzyme.

CLINICAL RAMIFICATIONS

The clinical significance of these observations includes the likelihood that a large number of pharmacological, environmental, and dietary agents negatively influence the sensitivity of malignant, and perhaps normal, cells, to cyclophosphamide and related compounds, and to the by-products of lipid peroxidation. Given that, in addition to that of cytosolic class-3 aldehyde dehydrogenase, Ah receptor ligands, phenolic antioxidants and other ARE activators and, at least in some cases, cyclophosphamide and related compounds, "coordinately" induce the overexpression of a number of other enzymes, and that a number of anticancer agents are known to be either activated or inactivated as a consequence of the catalytic action of these enzymes, it is likely that these agents induce resistance (or collateral sensitivity) to not only cyclophosphamide and related compounds, but also to a number of additional anticancer drugs so that they, in fact, effect multidrug resistance/collateral sensitivity (see Sreerama and Sladek [1994a] for further discussion). Of potential clinical significance in terms of diagnostic value, is that, at least some, neoplastic cells may express a tumor-specific variant of the cytosolic class-3 aldehyde dehydrogenase expressed in normal cells.

ACKNOWLEDGEMENT

Supported by USPHS Grant CA 21737 and Bristol-Myers Squibb Company Grant 100-R220.

REFERENCES

Arellano, L. O., Wang, X., and Safe, S., 1993, Effects of cycloheximide on the induction of *CYP1A1* gene expression by 2,3,7,8-tetrachlorodibenzo-*p*-dioxin (TCDD) in three human breast cancer cell lines, *Carcinogenesis*, 14:219.

Asman, D. C., Takimoto, K, Pitot, H. C., Dunn, T. J., and Lindahl, R., 1993, Organization and characterization of the rat class-3 aldehyde dehydrogenase gene, *J. Biol. Chem.*, 268:12530.

Belinsky, M., and Jaiswal, A. K., 1993, NAD(P)H:quinone oxidoreductase₁ (DT-diaphorase) expression in normal and tumor tissues, *Cancer Metastasis Rev.*, 12:103.

Benet, L. Z., and Sheiner, L. B., 1990, Pharmacokinetics: the dynamics of drug absorption, distribution, and elimination, *in* Goodman and Gilman's The Pharmacological Basis of Therapeutics. Eighth Edition, Edited by A. G. Gilman, T. W. Rall, A. S. Neis, and P. Taylor, p. 3, Pergaman Press, New York.

Bunting, K. D., and Townsend, A. J., 1993, Mafosfamide sensitivity in human MCF-7 breast carcinoma cell lines expressing transfected rat class-3 aldehyde dehydrogenase ("tumor ALDH"), *Proc. Am. Assoc. Cancer Res.*, 34:270.

DeMaster, E. G., Shirota, F. N., and Nagasawa, H. T., 1984, The metabolic activation of cyanamide to an inhibitor of aldehyde dehydrogenase is catalyzed by catalase, *Biochem. Biophys. Res. Commun.*, 122:358.

Downes, J. E., Swann, P. G., and Holmes, R. S., 1994, Differential corneal sensitivity to ultraviolet light among inbred strains of mice, *Cornea*, 13:67.

Engel, L. W., and Young, N. A., 1978, Human breast carcinoma cells in continuous culture: a review, *Cancer Res.*, 38:4327.

Favreau, L. V., and Pickett, C. B., 1991, Transcriptional regulation of the rat NAD(P)H:quinone reductase gene, *J. Biol. Chem.*, 266:4556.

Forrest, G. L., Akman, S., Krutzik, S., Paxton, R. J., Sparkes, R. S., Doroshow, J., Felsted, R. L., Glover, C. J., Mohandas, T., and Bachur, N. R., 1990, Induction of a human carbonyl reductase gene located on chromosome 21, *Biochim. Biophys. Acta*, 1048:149.

Friling, R. S., Bensimon, A., Tichauer, Y., and Daniel, V., 1990, Xenobiotic-inducible expression of murine glutathione S-transferase Ya subunit gene is controlled by an electrophile-responsive element, *Proc. Natl. Acad. Sci. USA*, 87:6258.

Hales, B. F., 1982, Comparison of the mutagenicity and teratogenicity of cyclophosphamide and its active metabolites, 4-hydroxycyclophosphamide, phosphoramide mustard, and acrolein, *Cancer Res.*, 42:3016.

Jaiswal, A. K., 1991, Human NAD(P)H:quinone oxidoreductase (NQO₁) gene structure and induction by dioxin, *Biochemistry*, 30:10647.

Jaiswal, A. K., 1994, Jun and Fos regulation of NAD(P)H: quinone oxidoreductase gene expression, *Pharmacogenetics*, 4:1.

Jaiswal, A. K., Gonzalez, F. J., and Nebert, D. W., 1985, Human P₁450 gene sequence and correlation of mRNA with genetic differences in benzo[a]pyrene metabolism, *Nucl. Acids Res.*, 13:4503.

Li, Y., and Jaiswal, A. K., 1992, Regulation of human NAD(P)H:quinone oxidoreductase gene, *J. Biol. Chem.*, 267:15097.

Lindahl, R., 1992, Aldehyde dehydrogenases and their role in carcinogenesis, *Crit. Rev. Biochem. Mol. Biol.*, 27:283.

Meijer, J., and DePierre, J. W., 1987, Hepatic levels of cytosolic, microsomal and 'mitochondrial' epoxide hydrolases and other drug-metabolizing enzymes after treatment of mice with various xenobiotics and endogeneous compounds, *Chem.-Biol. Interactions*, 62:249.

Nebert, D. W., 1994, Drug-metabolizing enzymes in ligand-modulated transcription, *Biochem. Pharmacol.*, 47:25.

Nebert, D. W., and Jones, J. E., 1989, Regulation of the mammalian cytocrome P_1450 (CYP1A1) gene, *Int. J. Biochem.*, 21:243.

Nebert, D. W., Petersen, D. D., and Fornace, A. J. Jr., 1990, Cellular responses to oxidative stress: the [*Ah*] gene battery as a paradigm, *Environ. Health Perspect.*, 88:13.

Neuhold, L. A., Gonzalez, F. J., Jaiswal, A. K., and Nebert, D. W., 1986, Dioxin-inducible enhancer region upstream from the mouse P_1450 gene and interaction with a heterologous SV40 promoter, *DNA*, 5:403.

Prochaska, H. J., and Talalay, P., 1988, Regulatory mechanisms of monofunctional and bifunctional anticarcinogenic enzyme inducers in murine liver, *Cancer Res.*, 48:4776.

Rekha, G. K., Sreerama, L., and Sladek, N. E., 1994, Intrinsic cellular resistance to oxazaphosphorines exhibited by a human colon carcinoma cell line expressing relatively large amounts of a class-3 aldehyde dehydrogenase, *Biochem. Pharmacol.*, 48:xxx (in press).

Rushmore, T. H., and Pickett, C. B., 1990, Transcriptional regulation of the rat glutathione S-transferase Ya subunit gene, *J. Biol. Chem.*, 265:14648.

Rushmore, T. H., King, R. G., Paulson, E. K., and Pickett, C. B., 1990, Regulation of glutathione S-transferase Ya subunit gene expression: identification of a unique xenobiotic-responsive element controlling inducible expression by planar aromatic compounds, *Proc. Natl. Acad. Sci. USA*, 87:3826.

Rushmore, T. H., Morton, M. R., and Pickett, C. B., 1991, The antioxidant responsive element, *J. Biol. Chem.*, 266:11632.

Russo, J. E., Hauquitz, D., and Hilton, J., 1988, Inhibition of mouse cytosolic aldehyde dehydrogenase by 4-(diethylamino)benzaldehyde, *Biochem. Pharmacol.*, 37:1639.

Safe, S., Astroff, B., Harris, M., Zacharewski, T., Dickerson, R., Romkes, M., and Biegel, L., 1991, 2,3,7,8-Tetrachlorodibenzo-*p*-dioxin (TCDD) and related compounds as antiestrogens: characterization and mechanism of action, *Pharmacol. Toxicol.*, 69:400.

Sreerama, L., and Sladek, N. E., 1993a, Identification and characterization of a novel class 3 aldehyde dehydrogenase overexpressed in a human breast adenocarcinoma cell line exhibiting oxazaphosphorine-specific acquired resistance, *Biochem. Pharmacol.*, 45:2487.

Sreerama, L., and Sladek, N. E., 1993b, Overexpression or polycyclic aromatic hydrocarbon-mediated induction of an apparently novel class 3 aldehyde dehydrogenase in human breast adenocarcinoma cells and its relationship to oxazaphosphorine-specific acquired resistance, *Adv. Exp. Med. Biol.*, 328:99.

Sreerama, L., and Sladek, N. E., 1994a, Identification of a methylcholanthrene-induced aldehyde dehydrogenase in a human breast adenocarcinoma cell line exhibiting oxazaphosphorine-specific acquired resistance, *Cancer Res.*, 54:2176.

Sreerama, L., and Sladek, N. E., 1994b, Identification of the class-3 aldehyde dehydrogenases present in human MCF-7/0 breast adenocarcinoma cells and normal human breast tissue, *Biochem. Pharmacol.*, 48:617.

Sreerama, L., Rekha, G. K., and Sladek, N. E., 1994a, Phenolic antioxidant-induced overexpression of class-3 aldehyde dehydrogenase and oxazaphosphorine-specific resistance, *Biochem. Pharmacol.*, XX:xxx, (submitted for publication).

Sreerama, L., Hedge, M., and Sladek, N. E., 1994b, Identification of a class-3 aldehyde dehydrogenase in human saliva and overexpression of this enzyme by a mucoepidermoid carcinoma and Warthin tumors of the paratid gland, *Proc. Am. Assoc. Cancer Res.*, 35:84.

Talalay, P., 1994, Enzyme regulation: a major mechanism for chemical protection against cancer, *Proc. Am. Assoc. Cancer Res.*, 35:697.

Talalay, P., DeLong, M. J., and Prochaska, H. J., 1987, Molecular mechanisms in protection against carcinogenesis, *in* Cancer Biology and Therapeutics. Edited by J. G. Cory and A. Szentivani, p. 197, Plenum Press, New York.

Taylor-Papadimitriou, J., Berdichevsky, F., D'Souza, B., and Burchell, J., 1993, Human models of breast cancer. *Cancer Surveys*, 16:59.

Vasiliou, V., and Marselos, M., 1989, Changes in the inducibility of a hepatic aldehyde dehydrogenase by various effectors, *Arch. Toxicol.*, 63:221.

Vickers, P. J., Dufresne, M. J., and Cowan, K. H., 1989, Relation between cytochrome P450IA1 expression and estrogen receptor content of human breast cancer cells, *Mol. Endocrinol.*, 3:157.

Williams, R. T., 1967, Comparative patterns of drug metabolism, *Fed. Proc.*, 26:1029.

15

METABOLISM OF CYCLOPHOSPHAMIDE BY ALDEHYDE DEHYDROGENASES

Dharam P. Agarwal, Ulrich v. Eitzen, Doris Meier-Tackmann and
H. Werner Goedde

Institute of Human Genetics
University of Hamburg
Germany

INTRODUCTION

Cyclophosphamide (Endoxan) and other oxazaphosphorines such as 4-hydroperoxy-cyclophosphamide, ifosfamide, and mafosfamide are widely used as antineoplastic drugs (Sladek, 1988; Lindahl, 1992). The cytotoxic effect is caused by alkylation reaction of these drugs with DNA and proteins inhibiting the cell proliferation. These chemotherapeutic agents are also extensively applied as immunosuppressants during bone marrow transplantation, and in autoimmune diseases. Cyclophosphamide is pharmacologically inactive, and needs to be biotransformed to its cytotoxic metabolite phosphoramide mustard via an intermediate metabolite 4-hydroxycyclophosphamide (Borch et al., 1984; Hill et al., 1973). The latter compound exists in equilibrium with aldophosphamide which can get converted to a non-cytotoxic compound carboxy-phosphamide through irreversible oxidation of the aldehyde group catalyzed by one or more forms of aldehyde dehydrogenases (Hilton et al., 1984; Sladek et al., 1989; Kastan et al., 1990; Dockham et al., 1992; Moreb et al., 1992). This enzymatic pathway leads to the detoxification of cyclophosphamide affecting its therapeutic efficiency. Therefore, induction or overexpression of one or more of the relevant ALDH forms in target cells might primarily account for the cyclophosphamide-specific acquired resistance exhibited by many neoplastic cells.

A number of aldehyde dehydrogenase isozymes have been identified based on their physicochemical characteristics, tissue and subcellular distribution, and enzymatic properties (Agarwal and Goedde, 1990). These isozymes are broadly subdivided into three classes, viz., 1, 2 and 3 (Lindahl and Hempel, 1991). Though the cytosolic class 1 ALDH isozyme, ALDH1, has been shown to be particularly important in the metabolism of aldophosphamide (Dockham et al., 1992; Yoshida et al., 1993), the contribution of other aldehyde dehydrogenases is not precisely known. In earlier studies, a stomach specific class 3 ALDH isozyme (ALDH3) was found to be overexpressed in human hepato-cellular carcinoma (Agarwal et al., 1989; Eckey et al, 1991) as well as in certain tumor cell lines with and without induction by polycyclic aromatic hydrocarbons such as 3-methylcholanthrene (Meier-Tackmann et al., 1993; Sreerama and Sladek, 1993). However, aldophosphamide was found to be a poor substrate for ALDH3 isolated from a mafosfamide resistant human breast adenocarcinoma cell line (Sreerama and Sladek, 1994). Moreover, in *in vitro* studies, aldophosphamide was not oxidized by ALDH3 (both intrinsic and inducible forms) isolated from human stomach mucosa and certain tumor cell lines (von Eitzen et al., 1994).

The purpose of the present study was to examine whether under *in vivo* conditions, normal and neoplastic cell lines expressing unequal levels of class 1 and class 3 ALDH also exhibit differential sensitivity to aldophosphamide-related cytotoxicity.

MATERIALS AND METHODS

Reagents

4-Hydroperoxycyclophosphamide and mafosfamide (ASTA Z 7557), a cyclophosphamide-mesna conjugate, were kindly provided by Dr. Jörg Pohl, ASTA Medica AG (Frankfurt a.M., F.R.G.). The preparation of aldophosphamide from 4-hydroperoxycyclophosphamide was performed as described by Manthey et al (1990).

Cell Lines

UMSCC2 cells isolated from a human pharynx carcinoma (originally obtained from Dr. T. Carey, University of Michigan) were provided by Dr. E. Schuuring (The Netherland Cancer Institute, Amsterdam). Breast adenocarcinoma cells (SK-BR-3), cells derived from a human lung carcinoma (A549), human hepatocellular carcinoma cells (Hep G2), and fibrosarcoma cells (HT-1080) were obtained from American Type Culture Collection (ATCC), Rockville, MD. The skin fibroblast cell line (Fib 236) was established from a healthy donor by Dr. I. Willers (Institute of Human Genetics, University of Hamburg).

The tumor cell lines were grown in minimal essential medium (MEM, Gibco) supplemented with 10% fetal calf serum (FCS), 2.2 g/l $NaHCO_3$, and 62.5 mg/l gentamycin. The cultures were kept at 37°C, 5% CO_2 in air atmosphere, and 90% relative humidity. The fibroblasts were grown in the same medium supplemented with 15% FCS and kept at 37°C in humid air atmosphere with 5% CO_2 and 5% O_2. Cells in asynchronous exponential growth were washed in saline solution (0.9% NaCl) and harvested by scrapping the cells in the same solution. Washed and packed cells were stored at -20°C until used. Cell extracts were prepared by thawing and homogenization in distilled water (1 ml packed cells/ml H_2O) using a micro-Potter homogenizer followed by centrifugation at 27,000g for 10 minutes.

Drug Exposure Studies

Tumor cells were harvested and seeded into Falcon tissue culture plates with 24 flat-bottom wells (Becton Dickinson, Oxnard) containing the growth medium and allowed to grow for 48 hours. The cells were exposed to aldophosphamide (0-100 µg/ml culture medium) for 30 min at 37°C after which they were washed free of drug with sterile phosphate buffered saline (PBS, Gibco) maintained at 37°C and cultured in drug-free growth medium for another 48 hours; the cells were washed with saline to remove the culture medium. A 0.3 M NaOH solution was layered over the cells for 24 hours to extract the soluble proteins. Assay of soluble proteins was carried out according to Lowry et al (1951).

Determination of IC_{50} Values

The inhibition of cellular proliferation as judged by inhibition concentration, IC_{50} (inhibition concentration of aldophosphamide required to reduce the amount of layered cells to 50% as compared to the untreated control cells) served as end point of the cytotoxic effect of aldophosphamide. Therefore, the amount of protein present in each well was indicative of the amount of the cells which remained resistant to aldophosphamide treatment. To calculate the IC_{50} values, mean average protein values from triplicate cultures at each aldophosphamide concentration were plotted (as % of control) against aldophosphamide concentration (µg aldophosphamide/ml culture medium).

Enzyme Activity Assay

ALDH activity was determined spectrophotometrically at 340 nm and 25°C in a 0.1 M pyrophosphate buffer, pH 9.5, using either 5 mM acetaldehyde or 0.25 mM 3-nitrobenzaldehyde and 2 mM NAD^+ or $NADP^+$ respectively. The preparation of aldophosphamide from 4hydroperoxycyclophosphamide was carried out as described by Manthey et al (1990). Accordingly, the concentration of aldophosphamide in the stock solution was assumed to be 3.2 mM. In the ALDH activity assay mixture the final concentration of aldophosphamide was thus calculated to be 0.16 mM. A stock solution of mafosfamide (80 mM) was prepared in a 60 mM sodium phosphate buffer, pH 6.0. In aqueous solution mafosfamide hydrolyzes spontaneously to 4-hydroxycyclophosphamide and aldophosphamide (Niemeyer et al., 1989). Though the equilibrium concentration of mafosfamide-derived aldophoshamide was not known, in ALDH activity assay system a mafosfamide concentration of 4 mM was used which gave a substrate saturation. The protein content of cell and tissue extracts as well as in partially purified ALDH preparations was determined according to the method of Marcart and Gerbaut (1982) using bovine serum albumin as the standard.

Isoelectric Focusing (IEF)

IEF was performed on 0.75% agarose gels containing 12% D-sorbitol, 1% polyethylene glycol 6000, and 7% carrier ampholytes, pH range 3.5 to 9.5. Focusing was done at a constant voltage of 400 V and a maximum of 20 W for 2 hours. The gels were stained for ALDH activity bands with 3-nitrobenzaldehyde and NAD^+ s substrate and cosubstrate, respectively.

RESULTS

Only cell lines A549, UMSCC2 and Hep G2 exhibited a substantial class 3 ALDH isozyme (ALDH3) activity. These cells also contained class 1 ALDH (ALDH1) activity, albeit in very low amounts (Fig. 1). Whereas Fib 236 and HT-1080 cells lacked in both ALDH1 and ALDH3 activities, no detectable level of ALDH3 was found in SK-BR-3 cells.

Specific ALDH activities, using various aldehyde substrates, from human tumor cells, stomach mucosa, and liver are presented in Table 1. Whereas the cytosolic class 1 ALDH (ALDH1) utilized various aldehydes as substrate, class 3 ALDH (ALDH3) from stomach mucosa as well as from lung and pharynx carcinoma cells oxidized only acetaldehyde and 3-nitrobenzaldehyde. ALDH3 isozyme, independent of the source, did not use aldophosphamide as substrate, neither with NAD^+ nor with $NADP^+$.

As shown in Fig. 2, based upon the IC_{50} values, Fib 236 cells and HT-1080 cells lacking in both class 1 and class 3 ALDH exhibited a greater sensitivity to aldophosphamide

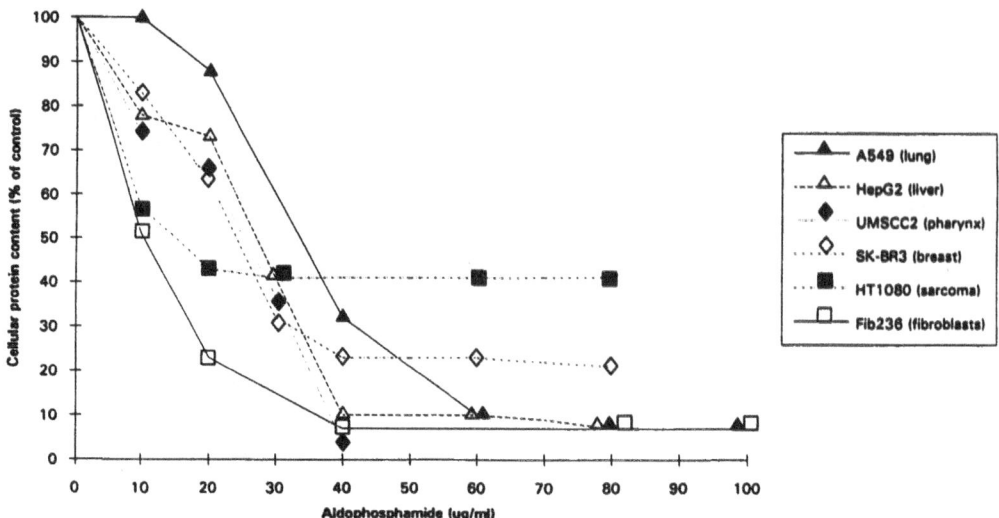

Figure 1. Sensitivity of human tumor and fibroblast cell lines to increasing aldophosphamide concentration. Cells were exposed to aldophosphamide for 30 min, washed, and grown in drug-free culture medium. Protein content of the surviving cells was determined. Each point is the mean of measurements on triplicate cultures.

toxicity than the cells (A549, UMSCC2, and Hep G2) possessing both ALDH1 and ALDH3 isozymes. Also, SK-BR-3 cells which lacked in ALDH3 activity were found to be very sensitive to aldophosphamide.

IC_{50} values for different cell lines are shown in Table 2. The number of cells seeded per ml culture medium in each well is also shown in the table. Cell lines having no ALDH3 activity were least resistant to aldophosphamide toxicity than those possessing both ALDH1 and ALDH3 activities; order of increasing aldophosphamide sensitivity: A549>Hep G2>UMSCC2>SK-BR-3>HT-1080>Fib 236.

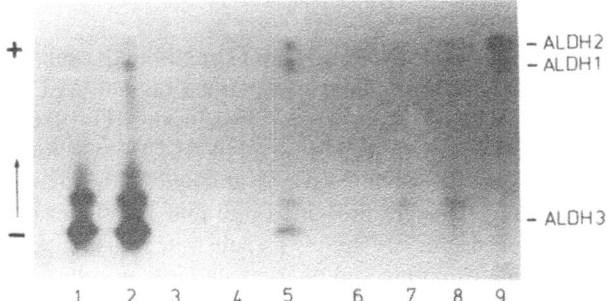

Figure 2. Isoelectric focusing pattern of aldehyde dehydrogenase isozymes from human tissue and cell extracts. Lane 1: UMSCC2; lane 2: A549; lane 3: HT-1080; lane 4: SK-BR-3; lane 5: Hep G2; lane 6: fibroblasts; lane 7 to 8: stomach; 9: liver.

Table 1. ALDH activity with different substrates and co-substrates***

Substrate	Co-substrate	Specific ALDH activity (mU/mg protein)			
		ALDH1 (liver)	ALDH3 (stomach)	ALDH3 (A549)	ALDH3 (UMSCC2)
Acetaldehyde (5 mmol)	NAD$^+$	770	80	578	58
3-Nitrobenzaldehyde (0.25 mmol)	NADP$^+$	45	1083	7345	871
Aldophosphamide* (0.16 mmol)	NAD$^+$	123	0	0	0
Aldophosphamide* (0.16 mmol)	NADP$^+$	38	0	0	0
Mafosfamide** (4 mmol)	NAD$^+$	108	0	0	0
Mafosfamide** (4 mmol)	NADP$^+$	31	0	0	0

* Aldophosphamide derived from 4-hydroperoxycyclophosphamide. The concentration of aldophosphamide was calculated according to Manthey et al. (1990).

** Aldophosphamide derived from mafosfamide; using 4 mM mafosfamide, a substrate saturation for aldophosphamide in the activity assay mixture was achieved.

*** Data taken from von Eitzen et al. (1994) with permission of the publisher.

DISCUSSION

The results shown above clearly demonstrate that under *in vivo* conditions, cell lines lacking in ALDH1 and/or ALDH3 were most sensitive to aldophosphamide toxicity than cell lines which expressed both the isozymes. The breast adenocarcinoma cells (SKBR-3) completely lacking in ALDH3 activity were found to be relatively less resistant to aldophosphamide-related cytotoxicity. Sreerama and Sladek (1994), too, observed very small amounts of ALDH activity (NADP-linked benzaldehyde oxidation) in the SK-BR-3 cell line which was also relatively sensitive to mafosfamide. Moreover, these cells did not become resistant to mafosfamide after treatment with 3-methylcholanthrene, and did not exhibit any induced aldehyde dehydrogenase activity. In an earlier study, Sreerama and Sladek (1993) reported an overexpression of a novel class 3 ALDH activity in a mafosfamide-resistant human breast adenocarcinoma cell line (MCF-7/OAP). These cells exhibited an elevated, albeit minimal, NAD$^+$ dependent enzymatic oxidation of aldophosphamide. Furthermore, ALDH3 activity of these cells was found to be inducible with 3-methylcholanthrene associated with a loss of sensitivity to mafosfamide by these cells (Sreerama and Sladek, 1994).

Table 2. IC_{50} values (mg aldophosphamide/ml culture medium) and number of cells seeded per well

cell line	IC_{50}	Seeded cells/well $\times\ 10^4$
A549	33.5	1.3
Hep G2	27.0	2.4
UMSCC2	24.5	2.2
SK-BR-3	24.0	2.6
HT-1080	15.0	1.0
Fib 236	10.5	0.8

Taken together, these observations imply that class 3 ALDH activity (ALDH3) may indeed represent an important determinant in the detoxification of oxazaphosphorines. However, both constitutive and inducible ALDH forms were found not to catalyze the oxidation of aldophosphamide derived from 4-hydroperoxycyclophosphamide or mafosfamide under *in vitro* conditions (von Eitzen et al., 1994). Similarly, no correlation was found between the level of class 3 ALDH (benzaldehyde-related ALDH activity) and cyclophosphamide sensitivity of several rat hepatoma cell lines (Lin and Lindahl, 1987). Thus, it remains to be shown in which way the class 3 ALDH is responsible for the development of oxazaphosphorine-specific resistance encountered frequently in neoplastic cells. On the other hand, the contribution of the constitutive class 1 ALDH1 isozyme in the oxidation of aldophosphamide, in normal as well as in tumor cells, also remains to be equivocally established. Whereas several fold increase of cytosolic ALDH1 was observed in a human ovarian carcinoma cell line made moderately resistant to cyclophosphamide, a human colon carcinoma cell line which became resistant to cyclophosphamide, showed no significant increase in ALDH1 activity (Yoshida et al., 1993). Moreover, the tumor cell lines investigated in the present study showed a very low ALDH1 activity as compared to the ALDH3 activity.

CONCLUDING REMARKS

The results of the present study though support the notion that ALDH3 (constitutive and/or inducible form) may be an important determinant in the development of cellular resistance to cyclophosphamide treatment, it remains to be proven unequivocally if the ALDH3 activity induction in target cells can indeed account for the detoxification of aldophosphamide leading to the development of acquired resistance to oxazaphosphorines. Thus further studies are necessary to ascertain the precise role of specific ALDH forms in cyclophosphamide-cytotoxicity.

REFERENCES

Agarwal, D.P., Eckey, R., Rudnay A.-C., Volkens, T., and Goedde, H.W., 1989, "High-Km"aldehyde dehydrogenase isozymes in human tissues: constitutive and tumor associated forms. In: Progress in Clin. Biol. Res., vol. 290: Enzymology and Molecular Biology of Carbonyl Metabolism 2, pp. 119-131. Editors: H. Weiner and T.G. Flynn. Alan R Liss, Inc., New York.

Agarwal, D.P., and Goedde, H.W., 1990, Alcohol Metabolism, Alcohol Intolerance, and Alcoholism. Biochemical and Pharmacogenetic Approaches. Springer-Verlag, Berlin, Heidelberg, New York.

Borch, R.F., Hoye, T.R ., and Swanson, T.A., 1984, In situ preparation and fate of cis-4-hydroxycyclophosphamide and aldophosphamide: ^{1}H and ^{31}P NMR evidence for equilibration of cis- and trans-4-hydroxycyclophosphamide with aldophosphamide and its hydrate in aqueous solution. J. Med. Chem., 27, 490-494.

Dockham, P.A., Lee, M.-O., and Sladek, N.E., 1992, Identification of human liver aldehyde dehydrogenases that catalyze the oxidation of aldophosphamide and retinaldehyde. Biochem. Pharmacol., 43, 2453-2469.

Eckey, R., Timmann, R., Hempel, J., Agarwal, D.P., and Goedde, H.W., 1991, Biochemical, immunological and molecular characterization of a "high Km" aldehyde dehydrogenase. In: Advances in experimental medicine and biology, vol. 284: Enzymology and molecular biology of carbonyl metabolism 3, pp. 43-52. Editors: H. Weiner, B. Wermuth and D.W. Crabb. Plenum Press, New York.

Hill, D.L., Laster, W,R, Jr., Kirk, M.C., ElDareer, S., and Struck, R.F., 1973, Metabolism of phosphamide (2-(2-chloroethylamino)3(2-chloroethyl)-tetrahydro 2H-1,3,2,-oxazaphosphorine 2-oxide) and production of a toxic phosphamide metabolite. Cancer Res., 33, 1016-1022.

Hilton, J., 1984, Role of aldehyde dehydrogenase in cyclophosphamideresistant L-1210 leukemia. Cancer Res. 44, 5156-5160.

Kastan, M.B., Schlaffer, E., Russo, J.E., Colvin, O.M., Civin, C.I. and Hilton, J., 1990, Direct demonstration of elevated aldehyde dehydrogenase in human hematopoietic progenitor cells. Blood, 75, 1947-1950.

Lin, K.-H., and Lindahl, R., 1987, Role of aldehyde dehydrogenase activity in cyclophosphamide metabolism in rat hepatoma cell lines. Biochem. Pharmacol., 36, 3305-3307.

Lindahl, R., 1992, Aldehyde dehydrogenases and their role in carcinogenesis. Critical Reviews in Biochemistry and Molecular Biology, 27, 283-334.

Lindahl R., and Hempel, H., 1991, Aldehyde dehydrogenases: what can be learned from a baker's dozen sequences? In: Advances in experimental medicine and biology, vol. 284: Enzymology and molecular biology of carbonyl metabolism 3, pp. 1-8. Editors: H. Weiner, B. Wermuth and D.W. Crabb. Plenum Press, New York.

Lowry, O.H., Rosebrough, N.J., Farr, A.L., and Randall, R.J., 1951, Protein measurement with the Folin phenol reagent. J. Biol. Chem. 193, 265-275.

Manthey, C.L., Landkamer, G.J., and Sladek, N.E., 1990, Identification of the mouse aldehyde dehydrogenases important in aldophosphamide detoxification. Cancer Res., 50, 4991-5002.

Marcart, M., and Gerbaut, L., 1982, An improvement of the Coomassie blue dye binding method allowing an equal sensitivity to various proteins: application to cerebrospinal fluid. Clin. Chim. Acta, 122, 93-101.

Meier-Tackmann, D., Eckey, R., Wolff, C., von Eitzen, U., Agarwal, D.P., and Goedde, H.W., 1993, Tumor-associated aldehyde dehydrogenase (ALDH3): expression in different human tumor cell lines with and without treatment with 3-methylcholanthrene. In: Advances in Experimental Medicine and Biology, vol. 328: Enzymology and Molecular Biology of Carbonyl Metabolism 4, pp. 115-122. Editors: H. Weiner, D.W. Crabb and T.G. Flynn. Plenum Press, N.Y.

Moreb, J., Zucali, J.R., Zhang, Y., Colvin, M.O., and Gross, M.A., 1992, Role of aldehyde dehydrogenase in the protection of hematopoietic progenitor cells from 4-hydroperoxy cyclophosphamide by interleukin 1ß and tumor necrosis factor. Cancer Res., 52, 1770-1774.

Niemeyer, U., Engel, J., Hilgard, P., Peukert, M., Pohl, J., and Sindermann, H., 1989, Mafosfamide - a derivative of 4hydroxycyclophosphamide. Progress Clin. Biochem. Med. 9, 35-60.

Sladek, N.E., 1988, Metabolism of oxazaphosphorines. Pharmacol. Ther. 37, 301-355.

Sladek, N.E., Manthey, C.L., Maki, P.A., Zhang, Z., and Landkammer, G.J., 1989, Xenobiotic oxidation catalyzed by aldehyde dehydrogenases. Drug Metabolism Review 20, 697-720.

Sreerama, L., and Sladek, N.E., 1993, Identification and characterization of a novel class 3 aldehyde dehydrogenase overexpressed in a human breast adenocarcinoma cell line exhibiting oxazaphosphorine-specific acquired resistance. Biochem. Pharmacol., 45, 2487-2505.

Sreerama, L., and Sladek, N.E., 1994, Identification of a methylcholanthrene-induced aldehyde dehydrogenase in a human breast adenocarcinoma cell line exhibiting oxazaphosphorinesepecific acquired resistance. Cancer Research 54, 2176-2185.

von Eitzen, U., Meier-Tackmann, D., Agarwal, D.P., and Goedde, H.W., 1994, Detoxification of cyclophos-
phamide by human aldehyde dehydrogenase isozymes. Cancer Lett, 76, 45-49.
Yoshida, A., Davé, V., Han, H., and Scanlon, K.J., 1993, Enhanced transcription of the cytosolic ALDH gene
in cyclosphosphamide resistant human carcinoma cells. In: Advances in Experimental Medicine and
Biology, vol. 328: Enzymology and molecular biology of carbonyl metabolism 4, pp. 63-72. Editors:
H. Weiner, B. Wermuth and D.W. Crabb, Plenum Press, New York.

16

TISSUE-SPECIFIC EXPRESSION AND PRELIMINARY FUNCTIONAL ANALYSIS OF THE 5' FLANKING REGIONS OF THE HUMAN MITOCHONDRIAL ALDEHYDE DEHYDROGENASE (*ALDH2*) GENE

Katrina M. Dipple, Mark J. Stewart, and David W. Crabb

Departments of Medicine and of Biochemistry and Molecular Biology
IB424 Medical Research and Library Building
975 West Walnut Street
Indiana University School of Medicine
Indianapolis, IN 46202-5121

INTRODUCTION

Mitochondrial aldehyde dehydrogenase (ALDH2) is a member of a large group of enzymes that catalyze the irreversible oxidation of aldehydes to carboxylic acids. The ALDH2 enzyme has a very low K_m (<1 µM) for short aliphatic aldehydes such as acetaldehyde and propionaldehyde. In human beings, this isozyme appears to play a major role in the removal of acetaldehyde generated from the oxidation of ethanol. This conclusion is based on the study of individuals with a genetically determined deficiency in ALDH2 activity. Liver or hair root extracts from these individuals show the absence of the ALDH2 activity band on starch gels or isoelectric focusing gels (also known in the literature as the ALDH I or E2 band) (Harada *et al.*, 1980). The deficiency results from a point mutation that substitutes a lysine for glutamate at position 487 (Yoshida *et al.*, 1984; Crabb *et al.*, 1989). Individuals with the deficient phenotype experience alcohol-induced flushing that is secondary to high levels of circulating acetaldehyde (Harada *et al.*, 1981; Enomoto *et al.*, 1991b). This is commonly referred to as the Oriental alcohol flush reaction. Assays of liver extracts from individuals carrying a deficient allele have about half the aldehyde dehydrogenase activity of extracts from subjects without the deficiency (S.-J. Yin, personal communication). Thus, ALDH2 activity makes a major contribution to the total aldehyde oxidizing capacity of the liver.

The alcohol flush reaction that is associated with ALDH2 deficiency has a strong protective effect against the development of alcoholism in most individuals; however, the flush reaction is remarkably variable (Enomoto *et al.*, 1991b), and some ALDH2-deficient patients drink heavily enough to suffer from alcoholism and its medical complications

Enzymology and Molecular Biology of Carbonyl Metabolism 5
Edited by H. Weiner *et al.*, Plenum Press, New York, 1995

(Enomoto *et al.,* 1991a). It is possible that some of the variability in the flush reaction is due to between-individual differences in the expression of the gene. It has been suggested that the ALDH2 enzyme is a constitutively expressed protein, since the enzyme activity can be detected in many tissues when examined by gel electrophoresis or isoelectric focusing and activity staining of the gels, and the enzyme does not appear to be inducible by hormones, drugs, alcohol, or carcinogens. However, it seemed unlikely that the enzyme would be expressed at uniformly high levels in various tissues when the enzymes that generate acetaldehyde are rather specifically expressed in the liver. These experiments were undertaken to quantitate the expression of ALDH2 in various tissues and to initiate studies on the promoter activity of the 5' flanking region of the *ALDH2* gene.

METHODS

Materials

Most chemicals and supplies were purchased from Sigma Chemical Company (St. Louis, MO). Sprague-Dawley rats were obtained from Harlan Sprague-Dawley (Indianapolis, IN). Cell lines were obtained from the American Type Culture Collection. Nitrocellulose was obtained from Schleicher and Schuell, Inc. (Keene, NH). Guanidinium isothiocyanate was from Fluka Biochemika (Buchs, Switzerland). Organic solvents, acids, and alcohols were from Mallinckrodt Specialty Chemicals Co. (Paris, KY). Agarose, trypsin, all restriction endonucleases, DNA modifying enzymes, and all tissue culture media and serum were purchased from GIBCO BRL (Gaithersburg, MD). All radioisotopes were purchased from DUPONT NEN Research Products Inc. (Boston, MA). Taq polymerase was from Perkin Elmer Cetus (Norwalk, CT). Oligonucleotides were synthesized by the Biotechnology Core Facility of the Department of Biochemistry and Molecular Biology on an Applied Biosystems model 380B DNA synthesizer. Dr. Henry Weiner, Department of Biochemistry, Purdue University, kindly provided the plasmid containing the rat ALDH2 cDNA (Farres *et al.,* 1989).

Isolation of Rna and Northern Blotting

Total cellular RNA was prepared from various tissues by a single extraction with guanidinium isothiocyanate by standard procedures (Chomczynski and Sacchi, 1987). The RNA pellet was dissolved in 4 M LiCl to remove polysaccharides and the RNA was recovered by centrifugation. It was quantified by measuring the absorbance at 260 nm and stored at -70°C. Twenty µg of total RNA was electrophoresed on a formaldehyde-agarose gel and transferred to a Nytran membrane using standard techniques (Rosen *et al.,* 1990). The baked filters were prehybridized and hybridized at 42°C in 50% formamide-containing buffers as previously described (Qulali *et al.,* 1991), with random hexamer radiolabeled rat ALDH2 cDNA. Blots were washed 3 times in 0.1X SSC and 0.5% SDS at 55°C for 20 minutes. The membranes were exposed to X-ray film with two intensifying screens at -70°C.

Dna Preparation and Southern Blotting

Rat liver, kidney, and spleen were quick-frozen in liquid nitrogen, then pulverized in a chilled mortar and pestle. 100 mg of the powder was then digested in Tris buffer containing 0.5% sodium dodecylsulfate and 100 µg/ml of proteinase K overnight at 50°C. The solution was then extracted with phenol and chloroform, and the high molecular weight DNA was spooled from a solution that was made 2.5 M in ammonium acetate and 70% in ethanol. Ten µg of the DNA was then subjected to digestion with Msp I or Hpa II or mock digested overnight at 37°C. The digests were fractionated on a 1.2% agarose gel, Southern blotted to nitrocellulose, and hybridized with the radiolabelled human pALDH600 fragment. The final wash was 0.5x SSC, 0.05% sodium dodecylsulfate at 55°C. The filter was then exposed to x-ray film with one intensifying screen at -70°C.

Polymerase Chain Reactions

Polymerase chain reactions (PCR) were performed in reactions using 2.5 U Taq polymerase, 0.25 mM of each dNTP, 10 mM ß-mercaptoethanol, 1.5 mg/ml bovine serum albumin, 1% dimethylsulfoxide, 20-60 pmole of each primer, and 1 µg template DNA in 100 µl of amplification buffer (67 mM Tris-HCl, pH 8.4, 6.7 mM $MgCl_2$, 16.7 mM $(NH_4)_2SO_4$). Samples were subjected to 30 cycles consisting of denaturation (at 94°C for 2 min), annealling of primers to the template (at 5°C less than the melting temperature (T_m) of the oligonucleotides for 2 min), and extension (at 72°C for 2 min).

Plasmid Constructions

Two oligonucleotide primers (DWC 22 and DWC 25) containing Bam HI and Hind III sites were synthesized to amplify the human *ALDH2* gene upstream sequences from -1 to -600 from the ATG translation start site (Hsu *et al.,* 1988). The 600 bp amplification product was digested with Bam HI and Hind III, ligated into pUC19, and the resultant plasmid was designated pALDH600. pALDH600-luc was constructed by digesting pALDH600 with HindIII and BamHI and ligating the 600 bp fragment into pXP2 (de Wet *et al.,* 1987). pALDH600-CAT was created by ligating the same *ALDH2* promoter fragment into the respective sites of pUC-CAT. All plasmid sequences were confirmed by restriction endonuclease digestion and double stranded DNA sequencing (Chen and Seeburg, 1985).

Transfection of Tissue-Culture Cells

All cells were grown in modified Eagle's medium (MEM) supplemented with 10% fetal bovine serum (FBS), 100 µg/ml streptomycin, and 63 µg/ml penicillin G. The day before transfection, the cells were plated at a density determined to give approximately 90% confluency at the time of preparing the cell extract (10^6 cells/100 mm dish). HeLa, H4IIE-C3, and HepG2 cells were transfected with 10 µg of reporter plasmid and an internal control vector (pSV2CAT) by calcium phosphate precipitation (Graham and van der Eb, 1973). Four hours later the cells were exposed to PBS containing 15% glycerol for 3 min. The cells were rinsed twice with PBS and fresh MEM with serum was added.

Twenty-four to forty-eight hours after transfection, cells were washed twice with phosphate buffered saline and lysed by resuspending cells in 150 µl of a buffer containing 25 mM Tris, pH 7.8, 2 mM EDTA, 20 mM dithiothreitol, 10 % glycerol, and 1% Triton X-100. Fifty µl of cell extract was incubated with luciferase assay reagent (20 mM Tricine, 1.07 mM $(MgCO_3)_4Mg(OH)_2$ $5H_2O$, 2.67 mM $MgSO_4$, 0.1 mM EDTA, 33.3 mM DTT, 270 µM coenzyme A, 470 µM luciferin, 530 µM ATP, pH 7.8) based on the original protocol of deWet (de Wet *et al.,* 1987). The activity of luciferase in relative light units was determined for each cell line based on a 10 sec delay and a 10 sec incubation for HeLa extracts and a 60 sec incubation for H4IIE-C3 and HepG2 extracts. To control for transfection efficiency, CAT activity was measured by incubating 50 µl of the extract with 1 mM acetyl CoA and 0.2 µCi of [C^{14}] chloramphenicol. HeLa cell extracts were incubated for 1 hour and H4IIE-C3 and HepG2 cell extracts were incubated for 4 hours. Chloramphenicol and its acetylated forms were extracted in ethyl acetate and separated by thin layer chromatography. Percent conversion of chloramphenicol to its acetylated products was quantified on an AMBIS β-scanner.

RESULTS AND DISCUSSION

Previous analyses of ALDH2 expression in various tissues has been complicated by the multiplicity of enzymes able to oxidize short chain aldehydes, even at low concentrations. For example, cytosolic ALDH1 has a K_m for acetaldehyde of about 30 μM, and hence might be active at 25% of V_{max} under the conditions of some ALDH2 assays. Thus, enzyme assays for "low K_m" ALDH will tend to overestimate the activity of ALDH2, although indeed, most tissues with mitochondria express low levels of ALDH2. Immunologic assays are complicated by the fact that the ALDH enzyme family contains a large number of members with significant sequence conservations, including strict conservation of two motifs (Lindahl, 1992). Polyclonal antisera are therefore likely to cross-react with several ALDHs. For instance, ALDH1, ALDH2, and $ALDH_x$ (ALDH5) share at least 65% sequence identity and must share many epitopes (Hsu and Chang, 1991). We therefore approached the question of tissue-specific expression by determining the steady state levels of ALDH2 mRNA in a variety of rat tissues (Figure 1) and in cell lines of rat and human origin.

By far the highest mRNA level was found in liver, with lower levels in the kidney, lung, and heart. mRNAs of the expected size for ALDH2 were also detected on Northern blots of several hepatoma cell lines. The highest levels were found in rat H4IIE and H4IIE-C3 hepatoma cells, with lower levels in the human HepG2 hepatoblastoma cells. No detectable

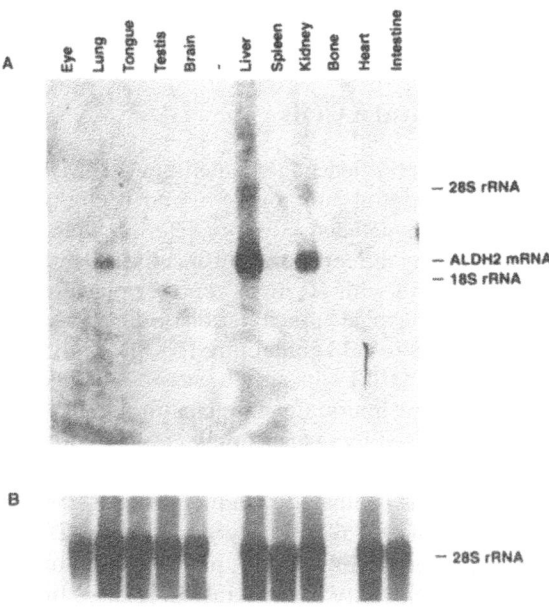

Figure 1. Total RNA was purified from the tissues noted and fractionated in a formaldehyde agarose gel. The RNA was blotted to nylon membranes and hybridized cDNAs. Panel A. Filter hybridized with a rat liver ALDH2 cDNA. The location of the 28 and 18S rRNAs are shown in the right margin. The dash indicates an empty lane. Panel B. The filter shown in panel A was stripped and rehybridized with a cDNA corresponding to the rat 28S rRNA to demonstrate that the amounts of RNA transfered to the filter were similar for each tissue, except for bone and eye. From (Dipple and Crabb, 1993), used with permission.

Human ALDH2 genomic sequence

```
GTCAACTGGG CTCCATTCAT TCTTTCCATT TCTCTAACAC GTGCCAGGTG GTCTCATCTC

CGGGCCTTTG CCCTTGCTGT TCCCTGTCAT CATTCAGGTC TCACTTGTCA TTTCCTGACC

ATGGTACTTA TAAAAGCAGT GCCGTCTGCC CCATCCATGT CACCTCGTTC ATCTCCTTCA

CCTCCGAAAT GATCTCGCTT TTCCCTTTAC GGCCGGTCTC TTCACCTGGA GCATCAGCCG

GGGAGGTCAG GGTCCCCTGG CTCGGGCCTG TTCACATTGG GGTCAAAGGC ACACATTGGG

GGCTCAACCA AGGCGAGCTC GTTCGCGGGG CCGGGTCTTT CCGCACAGGC GGAGGGCGGT

GGCGGGCGCG GAGGCGTCGC GCGAGCCAGG GGGCACGCAC GGGCCGGGGG TACCTAGCGC

CACCCGCTTC GCTTGCATCA GCTGCGCGCC CCATCCCTAT TAATGGTAGA GGCAGCCCGC

CCCCGGCCCG CCCCCGCCTT TCCATTGGCT GCCGCGCGGG GCGGGGAGCG GGGTCGGCTC
```

```
              AGTGGCCCTG AGACCCTAGC TCTGCTCTCG GTCCGCTCGC TGTCCGCTAG CCCGCTGCG ATG
Human cDNA                            G GTCCCCTCCC TGTCCGCTAG CCCGCTGCG ATG
Human S1 site                      TCTCG GTCCGCTCGC TGTCCGCTAG CCCGCTGCG ATG
Rat cDNA                           GCTTT ATC-GCTAAG -CTCCGCTCAG--TTCAGC- ATG
Bovine       A GTCAGGCTCC AGCCGCCCACTCGA GCCCCCTTGT CCCAGCTCCG GCCTCGGC- ATG
cDNA
```

Figure 2. The DNA sequence of the ALDH2 gene as determined by Hsu and coworkers (Hsu *et al.*, 1988). Lines above the sequence denote the location of GpC dinucleotides and the stippled pattern denotes the location of CpG dinucleotides. Below the human gene sequence, the 5' untranslated regions of the longest human, rat, and bovine cDNAs are shown, as well as the sequence up to a site determined by S1 nuclease analysis of human liver mRNA. References to the cDNA sequences and the human S1 cleavage site are given in the text.

ALDH2 mRNA was found in the human HuH7 and PLC/PRF/5 hepatoma cell lines, HeLa cells, or in simian CV-1 cells. The H4IIE, H4IIE-C3, and HepG2 cells expressed enzyme protein on Western blots at levels that paralleled the mRNA levels (not shown), and they had detectable low K_m ALDH activity in enzyme assays. Thus, this gene was not expressed constitutively in either rat tissues or in cultured cell lines, indicating the likelihood that tissue-specific factors exist that govern the expression of this gene.

The start sites of the ALDH2 genes are not known unequivocally (Figure 2). The rat cDNA extended to 30 bp 5' to the initiation ATG codon (Farres *et al.*, 1989), while the bovine cDNA had the longest 5' untranslated region of 53 nucleotides (Farres *et al.*, 1989). The bovine ALDH2 start site was mapped by primer extension to a site corresponding to either -53 or -57 bp upstream from the ATG (Guan and Weiner, 1990). In the case of the human ALDH2 gene, one laboratory cloned a cDNA from a fetal muscle library that extended to 30 bp 5' to the start codon (Braun *et al.*, 1987). Analysis of human liver ALDH2 mRNA by S1 nuclease protection assays indicated the presence of an S1 cleavage site that would be consistent with a 34 bp 5' untranslated region. A primer extension product of 240 bp that would correspond to a start site 94 bp longer than the site identified by S1 mapping was also noted (Hsu *et al.*, 1988). To help clarify this discrepancy, we performed primer extension assays using human liver poly A^+ mRNA and the same primer as used previously by others and detected extension products that would correspond to 5' untranslated regions of 52 or 77 bp in length (not shown). Thus, although it is not certain where transcription initiates in the human gene, several lines of evidence suggest that the transcription start site resides within the first 80 bp or so upstream of the initiation ATG. We therefore began studies on the possible promoter activity of the 600 bp of 5' flanking region of the human *ALDH2* gene. It remains possible that additional exons and an alternative promoter exist upstream of this region.

A prominent feature of the region of the *ALDH2* 5' flank is its high content of G+C and its high number of CpG dinucleotides (Figure 2). CpG dinucleotides are the substrates for vertebrate DNA methylases. The resulting methylated cytosines can spontaneously deaminate to form thymine residues that are not removed by DNA repair systems; thus, the frequency of CpG dinucleotides is generally lower in the genomic DNA than that predicted from the overall abundance of guanine and cytosine bases. The exception to this rule is the so-called HTF (*Hpa* II *tiny fragment*) or CpG island. These short regions (0.5-2.0 kb) are G+C rich and the CpGs are not methylated; thus, the CpG content has remained high. Constitutively expressed "housekeeping" genes almost always are associated with a CpG island at their 5' end (Larsen *et al.*, 1992). Tissue-specific genes are associated with these islands less often and the island may occur with the body or 3' regions of the gene instead of the 5' end.

To examine experimentally whether this region of the genome is in fact a CpG island, we prepared genomic DNA from three rat tissues that differed markedly in their expression of ALDH2 mRNA, namely liver, kidney, and spleen (see Figure 1). The DNAs were digested with two restriction enzymes that differ in their sensitivity to methylation of cytosines in the recognition sequence. Hpa II will not cleave the sequence 5'-CCGG-3' if the internal cytosine is methylated, while Msp I will cleave methylated or unmethylated sites. The digested DNA was then fractionated on an agarose gel, Southern blotted, and probed with the human *ALDH2* 5' flanking region contained in pALDH600 (since the rat *ALDH2* gene has not been cloned). As shown in Figure 3, both enzymes cut *ALDH2* sequences in each of

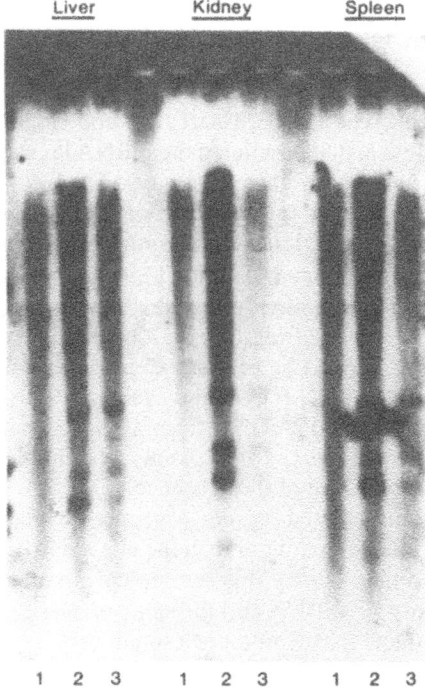

Figure 3. Genomic DNA isolated from liver, kidney, and spleen was digested with no enzyme (lanes 1), Msp I (lanes 2), or Hpa II (lanes 3). The DNA was fractionated on an agarose gel, transferred to nitrocellulose, and hybridized with the cloned human DNA fragment (pALDH600) shown in Figure 2. The bands generated by Msp I digestion were also generated by Hpa II digestion, indicating that the recognition sites were not heavily methylated. From (Dipple and Crabb, 1993), used with permission.

the tissues examined. This indicated that the sites involved were not (heavily) methylated, and supports the contention that the sites lie within a CpG island. It remains to be determined whether the human *ALDH2* 5' flank has similar properties; however, the human homologues of almost all rat or mouse genes that are associated with a CpG island are also associated with a CpG island, although the converse is not true (Aissani and Bernardi, 1991; Antequera and Bird, 1993).

Since it is somewhat unusual for genes associated with CpG islands to be expressed in as tissue-specific a fashion as *ALDH2* is, we have initiated promoter studies on the first 600 bp of 5' flanking region. The human *ALDH2* 5' flanking region was amplified by the polymerase chain reaction (PCR) based upon the sequence reported in the literature (Hsu *et al.*, 1988). When this fragment was ligated upstream of reporter genes (chloramphenicol acetyltransferase or luciferase), the fragment had substantial promoter activity (similar to that of the SV40 promoter in the case of the luciferase construct). The construct was active in a variety of cell lines, including HeLa, HepG2, and H4IIE-C3 cells. Thus, although it is difficult to directly compare the expression of promoter-reporter genes in different cell lines (because of differences in transfection efficiency), the promoter fragment was expressed in cells regardless of whether the endogenous *ALDH2* gene was active. In addition, preliminary studies showed that all tissues and cells surveyed to date contain nuclear factors that bind to the proximal *ALDH2* promoter. It is possible, therefore, that this proximal region, which falls within the CpG island, is in fact a constitutive promoter. Perhaps this constitutive promoter is responsible for the low level of enzyme activity that has been detected in many tissues by gel electrophoresis techniques. These initial observations suggested that additional regions of the gene, either upstream or downstream, control the proximal region to determine whether the gene is active in a given cell type. These additional regions are therefore likely to include transcriptional suppressors. This phenomenon may also be explained by differences in the methylation status of CpG islands between tissues and cultured cells. It has been shown that non-essential genes may be silenced in cultured cells and that their CpG islands become methylated (Antequera *et al.*, 1990). Whether this is the case and whether methylation might in part regulate this gene remains to be determined.

In conclusion, the *ALDH2* gene appears to be constitutively active in the liver, but is expressed at much lower levels in other tissues tested. The gene itself is associated with a CpG island. Preliminary results suggest that the 5' flanking region of the gene can act as a strong promoter in a variety of cell types, possibly through interactions with ubiquitous DNA binding proteins. It remains to be determined what other control mechanisms determine the highly tissue-specific expression of this gene.

ACKNOWLEDGEMENTS

The support of grants from the NIAAA to DWC (AA06434 and AA00081) and to KMD (AA05310), as well as a grant from the Indiana University School of Medicine Biomedical Research Committee to MJS are gratefully acknowledged. We also thank Dr. Henry Weiner for the rat ALDH2 cDNA and Dr. Rebecca Chan for helpful discussions.

REFERENCES

Aissani, B. and Bernardi, G. (1991) CpG islands: features and distribution in the genome of vertebrates. *Gene* **106**, 173-184.

Antequera, F. and Bird, A. (1993) Number of CpG islands and genes in humans and mouse. *Proc Nat Acad Sci USA* **90**, 11995-11999.

Antequera, F., Boyes, J., and Bird, A. (1990) CpG methylation in cultured cells. *Cell* **62**, 503-514.

Braun, T., Bober, E., Singh, S., Agarwal, D.P., and Goedde, H.W. (1987) Isolation and sequence analysis of a full length cDNA clone coding for human mitochondrial aldehyde dehydrogenase. *Nucleic Acids Research* **15**, 3179.

Chen, E.Y. and Seeburg, P.H. (1985) Supercoil sequencing: a fast and simple method for sequencing plasmid DNA. *DNA* **4**, 165-170.

Chomczynski, P. and Sacchi, N. (1987) Single-step method of RNA isolation by acid guanidinium thiocyanate-phenol-chloroform extraction. *Anal Biochem* **162**, 156-159.

Crabb, D.W., Edenberg, H.J., Bosron, W.F., and Li, T-K. (1989) Genotypes for aldehyde dehydrogenase deficiency and alcohol sensitivity. The inactive *ALDH2*2* allele is dominant. *J Clin Invest* **83**, 314-316.

de Wet, J.R., Wood, K.V., DeLuca, M., Helinski, D.R., and Subramani, S. (1987) Firefly luciferase gene: structure and expression in mammalian cells. *Mol Cell Biol* **7**, 725-737.

Dipple, K.M. and Crabb, D.W. (1993) The mitochondrial aldehyde dehydrogenase gene resides in an HTF island but is expressed in a tissue-sepcific manner. *Biochem. Biophys. Res. Comm.* **193**, 420-427.

Enomoto, N., Takase, S., Takada, N., and Takada, A. (1991a) Alcoholic liver disease in heterozygotes of mutant and normal dehydrogenase-2 genes. *Hepatology* **13**, 1071-1075.

Enomoto, N., Takase, S., Yasuhara, M., and Takada, A. (1991b) Acetaldehyde metabolism in different aldehyde dehydrogenase 2 genotypes. *Alcoholism: Clin Exp Res* **15**, 141-144.

Farres, J., Guan, K.-L., and Weiner, H. (1989) Primary structures of rat and bovine liver mitochondrial aldehyde dehydrogenases deduced from cDNA sequences. *Eur. J. Biochem.* **180**, 67-74.

Graham, F.L. and van der Eb, A.J. (1973) A new technique for the assay of infectivity of human adenovirus 5 DNA. *Virology* **52**, 456-467.

Guan, K. and Weiner, H. (1990) Sequence of the precursor of bovine liver mitochondrial aldehyde dehydrogenase as determined from its cDNA, its gene, and its functionality. *Arch Biochem Biophys* **277**, 351-360.

Harada, S., Agarwal, D.P., and Goedde, H.W. (1981) Aldehyde dehydrogenase deficiency as cause of facial flushing reaction to alcohol in Japanese. *Lancet* **ii**, 982.

Harada, S., Misawa, S., Agarwal, D.P., and Goedde, H.W. (1980) Liver aldehyde dehydrogenase in Japanese-isoenzyme variation and its possible role in alcohol intoxication. *Am J Hum Genet* **32**, 8-15.

Hsu, L.C., Bendel, R.E., and Yoshida, A. (1988) Genomic structure of the human mitochondrial aldehyde dehydrogenase gene. *Genomics* **2**, 57-65.

Hsu, L.C. and Chang, W.C. (1991) Cloning and characterization of a new functional human aldehyde dehydrogenase gene. *J Biol Chem* **266**, 12257-12265.

Larsen, F., Gundersen, G., Lopez, R., and Prydz, H. (1992) CpG islands as gene markers in the human genome. *Genomics* **13**, 1095-1107.

Lindahl, R. (1992) Aldehyde dehydrogenases and their role in carcinogenesis. *Crtitical Reviews in Biochem Mol Biol* **27**, 283-335.

Qulali, M., Ross, R.A., and Crabb, D.W. (1991) Estradiol induces class I alcohol dehydrogenase activity and mRNA in kidney of female rats. *Arch Biochem Biophys* **288**, 406-413.

Rosen, K.M., Lamperti, E.D., and Villa-Komaroff, L. (1990) Optimizing the Northern blot procedure. *Biotechniques* **8**, 398-403.

Yoshida, A., Huang, I-Y., and Ikawa, M. (1984) Molecular abnormality of an inactive aldehyde dehydrogenase variant commonly found in Orientals. *Proc Nat Acad Sci USA* **81**, 258-261.

17

TRANSGENESIS OF THE ALDEHYDE DEHYDROGENASE-2 (*ALDH2*) LOCUS IN A MOUSE MODEL AND IN CULTURED HUMAN CELLS

Cheng Chang, Jerry Mann and Akira Yoshida

Beckman Research Institute of the City of Hope
Duarte, CA 91010

INTRODUCTION

In 1657, Sir William Harvey stated: "For it has been found in almost all things, that what they contain of useful or applicable nature is hardly perceived unless we are deprived of them, or they become deranged in some way". Until recently, Harvey's approach for elucidation of biological roles of a given enzyme (or protein) was feasible only in cases of rare genetic disorders. It has become possible to control the expression of a given protein via genetic manipulation ("Transgenesis") *in vivo* as well as in cultured cells, and elucidate the ultimate roles of the protein. Developmental, physiological and behavioral roles of the proteins encoded by several genes were analyzed utilizing this technology.

Mammalian alcohol dehydrogenase (ADH) and aldehyde dehydrogenase (ALDH) systems are extremely complex. At the present time, seven non-allelic ADH genes and seven ALDH genes have been identified in humans. However, biological roles of the various isozymes encoded by these genes are obscure. We shall be able to answer the question through the transgenesis study. As the first subject of this line of study, we selected the *ALDH2* locus.

The severe genetic deficiency of the mitochondrial ALDH2 activity is related to alcohol sensitivity, and consequently to the low risk in developing alcoholic diseases in humans (Shibuya and Yoshida, 1988). Since the ALDH2 deficient individuals exhibit no adverse physiological problems, except for alcohol sensitivity, the ALDH2 may not be essential for survival. Alternatively, the low residual activity might be sufficient to fulfill the biological role(s) of ALDH2.

The mouse C57BL/6J strain exhibits a high alcohol preference; i.e. prefers to drink 10% ethanol rather than water in a free-choice setting (McClearn and Rodgers, 1961). We are undertaking to inactivate, totally and partially, the ALDH2 in the animal by genetic manipulation, and examine the change of biological and behavioral characteristics in the mutant mouse strains. In addition, we are currently undertaking to suppress the expression of ALDH2 in cultured human cells, and examine the consequence.

Enzymology and Molecular Biology of Carbonyl Metabolism 5
Edited by H. Weiner *et al.*, Plenum Press, New York, 1995

EXPERIMENTS AND RESULTS

Cloning and Structure of the Mouse ALDH2 Gene

Thus far, only the human *ALDH2* gene was cloned and analyzed (Hsu et al., 1988). The mouse *ALDH2* gene was cloned by screening a mouse (C57BL/6J) genomic library using the human *ALDH2* cDNA (Hsu et al., 1985) and synthetic oligonucleotides as probes. Nucleotide sequence analysis revealed an open reading frame of 1560 bp encoding 519 amino acid residues (Fig. 1). The mouse gene is composed of 13 exons and 12 introns and spans approximately 26 kbp. Organization of the *ALDH2* gene is highly similar between human and mouse (Table I). Except for a short signal peptide sequence (17 amino acid residues in human and 19 residues in the mouse) the amino acid identity is about 96% in human and mouse enzymes (both 500 amino acid residues).

Transgenesis in a Mouse System

There are two types of procedures, i.e., non-homologous integration of a mutant gene and targeting knockdown of an existing gene, for creating a mutant mouse.

In the non-homologous integration method, an artificial mutant gene is introduced into fertilized eggs to create a mutant mouse which contains the mutant gene in addition to

```
AACGCTGCAGTCGCCCTCCGATCGGCAAGGCTTCTCTCGGCTCCGTTCGGCTCGGCTCGCCCATTTCAGTTCAGTTCGGGTCAGTTAAGCTCCGCTCAGT   100
TCAGCATGCTGCGCGCCGCACTCACCACTGTCCGCCGCGGACCGCGCCTGAGCCGCCTGTTGTCCGCCGCCGCCACCAGCGCGGTGCCAGCCCCCAACCA   200
     M  L  R  A  A  L  T  T  V  R  R  G  P  R  L  S  R  L  L  S  A  A  A  T  S  A  V  P  A  P  N  H     32
                  1••2
TCAGCCTGAGGTCTTCTGCAACCAGATCTTCATTAACAATGAGTGGCACGACGCCGTCAGCAGGAAAACATTTCCCACCGTCAACCCTTCCACAGGGGAG   300
  Q  P  E  V  F  C  N  Q  I  F  I  N  N  E  W  H  D  A  V  S  R  K  T  F  P  T  V  N  P  S  T  G  E     65
                      2••3
GTCATCTGCCAGGTGGCCGAAGGGAACAAGGAGGACGTAGACAAGGCAGTGAAGGCTGCTCGTGCAGCCTTCCAGCTGGGCTCGCCCTGGCGCCGCATGG   400
 V  I  C  Q  V  A  E  G  N  K  E  D  V  D  K  A  V  K  A  A  R  A  A  F  Q  L  G  S  P  W  R  R  M      98
                                                     3••4
ATGCATCTGACCGGGGCCGGCTGTTGTACCGATTGGCGGATCTCATTGAACGGGACCGGACCTACCTAGCGGCCTTGGAGACCCTGGACAACGGCAAGCC   500
D  A  S  D  R  G  R  L  L  Y  R  L  A  D  L  I  E  R  D  R  T  Y  L  A  A  L  E  T  L  D  N  G  K  P    132
                                      4••5
TTATGTCATCTCGTACCTGGTGGATTTGGACATGGTCCTGAAATGTCTCCGCTATTACGCTGGCTGGGCTGACAAGTACCATGGGAAAACCATTCCCATC   600
  Y  V  I  S  Y  L  V  D  L  D  M  V  L  K  C  L  R  Y  Y  A  G  W  A  D  K  Y  H  G  K  T  I  P  I    165
                                               5••6
GACGGCGACTTCTTCAGCTATACCCGCCATGAGCCTGTGGGCGTGTGTGGACAGATCATTCCGTGGAACTTCCCGCTCCTGATGCAAGCATGGAAACTGG   700
 D  G  D  F  F  S  Y  T  R  H  E  P  V  G  V  C  G  Q  I  I  P  W  N  F  P  L  L  M  Q  A  W  K  L     198
                                                                    6••7
GCCCAGCCCTGGCAACCGGGAACGTGGTGGTGATGAAGGTGGCCGAGCAGACACCGCTCACCGCGCTCTACGTGGCCAACTTGATCAAGGAGGCAGGCTT   800
G  P  A  L  A  T  G  N  V  V  V  M  K  V  A  E  Q  T  P  L  T  A  L  Y  V  A  N  L  I  K  E  A  G  F    232
TCCCCCTGGCGTGGTCAATATCGTTCCCGGATTCGGCCCTACCGCCGGGGCTGCCATCGCATCCCATGAGGGTGTGGACAAAGTGGCGTTCACAGGCTCC   900
  P  P  G  V  V  N  I  V  P  G  F  G  P  T  A  G  A  A  I  A  S  H  E  G  V  D  K  V  A  F  T  G  S    265
   7••8
ACGGAGGTTGGTCACCTAATCCAGGTGGCCGCCGGGAGCAGCAACCTCAAGAGAGTAACCCTGGAGCTGGGGGGAAAGAGTCCCAACATCATCATGTCCG  1000
  T  E  V  G  H  L  I  Q  V  A  A  G  S  S  N  L  K  R  V  T  L  E  L  G  G  K  S  P  N  I  I  M  S    298
    8••9
ACGCTGACATGGACTGGGCTGTGGAGCAGGCCCACTTTGCCCTGTTCTTCAACCAGGGCCAGTGCTGCTGCGCAGGCTCCCGGACCTTCGTGCAGGAGAA  1100
D  A  D  M  D  W  A  V  E  Q  A  H  F  A  L  F  F  N  Q  G  Q  C  C  C  A  G  S  R  T  F  V  Q  E  N    332
                                                                       9••10
TGTGTATGACGAATTCGTGGAACGCAGCGTGGCTCGGGCCAAGTCTCGGGTGGTGGGGAACCCCTTCGACAGCCGGACGGAGCAGGGGCCTCAGGTGGAT  1200
  V  Y  D  E  F  V  E  R  S  V  A  R  A  K  S  R  V  V  G  N  P  F  D  S  R  T  E  Q  G  P  Q  V  D    365
GAAACTCAGTTTAAGAAGATCCTCGGCTACATCAAATCGGGACAACAAGAAGGGGCGAAGCTGCTGTGTGGTGGGGGCGCTGCCGCGGACCGTGGGTACT  1300
  E  T  Q  F  K  K  I  L  G  Y  I  K  S  G  Q  Q  E  G  A  K  L  L  C  G  G  G  A  A  A  D  R  G  Y    398
                                           10••11
TTATCCAGCCCACCGTGTTCGGGGACGTAAAAGACGGGATGACCATTGCCAAGGAGGAGATCTTTGGACCAGTGATGCAAATCCTCAAATTCAAGACCAT  1400
F  I  Q  P  T  V  F  G  D  V  K  D  G  M  T  I  A  K  E  E  I  F  G  P  V  M  Q  I  L  K  F  K  T  I    432
CGAGGAGGTTGTGGGGCGGGCCAATGATTCTAAGTATGGGCTGGCAGCCGCCGTCTTCACAAAGGACCTGGATAAAGCCAATTACCTGTCCCAAGCTCTG  1500
  E  E  V  V  G  R  A  N  D  S  K  Y  G  L  A  A  A  V  F  T  K  D  L  D  K  A  N  Y  L  S  Q  A  L    465
            11••12
CAGGCTGGCACTGTGTGGATCAACTGCTACGATGTGTTTGGGGCCCAGTCTCCATTTGGGGGCTATAAGATGTCAGGGAGTGGCAGGGAGCTGGGCGAGT  1600
  Q  A  G  T  V  W  I  N  C  Y  D  V  F  G  A  Q  S  P  F  G  G  Y  K  M  S  G  S  G  R  E  L  G  E    498
                       12••13
ATGGCCTGCAGGCGTACACAGAAGTGAAGACGGTTACTGTCAAAGTGCCACAGAAGAACTCGTAAAGCGGCATGCCTGCTTCCTCAGCCCGCACCCGAAA  1700
Y  G  L  Q  A  Y  T  E  V  K  T  V  T  V  K  V  P  Q  K  N  S  -                                       519
ACCCAACAAGATATACTGAGAAAAACCGCCACACACACTGCGCCTCCAAAGAGAAACCCCTTCACCAAAGTGTCTTGGGTCAAGAAAGAATTTTATAAAC  1800
AGGGCGGGGCTGGTGGGGGGAAAGCTCCTGATAAACTGGGTAGGGGATGAAGCTCAGTGCAGACCGATCACGCGTCCAGATGTGCAGGATGCTGCCTTC  1900
AACCTGCAGTCCCTAAGCAGCAAATGAGCAATAAAAATCAGCAGATCAAAGCCA                                                1954
```

Figure 1. Structure of mouse *ALDH2* gene and deduced amino acid sequence.

Table 1. Genomic Orgnaization of Human and Mouse *ALDH2* gene.

Exons	Amino Acid Coded Human	Mouse
1	-17-21	-19-21
2	22-56	22-56
3	57-103	57-103
4	104-130	104-130
5	130-167	130-167
6	168-210	168-210
7	211-248	211-248
8	249-283	249-283
9	283-344	283-344
10	344-399	344-399
11	400-452	400-452
12	452-490	452-490
13	491-500	491-500
Transcription site	-240 nt	-105 nt

Introns	Human	Mouse
1	15.4 kbp	12.7 kbp
2	1.4	0.9
3	2.3	1.2
4	4.5	1.7
5	0.5	0.4
6	0.8	0.7
7	0.6	1.7
8	0.6	0.5
9	5.0	0.5
10	1.6	0.9
11	3.7	1.7
12	5.7	0.8

the original normal gene. The genomic status is analogous to that of the heterozygous variant human subjects.

A more decisive way to inactivate a given gene includes a) construction of a gene replacement vector; b) transfection of embryonic stem cells with the vector; c) injection of the selected cells into blastocysts and implantation of the treated blastocysts into pseudo-pregnant mice; and d) breeding of chimeric male and female mice to produce homozygous and heterozygous mutant mouse strains (Joyner, 1993). The homozygous mutant mouse can no longer produce ALDH2 at all, while the ALDH2 activity of the heterozygous mutant mouse is expected to be 50% of that of the original strain. We used this method for creating mutant mouse strains.

The replacement vector (ca 12.5 kbp) used in the study contained exon 2, a partially deleted exon 3, a positive selection cassette (pPgk Neo), exon 4 with an artificial stop codon, exons 5, 6 and 7, and a negative selection cassette (pMCI-Tk) (Fig. 2).

Embryonic stem cells (approximately 10^7) transfected (electroporation method) with the linearized vector were screened by antibiotics (G418 and 5-bromodeoxyuridine). Genomic DNAs prepared from drug-resistant clones were analyzed by polymerase chain reaction using primers a, b, and c to distinguish random non-homologous integration from homologous recombination (Fig. 2). Out of 132 drug resistant clones

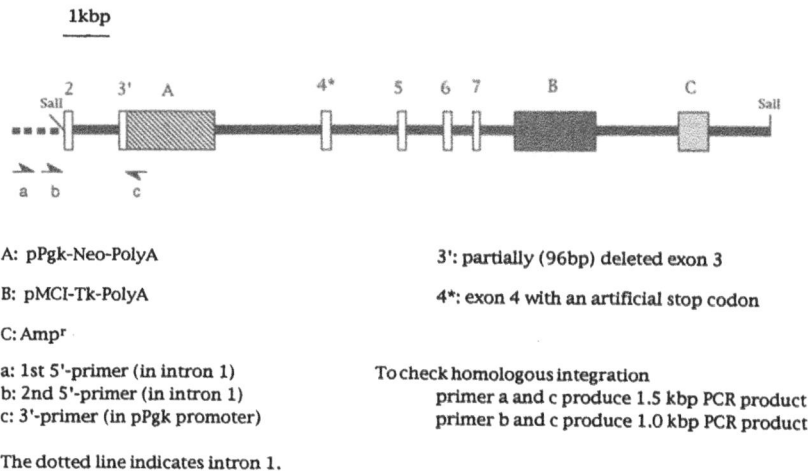

A: pPgk-Neo-PolyA

B: pMCI-Tk-PolyA

C: Amp^r

a: 1st 5'-primer (in intron 1)
b: 2nd 5'-primer (in intron 1)
c: 3'-primer (in pPgk promoter)

The dotted line indicates intron 1.

3': partially (96bp) deleted exon 3

4*: exon 4 with an artificial stop codon

To check homologous integration
    primer a and c produce 1.5 kbp PCR product
    primer b and c produce 1.0 kbp PCR product

Figure 2. Replacement Vector for Inactivation of Mouse *ALDH2* gene.

examined, eight had undergone homologous recombination at one of the *ALDH2* alleles (Fig. 3).

    The DNAs of these clones were digested by Hind III and subjected to Southern blot hybridization using the exon 2 - intron 2 fragment as a probe. All eight clones exhibited 8.0kb fragment which should be generated from the intact original gene and 1.7kb fragment which is expected to be produced from the homologous recombinant gene (Figure is not shown), i.e., these eight transformed embryonic stem cell clones contain one active *ALDH2* allele and one disrupted allele. These heterozygous mutant embryonic stem cells (C57BL/6J strain, black coat color) were injected into blastocysts of SW strain (white coat color). Transplantation of the blastocysts into surrogate mother mice yielded two chimeric male mice with a tiger-like coat pattern.

    The role of ALDH2 in alcohol preference, alcohol sensitivity and other biological and behavioral characteristics, can be elucidated by examining the heterozygous and homozygous mutant strain produced by breeding of chimeric mice with C57BL/6J strain.

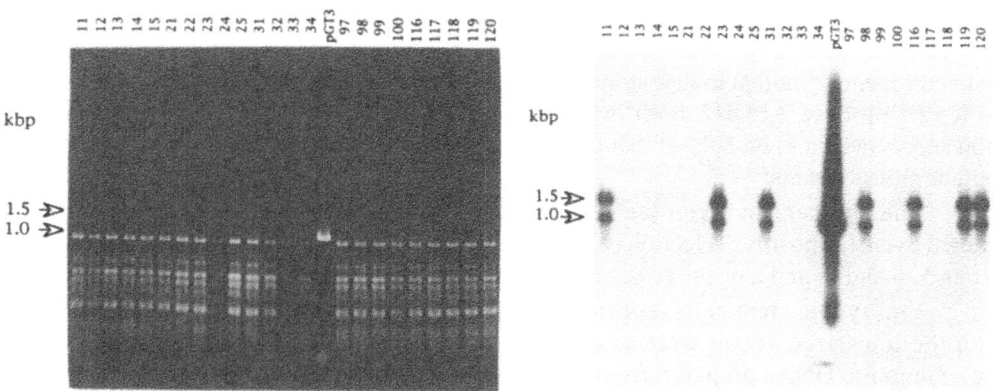

Figure 3. Identification of the clones with homologous recombination.

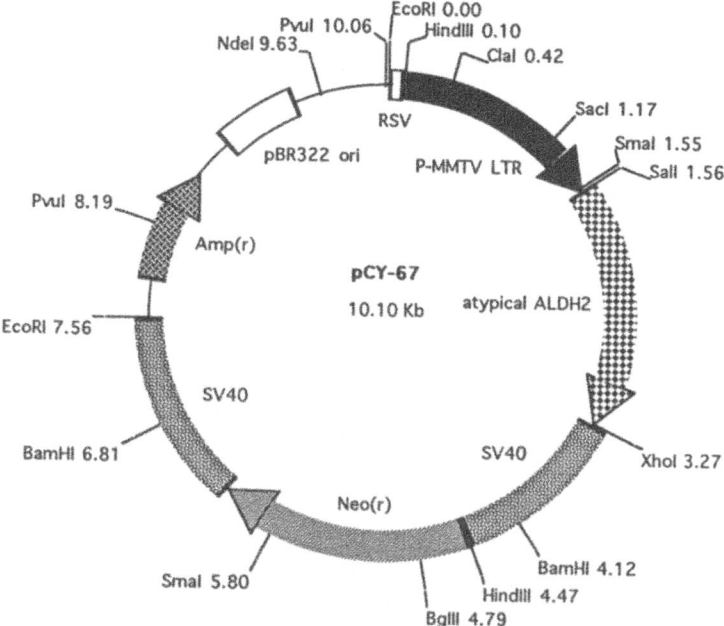

Figure 4. Expression construct of human atypical ALDH2^2 cDNA.

Transgenesis in Cultured Cells

In vivo transgenesis via germ line cells is not feasible in the human system. However, similar methods, i.e., non-homologous integration and homologous targeting knockdown, can be used in cultured somatic human (and other) cells.

Mutant cells which are analogous to the heterozygous atypical *ALDH2$^{1/}$ALDH2^2* status, can be created by the following procedures.

First, a full-length, atypical human ALDH2^2 cDNA, which has G → A nucleotide base change at nt 1510 counting from adenine of the chain initiation codon (Ikawa et al., 1983; Hsu et al., 1985) was created by site-specific mutagenesis of the full-length, normal human ALDH2^1 cDNA. The atypical ALDH2^2 cDNA (about 1.6 kbp) was incorporated into the eukaryotic expression vector pMAM-neo (about 8.5 kbp, Clontech Lab). The expression construct, pCY-67 (Fig. 4), was introduced into the human hepatoma cells (Hep G2 cell line) by the lipofection method (Felgner et al., 1987), and the transfected cells were screened in the presence of antibiotics G418. The PCR-based analysis indicated that the transfected cells contained the normal ALDH2^1 genes and the expression constructs with the atypical ALDH2^2 cDNA.

Complete inactivation of *ALDH2* gene in somatic human cells, hepatoma and other established cell lines is feasible by knocking down the gene using the replacement vector which is similar to that used in the mouse system described above. In order to inactivate both alleles existing in a pair of autosomal chromosome-12, in somatic cells, two types of replacement vectors, one with neo-resistant screening cassette and the other with Hyg-resistant cassette, were constructed. The procedures for screening and confirmation of the homologous integration are similar to that used for mouse embryonic stem cells, and two steps of gene inactivation have to be carried out.

DISCUSSION

At present, chimeric male mice were produced, and their off-springs were generated from C57BL/6J female mice. Breeding of these off-spring is required to obtain homozygous and heterozygous mutant strains, which will be used for the study of biological and behavioral characteristics. Since the chimeric mice exhibited no reproductive problem, the lack of the *ALDH2* gene in the hemizygous germ line cells does not seem to impair the reproductive process.

The naturally existing, severe ALDH2 deficiency in humans, i.e., the homozygous atypical $ALDH2^2/ALDH2^2$ and the heterozygous atypical $ALDH2^1/ALDH2^2$ subjects, is not associated with any biological problems, except for alcohol sensitivity. Growth rate and morphological changes were not observed in the hepatoma cells transformed with the atypical human ALDH2^2 cDNA construct. We shall be able to examine biological effects of alcohol and acetaldehyde using the cultured heterozygous mutant (with one active gene and one inactive gene) and homozygous mutant (with an inactive gene only) cells.

The approach described in the mouse system can be used for the study of biological roles of other aldehyde dehydrogenase and alcohol dehydrogenase isozymes. As indicated by the studies of many human genetic abnormalities, some genes (for example, the gene for serum cholinesterase and the gene for albumin) are not essential to the development, survival and reproduction (Nogueria et al., 1990; Ruttner et al., 1988). It is conceivable that some alcohol dehydrogenase and aldehyde dehydrogenase isozymes might not be essential for survival. Since the biological roles of most alcohol dehydrogenase and aldehyde dehydrogenase isozymes are not clear, the *in vivo* transgenesis in the animal model and the ex-vivo transgenesis of various types of cultured human cells have a merit to be pursued.

ACKNOWLEDGEMENTS

This study was supported in part by NIH Grant Number AA09727.

REFERENCES

Felgner, P.L., Gadek, T.R., Holm, M., Roman, R., Chan, H.W., Wenz, M., Northrop, J.P., Ringold, G.M., and Danielsen, M. (1987) Lipofection: A highly efficient, lipid-mediated DNA-transfection procedure. Proc. Natl. Acad. Sci. U.S.A. *84*, 7413.

Hsu, L.C., Bendel, R.E., and Yoshida, A. (1988) Genomic structure of the human mitochondrial aldehyde dehydrogenase gene. Genomics 2: 57.

Hsu, L.C., Tani, K., Fujiyoshi, T., Kurachi, K., and Yoshida, A., (1985) Cloning of cDNAs for human aldehyde dehydrogenase 1 and 2. Proc. Natl. Acad. Sci. U.S.A. 82: 3771.

Ikawa, M., Impraim, C.C., Wang, G., and Yoshida, A., (1983) Isolation and characterization of aldehyde dehydrogenase isozymes from usual and atypical human livers. J. Biol. Chem. 258: 6282.

Joyner, A.L. (1993) "Gene Targeting" Oxford Univ. Press, Oxford.

McClearn, G.E., and Rodgers, D.A. (1981) Genetic factors in alcohol preference of laboratory mice. J. Comp. Physiol. Psychol. 54: 116.

Nogueira, C.P., McGuire, M.C., Graeser, C., Bartels, C.F., Arpagaus, M., Van de Speik, A.F., Linghtstone, H., Lockridge, O., and LaDu, B.N. (1990) Identification of a frameshift mutation responsible for the silent phenotype of human serum cholinesterase, gly 117, (44T to GGAG) Am. J. Hum. Genet. 46: 934.

Ruffner, D.E., Dugaiczyk, A., (1988) Splicing mutation in human hereditary analbuminemia. Proc. Natl. Acad. Sci. U.S.A. *85*, 2125.

Shibuya, A., and Yoshida, A. (1988) Genotypes of alcohol metabolizing enzymes in Japanese alcoholic liver disease: A strong association of the usual caucasian type aldehyde dehydrogenase gene (*ALDH2¹*) with the disease. Am. J. Hum. Genet., *43: 744.*

18

CLASS 3 ALDEHYDE DEHYDROGENASE
A NORTHERN PERSPECTIVE IN THE LAND DOWN UNDER

Josette Feimer[1], Yiqiang Xie[1], Koichi Takimoto[2], David Asman[1],
Henry Pitot[2] and Ronald Lindahl[1]

[1]The University of South Dakota School of Medicine
Department of Biochemistry and Molecular Biology
414 E. Clark
Vermillion, SD 57069
[2]McArdle Laboratory for Cancer Research
Department, of Oncology and Pathology
University of Wisconsin Medical School
1400 University Ave.
Madison, WI 53706

INTRODUCTION

The mammalian Class 3 aldehyde dehydrogenase gene (*ALDH-3*) is an excellent model for studying tissue-specific, differential gene expression. Certain normal tissues, such as the cornea and stomach, constitutively express Class 3 ALDH at high levels. However, the *ALDH-3* gene is not expressed in normal mammalian liver. It is inducible in liver by a variety of aromatic hydrocarbon xenobiotics. The normally repressed *ALDH-3* gene is also activated during tumorigenesis in a number of tissues including the liver, colon and mammary tissue. All evidence to date indicates that both constitutive and inducible regulation of the *ALDH-3* gene occurs at the transcriptional level and involves use of both cis-acting elements and trans-acting factors.

The cloning and characterization of rat *ALDH-3* cDNA and gene, including several kilobases of it's 5' flanking region, have allowed us to initiate detailed studies of the regulation of this gene (Jones et al., 1988: Dunn et al., 1988; Asman et al., 1993; Takimoto et al., 1994). Computer analysis of the 5' flanking region of the gene indicates the existence of numerous cis-acting elements known to be involved in both the constitutive and inducible expression of other mammalian genes.

Studies of the inducible expression of Class 3 aldehyde dehydrogenase have been aided greatly by the availability of a number of rat and mouse hepatoma cell lines that express the *ALDH-3* gene at different levels (Lin et al., 1984, 1988; Takimoto et al., 1991; 1992; Vasiliou et al, 1992). Work employing cell lines has demonstrated that xenobiotic-induced expression of Class 3 ALDH can occur via an aromatic hydrocarbon (*Ah*) receptor pathway that involves the ligand-receptor-mediated activation of the *ALDH-3* gene via a cis-acting xenobiotic response element (XRE) (Takimoto et al, 1992; Vasiliou et al., 1993). However, studies of the regulation of constitutive expression of Class 3 ALDH has been limited by the lack of a suitable *in vitro* model system.

Here we report on two aspects of our more recent research: (1) preliminary fine structure analysis of regulation of the *ALDH-3* gene in liver and hepatoma cells; and (2) initial studies of regulation of *ALDH-3* gene expression in the cornea.

CLASS 3 ALDEHYDE DEHYDROGENASE GENE EXPRESSION

By deletion analysis we have identified three functional areas within 5.5 kb of the rat *ALDH-3* 5' flanking region (Takimoto et al., 1994). One area exhibits strong basal promoter activity (-0.0 to -0.4 kb). The remaining upstream regions control a variety of effects of transcription. From -0.4 to approximately -2.5 kb reside sequences that act to inhibit the strong basal promoter activity. The region centered around approximately -3.0 kb contains a positively acting xenobiotic response element surrounded by a unique 17 bp repetitive sequence. In transient transfection assays, this XRE and surrounding repetitive sequence is responsible for *Ah*-mediated *ALDH-3* induction by aromatic hydrocarbons. Both the XRE and repetitive sequence are necessary for maximal xenobiotic induction. Further upstream a second inhibitory may exist that down-regulates the *Ah*-mediated response. Neither the proximal or distal inhibitory regions have been studied in detail.

Initially, we have been interested in a detailed analysis of the basal promoter region. These experiments have been carried out using our battery of rat hepatoma cell lines differing in Class 3 ALDH expression as well as normal and 3-methyl-cholanthrene-induced rat liver samples. DNase I footprinting identifies 4 regions in the proximal 1 kb of the 5' flanking region that specifically bind proteins. These putative transcription factor binding sites include partially overlapping Sp1 and HNF4 sites, a putative Ap1 site and 2 putative NF1 sites (Figure 1). The Sp1/HNF4 region footprints in all cell lines and liver samples examined (HTC, high *ALDH-3* expression; 7777, low *ALDH-3* expression; Clone 9-no *ALDH-3* expression; normal and 3-MC-induced rat liver, no and high *ALDH-3* expression respectively). The putative Ap1 site footprints in all lines except 7777. The proximal NF1 site footprints in all lines tested. The distal NF1 site footprints only in *ALDH-3*-negative cell lines and in 3-MC-induced rat liver. Thus, some cell line and tissue-specificity is apparent, but no consistent footprinting pattern correlates with *ALDH-3* gene expression.

Gel mobility shift assays confirm that the Sp1/HNF4 sites are functional, as are the 2 NF1 sites. However, it is not yet clear the exact identity of the transcription factors interacting with these sites. Competition and/or antibody supershift assays for each site reveal that multiple proteins may be involved, especially at the Sp1/HNF4 site and the two

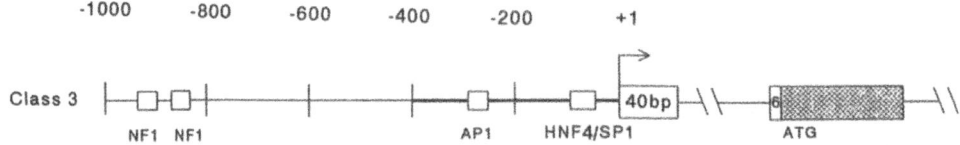

Figure 1. Organization of the Promoter Region of the Rat *ALDH-3* gene. Transcription start site is represented by the arrow. The ATG and hatched box represent the first protein-coding exon. The 5'-untranslated region is represented by the large box downstream from the transcription start site. The heavy line in the 5'-flanking region defines the basal promoter region. Location of the designated transcription factor binding sites in indicated by the small open boxes.

putative NF1 sites. For example, the affinity with which the Sp1 and HNF4 sites are bound by protein may differ significantly among cell lines. Likewise, although the putative Ap1 and NF1 sites are specifically shifted by proteins in nuclear extracts, supershift and competition experiments suggest that the complexing proteins may not be prototypical Ap1 or NF1. Taken together, the footprinting and gel retardation results suggest that although the cis-acting elements and transcription factors found in the basal promoter region are important, other elements and factors, yet to be identified, are also involved in regulating expression of the *ALDH-3* gene in liver.

CLASS 3 ALDEHYDE DEHYDROGENASE GENE EXPRESSION IN THE CORNEA

To begin creating a model for tissue-specific, differential expression of the Class 3 aldehyde dehydrogenase gene, we have developed a unique protocol for culturing rat corneal epithelium. The procedure involves excising 1 mm^2 explants from the corneal limbus and culturing them on Type I collagen-coated 30 mm^2 plastic culture dishes. Explants are grown in 1 ml of media at 37° C, in 5% CO_2. The culture medium is DMEM/F12 with high glucose, 10% fetal calf serum and a variety of growth factors including cholera toxin, EGF, and insulin. Medium is replaced after 24 hrs and daily thereafter. This culture system differs from large mammal corneal culture in that due to its small size, it is impossible to separate the rat corneal epithelium from the underlying stroma/endothelium as is typical for other systems.

In our hands, corneal epithelium proliferates down off the explant and forms a pavement of epithelial cells within 24-36 hours. Confluency is reached within 12-14 days when 3 explants are started per dish. We established that the cells growing out in culture were epithelium and not a heterogeneous population of endothelium, stroma and epithelium by immunoblotting of corneal culture extracts following SDS-gel electrophoresis using a

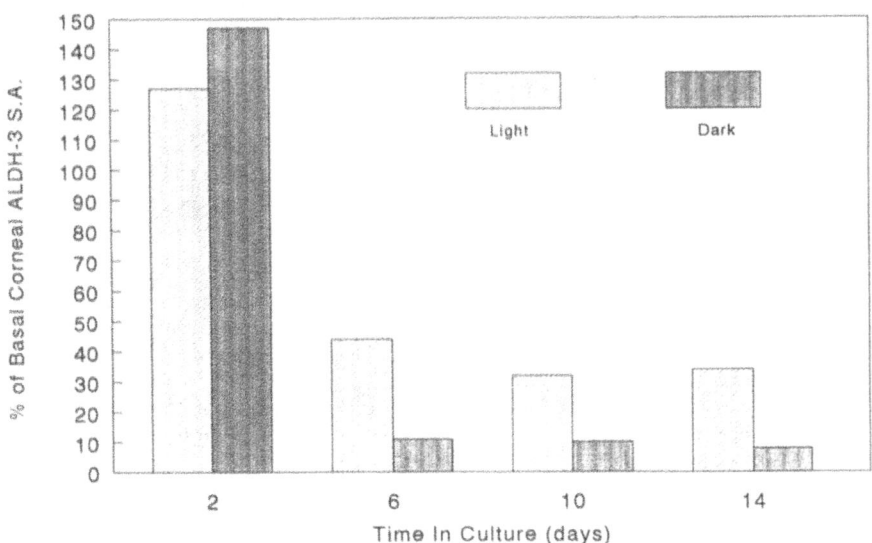

Figure 2. Class 3 Aldehyde Dehydrogenase Activity in Rat Corneal Epithelial Cultures. Corneas were explanted and cultured as described in the presence or absence of light. Enzyme activity was measured at the indicated times using benzaldehyde and NADP$^+$. Activity is expressed as the percent of corneal Class 3 ALDH specific activity at culture time zero.

monoclonal antibody against cytokeratins. Cytokeratins are produced by epithelial cells, but not endothelial or stromal cells. The results indicate that the corneal cultures are predominantly epithelial in composition.

By both histochemistry and specific activity determinations, corneal cultures express Class 3 ALDH activity for several days in culture. However, activity gradually declines from its initial extremely high levels (2000 mIU/mg protein) to near zero over a week in culture. Corneal Class 3- ALDH activity can be restored to high initial levels by treatment of 7-day cultures with 3-methylcholanthrene. This indicates that although the cultures lose Class 3 ALDH activity, they retain the ability to express the *ALDH-3* gene.

Because it is apparent that corneal cells alter *ALDH-3* expression by altering *ALDH-3* gene transcription, rather by some permanent change inherent to culturing the cells, as happens for liver, for example, we sought ways to maintain the high basal level of corneal *ALDH-3* expression in culture. Modifications to the culture system by supplementing the medium with other growth factors met with only minimal success.

Downes and Holmes (1992) reported that increased levels of corneal Class 3 ALDH were correlated with eye opening in newborn mice. They also reported that exposure of adult rabbit cornea to UV radiation *in vivo* resulted in a significant decrease long-term in Class 3 ALDH activity (Downes et al., 1993). Because of these observations, we cultured our corneal epithelial explants in the presence of continuous visible light using a fluorescent tube placed in the incubator. Control corneal cultures were maintained under identical incubation conditions in the dark. HTC and 7777 cells grown in the dark and light also served as controls. Light had no effect on the Class 3 ALDH activity of either hepatoma cell line. However, corneal epithelium maintained continuously in the light had significantly more Class 3 ALDH activity than comparable age cultures grown in the dark (Figure 2). Although we do not yet know the mechanism by which light maintains Class 3 ALDH activity in cornea and have not yet attempted to optimize the light regimen for maximal ALDH activity, we believe our preliminary results indicate a completely unexpected mechanism is operating to regulate constitutive *ALDH-3* expression in at least some tissues. The fact that liver cells do not respond in a similar fashion further suggests this response may be tissue-specific.

ACKNOWLEDGMENT

This work supported by NIH grant CA21103 to RL.

REFERENCES

Asman D. C., Takimoto, K., Pitot, H. C., Dunn, T. J., and Lindahl, R., 1993, Organization and characterization of the rat class 3 aldehyde dehydrogenase gene, *J. Biol. Chem.* 268;12530.

Downes, J., and Holmes, R., 1992, Development of aldehyde dehydrogenase and alcohol dehydrogenase in mouse eye: evidence for light-induced changes, *Biol. Neonate* 61:118.

Downes, J., E., Swann, P. G., and Holmes, R. S., 1993, Ultraviolet light-induced pathology in the eye: Associated changes in ocular aldehyde dehydrogenase and alcohol dehydrogenase activities, *Cornea* 12:241.

Dunn, T. J., Lindahl, R., and Pitot, H. C., 1988, Differential gene expression in response to 2,3,7,8-tetrachlorodibenzo-p-dioxin (TCDD). Noncoordinate regulation of a TCDD- inducible aldehyde dehydrogenase and cytochrome P450c in the rat, *J. Biol. Chem.* 263:10878.

Jones, D. E., Brennan, M. D., Hempel, J., and Lindahl, R., 1988, Cloning and complete nucleotide sequence of a full-length cDNA encoding a catalytically functional tumor associated aldehyde dehydrogenase, *Proc. Natl. Acad. Sci. U.S.A.* 85:1782.

Lin, K.-H., Winters, A. L., and Lindahl, R. 1984, Regulation of aldehyde dehydrogenase activity in five rat hepatoma cell lines, *Cancer Res.* 44:5219.

Lin, K.-H., Brennan, M.D., and Lindahl, R., 1988 Expression of tumor-associated aldehyde dehydrogenase gene in rat hepatoma cell lines, *Cancer Res.* 48:7009.

Takimoto, K., Lindahl, R., and Pitot, H. C., 1991, Superinduction of 2,3,7,8- tetrachlorodibenzo-p-dioxin-inducible expression of aldehyde dehydrogenase by the inhibition of protein synthesis, *Biochem. Biophys. Res. Comm.*, 180:953.

Takimoto, K., Lindahl, R., and Pitot, H. C., 1992, Regulation of 2,3,7,8- tetrachlorodibenzo-p-dioxin-inducible of aldehyde dehydrogenase in hepatoma cells, *Arch. Biochem. Biophys.*, 298:492.

Takimoto, K., Lindahl, R., Dunn, T. J., and Pitot, H. C., 1994, Structure of the 5' flanking region of class 3 aldehyde dehydrogenase gene in the rat, *Arch. Biochem. Biophys.*, In Press.

Vasiliou, V., Puga, A., and Nebert, D. W., 1992, Negative regulation of the murine cytosolic aldehyde dehydrogenase-3 (*ALDH-3C*) gene by functional CYP1A1 and CYP1A2 proteins, *Biochem. Biophys. Res. Comm.*, 187:413.

Vasiliou, V., Puga, A., and Nebert, D. W., 1993, Mouse Class 3 aldehyde dehydrogenases: Positive and negative regulation of gene expression, In *Enzymology and Molecular Biology of Carbonyl Metabolism 4*, Weiner, H., Crabb, D. W., and Flynn, T. G., eds., Plenum Press, New York, pp.131.

19

STUDIES ON THE INDUCTION OF RAT CLASS 3 ALDEHYDE DEHYDROGENASE

Perikles Pappas[1], Vasilis Vasiliou[2], Maria Karageorgou[1],
Panayiotis Stefanou[1] and Marios Marselos[1]

[1] Department of Pharmacology
Medical School
University of Ioannina
GR-451 10 Ioannina, Greece
[2] Department of Environmental Health
University of Cincinnati Medical Center
Cincinnati, OH, 45267-0056

INTRODUCTION

A Class 3 aldehyde dehydrogenase (ALDH3c; EC 1.2.1.3), present in rat liver cytosol is highly inducible by 2,3,7,8-tetrachlorodibenzo-p-dioxin (TCDD) or polycyclic hydrocarbons such as 3-methylcholanthrene (3MC) and aromatic amines (Törrönen et al., 1981; Vasiliou et al., 1988; Marselos and Vasiliou, 1991). Induction of ALDH3c has also been observed in a number of extrahepatic rat tissues including lungs, spleen, urinary bladder, heart, and brain (Dunn et al., 1988; Vasiliou and Marselos, 1989). Induction of ALDH3c by either TCDD or 3MC has also been found in cultures of human hepatoma cell lines (HepG2), normal human hepatocytes, mouse and rat hepatoma cells and (Marselos et al., 1987; Vasiliou et al., 1992, 1993a). This induction process requires a functional *a*romatic *h*ydrocarbon (Ah) receptor present in the cytosol (Vasiliou et al., 1992; 1993a; Nebert et al., 1993). After binding, the inducer-receptor complex translocates into the nucleus, forms a heterodimer with the Ah receptor nuclear translocator (ARNT), and binds to one or more *a*romatic *h*ydrocarbon-*r*esponsive *e*lements (AhREs) identified upstream of murine and rat ALDH3c genes (Asman et al., 1993; Vasiliou et al., 1994). In this chapter, we address questions regarding possible parameters regulating the induction of rat liver ALDH3c by 3MC, such as (i) glutathione depletion, (ii) sex hormones, (iii) non steroid anti-inflammatory drugs (NSAIDs) and (iv) the overall drug metabolizing capacity.

Enzymology and Molecular Biology of Carbonyl Metabolism 5
Edited by H. Weiner *et al.*, Plenum Press, New York, 1995

RESULTS AND DISCUSSION

Effects Of GSH-Depletion Chemicals on the ALDH3c Inducibility

Glutathione (GSH) is the major non-thiol protein in mammalian cells and tissues, and participates in such metabolic processes as the protection of cells against free radicals and toxic metabolites of endogenous and exogenous origin (Meister and Anderson, 1983; Ketterer, 1986). Inhibition of GSH synthesis is associated with conditions of oxidative stress and induces several enzymes including metalothionein glutathione transferase and UDP glucuronosyltransferase (Bauman et al., 1991; Manning and Franklin, 1990). Tissue GSH depletion can be induced by various chemicals which cause inhibition of the GSH-synthesis enzymes or react chemically with GSH. Among the most well known GSH-depletors are: **a.** buthionine sulfoximine (BSO) which inhibits gamma-glutamylcysteine synthetase activity and **b.** diethylmaleate (DEM) and phorone (PHR), which are electrophiles and conjugate with GSH.

Treatment of rats with BSO, PHR and DEM did not affect the ALDH3c activity but inhibited the mitochondrial low Km and high Km ALDH activities (Pappas et al., 1991, 1994). These data suggest that the induction does not occur *in vivo* as a direct effect of GSH-depletion. It is noteworthy that in some cases high levels of ALDH3c are associated with elevated levels of GSH. SV-40 transformed cell lines derived from the 14CoS/14CoS mutant mouse (homozygous for a 1.2 cM deletion on chromosome 7) have elevated ALDH3c activity and GSH levels (Vasiliou et al., 1993a, b; Liu et al., 1993). In addition, treatment with phenolic antioxidants increase ALDH3c activity and GSH levels in mouse cell cultures (Vasiliou et al., 1993a,c; Liu et al., 1994).

When the GSH-depleting agents were co-administered with 3MC, we found a significant inhibition of the ALDH3c induction (45-75% inhibition), compared with the 3MC-treated group (Table 1). Furthermore, the inhibitory effects of BSO and DEM treatment on the ALDH3c induction were dose-dependent. This interference does not probably involve an altered metabolism of the inducer, since the GSH-depleting agents were administered in sufficient doses to deplete GSH-content without apparently affecting biotransformation enzyme activities (Gerard-Monnier et al., 1992).

The results were different when the GSH-depleting agents were administered to animals after a prolonged period of 3MC treatment (20 mg/kg, i.p., for 7 days). In this case,

Table 1. Effects of GSH-depleting agents on the induction of ALDH3c by 3MC
in rat liver

Co-treatment with 3MC	Dose (mg/kg/d)	% Reduction of ALDH3c induction
BSO	50	35
BSO	100	60
DEM	100	44
DEM	200	55
PHR	50	75

Wistar rats (N=5 in each group) were treated with the GSH-depleting agents for 4 days. Five hours after first injection the animals received a single dose of 3MC (50 mg/kg), and were sacrificed 72 hrs after 3MC-dosing and no less than 24hrs since the last GSH-depleting agent's injection. All chemicals were administered intraperitoneally, and dissolved in olive oil (DEM, PHR, 3MC) or saline (BSO). The control groups received olive oil or saline, only. All the reduced values represent statistically significant inhibition of ALDH3c induction. Enzyme activities and protein were measured according to the methods described by Vasiliou and Marselos, 1989 and Gornall et al., 1949.

Table 2. Effects of GSH-depleting agents on the induced by 3MC
activity of ALDH3c in rat liver

Compound (days of treatment)	% of ALDH3c Activity
3MC (7d) and then olive oil (4d)	100
3MC (7d) and then BSO (4d)	97
3MC (7d) and then DEM (4d)	95
3MC (7d) and then PHR (4d)	85

All the animals were treated with 3MC (25 mg/Kg, i.p.) for 7 days. Forty-eight hrs after the last injection, the GSH-depleting agents were injected i.p. at doses described in Table 1 for 4 days; the control group received olive oil. All the animals were sacrificed 24 hrs after the last injection. ALDH activity and protein were measured as described in Table 1. ALDH3c specific enzyme acitivity (nmol NADPH/min/mg protein) was for the untreated animals: 3.9 ± 0.7 and for the 3MC-treated: 1,469 ± 98 .

none of the GSH-depleting chemicals affected the already highly induced ALDH3c activity (Table 2). Furthermore, glutathione depletion did not alter the induction of ALDH1 by phenobarbital in a responsive rat substrain (Wistar/Mol/Io/RR, P. Pappas and M. Marselos unpublished). Our results implicate the physiological levels of GSH as an important factor in the process of ALDH3c induction by PAHs.

Sex Hormones Affect the Inducibility of ALDH3c by 3MC

Many factors may affect the rate of drug metabolism. The influence of age and sex on drug metabolism has been well documented in various experimental animals. Male and female animals of the same strain and species could exhibit differences in susceptibility to toxic agents. Similarly, the induction of the microsomal enzymes is often more pronounced in one sex compared to the other (Burke et al., 1978). We have previously shown that induction of ALDH3c is differentially expressed in adult male and female rats indicating a sex hormone influence on the ALDH3c inducibility (Vasiliou, 1988; Karageorgou et al., 1993). In order to examine this hypothesis, we studied the effects of various sex hormones in the ALDH3c inducibility by 3MC. The basic ALDH3c enzyme activity was not affected by the treatment with sex hormones alone (data not shown). Co-treatment of rats with 3MC and each of the sex hormones shown in Table 3 and an estrogen receptor antagonist, tamoxifen, diminished the ALDH3c induction. Moreover, the 3MC pre-induced levels of

Table 3. Effects of sex hormones on the
induction of ALDH3c by 3MC in rat liver

Co-treatment with 3MC	Dose (mg/kg/d)	% Reduction of ALDH3c induction
β-Estradiol	1	77
Testosterone	1	80
Progesterone	4	83
Hydrocortisol	1	88
Diethylstilbestrol	1	82
Tamoxifen	1	63

Female rats (N=6 for each group) were treated with the sex hormones for 5 days. On the 4th day animals received a single dose of 3MC (50 mg/kg), and were sacrificed 96 hrs later. All the reduced values represent statistically significant inhibition of ALDH3c induction (P<0.001). Enzyme activity and protein were mesaured as described in Table 1.

Table 4. Effects of NSAIDs co-treatment on the induction of rat liver
ALDH3c by 3MC

NSAIDs co-injected with 3MC (dose)	% Reduction of the induced activity of		
	ALDH3c	AHH	GSTA1
Aspirin (100)	41	N.S.	N.S.
Diclofenac (25)	N.S.	23	N.S.
Ibuprofen (50)	81	52	12
Indomethacin (5)	76	23	15
Ketoprofen (20)	79	25	N.S.
Mefenamic Acid (80)	78	20	N.S.
Naproxen (20)	N.S.	N.S.	N.S.

Wistar rats (N=6 for each goup) were treated with the NSAIDs for 6
consecutive days, at pharmacologically equal doses. On the 3rd day of
treatment, animals received a single dose of 3MC (25 mg/kg), and were
sacrificed 96 hrs later and not less than 24 hrs after last NSAID injection. All
drugs, as well as 3MC were administered i.p. and were dissolved in olive oil
(ibuprofen, indomethacin, ketoprofen, mefenamic acid and 3MC) or in saline
(aspirin, diclofenac and naproxen). Enzyme activities were measured according
to the following methods: ALDH3c from Vasiliou and Marselos, 1989; AHH
from Nebert and Gelboin, 1968; GSTs from Habig et al., 1974. All the
reduced values except those indicated as N.S. (non significant) represent
statistically significant inhibition of ALDH3c induction (P < 0.001).

ALDH3c activity were not affected by post-treatment with β-estradiol or diethylstilbestrol
(data not shown).

It is well known that TCDD does not bind to the hormone receptors, nor do sex
hormones bind to the Ah receptor. Thomsen and co-workers (1994) have reported the
estrogen receptor antagonist ICI-1889 blocks the TCDD-induction of CYP1A1 in human
breast carcinoma cells. In addition ALDH3c or CYP1A1 were unaffected after treatment
with TCDD or 3MC in some cell lines which do have Ah receptor but lack estrogen receptor
(Thomsen et al., 1993; Sreerama and Sladek, 1994). One can conclude that there is a possible
functional interaction between the Ah receptor and the hormone receptor(s).

Effects of Non Steroid Anti-Inflammatory Drugs (NSAIDs)

Synthesis of the primary prostaglandins is accomplished by a ubiquitous complex of
microsomal enzymes, the first of which is referred to as fatty acid cyclooxygenase. The
precursor acids are oxygenated and form cyclic endoperoxide derivatives. These endoper-
oxides, which are chemically unstable and are possible physiological substrates for aldehyde
dehydrogenases, are farther isomerized creating prostaglandins and thromboxanes which
contribute to the genesis of the signs and symptoms of inflammation (Larsen and Henson,
1983; Lindahl, 1992.). It has been reported that cooxygenation of a PAH-diol to a corre-
sponding-diolepoxide occurs *in vitro* during prostaglandin biosynthesis. In addition pretreatment of the enzyme preparation with an inhibitor of cyclooxygenase pathway,
indomethacin, abolishes the oxygenation of arachidonic acid, PAH-7,8-diol, and the production of a mutagenic species, probably the 7,8-diol-9,10-epoxide product which is the eventual
inducer of the drug metabolizing enzyme activities. The aim of these series of experiments
was to examine if the induction of ALDH by PAHs may represents an atypical chemically-
produced inflammation of the hepatocyte. Various non steroid anti-inflammatory drugs were
tested for their ability to influence the inducibility of ALDH3c and some other drug
metabolizing enzymes such as benzopyrene hydroxylase (AHH) and glutathione transferase

(GSTA1). The drugs were different chemical products representatives of the following groups: the prototype aspirin, the methylated indole derivative (indomethacin), propionic acid derivatives (ibuprofen, ketoprofen and naproxen), the fenamate (mefenamic acid) and the phenylacetic acid derivative (diclofenac). Their therapeutic activity appears to depend on the inhibition of a defined biochemical pathway responsible for the biosynthesis of the prostaglandin and the related autacoid.

The NSAIDs administration without any co-treatment did not affect the ALDH3c enzyme activity (data not shown). On the contrary, the 100-fold induction of this enzyme activity after a single dose of 3MC, was blocked in animals co-treated with NSAIDs and 3MC (Table 5). Less ALDH3c induction (76-81%) was observed in animals treated with ibuprofen, indomethacin, ketoprofen and mefenamic acid. The induction of microsomal benzopyrene hydroxylase activity (AHH) was also reduced (Table 4). The results were somehow different when diclofenac and naproxen were administered to animals; although both drugs did not affect the ALDH3c induction, diclofenac reduced the AHH induction. Finally, co-treatment with aspirin and 3MC decreased the induction of ALDH3c (41% reduction) without affecting the AHH (Table 4).

Our results indicate that NSAIDs used in this study can regulate the induction of ALDH3c possibly by blocking the cyclooxygenase pathway of prostaglandin and thromboxane synthesis. Considering the inhibition of the AHH induction, it is possible that a decreased prostaglandin synthesis intervene in the inducer's metabolism (Marnett et al., 1982), reducing the metabolic activation of the PAH-inducer. An influence of the accumulated arachidonic acid or leukotrienes on the PAH induction of various enzymes including ALDH3c is also possible. It has been shown that leukotriene B4 activates the transcription of the inflammatory cytokine interleukin 6 (IL6) in human blood monocytes (Brach et al., 1992). Furthermore IL6 blocks the inducibility of the AHH by TCDD in HepG2 cells (Fukada and Sassa, 1994).

The Overall Drug Metabolizing Capacity

In a series of experiments, we examined the hypothesis that substances known to affect drug metabolism cause a modification of the ALDH3c inducibility by 3MC. Phenobarbital and disulfiram were respectively used for the induction and inhibition of microsomal enzymes, cyclohexene as an inhibitor of the epoxide hydrolase and calcium cyanamide as inhibitor of ALDH (*References in* Vasiliou, 1988). Pretreatment with disulfiram or cyclohexene oxide caused a significant decrease of ALDH3c induction (Table 5). The ALDH3c inducibility was not affected when disulfiram or cyclohexene oxide were administered after 3MC (data not shown). On the other hand, pretreatment with calcium cyanamide, a strong inhibitor of ALDHs, did not affect the ALDH3c induction. On the contrary, the ALDH3c

Table 5. Effect of inhibition or induction of drug metabolizing capacity on the induction of rat liver ALDH3c by 3MC

Cotreatment with 3MC	Dose (mg/kg)	% of ALDH3c induction
Disulfiram	100	30*
Cyclohexene oxide	200	19*
Calcium cyanamide	5	100
Phenobarbital	80	200*

Wistar rats were treated with the disulfiram or phenobarbital or cycloexene oxide or calcium cyanamide for 4 days. On the 5th day animals received a single dose of 3MC (50 mg/kg, i.p.), and were sacrificed 96 hrs later. ALDH assay and protein determinations are described in Table 1. *, Statistically significant (P < 0.001).

induction was significantly higher in phenobarbital treated rats. In conclusion an altered drug metabolizing capacity affects the inducibility of ALDH3c by 3MC; however; it is still unclear if the changes in the inducibility of ALDH3c are due to an altered metabolism of 3MC or to a more elaborate mechanism involving transcription factors like Ah receptor.

ACKNOWLEDGMENTS

The excellent technical assistance of Mrs. O. Tsoumani and Mrs. A. Balomenou is gratefully acknowledged. This work was supported by a grant from the Commission of the European Communities (STRIDE/Hellas).

REFERENCES

Asman, D.C., Takimoto, K., Pitot, H.C., and Lindahl, R., 1993, Organization of the rat class 3 aldehyde dehydrogenase gene, *J. Biol. Chem.* 268:12530.

Bauman, J.W., Liu, J., Liu, Y.P, and Klaassen, C.D., 1991, Increase in metalothionein produced by chemicals that induce oxidative stress, *Tox. App. Pharmacol.* 110:347.

Brach, M.A., De Vos, S., Arnold C., Gruβ H-J., Mertelsmann, R., and Herrmann F., 1992, Leukotriene B4 trancriptionally activates interleukin-6 expression involving NF-κB and NF-IL6, *Eur. J. Immunol.* 22:2705.

Burke, M.D., and Mayer, R.T., 1975, Inherent specificities of purified cytochromes P-450 and P-448 toward biphenyl hydroxylation and ethoxyresorufin deethylation, *Drug Metab. Dispos.* 3:245.

Burke, M.D., Orrenious, S., and Gustafsson, J., 1978, Pituitary involvement in the sexual differentiation and 3-methylcholanthrene induction of rat liver microsomal monooxygenases, *Biochem. Pharmacol.* 27:1125.

Dunn, T.J., Lindahl, R., Pitot, H.C., 1988, Differential gene expression in response to TCDD. Noncoordinate regulation of a TCDD-induced aldehyde dehydrogenase and cytochrome P-450c in the rat, *J. Biol. Chem.* 236: 10878.

Fukada, Y., and Sassa, S., 1994, Supression of cytochrome P450IA1 by interlukin-6 in human HepG2 hepatoma cells, *Biochem. Pharmacol.* 47:1187.

Gerard-Monnier, D., Fougeat, S., and Chaudiere, J., 1992, Glutathione and cysteine depletion in rats and mice following acute intoxication with diethylmaleate, *Biochem. Pharmacol.* 43: 451.

Gornall, A.G., Bardawill, C.J., and David, M.M., 1949, Determination of serum proteins by means of the biuret reaction, *J. Biol. Chem.* 177:751.

Habig, W.H., Pabst, M.J., and Jakoby, W.B., 1974, Glutathione S-transferases. The first enzymatic step in mercapturic acid formation, *J. Biol. Chem.* 249:7130.

Karageorgou, M., Papadimitriou, C., and Marselos, M., 1993, Sexual differentiation in the induction of the class 3 aldehyde dehydrogenase, *Adv. Exp. Med. Biol.* 328: 123.

Ketterer, B., 1986, Detoxication reactions of glutathione and glutathione transferases, *Xenobiotica* 16: 957.

Larsen, G.L., and Henson, P.M., 1983, Mediators of inflammation, *Ann. Rev. Immunol.* 1:335.

Lindahl, R., 1992, Aldehyde dehydrogenases and their role in carcinogenesis, *CRC Crit. Rev. Biochem. Mol. Biol.* 27:283.

Liu, R-M., Sainsbury, M., Tabor, M.W. and Shertzer H.G., 1994, Mechanism of protection from menadione toxicity by 5,10-dihydroindeno[1,2b]indole in a sensitive and resistant mouse hepatocyte lines. *Biochem. Pharmacol.* 46: 1491.

Liu, R-M., Vasiliou, V., Duh, L., Puga, A., Nebert, D.W., Sainsbury, M. and Shertzer, H.G., 1994, Regulation of [*Ah*] gene battery enzymes and glutathione levels by 5,10-dihydroindeno[1,2b]indole in mouse hepatoma cell lines. *Carcinogenesis* in press

Manning, B.W., and Franklin, M.R., 1990, Induction of rat UDP- glucuronosyltransferase and glutathione S-transferase activities by L-buthionine-S,R- sulfoximine without induction of cytochrome P-450, *Toxicology* 65:149.

Marnett, L.J., Panthananickal, A., and Reed, G.A., 1982, Metabolic activation of 7,8-dihydroxy-7,8-dihy-drobenzo[a]pyrene during prostaglandin biosynthesis, *Drug Metabol. Rev.* 13:235.

Marselos, M., Strom, S., and Michalopoulos, G., 1987, Effect of phenobarbital and 3-methylcholanthrene on aldehyde dehydrogenase activity in cultures of HepG2 cells and normal human hepatocytes, *Chem-Biol. Interactions* 62:75.

Marselos, M., and Vasiliou, V., 1991, Effect of various chemicals on the aldehyde dehydrogenase activity of the rat liver cytosol, *Chem-Biol. Interactions* 36:79.

Meister, A., and Anderson, M.E., 1983, Glutathione, *Ann. Rev. Biochem.* 52:711.

Nebert, D.W., and Gelboin, H.V., 1968, Substrate-inducible microsomal aryl hydrocarbon hydroxylase in mammalian cell culture. Assay and properties of the induced enzyme, *J. Biol. Chem.* 243:6242.

Nebert, D.W., Puga, A., and Vasiliou, V., 1993, Role of Ah receptor and the dioxin-inducible [Ah] gene battery in toxicity, cancer and signal transduction, *Ann. N. Y. Acad. Sci.* 685:624.

Pappas, P., Vasiliou, V., Karageorgou, M., and Marselos, M., 1991, Changes in the inducibility of a hepatic aldehyde dehydrogenase, *in:* Alcoholism: A Molecular Perspective. Edited by T. N. Palmer, Plenum Press, N.Y., pp. 115.

Pappas, P., Vasiliou, V., Nebert, D.W., and Marselos, M., Lack of response of the rat liver Class 3 cytosolic aldehyde dehydrogenase to toxic chemicals, glutathione depletion, and other forms of stress, *Biochem. Pharmacol.,* in press.

Sreerama, L., and Sladek, N.E., 1994, Identification of a methylcholanthrene- induced aldehyde dehydrogenase in human breast adenocarcinoma cells exhibiting oxazaphosphorine-specific a aquired resistance, *Cancer Res.* 54: 2176.

Thomsen, J., Wang, X., Hines, R.N.,and Safe, S., 1993, Introduction of a functional human estrogen receptor restores the function of the Ah receptor in the human breast carcinoma cell line MDA-MB-231, *The Thoxicologist,* 13:34.

Thomsen, J., Wang, X., and Safe, S., 1994, Regulation of induced CYP1A1 gene expression in human breast cancer cellls by the estrogen receptors, *The Thoxicologist,* 14:51.

Törrönen, R., Nousiainen, U., and Hänninen, O., 1981, Induction of aldehyde dehydrogenase activity by polycyclic aromatic hydrocarbons, *Chem-Biol. Interactions* 36:33.

Vasiliou, V., 1988, The effects of xenobiotics on hepatic aldehyde dehydrogenase activity, Ph.D. thesis, Ioannina 1988.

Vasiliou, V., Athanasiou, K., and Marselos, M., 1988, The use of ALDH induction as a carcinogenic risk marker in comparison with a typical in vitro mutagenicity system, *in;* Biologically Based Methods for Canser Risk Assessment. Edited by C.C. Travis, Plenum Press, New York and London, pp. 231.

Vasiliou, V., and Marselos, M., 1989, Tissue distribution of inducible aldehyde dehydrogenase activity in the rat after treatment with phenobarbital or methylcholanthrene, *Pharmacol. Toxicol.* 64:39.

Vasiliou, V., Puga, A., and Nebert, D.W., 1992, Negative regulation of the murine cytosolic aldehyde dehydrogenase-3 (Aldh-3c) gene by functional CYP1A1 and CYP1A2 proteins, *Biochem. Biophys. Res. Commun.* 187:413.

Vasiliou, V., Puga, A., and Nebert, D.W., 1993a, Mouse class 3 aldehyde dehydrogenases: positive and negative regulation of gene expression, *Adv. Exp. Med. Biol.* 328: 131.

Vasiliou, V., Reuter, S., Kozak, C., and Nebert, D.W., 1993b, Mouse dioxin- inducible cytosolic aldehyde dehydrogenase-3: AHD4 cDNA sequence, genetic mapping, and differences in gene expression, *Pharmacogenetics* 3:281.

Vasiliou V., Liu R-M., Shertzer H.G. and Nebert D.W., 1993c, Response of the [*Ah*] battery genes to antioxidants, *The Toxicologist* 13: 135.

Vasiliou, V., Reuter, S., and Nebert, D.W., 1994b, Organization and characterization of the murine dioxin-inducible cytosolic aldehyde dehydrogenase (Ahd4) gene, *The Toxicologist* 14: 140.

20

MOUSE CLASS 3 ALDEHYDE DEHYDROGENASES

Vasilis Vasiliou[1], Steven F. Reuter[1], Christine A. Kozak[2] and
Daniel W. Nebert[1]

[1]Department of Environmental Health
University of Cincinnati Medical Center
P.O. Box 670056
Cincinnati, Ohio 45267-0056
[2]Laboratory of Molecular Microbiology
National Institute of Allergy & Infectious Diseases
Bethesda, Maryland 20892

INTRODUCTION

The Class 3 aldehyde dehydrogenases include the 2,3,7,8-tetrachlorodibenzo-*p*-dioxin (dioxin, TCDD)-inducible cytosolic form (ALDH3c) and the constitutive microsomal form (ALDH3m), which is also TCDD-inducible. Table 1 summarizes the ALDH nomenclature used in this chapter. It might be noted that the Committee on Standardized Genetic Nomenclature for Mice is adamant about keeping the mouse gene designations the same as they have been for several decades.

We have previously shown that both ALDH3c and ALDH3m are inducible by dioxin in mouse hepatoma Hepa-1 cell cultures (Vasiliou et al., 1992, 1993a). This induction process requires a functional *a*romatic *h*ydrocarbon *r*eceptor (AHR) present in the cytosol. It has been shown that, after binding, the inducer-receptor complex translocates into the nucleus and forms a heterodimer with the *Ah* receptor *n*uclear *t*ranslocator protein (ARNT). This heterodimer, probably with additional proteins, becomes the complex which binds to one or more *a*romatic *h*ydrocarbon-*res*ponse *e*lements (AhREs) and turns on the transcription of several genes (*reviewed in* Nebert et al., 1993). AhREs have been identified in the 5' flanking region of several dioxin-inducible genes, including the rat ALDH3c (Asman et al., 1993).

Using both the Hepa-1 mutant *c4* cell line--which lacks ARNT--and the stably transfected *c4* line containing an ARNT cDNA expression vector, we have confirmed that induction of either ALDH3c or ALDHm requires the ARNT protein, in addition to the functional AHR protein (V. Vasiliou and D.W. Nebert unpublished). On the other hand, the murine ALDH3c (*Ahd4* gene), but not the ALDH3m (*Ahd3* gene), is negatively regulated by functional CYP1A1 and/or CYP1A2 enzymes; this negative control by a P450 gene product has been designated the necessary criterion for classification of a gene as a member of the [*Ah*] gene battery (Vasiliou et al. 1992; Nebert et al., 1993).

Table 1. Nomenclature for the class 3 ALDHs

	Cytosolic	Microsomal
Mouse gene	*Ahd4*	*Ahd3*
Mouse mRNA or protein	AHD4 or ALDH3c	AHD3 or ALDH3m
Rat gene	*ALDH3c*	*ALDH3m*
Rat mRNA or protein	ALDH3c	ALDH3m

Using the rat ALDH3c as a probe, we have cloned and sequenced the murine ALDH3c (AHD4) and ALDH3m (AHD3) cDNAs (Vasiliou et al. 1993b; Nebert et al., 1994). In this chapter, we review the two murine class 3 ALDHs with respect to cDNA and deduced amino acid sequence and chromosomal mapping. Finally, the putative transcription factor-binding sites found in a 3.2-kb fragment upstream of exon 1 of the *Ahd4* gene are also described.

RESULTS AND DISCUSSION

Comparison of the Murine AHD4 and AHD3 cDNAS and Proteins

After screening mouse liver cDNA libraries with the rat ALDH3c cDNA as a probe, we isolated several clones which were subsequently identified as the murine AHD3 cDNA, as well as others subsequently identified as the murine AHD4 cDNA. Both full-length cDNAs were sequenced and their deduced protein sequences were compared.

The AHD3 clone was found to be 3,011 bp in length, with 5' and 3' nontranslated regions of 113 bp and 1,443 bp, respectively (Fig. 1). The AHD3 cDNA encodes a protein of 484 amino acids ($M_r = 53,942$). On the other hand, the mouse AHD4 cDNA is 1,722 bp in length, with 5' and 3' nontranslated regions of 174 and 186 bp, respectively. The AHD4 cDNA encodes a protein of 453 amino acids ($M_r = 50,466$). The murine AHD3 protein is distinguished from the AHD4 protein primarily by a hydrophobic 33-residue putative membrane anchor at the COOH-terminus (Fig. 1).

Residues Gly-245 and Gly-250, as well as Cys-302, have been identified as important amino acids in the coenzyme-binding sites of the Class 1 and 2 ALDHs (Hempel et al., 1993). In addition, a conserved 10-residue segment (265-274) in ALDH1 and ALDH2 includes the catalytically essential Glu-268. These three amino acids and this 10-residue segment are

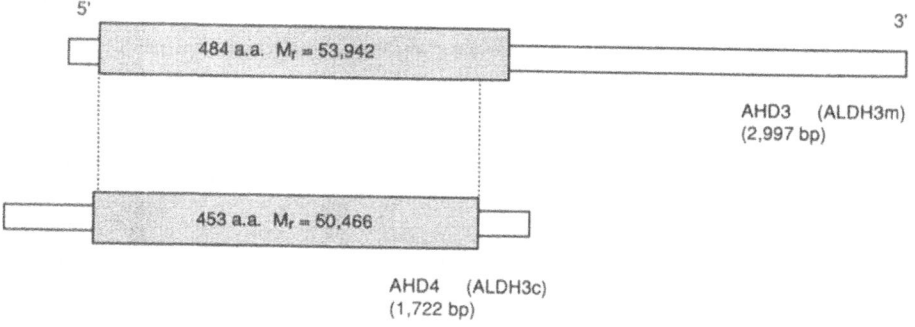

Figure 1. Comparison of the two murine Class 3 ALDH cDNAs and deduced proteins. The larger rectangles (*stippled*) represent the coding regions. Alignment of the two protein shows three unmatched NH$_2$-terminus residues in the AHD4 protein, compared with AHD3, and 33 additional residues in the COOH-terminus (Nebert et al., 1994).

remarkably conserved in both the murine and rat ALDH3m and ALDH3c proteins, suggesting an important role in the enzymic reaction.

Evolution of ALDH Genes

The mammalian ALDHs have been classified into three major classes, based on primary structures and also on the their biochemical and subcellular properties (***reviewed in*** Lindahl and Hempel, 1991; Lindahl, 1992). Class 1 includes the major hepatic cytosolic phenobarbital-inducible ALDH; the gene consists of 13 exons and encodes 501 amino acids. Class 2 is the major liver mitochondrial ALDH responsible for the atypical aldehyde dehydrogenase, a human pharmacogenetic disorder commonly seen in Asians (Yoshida,

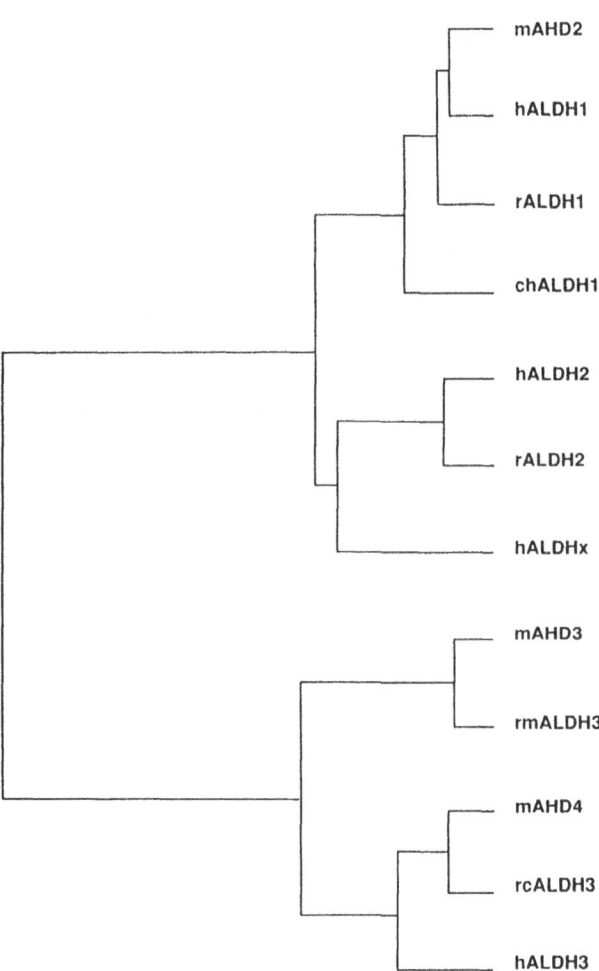

Figure 2. Clustering algorithm, showing the degrees of divergence for the proteins encoded by seven vertebrate ALDH genes: the cytosolic mouse mAHD2 (Rongnoparut and Weaver 1991); human hALDH1 (Hsu et al., 1989); rat phenobarbital-inducible rALDH1 (Dunn et al., 1989); chicken retina chALDH1 (McCaffery et al., 1991); human hALDH2 (Hsu et al., 1985, 1988); rat rALDH2 (Farres et al., 1989); human hALDHx (Hsu and Chang, 1991); murine microsomal mAHD3 (Nebert et al., 1994); rat microsomal rmALDH3 (Miyauchi et al., 1991); murine cytosolic mAHD4 (Vasiliou et al., 1993b); rat cytosolic rcALDH3 (Jones et al., 1988); and human hALDH3 (Hsu et al., 1992).

1992); this gene has 13 exons and encodes 517 amino acids. Class 3 includes the low constitutive ALDH present in stomach and elevated in hepatomas, and is dioxin-inducible; the gene has 11 exons and encodes 453 amino acids. The constitutive microsomal ALDH3m is also a member of Class 3, but the gene structure has not yet been reported. To date, more than 20 ALDH cDNA and/or genes have been isolated from various sources--including bacteria, plants, yeast and mammals. The fact that ALDH genes occur in bacteria (*reviewed in* Lindahl and Hempel, 1991) would suggest strongly that the ancestral gene was present on this planet more than 3.5 billion years ago, the time at which eocytes and eukaryotes are believed to have diverged from prokaryocytes, according to geologic and molecular biologic evidence (Knoll, 1992; Rivera and Lake, 1992).

Using the University of Wisconsin GCG Pileup program, we have aligned the protein sequences of twelve ALDHs encoded by vertebrate genes (Fig. 2). Pairwise alignments of these ALDHs reveal amino acid identities ranging from 95% to 22%. For example, the mouse AHD3 protein is 95% similar to the rat ALDH3m, about 65% identical to the mouse, rat and human ALDH3c, and <<28% similar to the rat ALDH1 and ALDH2 proteins. The murine AHD4 protein is 91% and 80% similar to the rat and human ALDH3c proteins, respectively, 64% identical to the rat ALDH3m protein, and <<28% similar to the rat ALDH1 and ALDH2 proteins.

From the Fig. 2 relationship dendrogram, one can see that two major gene families can be distinguished: (i) one which includes the Class 1 and Class 2 ALDHs, as well as the ALDHx, as two or possibly three subfamilies, and (ii) the other which includes the microsomal and the cytosolic forms of the Class 3 as two distinct subfamilies. If one is to be consistent with the gene nomenclature systems being proposed and used for dozens of gene superfamilies and being approved by the human and mouse gene nomenclature committees, therefore, it would appear that Class 1 and Class 2 ALDHs--although biochemically distinct from the standpoint of enzyme substrates and/or subcellular localizations--cannot be regarded as evolutionarily distant enough so as to warrant each of them as gene products from two distinct gene families. The longer the ALDH field waits before adopting standardized nomenclature for these genes, the more trivial names there will be assigned to the same gene, and the more confusing this field will become.

Chromosomal Mapping of the Ahd4 and Ahd3 Genes

Over the past two decades, several *Ahd* loci have been mapped in the mouse genome, by the use of electrophoretic and activity variants. These data confirm the gene multiplicity for the mammalian ALDHs (*reviewed in* Holmes et al., 1991). For example, *Ahd1* maps on mouse chromosome (Chr) 4, *Ahd2* on Chr 19, and both *Ahd4* and *Ahd6* on Chr 11.

Using our murine AHD4 and the AHD3 cDNAs separately, we have determined the chromosomal location of the corresponding genes. Parental DNAs of two multilocus crosses were screened for polymorphic fragments, using as the probes the full-length AHD4 cDNA and a 0.85-kb fragment of the AHD3 cDNA 5' end, respectively. Using the AHD4 probe, *Eco* RI digestion of DNA from the NSF/N mouse and *M. m. Musculus* produced fragments of 22.0 and 19.6 plus 3.0 kb, respectively; *Pst* I digestion of DNA from the NSF/N mouse and *M. spretus* gave fragments of 4.5 and 5.3 kb, respectively. Using the AHD3 probe, *Eco*

+1

ctctcttggctcttgcttattccaggagttCCAGGAGTTCCAGCTGCTGGAGAGGACTTG

Figure 3. Nucleotide sequence from -30 bp to +30 bp from the transcription start site (+1) of the murine *Ahd4* gene.

Table 2. Segregation of the *Ahd4* and *Ahd3* genes with alleles of the *Mgat1* and *Shbg* genes in two multilocus crosses

Recombination[a]:	*Locus pair*	*r/n*	*% recombination ± 1 SEM*
	Mgat1, Ahd4	11/224	4.9 ± 1.4
	Ahd4, Shbg	18/224	8.0 ± 1.8
	Mgat1; Ahd3,Ahd4	10/170	5.9 ± 1.8
	Ahd3, Ahd4	0/170	1.7
		0/280	1.3
	Ahd3, Ahd4; Shbg	17/170	10.0 ± 2.3

[a]Percentage recombination ± SEM between restriction fragments were calculated according to the method of Green (1981) from the numbers of recombinants (*r*) in a sample of size *n*.

RI digestion produced an NFS/N fragment of 15.2 kb and *M. m. musculus* fragments of 12.0 and 3.0 kb. Digestions of NFS/N and *M. spretus* DNA produced major *Pst* I fragments of 4.8 and 5.5 kb, respectively, and at least four smaller nonpolymorphic fragments. Inheritance of the polymorphic fragments in both crosses was compared with inheritance of more than 550 markers distributed on all 19 autosomes and the X chromosome. The *Ahd4* and *Ahd3* genes were both found to be linked to markers on Chr 11 (Table 2) between *Mgat-1* and *Shbg*. Interestingly, we identified no recombinants between the *Ahd4* and *Ahd3* genes among 208 mice. These data indicate that, at the 95% confidence level, the *Ahd4* and *Ahd3* genes must lie within 1.3 cM of each other and, in all likelihood, represent a gene duplication event that occurred less than 400 million years ago.

Ahd4 Gene Structure

Previously we have identified at least three mechanisms by which the murine *Ahd4* gene is regulated (*reviewed in* Nebert et al., 1993). (**i**) The gene is inducible by TCDD, a process dependent upon a functional AHR and ARNT and which presumes the presence of one or more AhREs in the regulatory region of the *Ahd4* gene (Vasiliou et al., 1992). The Ah receptor is encoded by the murine *Ahr* locus on proximal Chr 12, and the ARNT is encoded by the *Arnt* locus on Chr 3. (**ii**) Expression of the *Ahd4* gene is also elevated in untreated *14CoS/14CoS* murine cell lines, which have a deletion in both copies of a 1.2-cM region of Chr 7. This chromosome 7-mediated process has been shown to involve an electrophile response element (EpRE) (Nebert and Vasiliou, 1993). (**iii**) The *Ahd4* gene is repressed by a functional CYP1A1 or CYP1A2 enzyme. We have shown that AHD4 mRNA levels in the untreated *c37* mutant line are strikingly increased in the absence of the endogenous CYP1A1 enzyme (Vasiliou et al., 1992; 1993a). Introduction of a CYP1A1 or CYP1A2 functional enzyme activity--expressed on a transfected cDNA-containing plasmid--in *c37* stable transformants restores the wild-type phenotype, *i.e.* lowers AHD4 mRNA to negligible levels such as those found in the Hepa-1 *wt* cultures and restores the same amount of TCDD inducibility as that found in *wt* cells. This process is presumably dependent on a negative response element (NRE) in the regulatory region of the *Ahd4* gene (*reviewed in* Nebert et al., 1993).

To search for functional AhREs, EpREs and NRE in the 5' flanking region of the *Ahd4* gene, we isolated two overlapping genomic fragments which contain the entire *Ahd4* gene, plus extensive amounts of 5' and 3' flanking sequences (Vasiliou et al., 1994; V. Vasiliou and D.W. Nebert et al., in preparation). The *Ahd4* gene spans approximately 10 kb

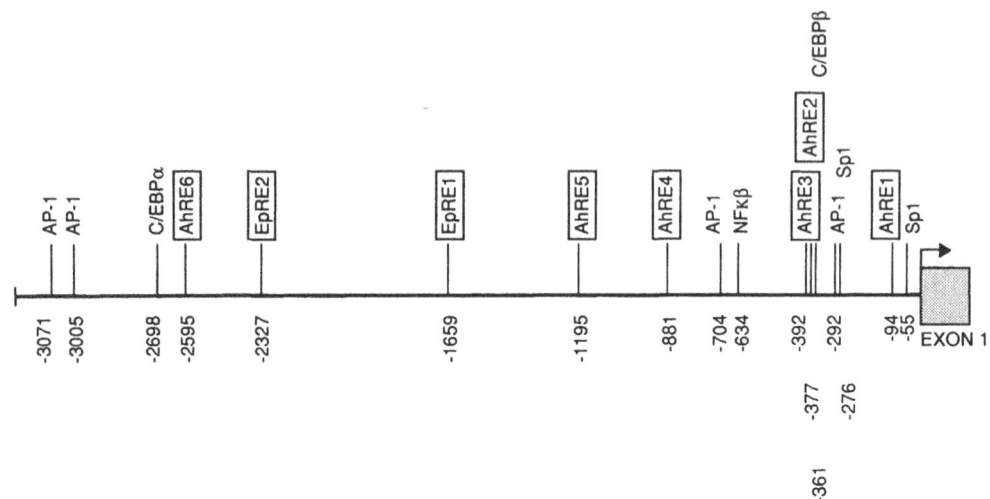

Figure 4. Putative transcription factor-binding sites in 3.2-kb of the 5' flanking region of the murine *Ahd4* gene.

and consists of 11 exons. The first exon comprises 169 bp, is noncoding, and is followed by a 3.2-kb first intron. A similar organization has been reported for the rat orthologous gene (Asman et al., 1993). It would appear that the human orthologous gene differs in structure from both the mouse *Ahd4* and rat *ALDH3c* genes. Specifically, the human stomach ALDH3 was reported not to contain a noncoding first exon, and putative CAATT box (-117 through -114) and TATA box (-79 through -76) regulatory elements were described (Hsu et al., 1992).

By primer extension analysis using extracts from several mouse cell lines following treatment with TCDD, we have determined the transcriptional start site of the *Ahd4* gene (Fig. 3). The *Ahd4* gene promoter contains no TATA box; these data are consistent with studies on the rat *ALDH3c* gene. It is possible that the rodent *ALDH3c* gene has multiple transcription start sites, which are tissue- specific and that the large first intron contains promoter activity--used for the low constitutive expression of ALDH activity.

Putative Dna Motifs in the Ahd4 5' Flanking Region

Sequencing of 3.2 kb of the 5' flanking region revealed several putative transcription factor-binding sites, which may be important in the positive and negative control of the *Ahd4* gene expression (Fig 4). Within the region of -55 and -300, we have identified two Sp1-binding sites. Six AhREs, which contain the core sequence of GCGTG, are located at -94, -377, -392, -881, -1195 and 2595. Two EpREs are found at -1659 and -2327. Four AP-1 sites were also identified. Finally, a putative C/EBPβ (NF-IL6) site was located at -361, and a C/EBPα site was found at -2698. Reporter gene assays are underway in order to determine which of these sites is functional and the degree to which they are functional.

ACKNOWLEDGMENTS

We appreciate the many hours of fruitful discussions with Alvaro Puga and his expert technical advice. We thank our colleagues for valuable discussions and a critical reading of this manuscript. V.V. was a Fogarty Scholar (1992-94) at the University of Cincinnati. This work was supported in part by NIH Grants R01 AG09235 and P30 ES06096.

REFERENCES

Asman, D.C., Takimoto, K., Pitot, H.C., and Lindahl, R., 1993, Organization of the rat class 3 aldehyde dehydrogenase gene, *J. Biol. Chem.* 268: 12530

Dunn, T.J., Koleske, J.E., Lindahl, R., and Pitot, H.C., 1989, Phenobarbital- inducible aldehyde dehydrogenase in the rat: cDNA sequence and regulation of the mRNA by phenobarbital in responsive rats, J. Biol. Chem. 264: 13507.

Farres, J., Guan, K-L., and Weiner H., 1989, Primary structures of rat and bovine liver mitochondrial aldehyde dehydrogenases deduced from cDNA sequences, *Eur. J. Biochem.* 180: 67.

Green, E.L., 1981, Genetics and probability in animal breeding experiments. New York: MacMillan Publishing Company, Inc.

Hempel, J., Nicholas, H., and Lindahl, R., 1993, Aldehyde dehydrogenases: widespread structural and functional diversity within a shared framework, *Protein Sci.* 2: 1890.

Holmes, R.S., von Oorschot, R.A.H, and Vandenberg, J.L., 1991, Aldehyde dehydrogenase (ALDH) isozymes in the gray short-tailed opossum (*Monodelphis domestica*): Tissue and subcellular distribution and biochemical genetics of ALDH3, *Biochem. Genet.* 29:163.

Hsu, L.C., Tani, K., Fujiyoshi, T., Kurachi, K., and Yoshida, A., 1985, Cloning of cDNAs for human aldehyde dehydrogenases 1 and 2, *Proc. Natl. Acad. Sci. U.S.A.* 82: 377.

Hsu, L.C., Bendel, R.E., and Yoshida, A., 1988, Genomic structure of the human mitochondrial aldehyde dehydrogenase gene, *Genomics* 2: 57.

Hsu, L.C., Chang, W-C., and Yoshida, A., 1989, Genomic structure of the human cytosolic aldehyde dehydrogenase gene, *Genomics* 5: 857.

Hsu, L.C., Chang, W-C., Shibuya, A., and Yoshida, A., 1992, Human stomach aldehyde dehydrogenase cDNA and genomic cloning, primary structure, and expression in *Escherichia coli, J. Biol. Chem.* 267: 3030.

Jones, D.E., Brennan, M.D., Hempel, J., and Lindahl, R., 1988, Cloning and complete nucleotide sequence of a full-length cDNA encoding a catalytically functional tumor-associated aldehyde dehydrogenase, *Proc. Natl. Acad. Sci. U.S.A.* 85: 1782.

Knoll, A.H., 1992, The early evolution of eukaryotes: A geological perspective, *Science* 256: 622.

Lindahl R, and Hempel, J., 1991, Aldehyde dehydrogenases: what can be learned from a baker's dozen sequences?, *Adv. Exp. Med. Biol.* 284: 1.

Lindahl, R., 1992, Aldehyde dehydrogenases and their role in carcinogenesis, *CRC Crit. Rev. Biochem. Mol. Biol.* 27: 283.

McCaffery, P., Tempst, P., Lara, G., and Drager, U., 1991, Aldehyde dehydrogenase is a positional marker in the retina, *Development* 112: 693.

Miyauchi, K., Masaki, R., Taketani, S., Yamamoto, A., Akayama, M., and Tashiro, Y., 1991, Molecular cloning, sequencing, and expression of cDNA for rat liver microsomal aldehyde dehydrogenase, *J. Biol. Chem.* 266: 19536.

Nebert, D.W., Puga, A., and Vasiliou, V., 1993, Role of Ah receptor and the dioxin-inducible [*Ah*] gene battery in toxicity, cancer and signal transduction, *Ann. N. Y. Acad. Sci.* 658: 624.

Nebert, D.W., Kozak, C.A., Lindahl, R., and Vasiliou, V., 1994, Mouse microsomal Class 3 aldehyde dehydrogenase: AHD3 cDNA sequence, chromosomal mapping, and dioxin inducibility, *The Toxicologist* 14:409.

Nebert, D.W. and Vasiliou, V., 1993, Involvement of the electrophile-responsive element (EpRE) in the murine chromosome 7-mediated derepression of the Phase II [*Ah*] battery genes, *The Toxicologist* 13: 134.

Rivera, M.C. and Lake, J.A., 1992, Evidence that eukaryotes and eocyte prokaryotes are immediate relatives, *Science* 257: 74.

Rongnoparut, P., and Weaver S., 1991, Isolation and characterization of a cytosolic aldehyde dehydrogenase-encoding cDNA from mouse liver, *Gene* 101: 261.

Vasiliou, V., Puga, A., and Nebert, D.W., 1992, Negative regulation of the murine cytosolic aldehyde dehydrogenase-3 (Aldh-3c) gene by functional CYP1A1 and CYP1A2 proteins, *Biochem. Biophys. Res. Commun.* 187: 413.

Vasiliou, V., Puga, A., and Nebert, D.W., 1993a, Mouse class 3 aldehyde dehydrogenases: positive and negative regulation of gene expression, *Adv. Exp. Med. Biol.* 328: 131.

Vasiliou V., Reuter S.F., Kozak C.A., and Nebert, D.W., 1993b, Mouse dioxin- inducible cytosolic aldehyde
    dehydrogenase: AHD4 cDNA sequence, genetic mapping, and differences in gene expression,
    Pharmacogenetics 3: 281.
Vasiliou, V., Reuter, S.F., and Nebert, D.W., 1994, Organization and characterization of the murine cytosolic
    TCDD-inducible aldehyde dehydrogenase (***Ahd4***) gene, *The Toxicologist* 14: 410.

21

CLONING AND CHARACTERIZATION OF GENES ENCODING FOUR ADDITIONAL HUMAN ALDEHYDE DEHYDROGENASE ISOZYMES

Lily C. Hsu, Wen-Chung Chang, Sharon W. Lin and Akira Yoshida

Department of Biochemical Genetics
Beckman Research Institute of the City of Hope
Duarte, California

INTRODUCTION

Aldehyde dehydrogenases (ALDH; aldehyde:NAD+ oxidoreductase, EC 1.2.1.3) are a group of isozymes with broad substrate specificity. They catalyze the oxidation of various aliphatic and aromatic aldehydes to the corresponding acids (Pietruszko, 1983) and have been frequently considered as detoxifying enzymes which eliminate aldehydes *in vivo*. Metabolism of ethanol-derived acetaldehyde and toxic aldehyde in the foodstuff and from lipid peroxidation are examples of the detoxification role of the ALDH isozymes (Parrilla et al., 1974; Harrington et al., 1987; Mitchell et al., 1987; Jakoby et al., 1990). Recently, it has been demonstrated that the enzymes also play crucial roles in the metabolism of retinoic acid, biogenic amine and neurotransmitters (Ambroziak and Pietruszko, 1991, 1993; Yoshida et al., 1992, 1993).

ALDH isozymes can be distinguished based on their physico-chemical properties, enzyme properties, tissue/subcellular distributions and sequence identities (Yoshida et al., 1991). In humans, five liver ALDH isozymes have been purified and characterized. Cytosolic ALDH1 and mitochondrial ALDH2, the two major liver isozymes, and γ-aminobutyraldehyde dehydrogenase (GABALDH) exhibit low Km values (μM range) toward acetaldehyde, while ALDH3 and ALDH4 have high Km values (mM range) toward acetaldehyde. ALDH3 is expressed at a high level in stomach, lung, cornea, and carcinoma liver, but hardly in healthy liver, and utilizes benzaldehyde and medium-chain aliphatic aldehydes as optimal substrates (Yin et al., 1989; Wang et al., 1990; Hsu et al., 1992; King and Holmes, 1993). The ALDH3 isozyme can be induced in liver and urinary bladder during carcinogenesis (Lindahl, 1992), and by 2, 3, 7, 8-tetrachlorodibenzo-p-dioxin or polycyclic aromatic hydrocarbons in rat and mice (Lindahl and Evces, 1984; Koivusalo et al., 1989; Vasiliou et al., 1993). ALDH4 has been identified as glutamic γ-semialdehyde dehydrogenase (Forte-McRobbie and Pietruszko, 1986). Partial purifications of low Km human brain- and saliva-specific ALDH isozymes have been reported (Ryzlak and Pietruszko, 1989; Harada et al., 1989). In addition, multiple forms of ALDH isozymes have been demonstrated in the kidney tissue (Harada et al., 1980; Duley et al., 1985). However, detailed biochemical and structural analyses of these tissue-specific ALDHs are still lacking.

Human ALDH1 and ALDH2 are homotetramers consisting of subunits of 54 kDa. Their amino acid sequences (500-residue polypeptide) have been determined (Hempel et al., 1984, 1985). The cDNAs and genes of human ALDH1 and ALDH2 have been cloned, characterized (Hsu et al., 1988, 1989, 1992) and mapped (Raghunathan et al., 1988). Both genes consist of 13 exons spanning about 50 kbp in length. A partial amino acid sequence of human stomach ALDH3 has been reported (Yin et al., 1991). Unlike ALDH1 and 2, ALDH3 has a dimeric structure (Santisteban et al., 1985). The cDNA and gene of human ALDH3 have also been cloned and expressed (Hsu et al., 1992). The deduced sequence encodes 453 amino acid residues (aa). Although the kinetic properties of GABALDH have been extensively studied (Ambroziak and Pietruszko, 1991), its structural properties have only been reported recently. cDNAs containing the complete (Lin et al., 1993) and partial (Kurys et al., 1993) coding region have been obtained independently. One isozyme, ALDH5 (ALDHx), has been identified by reverse genetics (Hsu and Chang, 1991). Unlike other characterized *ALDH* genes, the structure of the *ALDH5* gene contains no intron in the coding region (Hsu and Chang, 1991). The *ALDH5* gene is expressed in various tissues including liver, brain, adrenal gland, testis, stomach and parotid gland as detected by reverse transcriptase-polymerase chain reaction (RT-PCR).

We used RT-PCR approach to clone and identify ALDH6 and ALDH7 cDNA from a human salivary gland and a human kidney library, respectively. *ALDH8* genomic clone was isolated during the process of the screening for the human *ALDH7* genomic clones. GABALDH cDNA was isolated from a human placenta library. We have further elucidated gene organization and assigned chromosomal localization of these genes. In this report, we also describe the sequence relationship and the conservation of the exon organizations among the three newly identified human *ALDH* genes, *ALDH6, 7* and *8,* and the previously characterized human *ALDH* genes.

RESULTS AND DISCUSSION

The Screening Probe

The sequences of degenerate oligonucleotide primers, primer-1 and primer-2, used in the RT-PCR procedure to prepare screening probe were derived from two highly conserved molecular regions among the known sequences of various ALDH isozymes (Hsu et al., 1992). The sequences of primer-1 and primer-2, corresponding to the sense and antisense sequences of the upstream and downstream conserved regions, are 5'-GAGGTACCCTGGAGCTKGGRGGWAARAGCCC-3'   and   5'-TCGAATTCACWGGYCCRAARATCTCCTC-3', respectively.

ALDH6. A single RT-PCR product band, about 420 bp, was obtained by amplifying human salivary gland total RNA with the primer-1 and -2. The gel-purified DNA was subcloned subsequently into pBluescript KS(+) vector. Fifty five clones were randomly selected for colony hybridization with mixed ALDH probes (including ALDH1, 2, 3 and 5 cDNA). Twenty-four clones with a weak hybridization signal were subjected to sequencing and the novel sequences were identified. Fifteen clones had the same nucleotide sequence between the two PCR primers and encoded 123 aa which were highly similar (73% identity) to that of the corresponding region of the human ALDH1 (aa positions 275-397. The aa position number used in this report for the ALDH1/2 isozymes is based on the 500-aa polypeptide of the human ALDH1/2, Hempel et al., 1984, 1985). One of these clones (135#9) was used for cloning of the ALDH6 cDNA from a human salivary gland library.

ALDH7. A 408 bp RT-PCR band, obtained from total human kidney RNA, was eluted and subcloned. Out of eight weakly hybridized clones, five clones had the same nucleotide sequence. The deduced aa sequence is 54% identical to that of the corresponding region of the human ALDH3 (aa positions 206-341. The aa position number used in this report for the ALDH3 isozyme is based on the deduced human ALDH3 sequence of 453-aa. Hsu et al., 1992). One of these clones (135#23) was used for cloning of the ALDH7 cDNA from a human kidney library.

ALDH8. The human *ALDH8* gene was identified during the process of the screening for the human *ALDH7* genomic clones. The ALDH8 cDNAs were obtained by screening a human salivary gland library with a ALDH8-specific oligomer.

GABALDH. A 440 bp PCR product was obtained by amplifying the purified recombinant DNA of the human placenta cDNA library with the primer-1 and -2. The subclones, which weakly hybridized with the ALDH mixed probes and had sequence homology with other ALDHs, were identified. One of such subclones was used as the probe to screen human placenta cDNA library.

The ALDH cDNAS

The structural features of the human *ALDH6, 7* and *GABALDH* cDNA are summarized in the Table 1. The ALDH6, 7 and GABALDH cDNA contain an open reading frame encoding for a polypeptide chain of 512, 468 and 493 aa, respectively. The sequence around the ATG initiation codon of the respective cDNA is similar to the consensus sequence (GCCA/GCCATGG) for higher eukaryotes (Kozak, 1987), and the polyadenylation signal (AATAAA) (Breathnach and Chambon, 1981) at the 3'-end is located within the consensus distance upstream from the poly(A) tract.

The ALDH8 nucleotide sequence of the coding region was derived from a *ALDH8* genomic clone based on the conservation between the *ALDH7* and *8* exon sequences. However, a termination codon was found in frame in the second coding exon of the only *ALDH8* genomic clone we obtained in this region. If the existence of this termination codon is not due to the cloning artifact, which is very unlikely, there are three possibilities. Firstly, this clone represents a variant *ALDH8* gene. Then, there is a wild type *ALDH8* gene which encodes for a 466-aa polypeptide chain. This possibility can be tested by analyzing the PCR products of the variation region obtained from the genomic DNA of unrelated persons. Secondly, the human ALDH8 uses a downstream ATG initiation codon. A shorter polypeptide chain of 385 aa will be the gene product, if the first downstream in-frame ATG is used. Thirdly, the human *ALDH8* gene is a pseudogene. We are currently characterizing ALDH8 cDNA clones isolated from a human salivary gland library. The cDNA sequence data can reveal the nature of the human *ALDH8* gene.

Analysis of ALDH mRNA

To determine the tissue distribution of ALDH6, 7, 8 and GABALDH, we analyzed the levels of the *ALDH* expression by Northern blotting and RT-PCR analysis with respective specific probe and primers. These two analyses gave consistent results which are summarized in the Table 1. For each of the ALDH, the Northern blot analysis revealed a single positive band which was always consistent with the size of the respective cloned ALDH cDNA (data not shown).

The *ALDH6* gene is expressed at low levels in many tissues and at higher levels in salivary gland, stomach and kidney. The *ALDH7* is expressed primarily in kidney and lung.

Table 1. Characteristics of the newly cloned human *ALDH* genes

	ALDH6	ALDH7	ALDH8	GABALDH
Full length cDNA	3457 nt	2791 nt	-	2392 nt
5'-UT	85 nt	>47 nt	-	>57 nt
ORF	1536 nt (512 aa)	1404 nt (468 aa)	1398 nt (466 aa)	1482 nt (493 aa)
3'-UT	1853 nt	1337 nt	-	853 nt
Gene structure				
Size	37 kbp	20 kbp	37 kbp	>50 kbp
Exons	13	10	9 (coding region)	10
Chromosomal Localizations	15q26	11q13	11q13	1q
Tissue distribution	Wide (Salivary gland, stomach and kidney)	Kidney, lung	Parotid gland	Wide (Kidney, heart, liver and muscle)

The expression of the *ALDH8* is found only in the parotid gland. The *GABALDH* gene is widely expressed. Higher level of expression was found in human kidney, heart, liver and muscle. The subcellular localization as well as the functional significance of these ALDHs are presently unclear.

The Sequence Identity

The length of the known primary or deduced mammalian ALDH polypeptides ranges from 435 to 520 aa. They have been grouped into three classes based on their structural relationships. No gap is required to align the sequences of the class 1 and 2 isozymes (cytosolic ALDH1 and mitochondrial ALDH2, respectively). However, a total of 7 gaps (14 nucleotide positions) is required to place into the class 1/2 and 3 isozyme structures for maximal positional identity, and the class 1/2 and 3 isozyme structures have the NH2- and

Table 2. Percentage identity between the known human ALDH primary and/or deduced amino acid sequences

	ALDH1	ALDH2	ALDH5	ALDH6	ALDH3	ALDH7	ALDH8
Length	501	517	517	512	453	468	466
ALDH1	(81-91)[b]	68	65	70	28	22	22
ALDH2		(93-99)	74	65	28	23	23
ALDH5			--	63	28	24	24
ALDH6				--	27	26	26
ALDH3					(80/64-91)	52	50
ALDH7						--	87
ALDH8							--

[a] The alignments are based on the maximal positional identity.

[b] The range of the percentage positional identity of currently available ALDH sequences within each class is indicated in the parenthesis.

COOH- terminal extension, respectively. That is, the class 1/2 and 3 ALDH structures have a common core of about 430-440 aa. The human ALDH6, 7 and 8 also have this common core region. The ALDH6 sequence has significant similarity along its entire length to the class 1/2 isozymes and ALDH5, but not to the class 3 isozymes. However, the ALDH7 and 8 sequences are more similar to the class 3 structure. Table 2 summarizes the percentage positional identity between all the currently known human ALDH aa sequences.

ALDH6. The degree of identify of ALDH6/1, 6/2 and 6/5 is 70%, 65% and 63%, respectively, whereas the degree of identity of ALDH6/3, 6/7 and 6/8 is about 26%. Although ALDH6 sequence shares relatively better % identity with ALDH1 structure, the % is lower than that of any other pair of the existing class 1 ALDH member (81%-91%) and is near the % between class 1 and 2 (68%).

ALDH7 and 8. Similarly, although ALDH7 and 8 are more related to the ALDH3 than the ALDH1/2, the % of identity between 7/3 or 8/3 (about 50%) is much lower than that of any other pair of the existing cytosolic class 3 ALDH3 structures (80-91%, not including the microsomal ALDH structure).

GABALDH. In order to align the human GABALDH with other human ALDHs, it is necessary to introduce a total of 10 gaps (19 nucleotide positions) to GABALDH sequence for maximal positional identity. The GABALDH sequence shares about 40% identity with that of the ALDH1 and 2, and 23% identity with that of the ALDH3.

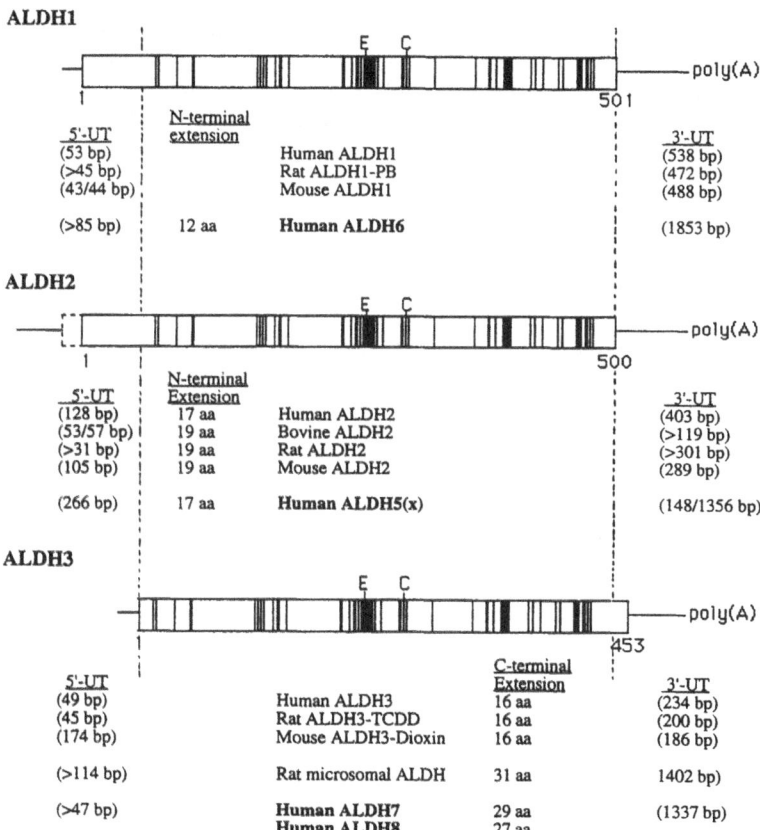

Figure 1. The alignment of the mammalian ALDH cDNAs.

The Conservation of the Amino Acid Residues of the Aldh Sequences

A total of 58 aa residues is conserved among the known mammalian ALDH isozymes including the three newly cloned human ALDHs, ALDH6, 7, and 8 (Fig. 1). Recently, Hempel et. al. reported the alignments of the primary sequences of 16 ALDH isozymes, including those with more narrow substrate preferences and those from bacteria, fungi and plants, by using UWGCG programs (Hempel et al., 1993). They found that most of the invariant residues are clustered with other nearly invariant residues (conserved in 14 or 15 of 16 aligned sequences) to form highly conserved segments. These invariant and nearly invariant residues are part of the 58 conserved aa as mentioned previously.

Among the 58 conserved aa, there are functionally important residues. Glu-268 and Cys-302 have been previously implicated in the active sites of the ALDH isozymes by the studies of selective chemical modifications (von Bahr-Lindstrm et al., 1985; Abriola et al., 1987). Furthermore, Gly-245/-250 have been suggested to be involved in the NAD-binding domain (Hempel et al., 1984).

The Genomic Structure

The restriction maps derived from the genomic clones of the human *ALDH6, 7, 8* and *GABALDH* were consistent with that obtained from Southern blot analysis of total genomic DNA digested by several restriction enzymes, respectively. Hence, these clones are able to represent the sequence organization of the cellular DNA. Table 1 summaries the genomic cloning results. The *ALDH6, 7, 8* and *GABALDH* gene are 37, 20, 37 and >>50 kbp long and contain 13, 10, 9 and 10 exons for the coding region, respectively. All exon/intron junctions of the respective *ALDH* gene conform to the consensus sequences for intronic donor and acceptor splice signals (Breathnach and Chambon, 1981). Most of the introns of a *ALDH* gene interrupt the coding sequence between the codons.

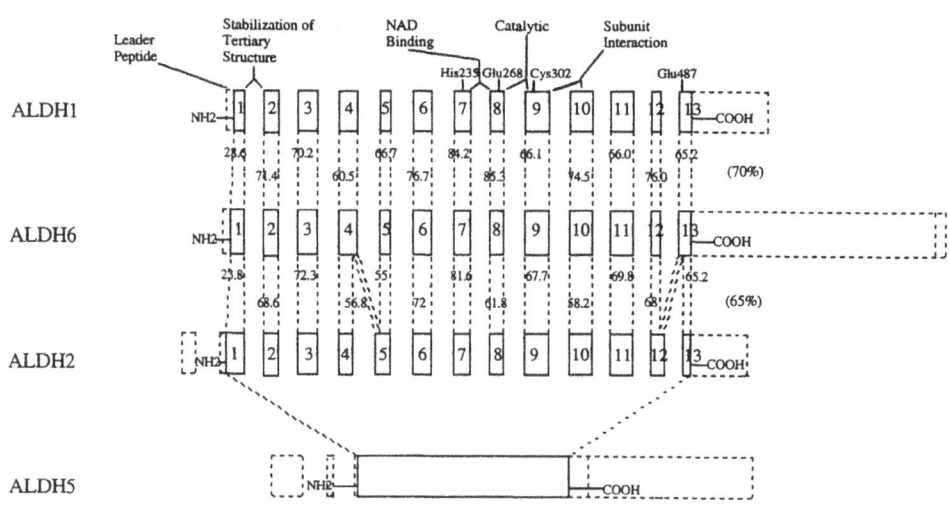

Figure 2. Conservation of exon organization and functionally important amino acid residues among the human *ALDH1, 2, 5,* and *6* genes.

The Conservation of the Exon Organization

Figure 2 and 3 demonstrate the conservation of exon organization and functionally important amino acid residues among the known human *ALDH* genes. It is interesting to note that there are two basic patterns of exon organization among the seven characterized human *ALDH* genes except *ALDH5. ALDH1, 2* and *6* contain 13 exons (Fig. 2), whereas *ALDH3, 7* and *8* have 9/10 exons for the coding region (Fig. 3). ALDH5 sequence shares 74% identity with ALDH2 and 65% identity with ALDH1 and 6, respectively (Table 2), its gene contains no introns in the coding region. This structure organization implicates its unique evolutionary event as to other *ALDH* genes.

ALDH1, 2 and 6. The boundaries of the individual exons are precisely conserved between the human *ALDH6* and *ALDH1* genes, whereas 10 out of the 12 introns of the *ALDH2* gene have the same exon-intron anatomy as that of the *ALDH1/6* gene (Fig. 2). Based on the facts of a higher % positional identity (Table 2) and the complete conservation of the exon organization between *ALDH6* and *1* gene, it is very likely that late gene duplications created *ALDH1* and *ALDH6* gene, whereas *ALDH1/ALDH6* and *ALDH2* genes were duplicated earlier during the evolution. Figure 2 also shows the current knowledge of the correlation between functional domains and exon organizations of the *ALDH* genes. Exon 1 contains signal sequence for final subcellular localization (Guan and Weiner, 1990; Jeng and Weiner, 1991; Thornton et al., 1993). Exon 1 and 2 encode sequences involved in the stabilization of the tertiary structures of the ALDH isozymes. It has been demonstrated that variants of ALDH1 and 2, lacking 56-residues in the NH2-terminal segment, formed dimeric instead of tetrameric structure (Hempel et al., 1985). A similar conclusion has also been obtained by the limited proteolysis (Loomes and Jörnvall, 1991). Exon 7 and 8 encode NAD-binding domain, while exon 8 and 9, which contain functionally important Glu-268 and Cys-302, respectively, encode active-site domain. A possible functional role of the exon 9 and 10 as the subunit interaction region has been proposed previously based on the different hydropathic properties between ALDH1 and 2 (Johansson et al., 1988).

ALDH3, 7 and 8. The comparison of the exon organization of the *ALDH3, 7 and 8* gene shows that the boundaries of the individual exons between the human *ALDH7* and *8* genes are precisely conserved (Fig. 3), whereas intron insertions or deletions were involved in the evolution of these genes.

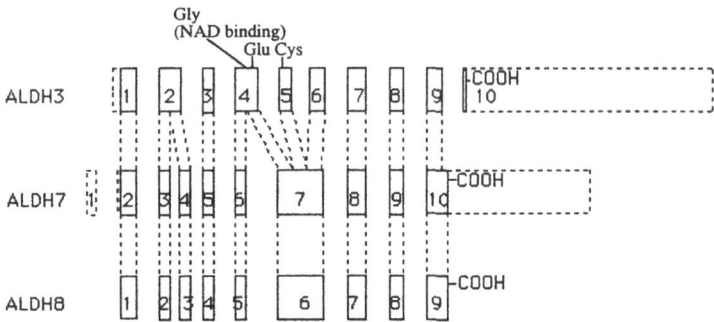

Figure 3. Conservation of exon organization and functionally important amino acid residues among the human *ALDH3, 7,* and *8* genes.

Chromosomal Localization

Chromosomal assignment of the human *ALDH6* , *7, 8* and *GABALDH* gene was determined by *in situ* hybridization. They were assigned to 15q26, 11q13, 11q13 and 1q, respectively. Southern blot analysis of the human genomic DNA treated with various restriction enzymes indicated that each of these genes is present as a single copy in the human genome (data not shown). It is known that the *ALDH1* gene was localized to chromosome 9q21, *ALDH2* gene to 12q24, and *ALDH5* gene to 9p13. However, there is no sufficient available data for the correlation of the duplication of the chromosomal segments among those closely related non-allelic *ALDH* genes (Cytogenetics and Cell Genetics: Human Gene Mapping vol. 58, 1991).

ACKNOWLEDGMENTS

This work was supported by NIH Grant AA05763.

REFERENCES

Abriola, D. P., Fields, R., Stein, S., Mackerell, A. D., Jr., and Pietruszko, R. (1987). Active site of human liver aldehyde dehydrogenase. *Biochemistry* **26**: 5679-5684.

Ambroziak, W., and Pietruszko, R. (1991). Human aldehyde dehydrogenase: Activity with aldehyde metabolites of monoamines, diamines and polyamines. *J. Biol. Chem.* **266**: 13011-13018.

Ambroziak, W., and Pietruszko, R. (1993). Metabolic role of aldehyde dehydrogenase. *Adv. in Exp. Med. Bio.* **328**: 5-15.

Breathnach, R., and Chambon, P. (1981). Organization and expression of eucaryotic split genes coding for proteins. *Annu. Rev. Biochem.* **50**: 349-383.

Duley, J. A., Harris, O., and Holmes, R. S. (1985). Analysis of human alcohol- and aldehyde-metabolizing isozymes by electrophoresis and isoelectric focusing. *Alcoholism: Clinical and Exp. Res.* **9**:263-271.

Forte-McRobbie, C. M., and Pietruszko, R. (1986). Purification and characterization of human liver "high Km" aldehyde dehydrogenase and its identification as glutamic-semialdehyde dehydrogenase. *J. Biol. Chem.* **261**: 2154-2163.

Guan, K., and Weiner, H. (1990). Sequence of the precursor of bovine liver mitochondrial aldehyde dehydrogenase as determined from its cDNA, its gene, and its functionality. *Arch. Biochem. Biophy.* **277**: 351-360.

Harada, S., Agarwal, D. P., and Goedde, H. W. (1980). Electrophoretic and biochemical studies of human aldehyde dehydrogenase isozymes in various tissues. *Life Sci.* **26**: 1773-1780.

Harada, S., Muramatsu, T., Agarwal, D. P., and Goedde, H. W. (1989). Polymorphism of aldehyde dehydrogenase in human saliva. *Prog. Clin. Biol. Res.* **290**: 133-139.

Harrington, M. C., Henehan, G. T. M., and Tipton, K. F. (1987). The roles of human aldehyde dehydrogenase isozymes in ethanol metabolism. *Prog. Clin. Biol. Res.* **232**: 111-125.

Hempel, J., von Bahr-Lindstrm, H., Jörnvall, H. (1984). Aldehyde dehydrogenase from human liver: Primary structure of the cytoplasmic isoenzyme. *Eur. J. Biochem.* **141**: 21-35.

Hempel, J., Kaiser, R., and Jörnvall, H. (1985). Mitochondrial aldehyde dehydrogenase from human liver: Primary structure, differences in relation to the cytosolic enzyme, and functional correlations. *Eur. J. Biochem.* **153**: 13-28.

Hempel, J., Nicholas, H., and Lindahl, R. (1993). Aldehyde dehydrogenases: Widespread structural and functional diversity within a shared framework. *Protein Science* **2**: 1890-1900.

Hsu, L. C., Bendel, R. E. and Yoshida, A. (1988). Genomic structure of the human mitochondrial aldehyde dehydrogenase gene. *Genomics* **2**: 57-65.

Hsu, L. C., Chang, W.-C., and Yoshida, A. (1989). Genomic structure of the human cytosolic aldehyde dehydrogenase gene. *Genomics* **5**: 857-865.

Hsu, L. C., and Chang W.-C. (1991). Cloning and characterization of a new functional human aldehyde dehydrogenase gene. *J. Biol. Chem.* **266**: 12257-12265.

Hsu, L. C., Chang, W-C., Shibuya, A., and Yoshida, A. (1992). Human stomach aldehyde dehydrogenase: DNA and genomic cloning, primary structure, and expression in *Escherichia coli. J. Biol. Chem.* **267**: 3030-3037.

Jakoby, W. B., and Ziegler, D. M. (1990). The enzymes of detoxication. *J. Biol. Chem.* **265**:20715-20718.

Jeng, J., and Weiner, H. (1991). Purification and characterization of catalytically active precursor of rat liver mitochondrial aldehyde dehydrogenase expressed in *Escherichia coli. Arch. Biochem. Biophy.* **289**: 214-222.

Johansson, J., von Bahr-Lindsrm, H. , Jeck, R., Woenckhaus, C., and Jrnvall, H., (1988). Mitochondrial aldehyde dehydrogenase from horse liver: Correlations of the same species variants for both the cytosolic and the mitochondrial forms of an enzyme. *Eur. J. Biochem.* **172**: 527-533.

King, G., and Holmes, R.S. (1993). Human corneal aldehyde dehydrogenase: purification, kinetic characterization and phenotypic variation. *Biochem. Mol. Biol. Int.* **31**: 49-63.

Koivusalo, M., Aarnio, M., Baumann, M., Rautoma, P. (1989). NAD(P)-Linked aromatic aldehydes preferring cytoplasmic aldehyde dehydrogenases in the rat. Constitutive and inducible forms in liver, lung, stomach and intestinal mucosa. *Prog. Clin. Biol. Res.* **290**: 19-33.

Kozak, M. (1987). An analysis of 5'−noncoding sequences from 699 vertebrate messenger RNAs. *Nucleic Acid Res.* **15**: 8125-8148.

Kurys, G., Shah, P. C., Kikonyogo, A., Reed, D. Ambroziak, W., and Pietruszko, R. (1993). Human aldehyde dehydrogenase: cDNA cloning and primary structure of the enzyme that catalyzes dehydrogenation of 4-aminobutyraldehyde. *Eur. J. Biochem.* **218**: 311-320.

Lin, S. W., Hsu, L. C., and Yoshida, A. (1993). Cloning and characterization of cDNA for human γ-aminobutyraldehyde dehydrogenase. *FASEB J.* 7:A1297.

Lindahl, R., and Evces, S. (1984). Rat liver aldehyde dehydrogenase II. Isolation and characterization of four inducible isozymes. *J. Biol. Chem.* **259**: 11991-11996.

Lindahl, R. (1992). Aldehyde dehydrogenases and their role in carcinogenesis. *Crit. Rev. Biochem Mol. Biol.* **27**:283-335.

Loomes, K., and Jörnvall, H. (1991). Structural organization of aldehyde dehydrogenases probed by limited proteolysis. *Biochemistry* **30**: 8865-8870.

Mitchell, D. Y., and Petersen, D. R. (1987). The oxidation of α,β-unsaturated aldehydic products of lipid peroxidation by rat liver aldehyde dehydrogenases. *Toxicol. Appl. Pharmacol.* **87**: 403-410.

Parrilla, R., Okhawa, K., Lindros, K. O., Zimmerman, U.-I. P., Kobayashi, K., and Williamson, J. R. (1974). Functional compartmentation of acetaldehyde oxidation in rat liver. *J. Biol. Chem.* **249**: 4926-4933.

Pietruszko, R. (1983). Aldehyde dehydrogenase isozymes. *Isozymes: Current Topics in Biol. Med. Res.* **8**: 195-217.

Raghunathan, L, Hsu, L. C., Klisak, I., Sparkes, R. S., Yoshida, A., and Mohandas, T. (1988). Regional localization of the human genes for aldehyde dehydrogenase-1 and aldehyde dehydrogenase-2. *Genomics* **2**: 267-269.

Ryzlak, M. T., and Pietruszko, R. (1989). Human brain glyceraldehyde-3-phosphate dehydrogenase, succinic semialdehyde dehydrogenase and aldehyde dehydrogenase isozymes: substrate specificity and sensitivity to disulfiram. *Alcohol Clin. Exp. Res.* **13**: 755-761.

Santisteban, I., Povey, S., West, L. F., Parrington, J. M., and Hopkinson, D. A. (1985). Chromosome assignment, biochemical and immunological studies on a human aldehyde dehydrogenase, ALDH3. *Ann. Hum. Genet.* **49**: 87-100.

Thornton, K., Wang, Y., Weiner, H., and Gorenstein, D. G. (1993). Import, processing, and two-dimensional NMR structure of a linker-deleted signal peptide of rat liver mitochondrial aldehyde dehydrogenase. *J. Biol. Chem.* **268**: 19906-19914.

Vasiliou, V., Reuter, S. F., Kozak, C. A., and Nebert, D. W. (1993). Mouse dioxin-inducible cytosolic aldehyde dehydrogenase-3: AHD4 cDNA sequence, genetic mapping, and differences in mRNA levels. *Pharmacogenetics* 3:281-290.

von Bahr-Lindstrm, H., Jeck, R., Woenckhaus, C., Sohn, S., Hempel, J., and Jrnvall, H. (1985). Characterization of the coenzyme binding site of liver aldehyde dehydrogenase: Differential reactivity of coenzyme analogues. *Biochemistry* **24**: 5847-5851.

Wang, S.-L., Wu, C.-W., Cheng, T.-C., and Yin, S.-J. (1990). Isolation of high-Km aldehyde dehydrogenase isoenzymes from human gastric mucosa. *Biochem. Int.* **22**: 199-204.

Yin, S.-J., Liao, C.-S., Wang, S.-L., Chen, Y.-J., and Wu, C.-W. (1989). Kinetic evidence for human liver and stomach aldehyde dehydrogenase-3 representing a unique class of isozymes. *Biochem. Genet.* **27**. 321-331.

Yin, S.-J. Vagelopoulos, N., Wang, S.-L., and Jrnvall, H. (1991). Structural features of stomach aldehyde dehydrogenase distinguish dimeric aldehyde dehydrogenase as a "variable" enzyme. "Variable" and "constant" enzymes within the alcohol and aldehyde dehydrogenase families. *FEBS Lett.* **283**: 85-88.

Yoshida, A., Hsu, L.C., and Yasunami, M. (1991). Genetics of human alcohol-metabolizing enzymes. *Prog. Nuc. Acid Res. and Mol. Biol.* **40**. 255-287.

Yoshida, A., Hsu, L.C., and Dav, V. (1992). Retinal oxidation activity and biological role of human cytosolic aldehyde dehydrogenase. *Enzyme* **46**: 239-244.

Yoshida, A., Hsu, L.C., and Yanagawa, Y. (1993). Biological role of human cytosolic aldehyde dehydrogenase I: Hormonal response, retinal oxidation and implication in testicular feminization. Adv. in Exp. Med. Bio. **328**: 37-44.

22

NEW HUMAN ALDEHYDE DEHYDROGENASES

Regina Pietruszko, Pritesh C. Shah, Alexandra Kikonyogo,
Ming-Kai Chern and Teresa Lehmann

Center of Alcohol Studies
Rutgers University
Piscataway, New Jersey 08855-0969

Electrophoretic separation techniques of alcohol and aldehyde dehydrogenases have been generally employed for identification of the individual isozyme components. The electrophoretic patterns of alcohol dehydrogenase, a largely cathodal enzyme, are relatively easy to interpret because only a few other proteins migrate towards cathode and different isozymes have different isoelectric points. Aldehyde dehydrogenase, however, is largely an anodal enzyme, which migrates in the direction of the majority of other proteins, large numbers of which superimpose with aldehyde dehydrogenase. Some of these proteins produce gel bleaching during nitroblue tetrazolium / phenazine methosulfate color development and thereby mask some weaker aldehyde dehydrogenase bands. This is the reason why only the major bands of aldehyde dehydrogenase are usually visualized. In addition, there are several different aldehyde dehydrogenases that have the same pI as the major bands and therefore superimpose on isoelectric focusing gels. The superimposals that are now known to occur as a result of work of a large number of investigators (Greenfield and Pietruszko, 1977; Impraim et al., 1982; Jones and Teng, 1983; Palmer and Jenkins, 1985; Santisteban et al., 1985; Forte McRobbie and Pietruszko, 1986; Ryzlak and Pietruszko, 1988 a, b; Kurys et al., 1989; Wang et al., 1990) are shown in Table 1. The best known superimposals occur at pI 5.3 where the human E1 isozyme, human E3 isozyme and the Oriental E2 isozyme are found.

There are several other aldehyde dehydrogenases in the human liver and brain that have recently been purified in our laboratory. Some also superimpose with known aldehyde dehydrogenases on isoelectric focusing (Table 1).

The new enzymes, whose pI values are shown in Table 1, have been obtained homogeneous by employing preliminary purification procedures via ion exchange chromatography, on CM-Sephadex and DEAE Sephadex, followed by affinity chromatography on 5'AMP Sepharose 4B, Blue Sepharose, or Acetophenone Sepharose (Ghenbot and Weiner, 1992). Properties of two homogeneous enzymes from human liver, currently available in our laboratory, are listed in Table 2. The first one (pI = 5.5) could be the same as that previously described by Jones and Teng (1983) and by Palmer and Jenkins (1985). It is distinct from the E3 isozyme minor component (Kurys et al., 1989) because it has no activity

Table 1. Relationship Between Human Aldehyde Dehydrogenases and Their Isoelectric Points

pI	Enzymes Previously Identified (Reference)	New Enzymes
>9.0		New cathodic enzyme.
6.8 - 9	High Km for acetaldehyde , utilizing glyceraldehyde-3-phosphate as substrate (Ryzlak and Pietruszko, 1988 a).	
6.8 - 6.9	High Km for acetaldehyde, utilizing glutamic -γ -semialdehyde as substrate (Forte-McRobbie and Pietruszko, 1986).	
6.3 - 7.2	High Km for acetaldehyde, utilizing succinic semialdehyde as substrate (Ryzlak and Pietruszko, 1988 b).	
6.5	ALDH 3 - high Km for acetaldehyde (Santisteban et al., 1985; Wang et al., 1990).	
5.5	Enzyme described by Jones and Teng (1983) and by Palmer and Jenkins (1985); also a minor component of the E3 isozyme (Kurys et al., 1989).	New enzyme purified from human liver.
5.3	E1 isozyme (Greenfield and Pietruszko, 1977; Oriental E2 (Impraim et al., 1982; E3 isozyme (Kurys et al., 1989).	New enzyme from brain; enzyme like E1 (liver).
5.1	E2 isozyme (Greenfield and Pietruszko, 1977).	Enzyme like E2 (liver).

with 4-aminobutyraldehyde. Its Km for acetaldehyde is considerably larger than that of the E1 isozyme, while its Km for NAD resembles that of the E2 isozyme. Its subunit MW resembles those of E1 and E2 isozymes. The second enzyme from liver with cathodic (pH > 9) pI appears to be an aminoaldehyde dehydrogenase which efficiently utilizes aminoacetaldehyde as substrate (Km = 15 μM). Its Km for 4-aminobutyraldehyde is 40 μM. Its Km with acetaldehyde is greater than 1 mM while its Km for NAD is similar to that of the E2 isozyme. The enzyme is a dimer with subunit MW of ca. 46,000 daltons.

An enzyme from rat and bovine brain capable of metabolizing 4-aminobutyraldehyde was previously described (Abe et al., 1990; Lee and Cho, 1992). The reported Km for 4-aminobutyraldehyde of the enzyme from rat brain was 151 μM; while that from bovine brain was 154 μM, one to two orders of magnitude greater than that of the human liver E3 isozyme. The brain enzyme is of interest because metabolism of putrescine is different in brain from that in liver (see Figure 1). In the liver putrescine

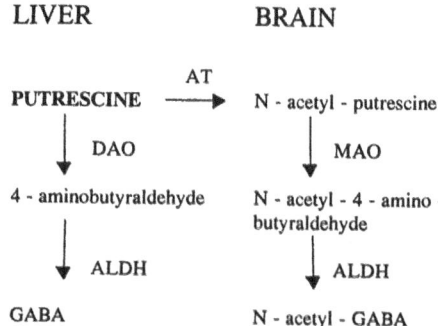

Figure 1. Metabolic pathway of putrescine to 4-aminobutyric acid in liver and in brain. AT = N-acetyl transferase; DAO = diamine oxidase; MAO = monoamine oxidase; ALDH = aldehyde dehydrogenase; GABA = 4-aminobutyric acid.

Table 2. New Aldehyde Dehydrogenases and Their Properties

Enzyme Source and pI		Subunit MW (daltons)	Native MW (daltons)	Km (μM)		
				Acet.	4-Abal.	NAD
Liver	5.5	54,000	-	900	inactive	80
Liver	>9	46,000	90,000	>1,000	40	70
Brain	5.3	54,000	-	140	7	4

Km values determined at pH 7.0 in 30 mM phosphate buffer, at 25°C. Acet. = acetaldehyde; 4-Abal. = 4-aminobutyraldehyde.

is converted to 4-aminobutyraldehyde via diamine oxidase and then to 4-aminobutyric acid via aldehyde dehydrogenase. In the brain putrescine is first acetylated by N-acetyl transferase and then converted via monoamine oxidase and aldehyde dehydrogenase to N-acetyl-4-aminobutyric acid. Deacylation is a necessary step in brain to form 4-aminobutyric acid (Caron et al., 1987). Diamine oxidase which deaminates putrescine in liver is absent from mammailan brain.

Employing the purification procedure used for the bovine brain enzyme (Lee and Cho, 1992) with some modifications, we purified from human brain the enzyme catalyzing dehydrogenation of 4-aminobutyraldehyde. The result of its preliminary characterization is shown in Table 2. It can be seen that the Km value for 4-aminobutyraldehyde is low as is Km for NAD; both are similar to those of the E3 isozyme from human liver (Kurys et al, 1989; Ambroziak and Pietruszko, 1991). Subunit MW as well as pI value are also similar. These suggest that the major 4-aminobutyraldehyde-metabolizing enzyme from human brain is the E3 isozyme. There are, however, some differences: (1) in Km for acetaldehyde, which is ca. 3 times larger; at the present time it is not clear whether this difference is real or a result of an experimental error; (2) stability of the brain enzyme was much greater than that of the liver E3 isozyme and the minor component was never observed. Attempts are being made to determine if the brain and liver enzyme are products of the same or different gene.

In addition to the new enzymes listed in Table 2, other differences were noted during enzyme purification. Some human livers contain a large amount of an enzyme that has properties similar to those of the E2 isozyme. Its Km for acetaldehyde is 1 μM, and its pI value is ca pH 5.1 but it does not bind to DEAE Sephadex (the E2 isozyme readily binds to DEAE Sephadex in our purification conditions). The same livers also contain an enzyme that behaves like E1 isozyme (the same Km and pI) and also does not bind to DEAE Sephadex. All the above enzymes are NAD-linked and appear to have broad substrate specificity. Information about their catalytic properties and structural relationship to E1, E2 and E3 isozymes is being currently obtained.

The information presented in this paper, in all probability, is far from completing identifying and characterizing aldehyde dehydrogenases present in the human organism. Our previous experiments with human brain aldehyde dehydrogenases showed preliminary evidence of great complexity (Pietruszko et al., 1987). Isoelectric focusing of human brain homogenate showed only two activity bands, however, following pre-purification of the same homogenate on 5'AMP Sepharose 4B column eleven aldehyde dehydrogenase activity bands were visualized. Their pI values ranged from pI 5-10; many of those, located in the pI 7-9 region, have not yet been characterized.

ACKNOWLEDGEMENTS

Financial support of USPHS Grant 1R01AA00186 and Research Scientist Award K05AA00046 from NIAAA is gratefully acknowledged.

REFERENCES

Abe, T., Takada, K., Ohkawa, K., and Matsuda, M. (1990) Purification and characterization of rat brain aldehyde dehydrogenase able to metabolize γ-aminobutyraldehyde to γ-aminobutyric acid. Biochem. J. 269, 25-29.

Ambroziak, W., and Pietruszko, R. (1991) Human aldehyde dehydrogenase activity with aldehyde metabolites of monoamines, diamines and polyamines. J. Biol. Chem. 266, 13011-13018.

Caron, P.C., Kremzner, L.T., and Cote, L.J. (1987) GABA and its relationship to putrescine metabolism in the rat brain and pancreas. Neurochem. Int. 10, 219-229.

Forte-McRobbie, C.M., and Pietruszko, R. (1986) Purification and characterization of human liver "high Km" aldehyde dehydrogenase and its identification as glutamic-γ-semialdehyde dehydrogenase. J. Biol. Chem. 261, 2154-2163.

Ghenbot, G., and Weiner, H. (1992) Purification of liver aldehyde dehydrogenase by p-hydroxyacetophenone-Sepharose affinity matrix and the coelution of chloramphenicol acetyl transferase from the same matrix with recombinantly expressed aldehyde dehydrogenase. Prot. Express. Purif. 3, 470-478.

Greenfield, N.J., and Pietruszko, R. (1977) Two aldehyde dehydrogenases from human liver. Isolation via affinity chromatography and characterization of the isozymes. Biochim. Biophys. Acta 483, 35-45.

Impraim, C., Wang, G., and Yoshida, A. (1982) Structural mutation in a major human aldehyde dehydrogenase gene results in loss of enzyme activity. Am. J. Hum. Genet. 34, 837-841.

Jones, G.L., and Teng, Y.S. (1983) A chemical and enzymological account of human liver aldehyde dehydrogenase. Implications for ethnic differences in alcohol metabolism. Biochim. Biophys. Acta. 745, 162-174.

Kurys, G., Ambroziak, W., and Pietruszko, R. (1989) Human aldehyde dehydrogenase. Purification and characterization of a third isozyme with low Km for γ-aminobutyraldehyde. J. Biol. Chem. 264, 5715-5721.

Lee, J.E., and Cho, Y.D. (1992) Purification and characterization of bovine brain γ-aminobutyraldehyde dehydrogenase. Biochem. Biophys. Res. Commun. 189, 450-454.

Palmer, K.R., and Jenkins, W.J. (1985) Aldehyde dehydrogenase in alcoholic subjects. Hepatology 5, 260-263.

Pietruszko, R., Ryzlak, M.T., and Forte-McRobbie, C.M. (1987) Multiplicity and identity of human aldehyde dehydrogenases. Advances in Biomedical Alcohol Research pp 175-179, Lindros, K.O., Ylikahri, R., and Kiianmaa, K. Eds. Pergamon Press.

Ryzlak, M.T., and Pietruszko, R. (1988 a) Heterogeneity of glyceraldehyde-3-phosphate dehydrogenase from human brain. Biochim. Biophys. Acta 954, 309-324.

Ryzlak, M.T., and Pietruszko, R. (1988 b) Human brain high Km aldehyde dehydrogenase: purification and identification as NAD$^+$-dependent succinic semialdehyde dehydrogenase. Arch. Biochem. Biophys. 266, 386-396.

Santisteban, I., Powey, S., West, L.F., Parrington, J.M., and Hopkinson, D.A. (1985) Chromosome assignment, biochemical and immunological studies on human aldehyde dehydrogenase, ALDH3. Ann. Hum. Genet. 49, 87-100.

Wang, S.L., Wu, C.W., Cheng, T.C., and Yin, S.J. (1990) Isolation of high Km aldehyde dehydrogenase isozymes from human gastric mucosa. Biochem. Int. 22, 199-204.

23

RETINOIC ACID SYNTHESIZING ENZYMES IN THE EMBRYONIC AND ADULT VERTEBRATE

Peter McCaffery and Ursula C. Dräger

E. Kennedy Shriver Center and
Department of Psychiatry
Harvard Medical School
Waltham MA 02254

INTRODUCTION

The oxidation of retinaldehyde to retinoic acid (RA) provides the retinoid form of highest potency for a variety of cellular systems. RA has been implicated in many processes, such as growth and differentiation of epithelia in the adult organism (De Luca 1991), and determination of the antero-posterior axis for the limb bud (Eichele and Thaller 1987; Tickle et al. 1982) and the entire body of the vertebrate embryo (Durston et al. 1989; Hogan, Thaller, and Eichele 1992). In addition, RA is thought to promote neuronal survival, differentiation and neurite outgrowth (Haskell et al. 1987; Quinn and De Boni 1991; Wuarin, Sidell, and De Vellis 1990). RA exerts its effects by binding to specific nuclear receptors that regulate transcription. The diversity in RA actions is commonly attributed to differences in local expression patterns of different receptors and cytoplasmic binding proteins that modify the availability of intracellular RA (Giguère 1994). In addition, however, retinoid metabolism may contribute significantly to local diversity in RA actions. Retinoid metabolism includes the processes of precursor circulation and cellular uptake mediated by binding proteins, the reversible oxidation of retinol to retinaldehyde, the irreversible oxidation of retinaldehyde to RA, and RA degradation. Here we focus on the enzymes that mediate the oxidation of retinaldehyde to RA.

The synthesis of RA has been studied in a variety of tissues and cell lines (Napoli 1986; Napoli and Race 1987), and both NAD-dependent dehydrogenases and flavoprotein-linked oxidases have been implicated in retinaldehyde oxidation, depending on cell type and species (Bhat, Poissant, and Lacroix 1988; Lee, Manthey, and Sladek 1991; McCaffery et al. 1992). In adult mouse liver, at least 90% of RA synthesis is NAD-dependent, being mediated by the two class-1 aldehyde dehydrogenases AHD2 and AHD7, as determined by HPLC assay (Lee, Manthey, and Sladek 1991). In the course of studies on the retina of the embryonic mouse, we found evidence for NAD-dependent retinaldehyde oxidation by unknown enzymes. For their characterization we developed a technique based on RA reporter

Enzymology and Molecular Biology of Carbonyl Metabolism 5
Edited by H. Weiner *et al.*, Plenum Press, New York, 1995

cells (Wagner, Han, and Jessell 1992), a zymography bioassay of charge separated proteins which allows to test minute amounts of embryonic tissue that cannot easily be analyzed by any other method (McCaffery et al. 1992). The present study presents an overview of retinaldehyde oxidizing enzymes detected by the zymography bioassay in embryonic and adult vertebrates.

RETINALDEHYDE DEHYDROGENASES IN THE EYE

The vertebrate retina derives in the early embryo as an outgrowth from the diencephalon and develops into a layered neuronal sheet that appears completely homogeneous throughout its extent: there are no obvious anatomical or biochemical differences between different regions of the developing retina. Nevertheless, the existence of some biochemical differences has been postulated in order to account for the phenomenon of retinotopic projections: optic axons project to visual areas in the brain in a topographical manner, creating central neuronal representations of the visual world. The general lay-out of these topographical maps, which is identical in all vertebrates, is thought to be created by a biochemically-based Cartesian coordinate system of positional information established in the early embryo, one in the antero-posterior, the other in the dorso-ventral dimension of the map. In a search for the biochemical nature of such positional determinants, we detected a very abundant cytosolic protein in the dorsal half of the embryonic mouse retina, and through microsequencing we identified it as AHD2, the major RA-producing retinaldehyde dehydrogenase (Lee, Manthey, and Sladek 1991; McCaffery et al. 1991).

Comparative estimates of RA levels in dorsal and ventral halves of embryonic mouse retinas, however, showed substantially higher RA levels in the ventral halves that lack AHD2 (McCaffery et al. 1992). These RA estimates were done by co-culturing pieces of embryonic retinas with the RA reporter cells that respond to different RA levels by proportional synthesis of β-galactosidase (Wagner, Han, and Jessell 1992). In order to identify the source of the high RA levels in ventral embryonic retina, extracts of dorsal and ventral retina halves were separated by agarose-gel isoelectric focusing, and gel slices were assayed for RA synthesis from retinaldehyde with the reporter cells (McCaffery et al. 1992). Figure 1 (left) shows such zymography-bioassay traces for the retina of the day-13.5 mouse embryo: the dorsal retina contains the basic AHD2 (pI 7.6), and the ventral retina three acidic activities, labelled V1, V2, and V3. Like AHD2, the acidic activities require NAD as cofactor. The V1 activity at pH 5.7 represents the major ventral-specific dehydrogenase; at a protein level comparison, V1 is at least two orders of magnitude more efficient in RA synthesis than AHD2 (McCaffery and Dräger 1993; McCaffery et al. 1992). The V3 peak is localized at the point of sample loading and likely consists of membrane-trapped or precipitated V1 enzyme. The V2 activity, barely visible in Figure 1A, probably originates in the retinal pigment epithelium, a layer adjacent and continuous with the embryonic retina.

An asymmetrical distribution of different retinaldehyde dehydrogenases in the retina is conserved between divergent species. In the embryonic chick (Figure 1, middle) a basic AHD2 homolog (D) is present in the dorsal half of the retina and a slightly more acidic dehydrogenase (V1) in the ventral half. In addition, a ventral specific dehydrogenase peak is located at the gel origin, similar to the mouse V3 peak. The retina of the embryonic zebrafish (Marsh-Armstrong et al. 1994) contains single dorsal (D) and ventral (V) dehydrogenase activities (Figure 1, right). In contrast to the mouse retina where overall levels of retinaldehyde dehydrogenases decrease dramatically between embryonic and adult age (McCaffery and Dräger 1993; McCaffery et al. 1993), in zebrafish high enzymes levels persist throughout life, consistent with the continued retinal growth in adult amphibians (Marsh-Armstrong et al. 1994). In the embryonic retinas of all three vertebrate classes,

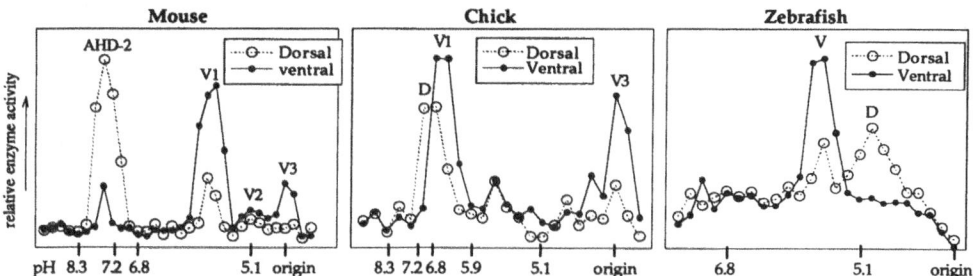

Figure 1. Zymography bioassays comparing retinaldehyde dehydrogenases in dorsal and ventral halves of retina from embryonic day-13.5 (E13.5) mouse (left), E6 chicken (middle) and 2.5-day zebrafish embryos (right). The enzyme activity measurements represent arbitrary units based on colorimetric readings of the β-galactosidase activity induced in the reporter cells. The samples were processed in parallel with isoelectric focusing standards (Sigma) as indicated.

including mammals, birds and fish, the asymmetric distribution of retinaldehyde dehydrogenases results in substantially higher RA levels in ventral than dorsal retina (Marsh-Armstrong et al. 1994; McCaffery et al. 1992).

Retinaldehyde dehydrogenases are not confined to the neural retina but are found also in other ocular tissues; different tissues express different enzymes or combinations of enzymes, and levels and patterns of expression vary with developmental age. A comparison of retinaldehyde dehydrogenases in the eye of the day-14 mouse embryo is shown in Figure 2. The neural retina contains the AHD2, V1 and V3 activities, the cornea has only V1 activity, and the lens expresses both AHD2 and V1 activities. In lenses of several adult mammals, cytosolic aldehyde dehydrogenases have been implicated as crystallins (Wistow and Kim 1991). The retinal pigment epithelium (RPE) shows both V1 and V2 activities in the sample assayed here. As in samples from older embryos we find only V2 (not shown), it is possible that the V1 detected here originated from the region of retina/RPE transition at the optic fissure in the ventral eye. The assayed ocular tissue samples were not normalized for protein

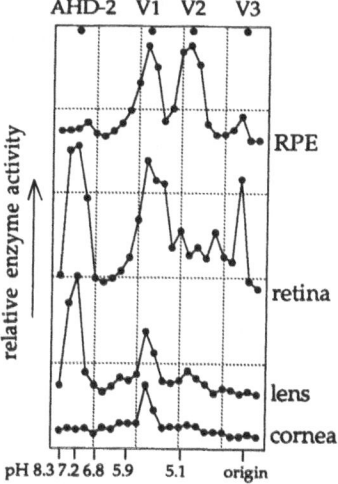

Figure 2. Zymography bioassays comparing retinaldehyde dehyrogenases in E14 retinal pigment epithelium (RPE), retina, lens and cornea. The samples were not normalized for protein content.

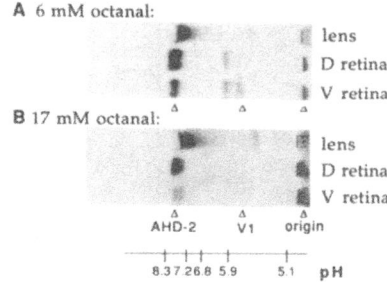

Figure 3. Standard zymography of aldehyde dehydrogenases present in lens, and dorsal (D) and ventral (V) retina halves of day-14 mouse embryos. Enzymes were visualized using octanal as substrate at concentrations of 6 mM (A) and 17 mM (B), NAD as cofactor and phenazine methosulfate and nitroblue tetrazolium as color reagents responsive to oxidation. Reaction was performed in 30 mM phosphate buffer at pH 8.9. Sigma isoelectric focusing standards are shown at the bottom of the figure.

content, in order to facilitate enzyme type comparisons. At E14, RA synthesis in the neural retina far outweighs synthesis in other ocular tissues, but in the newborn mouse, relatively highest levels of RA synthesis undergo a shift: postnatally the RPE represents the by far RA-richest tissue in the eye (unpublished observation). As during the postnatal period the RA-rich RPE is located directly adjacent to rod photoreceptors undergoing final mitosis, such proximity is likely to influence rod determination, a process known to be regulated by RA (Kelley and Reh 1993).

CHARACTERIZATION OF THE V1 RETINALDEHYDE DEHYDROGENASE

Two of the retinaldehyde dehydrogenases identified in the embryonic eye, AHD2 and V2, have been characterized in terms of substrate specificities and amino acid sequence (McCaffery et al. 1992; Rongnoparut and Weaver 1991; Zhao et al. 1994): while AHD2 can oxidize a broad range of substrates, V2 is retinaldehyde-specific, and the two enzymes share 70% homology at the protein level. By contrast, the V1 dehydrogenase is mainly identified by its isoelectric point. To characterize this enzyme further, we analyzed dorsal and ventral embryonic retina halves by standard isoelectric focusing gel zymography with substrates other than retinaldehyde. No enzyme of the correct isoelectric point was evident in the ventral retina using the substrates acetaldehyde, benzaldehyde or propionaldehyde (not shown). However, when 6mM octanal is used as substrate, a ventral specific enzyme with isolectric point of 5.7 becomes apparent, which is very likely the V1 dehydrogenase (Figure 3A). With higher octanal concentrations (17mM), this enzyme can no longer be visualized (Figure 3B), indicating that the characteristic substrate inhibition of V1, observed previously with retinaldehyde (McCaffery et al. 1992), extends to other substrates.

To determine the native size of the V1 dehydrogenase, we separated proteins by non-denaturing gradient polyacrylamide electrophoresis and analyzed this gel with the zymography bioassay technique. The embryonic dorsal and ventral retinas each express a single enzymatic peak (Figure 4) which are likely to represent V1 and AHD2, the major soluble ventral and dorsal retinaldehyde dehydrogenases here. The V1 peak runs with a mobility slightly lower than the 272 kDa molecular weight marker (Figure 4) and similar to, but slightly higher than other aldehyde dehydrogenases in the retina identified by im-munoblotting (Figure 5); most cytosolic aldehyde dehydrogenases are known to have a

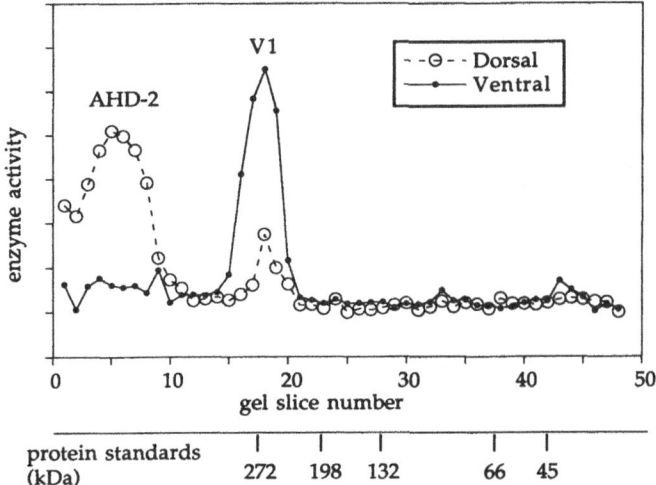

Figure 4. Zymography bioassay of of E13.5 dorsal and ventral retina halves separated by 3-27% polyacrylamide gel nondenaturing electrophoresis. Electrophoresis was performed overnight at 70 volts. Gels were sliced, and each slice assayed for synthesis of retinoic acid from retinaldehyde similar to the zymography bioassay of agarose gels. Migration of nondenaturing molecular weight standards (Sigma) is shown underneath.

native molecular weight of 240 kDa. When the retina was compared by nondenaturing gel analysis/immunoblotting to liver, kidney and testes, all four tissues showed aldehyde dehydrogenases of identical mobility (not shown). This suggests that V1 occurs as a tetrameric complex like other cytoplasmic aldehyde dehydrogenases. It is interesting to note that AHD2 runs with an anomalously low mobility (dorsal traces, Figures 4 and 5); this is likely due to the low negative charge of this basic enzyme in the pH 8.9 buffer system used for the gradient gel.

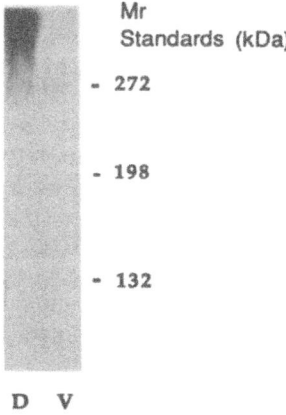

Figure 5. Immunoblot of E13.3 dorsal and ventral retina halves, separated by 3-27% polyacrylamide gel nondenaturing electrophoresis. The blot was probed with an antiserum to rat class I aldehyde dehydrogenase (Lindahl and Evces 1984). This antiserum recognizes several cytoplasmic aldehyde dehydrogenases but not the V1 dehydrogenase.

RETINALDEHYDE DEHYDROGENASES IN ADULT NON-NEURONAL TISSUES

While some form of RA-mediated transcriptional regulation probably takes place in most or all embryonic and adult cells, several adult organs have been particularly tightly implicated in RA function and metabolism (De Luca 1991). In the skin RA is important for keratinocyte differentiation (Asselineau et al. 1989), in the testes it is required for spermatogenesis (Lufkin et al. 1993), and both kidney and liver are involved in retinoid metabolism and storage (Blomhoff et al. 1990); all four tissues have been reported to be capable of NAD-dependent retinoic acid synthesis (Conner and Smit 1987; Napoli and Race 1987). Zymographical determinations of retinaldehyde dehydrogenase profiles in these tissues (Figure 6A) show at least four dehydrogenase activities, focussing at pH 8.0, 6.4, 5.7 and 5.0; these include AHD2 (pI 8.0), V1 (pI 5.7) and V2 (pI 5.0).

The pituitary has not been reported as a region of RA synthesis, but it contains high levels of an otherwise rare RA receptor, an RXRγ (Mangelsdorf et al. 1992). It expresses a major retinaldehyde dehydrogenase at pH 5.0, possibly in addition to some minor activity obscured at the acidic shoulder of the major activity. All pituitary activity is restricted to the anterior lobe that secretes trophic peptide hormones, including thyroid-stimulating hormone and gonadotrophins. RA, thyroid and steroid hormones bind to different members of the same nuclear receptor superfamily (Evans 1988), and the anterior lobe of the pituitary may be a region in which RA can regulate the release of thyroid and steroid hormones.

Like in the retina, retinaldehyde dehydrogenase profiles and levels change with developmental stage of peripheral organs, as illustrated here in a comparison of the embryonic and adult liver (Figure 6B); samples were normalized for protein content. While in the day-15 embryo, very little AHD2 is detectable, this dehydrogenase is prominent in the adult. In addition there is an increase in overall enzyme activity levels in this organ. This contrasts

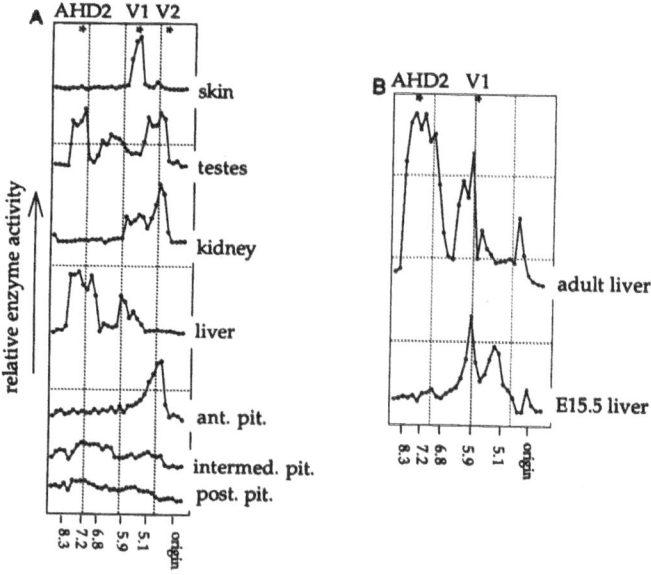

Figure 6. A: Zymography bioassays comparing retinaldehyde dehydrogenases in adult skin, testes, kidney, liver and anterior, intermediate and posterior pituitary; no activity is located at the gel origin, as the samples were high-speed supernatants, spun in an airfuge. B: Comparison of retinaldehyde dehydrogenases in embryonic and adult liver.

with the nervous system, such as retina and spinal cord, where retinaldehyde dehydrogenase levels are high in the embryo but substantially lower in the adult (McCaffery and Dräger 1994b; McCaffery et al. 1993).

RETINOIC ACID SYNTHESIS IN THE ADULT BRAIN

Overall levels of RA and retinaldehyde dehydrogenases in the adult central nervous system are rather low in comparison to levels in adult liver, testes and kidney. Longitudinal comparisons of the developing retina and spinal cord show a temporal and spatial correlation of high levels with regions of neuronal differentiation but not adult function (Marsh-Armstrong et al. 1994; McCaffery and Dräger 1994b; McCaffery et al. 1993). Nevertheless, different lines of evidence point to a significant role of RA in the adult brain: (1) Cells of the blood-brain barrier express high levels of retinoid binding proteins probably involved in the control of retinoid entry into the brain (MacDonald, Bok, and Ong 1990). (2) Several regions of the brain are capable of synthesizing RA from retinol and retinaldehyde (Dev, Adler, and Edwards 1993). (3) The adult brain expresses distinct patterns of RA receptors as well as retinol and RA binding proteins (Dollé et al. 1994; Ruberte et al. 1993). With the zymography assay we find several retinaldehyde dehydrogenases in the adult brain, with pronounced regional differences in enzyme types and overall levels. Parenchymal regions tend to express rather low levels of retinaldehyde dehydrogenases, and the pial/ meningeal region, including cortical layer I, exhibit high levels of the V2 dehydrogenase (Figure 7). The hippocampus has very low levels, and the olfactory bulb is one of the enzyme-richest regions in the brain (Figure 7), expressing mainly V2 in addition to minor, less acidic activities. The olfactory bulb is unique in the adult brain for having to accommodate the new axons of olfactory receptors which undergo continued cell division throughout life. The corpus striatum, a region of the basal ganglia involved in involuntary motor control, is the only brain region expressing high levels of AHD2, all of which is localized in axons and synaptic terminals of a subpopulation of dopaminergic neurons projecting from the substantia nigra and adjacent ventral tegmental regions (McCaffery and Dräger 1994a). The AHD2-containing dopaminergic neurons seem to be the homologous population to the dopaminergic neurons that degenerate in Parkinson's disease. Moreover, incidence of Parkinson's disease has been reported as a neurotoxic side effect of the AHD2 inhibitor

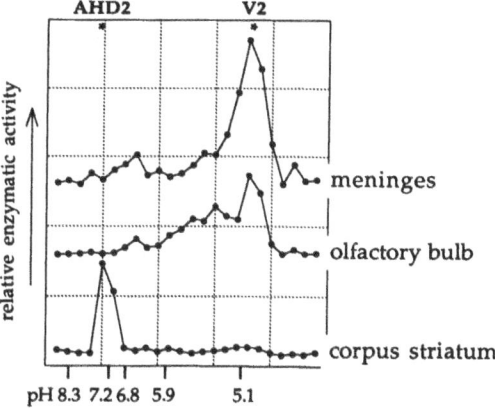

Figure 7. Zymography bioassays comparing retinaldehyde dehydrogenases in meninges from adult visual cortex, olfactory bulb and corpus striatum.

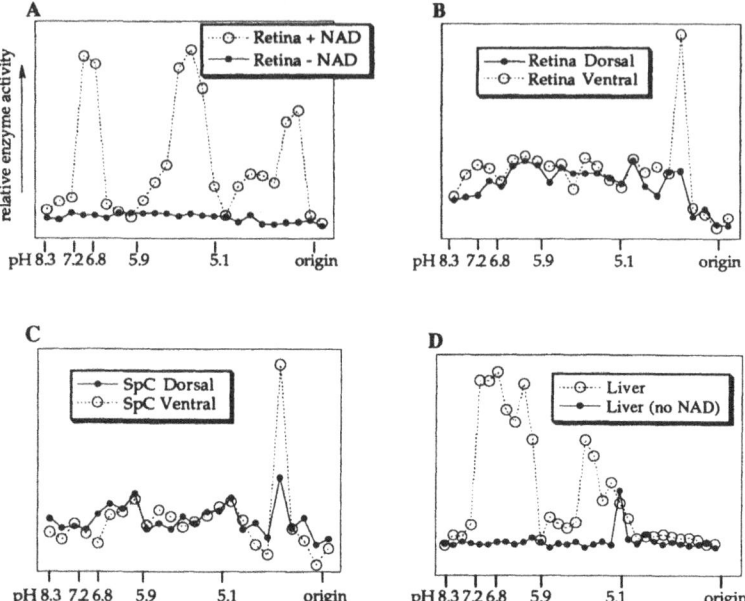

Figure 8. Zymography bioassays run in the absence of NAD, to test for retinaldehyde oxidases. No oxidase activity is detectable in the E12 retina (A), but by E13.5 (B) the ventral retina shows a small but distinct peak. The ventral retina activity is identical in charge to an oxidase activity located predominantly in ventral halves of E13 spinal cord (C), but it is different from an oxidase activity in adult liver (D).

disulfiram, given as aversive treatment in alcoholism (Fisher 1989; Krauss et al. 1991; Laplane et al. 1992).

ALDEHYDE OXIDASES

While most of the zymographically detectable retinaldehyde oxidation in embryonic and adult tissues requires NAD, in some locations we find NAD-independent activities indicative of FAD-linked oxidases. In the embryonic day-12 retina, no retinaldehyde oxidation can be detected when the zymography bioassay method is performed in the absence of NAD (Figure 8A), but by embryonic day 13 an NAD-independent retinaldehyde oxidase is evident (Figure 8B). This oxidase is localized in the ventral retina, and a similar oxidase is also present in the embryonic ventral spinal cord (Figure 8C). The embryonic oxidase differs in charge from an oxidase activity detectable in the adult liver (Figure 8D). In general, the oxidases seem to be of much lower activity than the dehydrogenases in the same tissue, and their visibility is obscured by the dehydrogenases when the zymographs are run in the presence of NAD. They may be important in maintaining RA synthesis under conditions in which NAD is limiting.

WHAT IS THE REASON FOR THE USE OF DIFFERENT RETINALDEHYDE OXIDIZING ENZYMES?

We have shown here that RA can be generated *in vitro* by several enzymes, expressed in distinct spatial and developmentally regulated patterns *in vivo*. The expression patterns are reminiscent to, but distinct in detail from expression patterns of retinol dehydrogenases, RA receptors and retinoid binding proteins (Dollé et al. 1994; Dollé et al. 1990; Duester et al. 1991; MacDonald, Bok, and Ong 1990; Mangelsdorf et al. 1992; Ruberte et al. 1993; Zgombic-Knight, Satre, and Duester 1994). It is likely that all of the enzymes contribute to RA synthesis *in vivo*, because for those expressed as the only enzymes in some locations, *in-vivo* RA synthesis has been demonstrated: AHD2 generates RA in the dorsal retina and the corpus striatum, V1 generates very high RA levels in ventral embryonic retina, and V2 creates RA-rich foci in the embryonic spinal cord (McCaffery and Dräger 1994a; McCaffery and Dräger 1994b; McCaffery et al. 1992). The use of different RA generating enzymes allows for separate forms of regulation: distinct modes of transcriptional control are likely to apply to AHD2, V1, V2, and the oxidases encoded by different genes (Rongnoparut and Weaver 1991, Zhao et al. 1994). Moreover, a possible role in translational control is suggested for AHD2, expressed in some locations at levels far exceeding ordinary enzyme levels, by its correlation with a ribosome-associated translation factor (McCaffery et al. 1991; Rabacchi, Neve, and Dräger 1990); RNA-binding through the NAD-binding sites has been reported for similar enzymes (Hentze 1994).

In addition to differences in expressional regulation, the enzymes differ in enzymatic characteristics and probably in ranges of reaction products. AHD2 can oxidize a broad range of substrates, V1 acts as a highly efficient retinaldehde dehydrogenase but is not completely specific (Figure 3), and for V2 so far no substrate other than retinaldehyde was found (Zhao et al. 1994). The enzymes may also differ in affinity for retinaldehyde isomers and production of RA isomers (Akawi and Napoli 1994). The all-*trans* and 9-*cis* isomers of RA are known to activate different classes of RA receptors and modulate the transcription of different sets of genes (Allenby et al. 1993). The highly conserved expression patterns of the different RA generating enzymes make it likely that the different enzymes are not just a redundant back-up system but that they have distinct morphogenetic roles in the embryo and serve distinct functions in the adult.

ACKNOWLEDGEMENTS

We thank M. Wagner and T. Jessell for the retinoic-acid reporter cells and R. Lindahl for the aldehyde dehydrogenase antiserum. This work was supported by grant R01 EY01938 from the National Eye Institute and a gift from Johnson & Johnson.

REFERENCES

Akawi, Z.E., and J.L. Napoli. 1994. Rat liver cytosolic retinal dehydrogenase: Comparison of 13-*cis*, 9-*cis*-, and all-*trans*-retinal as substrates and effects of cellular retinoid-binding proteins and retinoic acid on activity. *Biochemistry* 33 : 1938-1943.

Allenby, G., M. Bocquel, M. Saunders, S. Kazmer, J. Speck, M. Rosenberger, A. Lovey, P. Kastner, J.F. Grippo, P. Chambon, and A.A. Levin. 1993. Retinoic acid receptors and retinoid X receptors: Interactions with endogenous retinoic acids. *Proc. Natl. Acad. Sci. USA* 90 : 30-34.

Asselineau, D., B.A. Bernard, C. Bailly, and M. Darmon. 1989. Retinoic acid improves epidermal morpho-genesis. *Dev. Biol.* 133 : 322-335.

Bhat, P.V., L. Poissant, and A. Lacroix. 1988. Properties of retinal-oxidizing enzyme activity in rat kidney. *Biochim. Biophys. Acta* 967 : 211-217.

Blomhoff, R., M.H. Green, T. Berg, and K.R. Norum. 1990. Transport and storage of vitamin A. *Science* 250 : 399-404.

Conner, M.J., and M.H. Smit. 1987. Terminal-group oxidation of retinol by mouse epidermis. Inhibition in vitro and in vivo. *Biochem. J.* 244 : 489-492.

De Luca, L.M. 1991. Retinoids and their receptors in differentiation, embryogenesis and neoplasia. *FASEB J.* 5 : 2924-2933.

Dev, S., A. J. Adler, and R. B. Edwards. 1993. Adult rabbit brain synthesizes retinoic acid. *Brain Res.* 632 : 325-328.

Dollé, P., V. Fraulob, P. Kastner, and P. Chambon. 1994. Developmental expression of murine retinoid X receptor (RXR) genes. *Mech. Develop.* 45 : 91-104.

Dollé, P., E. Ruberte, P. Leroy, G. Morriss-Kay, and P. Chambon. 1990. Retinoic acid receptors and cellular retinoid binding proteins. I. A systematic study of their differential pattern of transcription during mouse organogenesis. *Development* 110 : 1133-1151.

Duester, G., M.L. Shean, M.S. McBridge, and M.J. Steward. 1991. Retinoic acid response element in the human alcohol dehydrogenase gene ADH3: implications for regulation of retinoic acid synthesis. *Molec. Cell. Biol.* 11 : 1638-1646.

Durston, A.J., J.P.M. Timmermans, W.J. Hage, H.F.J. Hendriks, N.J. de Vries, M. Heideveld, and P.D. Nieuwkoop. 1989. Retinoic acid causes an anteroposterior transformation in the developing central nervous system. *Nature* 340 : 140-144.

Eichele, G., and C. Thaller. 1987. Characterization of concentration gradients of morphogenetically active retinoid in the chick limb bud. *J. Cell. Biol.* 105 : 1917-1923.

Evans, R.M. 1988. The steroid and thyroid hormone receptor superfamily. *Science* 240 : 889-895.

Fisher, C.M. 1989. 'Catatonia' due to disulfiram toxicity. *Arch. Neurol.* 46 : 798-804.

Giguère, V. 1994. Retinoic acid receptors and cellular retinoid binding proteins: complex interplay in retinoid signaling. *Endocr. Rev.* 15 : 61-79.

Haskell, B. E., R.W. Stach, K. Werrbach-Perez, and J.R. Perez-Polo. 1987. Effect of retinoic acid on nerve growth factor receptors. *Cell Tiss. Res.* 247 : 67-73.

Hentze, M.W. 1994. Enzymes as RNA-binding proteins: a role for (di)nucleotide-binding domains? *TIBS* 19 :101-103.

Hogan, B.L., C. Thaller, and G. Eichele. 1992. Evidence that Hensen's node is a site of retinoic acid synthesis. *Nature* 359 : 237-241.

Kelley, M.W., and T.A. Reh. 1993. Retinoic acid influences the differentiation of photoreceptor cells in embryonic rat retina in vitro. *Soc. Neurosci. Abstr.* 19 : 1288.

Krauss, J.K., M. Mohadjer, A.K. Wakloo, and F. Mundinger. 1991. Dystonia and akinesia due to pallidoputaminal lesions after disulfiram intoxication. *Movement Disorders* 6 : 166-177.

Laplane, D., N. Attal, B. Sauron, A. de Billy, and B. Dubois. 1992. Lesions of basal ganglia due to disulfiram neurotoxicity. *J. Neurol. Neurosurg. Psychiatry* 55 : 925-929.

Lee, M.-O., C.L. Manthey, and N.E. Sladek. 1991. Identification of mouse liver aldehyde dehydrogenases that catalyze the oxidation of retinaldehyde to retinoic acid. *Biochem. Pharmacol.* 42 : 1279-1285.

Lindahl, R., and S. Evces. 1984. Rat liver aldehyde dehydrogenase. I. Isolation and characterization of four inducible isozymes. *J. Biol. Chem.* 259 : 11991-11996.

Lufkin, T., D. Lohnes, M. Mark, A. Dietrich, P. Gorry, M.-P. Gaub, M. LeMeur, and P. Chambon. 1993. High postnatal lethality and testis degeneration in retinoic acid receptor a mutant mice. *Proc. Natl. Acad. Sci. USA* 90 : 7225-7229.

MacDonald, P.N., D. Bok, and D.E. Ong. 1990. Localization of cellular retinol-binding protein and retinol-binding protein in cells comprising the blood-brain barrier of rat and human. *Proc. Natl. Acad. Sci. USA* 87 : 4265-4269.

Mangelsdorf, D.J., U. Borgmeyer, R.A. Heyman, J.Y. Zhou, E.S. Ong, A.E. Oro, A. Kakizuka, and R.M. Evans. 1992. Characterization of three RXR genes that mediate the action of 9-*cis* retinoic acid. *Genes Dev.* 6 : 329-344.

Marsh-Armstrong, N., P. McCaffery, W. Gilbert, J.E. Dowling, and U.C. Dräger. 1994. Retinoic acid is necessary for development of the ventral retina in zebrafish. *Proc. Natl. Acad. Sci. USA* 91:

McCaffery, P., and U.C. Dräger. 1993. Retinoic acid synthesis in the developing retina. In *Enzymology and Molecular Biology of Carbonyl Metabolism.* Vol IV. Edited by H. Weiner, D. W. Crabb and T. G. Flynn. 181-190. New York: Plenum Press.

McCaffery, P., and U.C. Dräger. 1994a. High levels of a retinoic-acid generating dehydrogenase in the meso-telencephalic dopamine system. *Proc. Natl. Acad. Sci. USA* 91:

McCaffery, P., and U.C. Dräger. 1994b. Hotspots of retinoic acid synthesis in the developing spinal cord. *Proc. Natl. Acad. Sci. USA* 91:

McCaffery, P. , M.-O. Lee, M.A. Wagner, N.E. Sladek, and U.C. Dräger. 1992. Asymmetrical retinoic acid synthesis in the dorso-ventral axis of the retina. *Development* 115 : 371-382.

McCaffery, P., K.C. Posch, J.L. Napoli, L. Gudas, and U.C. Dräger. 1993. Changing patterns of the retinoic acid system in the developing retina. *Dev. Biol.* 158 : 390-399.

McCaffery, P., P. Tempst, G. Lara, and U.C. Dräger. 1991. Aldehyde dehydrogenase is a positional marker in the retina. *Development* 112 : 693-702.

Napoli, J.L. 1986. Retinol metabolism in LLC-PK₁ cells. *J. Biol. Chem.* 261 : 13592-13597.

Napoli, J.L., and K.R. Race. 1987. The biosynthesis of retinoic acid from retinol by rat tissues in vitro. *Arch. Biochem. Biophys.* 255 : 95-101.

Quinn, S.D.P., and U. De Boni. 1991. Enhanced neuronal regeneration by retinoic acid of murine dorsal root ganglia and of fetal murine and human spinal cord in vitro. *In Vitro Cell. Dev. Biol.* 27A : 55-62.

Rabacchi, S.A., R.L. Neve, and U.C. Dräger. 1990. A positional marker for the dorsal retina is homologous to the 68kD-laminin receptor. *Development* 109 : 521-531.

Rongnoparut, P., and S. Weaver. 1991. Isolation and characterization of a cytosolic aldehyde dehydrogenase-encoding cDNA from mouse liver. *Gene* 101 : 261-265.

Ruberte, E., V. Friederich, P. Chambon, and G. Morriss-Kay. 1993. Retinoic acid receptors and cellular retinoid binding proteins. III. Their differential transcript distribution during mouse nervous system development. *Development* 118 : 267-282.

Tickle, C., B. Alberts, L. Wolpert, and J. Lee. 1982. Local application of retinoic acid to the limb bud mimics the action of the polarizing region. *Nature* 296 : 564-566.

Wagner, M., B. Han, and T.M. Jessell. 1992. Regional differences in retinoid release from embryonic neural tissue detected by an in vitro reporter assay. *Development* 116 : 55-66.

Wistow, G., and H. Kim. 1991. Lens protein expression in mammals: Taxon-specificity and the recruitment of crystallins. *J. Mol. Evol.* 32 : 262-269.

Wuarin, L., N. Sidell, and J. De Vellis. 1990. Retinoids increase perinatal spinal cord neuronal survival and astroglial differentiation. *Int. J. Devl. Neurosci.* 8 : 317-326.

Zgombic-Knight, M., M.A. Satre, and G. Duester. 1994. Differential activity of the promoter for the human alcohol dehydrogenase (retinol dehydrogenase) gene ADH3 in neural tube of transgenic mouse embryos. *J. Biol. Chem.* 269 : 6790-6795.

Zhao, D., P. McCaffery, R.L. Ivins, R.L. Neve, P. Hogan, W.W. Chin, and U.C. Dräger. 1994. Molecular identification of a major retinoic-acid synthesizing enzyme: a retinaldehyde-specific dehydrogenase. *Submitted for publication*

24

RETINOIC ACID SYNTHESIS IN THE DEVELOPING SPINAL CORD

Ursula C. Dräger and Peter McCaffery

E. Kennedy Shriver Center and
Department of Psychiatry
Harvard Medical School
Waltham MA 02254

A MICRO METHOD FOR DETECTION OF RETINALDEHYDE DEHYDROGENASES

The lipid retinoic acid is the activating ligand for a diverse group of transcription factors, the retinoic acid receptors, which are members of the superfamily of nuclear receptors that include, in addition, the receptors for thyroid hormone, steroids and vitamin D (Chambon et al. 1991; Mangelsdorf 1993). Expression of more than 100 proteins is known to be regulated by retinoic acid (Chytil and ul-Haq 1990). While much has been learned in recent years about the mechanisms of retinoic-acid mediated transcriptional regulation, relatively little was known about the enzymes that catalyze the oxidation of retinaldehyde to retinoic acid in the developing embryo. We previously developed a technique for the detection of such enzymes based on retinoic acid reporter cells (Wagner, Han, and Jessell 1992): tissue extracts are separated by isoelectric focussing gel electrophoresis, the gel is cut into thin consecutive slices, from which the proteins are eluted and assayed for capacity to synthesize retinoic acid from added retinaldehyde (McCaffery et al. 1992). This technique requires only minute amounts of tissue, and it makes it possible to analyze detailed patterns of retinoic acid synthesis in the developing embryo.

CONSERVED DIVERSITY OF RETINALDEHYDE DEHYDROGENASES

In the mouse embryo the final step in retinoic acid synthesis is mediated primarily by several NAD-dependent dehydrogenases, which are arranged in distinct spatial and developmentally regulated patterns. For instance the embryonic retina, one of the retinoic-acid richest organs, contains in its dorsal part the class-1 aldehyde dehydrogenase AHD2, the only class previously identified as mediating retinoic acid synthesis (Lee, Manthey, and Sladek 1991), and the ventral embryonic retina contains a novel dehydrogenase, provisionally named V1. The spinal cord contains a third dehydrogenase, the V2 activity (McCaffery

Enzymology and Molecular Biology of Carbonyl Metabolism 5
Edited by H. Weiner *et al.*, Plenum Press, New York, 1995

and Dräger 1993; McCaffery et al. 1992). V2 is a novel dehydrogenase with 70% homology to class 1 aldehyde dehydrogenase; it is highly effective in retinaldehyde oxidation, and it seems specific for this substrate (Zhao et al. 1994).

In the zymograph of Figure 1 a similar diversity of retinoic acid generating dehydrogenases is illustrated for the day-6 chick embryo. The dorsal chick retina (Figure 1, upper trace) contains a basic dehydrogenase activity peak. This is likely to represent the chick homolog to mammalian class-1 aldehyde dehydrogenase, because a homologous mRNA has been found at abundant levels in the embryonic chick retina (Godbout 1992), and because in chick, as in other vertebrate embryos, retinoic acid synthesis in dorsal retina is much more susceptible to the class-1 aldehyde dehydrogenase inhibitor disulfiram than synthesis in ventral retina (McCaffery et al. 1993). The ventral chick retina (Figure 1, middle trace) contains two dehydrogenase peaks: a basic activity, slightly more acidic than the dorsal enzyme, and dehydrogenase activity precipitated at the origin of the isoelectric focussing gel. These two peaks are likely to represent the same enzyme, as variable enzyme precipitation/ membrane trapping is also a characteristic of the corresponding enzyme in the mouse (McCaffery et al. 1992). The chick spinal cord (Figure 1, lower trace) contains an acidic retinaldehyde dehydrogenase whose charge (~pI 5.7) is similar to the charge of the spinal cord dehydrogenase V2 in the mouse (pI 5.0).

SPATIAL PATTERNS OF RETINALDEHYDE DEHYDROGENASES IN THE DEVELOPING SPINAL CORD

The zymography bioassay (Figure 1) measures only the capacity of gel-separated enzymes to generate retinoic acid from added retinaldehyde, but it does not indicate whether such synthesis occurs *in vivo*. Moreover, a commonly held notion assumes that most retinoic acid originates in the liver, and that local differences in tissue levels are due to local differences in uptake, binding and breakdown of blood-born retinoic acid. In order to address the question in how far local levels of retinaldehyde dehydrogenase are indicative of local retinoic acid levels, we analyzed the spinal cord of the developing mouse, an anatomically

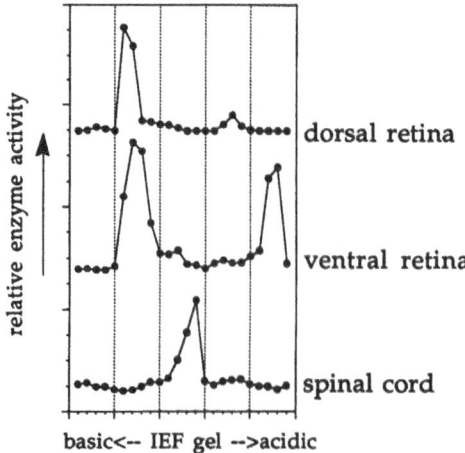

Figure 1. Zymographical comparison of retinaldehyde dehydrogenases in dorsal and ventral retinas and the spinal cord of the day-6 chick embryo. The range of ampholytes used for this and other zymographs shown here covered pH 3-10.

homogeneous system containing only the V2 activity (McCaffery and Dräger 1994). Spinal cords were dissected into small consecutive pieces, cultured overnight, and the amounts of retinoic acid released from the pieces were assayed with the reporter cells in supernatant volumes normalized for tissue protein. Then retinaldehyde dehydrogenase activities were determined zymographically in the same tissue samples. We found an excellent agreement between the two parameters, indicating that the major determinant for local retinoic acid levels in the developing spinal cord is local synthesis. Both levels of retinoic acid and retinaldehyde dehydrogenase vary dramatically along the rostro-caudal extent of the embryonic neural tube: hindbrain/medulla levels are low, and the spinal cord shows two foci of very high synthesis; these are localized in the ventral cord at the regions that give rise to the limb innervations. *In situ* hybridizations show high levels of V2 dehydrogenase mRNA in the embryonic limb motorneurons (Zhao et al. 1994).

The rostro-caudal pattern in retinoic acid levels for the 13-day old mouse embryo (McCaffery and Dräger 1994) is compared in Figure 2 to the pattern of retinaldehyde dehydrogenase levels from the 6-day old chick: as in the mouse, there are two maxima approximately corresponding to the spinal cord regions innervating the developing limbs. This makes it very likely that the distribution analyzed in detail for the mouse applies also to the embryonic chick.

In the developing mouse, the rostro-caudal retinoic acid variations are pronounced during embryonic stages, persist for several days postnatally, and are no longer detectable in the spinal cord of the two-week old pup (McCaffery and Dräger 1994). In the younger embryo (such as embryonic day 13, see Figure 2A), the rostral maximum is highest, and later the caudal maximum predominates, a shift consistent with a rostro-caudal developmental gradient: hindlimb maturation is known to lag behind forelimb maturation. In addition, overall retinoic acid levels are highest in the early embryonic spinal cord and decrease substantially with age (McCaffery and Dräger 1994). These spatial and temporal correlations point to a role of retinoic acid in neuronal differentiation in the spinal cord: highest levels

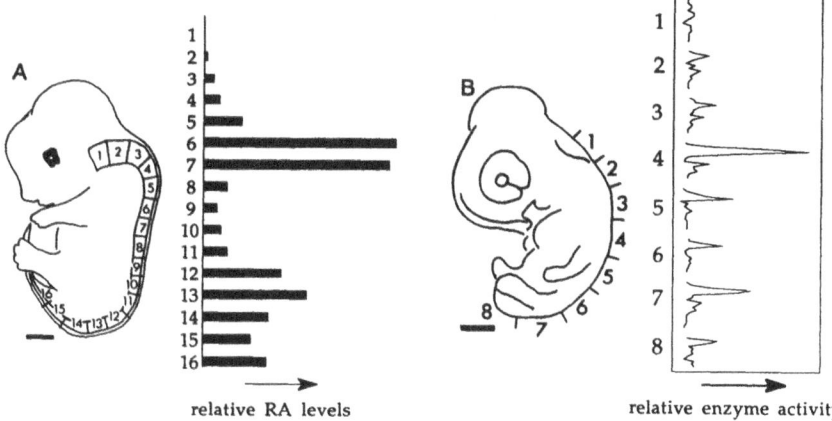

relative RA levels   relative enzyme activity

Figure 2. Comparisons of rostro-caudal distribution of released retinoic acid amounts in the spinal cord of the day-13 mouse embryo (McCaffery and Dräger 1994) in (A) and retinaldehyde dehydrogenase levels in the day-6 chick spinal cord in (B). For the mouse, a colocalization of the retinoic acid maxima with the origins of the limb innervations in the spinal cord was demonstrated by retrograde labeling from the limbs. The measured values given here represent colorimetric readings obtained in an ELISA reader for samples normalized for tissue protein; doubling of the values corresponds to a 10-100 fold increase in retinoic acid levels (McCaffery and Dräger 1994). Scales: (A) 100μm; (B) 200μm.

coincide with regions and periods of growth, decreased morphogenetic cell death and maximal neurite formation (see below).

EXPRESSION OF THE V2 RETINALDEHYDE DEHYDROGENASE BEGINS AT HENSEN'S NODE

The earliest age for which we had previously analyzed retinoic acid generating enzymes in the trunk region was the day-8.5 mouse embryo, when V2 is the only detectable retinaldehyde dehydrogenase and its levels are already very high (McCaffery et al. 1993). In order to determine the onset of V2 expression, we tested one-day younger embryos. Due to the natural age variations within one litter and the small sizes, we had to combine embryos from several litters that were nominally 7.5 days old. The embryos were dissected and sorted into five representative size groups (i.e. excluding outliers) labeled for brevity E7.0, E7.2, E7.4, E7.6 and E7.8 in Figure 3. These correspond approximately to the following stages of the gastrulating mouse embryo: mid-primitive streak to late-streak stages (E7.0-7.2), neural plate stages (E7.4-7.6), and early headfold stage (E7.8) (Downs and Davies 1993). From five embryos in each group we dissected the distal tip of the egg cylinder (=prospective node region) or the region containing a visible node, in addition to the base of the egg cylinder or the beginning headfold regions; tissue amounts for each sample were barely visible to the naked eye. Zymographs of these samples (Figure 3) show appearance of a retinaldehyde dehydrogenase in the node region of the third size sample ('E7.4'), which represents the early neural plate stage. The headfold/future eye region even of the most advanced group ('E7.8') shows no detectable dehydrogenase activity yet. The node enzyme has the same

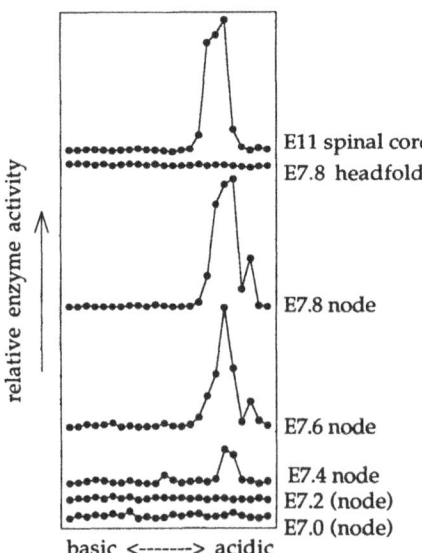

Figure 3. Zymographical comparisons of the appearance of retinaldehyde dehydrogenase activity in gastrulating mouse embryos from a B6/D2 outbred colony, nominally at day 7.5 of development, but sorted into five size groups. The first activity is detectable in the node region of early neural-plate stage embryos ('E7.4'), a stage when the node could be clearly identified under the dissecting microscope. In the smaller embryos, the prospective node regions '(node)' had no detectable activity. The headfold region of the largest embryo group ('E7.8') is still negative; we previously found the first activity here at day 8.0 (McCaffery et al. 1993). The node enzyme seems to be identical to the V2 activity expressed in the embryonic day-11 (E11) spinal cord.

charge as the V2 dehydrogenase in the day-11 (E11) embryonic spinal cord processed in parallel. These observations show that the earliest detectable retinoic acid synthesis in the vertebrate embryo in the node/ organizer region (Chen et al. 1992; Hogan, Thaller, and Eichele 1992) is mediated by the major spinal cord retinaldehyde dehydrogenase.

OTHER RETINOIC ACID GENERATING ENZYMES IN THE SPINAL CORD

Although in most spinal cord preparations V2 seemed to be the only detectable retinaldehyde dehydrogenase, we had noted some additional enzymatic activity in particular in older embryos and in hastily dissected spinal cord samples that contained some of the adhering pial/meningeal membranes. In order to sort out this variability, we dissected pial membranes from developing murine spinal cords, cultured them overnight for determinations of retinoic acid tissue levels, and then analyzed them zymographically for enzyme content. In addition, samples from different ages and different anatomical locations were assayed for spatio-temporal variations in activity levels (Figure 4). Like the spinal cord explants, the pial samples released high amounts of retinoic acid commensurate with measured levels of retinaldehyde dehydrogenase in the same tissues (not shown). Apart from V2 dehydrogenase, the pial/meningeal samples showed at least two additional activities as a poorly resolved shoulder towards basic charge values (Figure 4A); their identity was not further analyzed here. Like the V2 levels within the spinal cord, overall retinaldehyde dehydrogenase levels in the pial/meningeal samples decrease with age (Figure 4A). In order to determine whether synthesis in the pial/meningeal membranes contributes to the rostro-caudal variations in retinoic acid levels, membrane samples were dissected from spinal cord

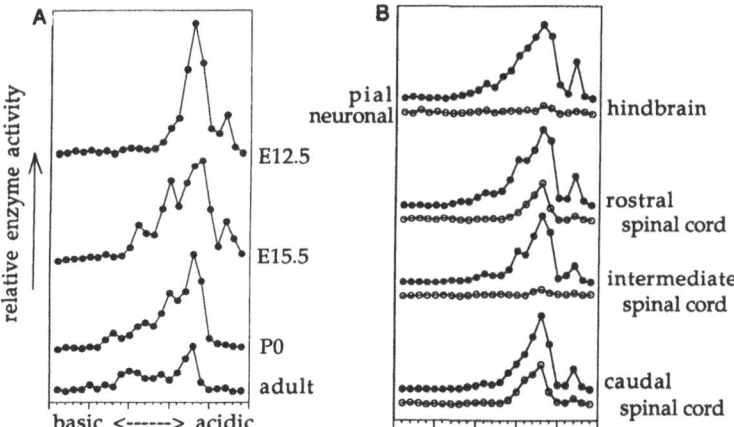

Figure 4. (A) Zymographical comparisons of retinoic acid generating enzymes in pial/ meningeal tissue samples dissected from spinal cords of embryonic day 12.5 (E12.5), E15.5, newborn (P0) and adult spinal cords, normalized for protein content. Note the presence of poorly resolved activities at the left shoulder of the main activity that represents V2. The small acidic peaks in the E12.5 and E15.5 samples represent activity trapped at the gel origin. (B) Comparisons of neural (open circles = lower paired traces) and pial/ meningeal tissue samples (closed circles = upper paired traces) dissected from the hindbrain/medulla, rostral, intermediate and caudal spinal cords of day-16 mouse embryos. As the samples were normalized for protein content, the very thin pial samples had to be collected from many spinal cords in order to match the protein content of the neuronal sample collected from one cord, which makes the pial activity appear larger than the neuronal activity. Note that the pia does not contribute to the rostro-caudal activity variations.

regions giving rise to the limb innervation and from the intermediate cord. Figure 4B, from a day-16 mouse embryo, shows a comparison of neuronal and pial samples along the rostro-caudal dimension: only the neuronal component varies, pial enzyme levels remain even.

ROSTRO-CAUDAL VARIATIONS IN DORSAL ROOT GANGLIA

The neuronal somata for the somatosensory innervation of the body are located within the dorsal root ganglia, neural-crest derived organs situated next to the spinal cord. Like neurons within the spinal cord, the dorsal root ganglia have to adjust to the peripheral innervation needs: dorsal root ganglia innervating the limbs contain more neurons than ganglia innervating the chest or belly (Hamburger and Levi-Montalcini 1949). To test whether retinoic acid might play a role in the generation of this rostro-caudal patterning of the somatosensory system, we dissected dorsal root ganglia from the limb and intermediate regions, normalized them for protein content and tested for zymographically detectable retinaldehyde dehydrogenase. Dorsal root ganglia from day-13 mouse embryos (Figure 5) show a single retinaldehyde dehydrogenase activity, whose levels are considerably higher in the limb-innervating samples than in the intermediate sample. Although by charge this dehydrogenase is indistinguishable from the V2 activity, its identity is not certain: live explants of the same dorsal root ganglia generate very little retinoic acid detectable by the reporter cells (not shown). As the reporter cells respond mainly to all-*trans* retinoic acid (McCaffery and Dräger 1994), a possible explanation for this discrepancy could be *in-vivo* conversion of all-*trans* retinoic acid, generated by V2, into a different isomer such as *9-cis* retinoic acid.

A POSSIBLE ROLE OF RETINOIC ACID IN SPINAL CORD MORPHOGENESIS

In the vertebrate retina, high levels of retinoic acid and retinoic acid generating enzymes correlate spatially and temporally with neuronal growth and differentiation, but not with adult function: in the mouse, high levels are restricted to embryonic and early postnatal stages, and adult levels are much lower (McCaffery and Dräger 1993; McCaffery et al. 1993);

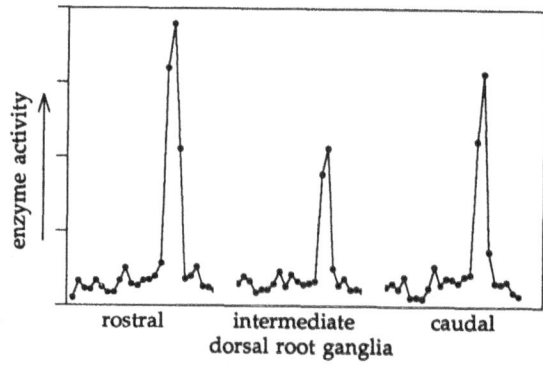

Figure 5. Rostro-caudal variations in activity levels of retinaldehyde dehydrogenase in dorsal root ganglia innervating the forelimb (rostral), chest/belly (intermediate) and hindlimb (caudal); samples taken from day-13 mouse embryos.

in the zebrafish, high levels persist in the adult retina, but are restricted to the peripheral retina containing the germinal zone for continued neuronal growth (Marsh-Armstrong et al. 1994). Similarly in the spinal cord overall levels of retinaldehyde dehydrogenase and retinoic acid levels decrease substantially between early embryonic stages to the adult (McCaffery and Dräger 1994). In addition, the variations along the rostro-caudal dimension point to a role in the patterning along this axis. Early in development the spinal cord starts out as a neural tube that appears morphologically homogeneous in the rostro-caudal extent, but later two localized enlargements form at the origins of the forelimb and hindlimb innnervations in the cervical and lumbar cord. These limb zones contain more neurons than intermediate cord regions due to higher proliferative rates and decreased morphogenetic cell death (Oppenheim, Cole, and Prevette 1989), and the neurons innervating the limbs have to sprout and maintain much longer axons than neurons innervating the trunk. The retinoic-acid rich zones probably provide a morphogenetic mechanism for the formation of the limb zones, because in cultures of dissociated embryonic spinal cords retinoic acid is known to increase the number of surviving neurons by two- to three-fold; in addition, such neurons show significant neurite outgrowth only in the presence of high retinoic-acid levels (Haskell et al. 1987; Quinn and De Boni 1991; Wuarin, Sidell, and De Vellis 1990).

The first indications for rostro-caudal variations in retinoic acid levels along the spinal cord (Colbert, Linney, and LaMantia 1993) came from retinoic acid indicator mice transgenic for a construct consisting of the response element from the retinoic acid receptor β (de Thé et al. 1990) driving β-galactosidase expression, a construct similar to the one in the retinoic acid reporter cells used here (Wagner, Han, and Jessell 1992). The transgenic embryos show two broad zones of high β-galactosidase expression in cervical and lumbar regions, including both ventral and dorsal cord (Colbert, Linney, and LaMantia 1993). As the only source of rostro-caudal variations in retinoic acid synthesis within the cord is in the motorneurons and not in the pial/ meningeal synthesis (Figure 4B), the broad zones must represent retinoic acid diffusing from the motorneurons. As diffusion of the lipid retinoic acid is rather little impeded by plasma membranes, localized synthesis foci will set up diffusion concentration gradients in the surrounding tissue which could result in the spatially sequential activation of genes with graded retinoic acid response thresholds (Boncinelli, Simeone, and Mavilio 1991; Simeone et al. 1990). Retinoic acid synthesized within the motorneurons is likely to influence transcription in the surrounding tissue in a graded manner determined by tissue diffusion halos.

ACKNOWLEDGMENTS

We thank M. Wagner and T. Jessell for the retinoic-acid reporter cells. This work was supported by grant R01 EY01938 from the National Eye Institute and a gift from Johnson & Johnson.

REFERENCES

Boncinelli, E., F. Simeone, and F. Mavilio. 1991. Hox gene activation by retinoic acid. *Trends Genet.* 7 : 329-334.

Chambon, P., A. Zelent, M. Petkovich, C. Mendelsohn, P. Leroy, A. Krust, P. Kastner, and N. Brand. 1991. The family of retinoic acid nuclear receptors. In *Retinoids: 10 Years On.* Edited by J.-H. Saurat. 10-27. Basel: Karger.

Chen, Y., L. Huang, A. F. Russo, and M. Solursh. 1992. Retinoic acid is enriched in Hensen's node and is developmentally regulated in the early chicken embryo. *Proc. Natl. Acad. Sci. USA* 89 : 10056-10059.

Chytil, F., and R. ul-Haq. 1990. Vitamin A mediated gene expression. *Crit. Rev. Eukaryot. Gene Expr.* 1 : 61-73.

Colbert, M.C., E. Linney, and A.-S LaMantia. 1993. Local sources of retinoic acid coincide with retinoid-mediated transgene activity during embryonic development. *Proc. Natl. Acad. Sci. USA* 90 : 6572-6576.

de Thé, H., M. del Mar Vivanco-Ruiz, P. Tiollais, H. Stunnenberg, and A. Dejean. 1990. Identification of a retinoic acid responsive element in the retinoic acid receptor beta gene. *Nature* 343 : 177-180.

Downs, K., and T. Davies. 1993. Staging of gastrulating mouse embryos by morphological landmarks in the dissecting microscope. *Development* 118 : 1255-1266.

Godbout, R. 1992. High levels of aldehyde dehydrogenase transcripts in the undifferentiated chick retina. *Exp. Eye Res.* 54 : 297-305.

Hamburger, V., and R. Levi-Montalcini. 1949. Proliferation, differentiation and degeneration in the spinal ganglia of the chick embryo under normal and experimental conditions. *J. Exp. Zool.* 111 : 457-501.

Haskell, B. E., R.W. Stach, K. Werrbach-Perez, and J.R. Perez-Polo. 1987. Effect of retinoic acid on nerve growth factor receptors. *Cell Tiss. Res.* 247 : 67-73.

Hogan, B.L., C. Thaller, and G. Eichele. 1992. Evidence that Hensen's node is a site of retinoic acid synthesis. *Nature* 359 : 237-241.

Lee, M.-O., C.L. Manthey, and N.E. Sladek. 1991. Identification of mouse liver aldehyde dehydrogenases that catalyze the oxidation of retinaldehyde to retinoic acid. *Biochem. Pharmacol.* 42 : 1279-1285.

Mangelsdorf, D.J., and Evans, R.M. 1993. Retinoid receptors as transcription factors. In *Transcriptional Regulation*. Edited by S. L. McKnight and K. R. Yamamoto. 1137-1167. Plainview, NY: Cold Spring Harbor Laboratory Press.

Marsh-Armstrong, N., P. McCaffery, J.E. Dowling, W. Gilbert, and U.C. Dräger. 1994. Retinoic acid is necessary for development of the ventral retina in zebrafish. *Proc. Natl. Acad. Sci. USA* 91

McCaffery, P., and U.C. Dräger. 1993. Retinoic acid synthesis in the developing retina. In *Enzymology and Molecular Biology of Carbonyl Metabolism*. Vol IV. Edited by H. Weiner, D. W. Crabb and T. G. Flynn. 181-190. New York: Plenum Press.

McCaffery, P., and U.C. Dräger. 1994. Hotspots of retinoic acid synthesis in the developing spinal cord. *Proc. Natl. Acad. Sci. USA* 91

McCaffery, P. , M.-O. Lee, M.A. Wagner, N.E. Sladek, and U.C. Dräger. 1992. Asymmetrical retinoic acid synthesis in the dorso-ventral axis of the retina. *Development* 115 : 371-382.

McCaffery, P., K.C. Posch, J.L. Napoli, L. Gudas, and U.C. Dräger. 1993. Changing patterns of the retinoic acid system in the developing retina. *Dev. Biol.* 158 : 390-399.

Oppenheim, R.W., T. Cole, and D. Prevette. 1989. Early regional variations in motoneuron numbers arise by differential proliferation in the chick embryo spinal cord. *Dev. Biol.* 133 : 468-474.

Quinn, S.D.P., and U. De Boni. 1991. Enhanced neuronal regeneration by retinoic acid of murine dorsal root ganglia and of fetal murine and human spinal cord in vitro. *In Vitro Cell. Dev. Biol.* 27A : 55-62.

Simeone, A., D. Acampora, L. Arcioni, P. W. Andrews, E. Boncinelli, and F. Mavilio. 1990. Sequential activation of Hox2 homeobox genes by retinoic acid in human embryonal carcinoma cells. *Nature* 346 : 763-766.

Wagner, M., B. Han, and T.M. Jessell. 1992. Regional differences in retinoid release from embryonic neural tissue detected by an in vitro reporter assay. *Development* 116 : 55-66.

Wuarin, L., N. Sidell, and J. De Vellis. 1990. Retinoids increase perinatal spinal cord neuronal survival and astroglial differentiation. *Int. J. Devl. Neurosci.* 8 : 317-326.

Zhao, D., P. McCaffery, K.J. Ivins, R.L. Neve, P. Hogan, W.W. Chin, and U. C. Dräger. 1994. Molecular identification of a major retinoic-acid synthesizing enzyme: a retinaldehyde-specific dehydrogenase. *Submitted for publication*

25

STRUCTURE AND MECHANISM OF ALDEHYDE REDUCTASE

T. Geoffrey Flynn[1], Nancy C. Green[1], Mohit B. Bhatia[1], and
Ossama El-Kabbani[2]

[1]Department of Biochemistry
Queen's University
Kingston, Ontario, Canada K7L 3N6
[2]Center for Macromolecular Crystallography
University of Alabama
Birmingham, Alabama 35294

Aldehyde reductase (ALR1, EC 1.1.1.2) and aldose reductase (ALR2, EC 1.1.1.21) catalyze the NADPH-dependent reduction of a wide range of aromatic and aliphatic aldehydes to their corresponding alcohols. Despite a recently expressed opinion that aldose reductase is of little consequence (Harding, 1992) the past few years have seen a great advancement in our knowledge of the structure and mechanism of both enzymes, and of aldose reductase in particular. The three-dimensional structure of pig (Rondeau et al., 1992) and human (Wilson et al., 1992) aldose reductase revealed that as an oxido-reductase the enzyme is unique in that it has a β/α-TIM barrel structure and is the first oxido-reductase known to possess such a structure and to not have a dinucleotide or Rossmann binding fold. Kinetic studies have shown that the enzyme operates by an ordered mechanism with NADPH binding first (Grimshaw et al., 1990; Kubiseski et al., 1992). The binding of coenzyme is very tight ($<<1\mu M$, see Grimshaw and Lai in this Proceedings) and following coenzyme binding there is a conformational change in the enzyme. This has been shown by fluorescence spectroscopy (Kubiseski et al., 1992); by a combination of chemical modification and X-ray crystallography (Kubiseski et al., 1994) and from a comparison of the three-dimensional structures of human and porcine aldehyde and aldose reductase (El-Kabbani et al., 1994). Structural studies have also revealed that an aldose reductase inhibitor (ARI), zopolrestat, binds in the active site (Wilson et al., 1993) and not at a site removed from the active site as suggested by the fact that all ARIs are uncompetitive or non-competitive inhibitors in the forward direction of the reaction (Sato & Kador, 1990). Most ARIs are in fact competitive with the alcohol product and binding of ARIs at the active site is, of course, feasible and understandable (Sato & Kador, 1990).

Knowledge of both the primary and tertiary structure has facilitated the use of site-directed mutagenesis to examine the role of various residues in the catalytic mechanism. Such studies have suggested that in ALR2 Tyr-48 donates a proton to the carbonyl group of the substrate following hydride transfer (Tarle et al., 1993; Bohren et al., 1994). Mutation

Enzymology and Molecular Biology of Carbonyl Metabolism 5
Edited by H. Weiner *et al.*, Plenum Press, New York, 1995

of His-110 also seriously impairs the activity of the enzyme (Tarle et al., 1993; Bohren et al., 1994) and classical pH-activity studies also suggest a group with a pKa of 6.5-7.0 being involved in catalysis (Liu et al., 1993).

Almost all of the recent crystallographic and mechanistic studies have been concerned with aldose reductase. The close similarity in amino acid sequence between aldose and aldehyde reductase (65% identity; Green et al., 1994) suggests that their three-dimensional structures will also be similar. To aid in the refinement of the crystallographic structure of pig aldehyde reductase we obtained the primary structure of the enzyme following cloning and sequencing of ALR1 cDNA. The crystal structure of both human and pig ALR1 was refined to 2.8 A and 2.48 A respectively. We show here that there is a high degree of conservation of both primary and tertiary structure between ALR1 and ALR2.

METHODS AND MATERIALS

Probe Generation

A fragment of the human ALR1 gene was generated using the polymerase chain reaction (PCR). PCR primers were made to portions of the hALR1 cDNA sequence (Bohren et al., 1989) that were specific for ALR1 but not for ALR2. The primer sequences were 5'ACTATATGCTACGACTCCACCCACTAC3' and 5'CCATCCTCCGTAAG-CATAGGCACAATATA3'. The oligonucleotides were synthesized at the Core Facility at Queen's University. Crude oligonucleotides were purified using Econo-Pac 10DG desalting columns (Biorad). PCR was performed using a Perkin-Elmer Cetus DNA thermal cycler. An aliquot of human genomic DNA (2μL) was amplified using 2.5 units of Taq DNA polymerase (Promega), 200 μM nucleotides, 1X Promega Mg free buffer, 3.5 mM $MgCl_2$ and 25 pmol of each primer in a total volume of 50 μL. The cycle consisted of 94°C for 45 seconds, 65°C for 45 seconds and 72°C for 60 seconds for 30 cycles followed by 10 minutes at 72°C. To remove excess primers and nucleotides, the PCR reaction mixture was passed through a 1 mL Sephadex G-50 spun column. The product was nick translated following manufacturer's protocols (BRL).

Library Screening

A pig brain λgt 10cDNA library (Clonetech) was titred and 200 000 pfu were plated at a density of 10 000 pfu/150mm plate following protocols outlined in Sambrook et al. (1989). Nitrocellulose filters (Millipore) were prehybridized in 6X SSC, 5X PVP/Ficoll, 0.1% SDS and 100 μg/mL herring testes DNA (Sigma) at 42°C overnight. The filters were hybridized at 42°C for 10 hours in fresh prehybridization fluid with probe added at a concentration of 1 x 10^6 cpm/mL. Filters were washed twice at 42°C in 2X SSC, 0.1% SDS for 30 minutes and exposed to Kodak XAR film for 2 days at -70°C. Insert sizes of putative clones were estimated by performing PCR using a 2 μL aliquot of phage stock and λgt10 leftward and rightward primers (Core Facility, Queen's University). Based on insert size, one clone was chosen for subcloning into the EcoRI site of the pGEM3Zf+ vector (Promega).

DNA Sequencing

The insert was sequenced on both strands using successive primers and a dideoxy DNA sequencing kit (Pharmacia) following manufacturer's protocols.

Crystallography Conditions

Monoclinic crystals of the enzyme were grown at 295 K by vapour diffusion using the hanging drop method as described in (El-Kabbani et al., 1991). Briefly, 2 μL of a 12 mg/mL ALR1 solution in 10mM PIPES, 2 mM β-mercaptoethanol, pH 6.5 was mixed with 2 μL each of 20% ammonium sulfate and 10 mM β-octylglucoside made up in 10 mM PIPES, 2mM β-mercaptoethanol, pH 6.5. The resulting droplets were vapour equilibrated against 1 mL of 35% ammonium sulfate in 50 mM PIPES, 2 mM β-mercaptoethanol, pH 6.5. X-ray diffraction data from one native monoclinic crystal were recorded on a Nicolet multiwire area detector and processed by the ZENGEN program package. A heavy atom derivative was obtained by adding a concentrated potassium platinum nitrate solution directly to the ALR1-containing droplet (final concentration 5 mM) and soaking for 8d.

Standard Enzymatic Assay

Aldehyde reductase was assayed by a modification of methods described previously (Kubiseski et al., 1992) based on the decrease in absorbance at 340 nm due to the consumption of NADPH. Assays were performed on a Hewlett Packard HP8452A diode array spectrophotometer. Absorbance measurements were taken every 1 sec over a period of 60 s and the measurement taken at 340 nm was subtracted from that measured at 600 nm. A 295 nm UV cut-off filter was mounted to the cell holder to prevent photodegradation of NADPH during the kinetic measurements. All assays were conducted at 19°C. A 1 mL reaction mixture contained 60 uM NADPH, 1 mM p-nitro benzaldehyde in 100 mM sodium phosphate, pH 7.0. The assay reaction was initiated by the addition of 1 ug of aldehyde reductase; this amount of protein provided a linear rate of NADPH consumption over a period of 1 min. One unit of enzyme is defined as the amount of ALR1 required to convert 1 umol of NADPH to $NADP^+$ per minute at 19°C. For the pH effect studies, a 3-component buffer system consisting of MES-TRIS-CHES was employed at an ionic strength of 0.3M at 19°C.

RESULTS

Fig. 1 shows the cDNA derived amino acid sequence of pig ALR1 in alignment with the amino acid sequence of pig ALR2 and human ALR1 and human ALR2. Pig ALR1 is 96% identical with human ALR1, 65% identical with pig ALR2 and 53% with human ALR2.. The remarkable primary sequence identity that exists between the primary structures of pig and human aldehyde reductase is also reflected in their tertiary structures which are virtually completely superimposable. The tertiary structures of pig ALR1 and human ALR2 are also very similar. Shown in Fig. 2 are the tertiary structures of NADPH bound pig ALR1 and human ALR2.

Like ALR2, pig aldehyde reductase consists of an eight-stranded β/α barrel with the coenzyme binding site located at the C-terminus of the ends of the barrel. Two short antiparallel β-strands at the N-terminus, three large exposed loops and two α-helices are additional structural components of the ALR1 and ALR2 molecules. There are minor differences in the overall conformation of ALR1 in comparison to ALR2 but there appear to be confined to the surface of the enzyme and not at the active site.

It is apparent from the primary and tertiary structure of ALR1 and ALR2 that several residues important to structure and function are conserved e.g. Tyr-48 (50 in ALR1) and His-110 (113 in ALR1). In preliminary experiments designed to examine the catalytic

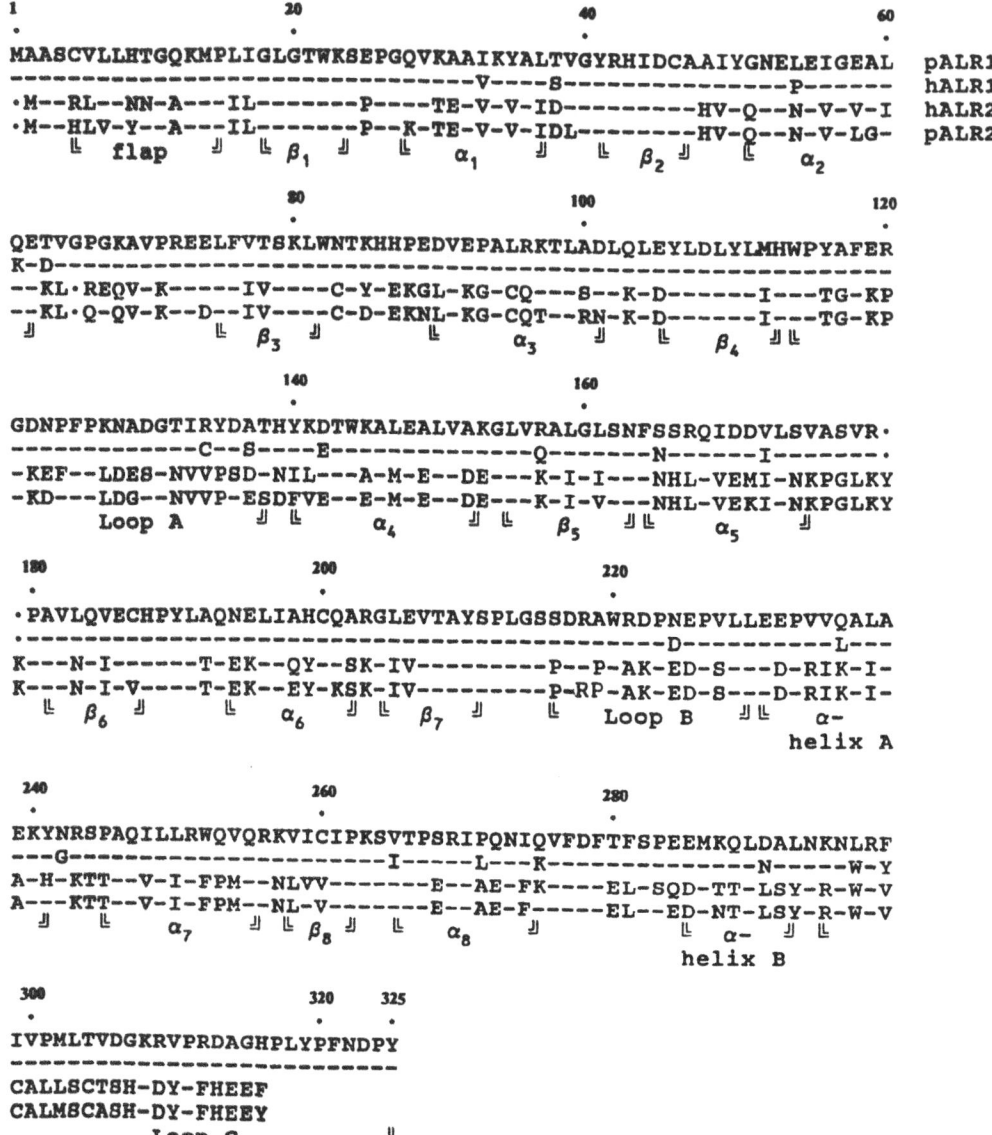

Figure 1. Comparison of primary sequence of pig ALR1, pig ALR2 (Kubiseski et al., 1993) human ALR1 (Bohren et al. 1989) and human ALR2 (Nishimura et al., 1990).

activity of residues at the active site of ALR1 we have examined various kinetic parameters as a function of pH.

EFFECT ON PH ON KINETIC PARAMETERS

The maximal velocity of the reduction of p-nitro benzaldehyde (pNB) decreases with high pH (pKa 6.5-7.0) (Fig 3). The k_{cat}/K_{pNB} also decreases with high pH (pKa 6.5-7.0) (Fig

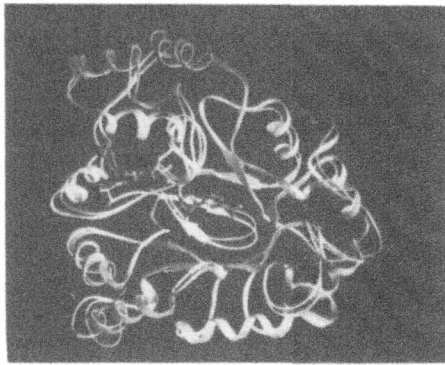

Figure 2. Comparison of the tertiary structures of NADPH-bound pig ALR1 and human ALR2. The light strand is pig ALR1 and the darker strand is human ALR2.

4). The effect is primarily an effect on V_{MAX}; the K_M for pNB shows no change in the pH range of 5.5-8.0. Thus, the pKa of 6.5-7.0 observed in these experiments may reflect either a catalytic step or that the protonation of an active site residue is important for substrate binding (rate-limiting product release step).

The dependence of k_{cat}/K_{NADPH} on pH was bell-shaped decreasing at both high and low pH (Fig 5). The K_I for NADP$^+$ as determined from competitive product inhibition studies displayed a similar bell-shaped dependence on pH (Fig 6). The lower pKa of 5.75-6.0 may reflect the pKa of the phosphate group of NADP$^+$. On the basis of ^{31}P NMR experiments, Sem & Kasper (1993) have determined the pKa of the phosphate group of NADP$^+$ to be 5.8. The higher pKa of 7-7.5 may reflect a residue in the active-site of aldehyde reductase that may interact with the coenzyme. Liu et al. (1993) have proposed that this may be lysine-262 (in aldose reductase).

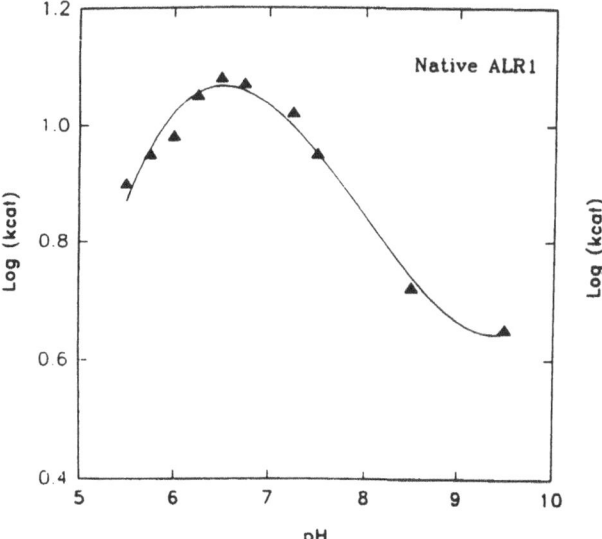

Figure 3. Effect of pH variation (5.5-8.0) on k_{cat} and k_{cat}/K_M^{pNB} in native aldehyde reductase.

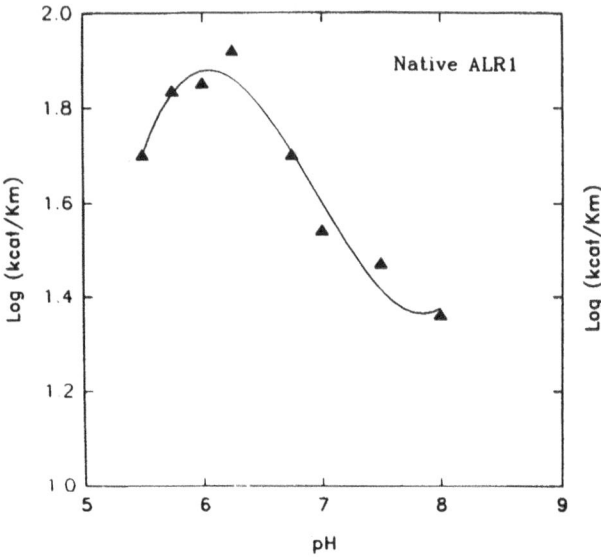

Figure 4. Effect of pH variation (5.5-8.0) on k_{cat} and k_{cat}/K_M^{pNB} in native aldehyde reductase.

DISCUSSION

The catalytic mechanism of aldehyde reduction by aldose or aldehyde reductase involves a stereospecific transfer of a pro(R) hydride from the C4 of NADPH to the carbonyl carbon of the aldehydic substrate and protonation of the oxygen of the aldehyde by a residue in the enzyme active site. Crystallographic studies on aldose reductase (ALR2) have

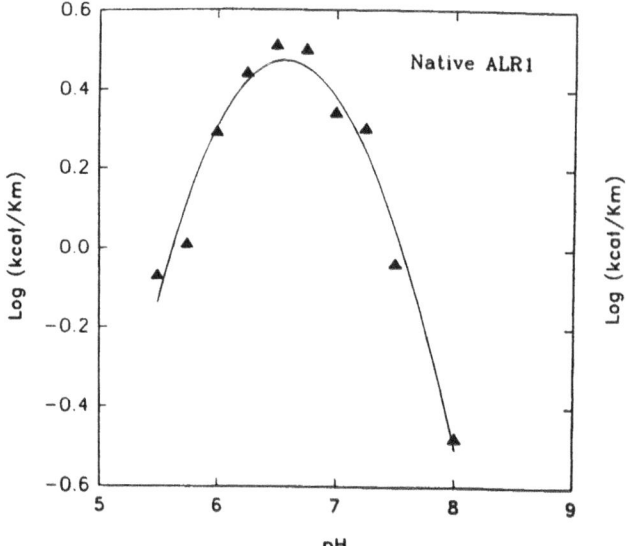

Figure 5. Effect of pH variation on k_{cat}/K_M for NADPH and K_I of NADP+ determined from competitive product inhibition studies.

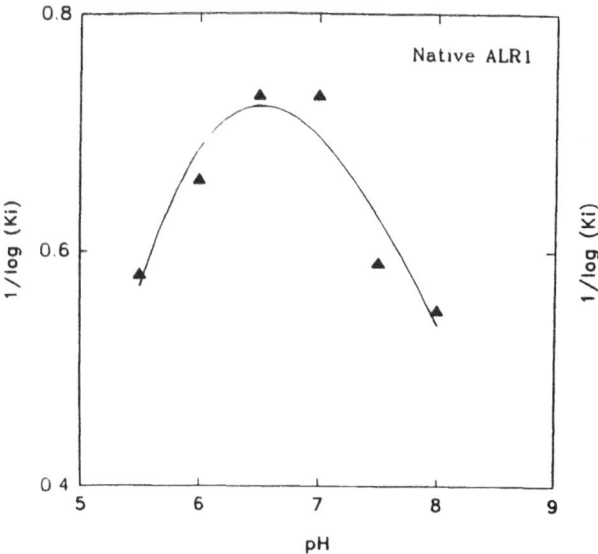

Figure 6. Effect of pH variation on k_{cat}/K_M for NADPH and K_I of NADP+ determined from competitive product inhibition studies.

implicated either Tyr-48 or His-110 as the final proton donor due to the proximity of these residues to the C4 of the nicotinamide ring of NADPH (Wilson et al., 1992). Wilson et al. (1992) favor Tyr-48 as the proton donating residue. Site-directed mutagenesis studies designed to investigate the role of His-110 and Tyr-48 in aldose reductase. (Tarle et al. 1993) have shown that the mutant Y48F (where Tyr-48 had been mutated to phenylalanine) aldose reductase has no detectable enzymatic activity. When His-110 was mutated to asparagine (H110N mutant) the enzyme showed measureable activity. On the basis of these results it was concluded that Tyr-48 was the proton donor. In a more detailed study Bohren et al. (1994) also constructed and characterized several mutant enzymes in which either Tyr-48 or His-110 was mutated (Y48F, Y48H, Y48S, H110Q and H110A) Based on their kinetic studies they concluded also that Tyr-48 was the proton donor and that His-110 was involved in substrate recognition. Recently, however Liu et al. (1993) have explored the pH dependence of the kinetic parameters of aldose reductase. They have concluded that the amino acid residue involved in catalysis has a pKa of 6.5-7.0 and that this is more characteristic of a histidine residue than a tyrosine. Thus, the nature of the proton donor in aldose reductase remains a subject of some debate.

In the case of aldehyde reductase, the structure and reaction mechanism are less studied. The three-dimensional structures of ALR1 and ALR2 are virtually superimposable and from the data available so far it is likely that in ALR1 the same residues are involved in binding and catalysis as in ALR2. From the crystallographic data now available, the corresponding residues for Tyr-48 and His-110 in aldehyde reductase are Tyr-50 and His-113 respectively. Like, aldose reductase, aldehyde reductase also proceeds via an ordered sequential kinetic mechanism with NADPH binding first and NADP+ being released last. In this work, we conducted a study of the effect of pH on the kinetic parameters. Our results indicate that a residue with a pKa of 6.5-7.0 is important in catalysis. This residue is most likely a histidine residue; in agreement with the results of Liu, et al. (1993). The pKa of 6.5-7.0 may represent the catalytic step or may represent a product off-rate. Although the kinetic mechanism for aldehyde reductase is known, the rate determining step remains to be determined. A more detailed investigation of the kinetic mechanism of ALR1 is warranted.

ACKNOWLEDGEMENT

This work was supported by a grant from the Medical Research Council of Canada (TGF) and the American Diabetes Association (O.E-K).

REFERENCES

Bohren, K.M., Bullock, B., Wermuth, B. & Gabbay, K.H., 1989, The aldo-keto reductase superfamily. cDNAs and deduced amino acid sequences of human aldehyde and aldose reductases, J. Biol. Chem. 264,9547-9551.

Bohren, K.M., Grimshaw, C.E., Lai, C-J., Harrison, D.H., Ringe, D., Petsko, G.A., & Gabbay, K.H., 1994, Tyrosine-48 is the proton donor and Histidine-110 directs substrate stereochemical selectivity in the reduction reaction of human aldose reductase: Enzyme kinetics and crystal structure of the Y48H mutant enzyme. Biochemistry, 33, 2021-2032.

Daly, A.K., & Mantle, T.J., 1982, The kinetic mechanism of the major form of ox kidney aldehyde reductase with D-glucoronic acid. Biochem. J., 205, 381-388.

Davidson, W., & Flynn, T.G., 1979, Kinetics and mechanism of action of aldehyde reductase from pig kidney. Biochem. J., 177, 595-601.

El-Kabbani, O., Narayana, S.V.L., Babu, Y.S., Moore, K.M., Flynn, T.G., Petrash, J.M., Westbrook, E.M., DeLucas, L.J., & Bugg, C.E., 1991, Purification, crystallization and preliminary crystallographic analysis of porcine aldose reductase. J. Mol. Biol. 218, 695-698.

El-Kabbani, O., Green, N.C., Lim, G., Carson, M., Narayana, S.V-L., Moore, K.M., Flynn, T.G. and deLucas, L.J., 1994, Structures of human and porcine aldehyde reductase. Acta Cryst D50 (in press)

Green, N.C., El-Kabbani, O., Flynn, T.G., 1994, Primary structure of pig aldehyde reductase and tertiary structures of pig and human enzymes. FASEB Journal, 8, ABSTR. 217

Grimshaw, C.E., Shahbaz, M., & Putney, C.G., 1990, Mechanistic basis for non-linear kinetics of aldehyde reduction catalyzed by aldose reductase. Biochemistry, 29, 9947-9955.

Harding, J.J., 1992, Aldose reductase (or is it?) TIBS, 17, 494.

Kubiseski, T.J., Hyndman, D.J., Morjana, N.A., & Flynn, T.G., 1992, Studies on pig muscle aldose reductase. Kinetic mechanism and evidence for a slow conformational change upon coenzyme binding. J. Biol. Chem., 267, 6510-6517.

Kubiseski, T.J., Green, N.C., & Flynn, T.G., 1993, Location of an essential arginine residue in the primary structure of pig aldose reductase. In Enzymology and Molecular Biology of Carbonyl Metabolism 4, Plenum Press, N.Y. 259-265.

Kubiseski, T.J., Green, N.C., Borhani, D.W. and Flynn, T.G., 1994, Studies on pig aldose reductase. Identification of an essential arginine in the primary and tertiary structure of the enzyme. J. Biol. Chem. 269: 2183.

Lui, S-Q, Bhatnagar, A, & Srivastava, S.K., 1993, Bovine lens aldose reductase: pH dependance of steady-state kinetic parameters and nucleotide binding. J. Biol. Chem., 268, 25494-25499.

Nishimura, C., Matsuura, Y., Kokai, Y., Akera, T., Carper, D., Morjana, N., Lyons, C. & Flynn, T.G., 1990, Cloning and expression of human aldose reductase, J. Biol. Chem. 265, 9788-9792.

Rondeau, J-M., Tete-Favier, F., Podjarny, A., Reymann, J-M., Barth, P., Biellman, J-F. and Moras, D., 1992, Novel NADPH binding domain revealed by the crystal structure of aldose reductase. Nature 355:469.

Sambrook, J., Fritsch, E.F. & Mamatis, T., 1989, in Molecular Cloning. A laboratory manual. Cold Spring Harbour Laboratory Press.

Sato, S., & Kador, P.F., 1990, Inhibition of aldehyde reductase by aldose reductase inhibitors. Biochemical Pharmacology, 40, 1033-1042.

Sem, D.S., & Kasper, C.B., 1993, Enzyme-substrate binding interactions of NADPH-cytochrome P-450 oxidoreductase characterized with pH and alternate substrate-inhibitor studies. Biochemistry, 32, 11539-11547.

Tarle, I., Borhani, D.W., Wilson, D.K., Quiocho, F.A, & Petrash, J.M., 1993, Probing the active site of human aldose reductase site-directed mutagenesis of Asp-43, Tyr-48, Lys-77 and His-110. J. Biol. Chem., 268, 25687-25693.

Wilson, D.K., Bohren, K.M., Gabbay, K.H., & Quiocho, F.A., 1992, Science, 257: 81-84 An unlikely sugar substrate site in the 1.6A structure of the human aldose reductase holoenzyme implicated in diabetic complications.

Wilson, D.K., Tarle, I., Petrash, J.M., & Quiocho, F.A., 1993, Refined 1.8 A structure of human aldose reductase complexed with the potent inhibitor zopolrestat. Proc. Natl. Acad. Sci., 90, 9847-9851.

26

EXPRESSION OF HUMAN AND RAT CARBONYL REDUCTASE IN E. COLI

COMPARISON OF THE RECOMBINANT ENZYMES

Bendicht Wermuth

Department of Clinical Chemistry
University of Berne
Inselspital, CH-3010 Berne

INTRODUCTION

Carbonyl reductase (secondary-alcohol:NADP$^+$ oxidoreductase, EC 1.1.1.184) is a cytosolic, monomeric oxidoreductase that catalyzes the reduction of a variety of endogenous and xenobiotic carbonyl compounds (Ahmed et al., 1978; Wermuth, 1981; Jarabak et al., 1983; Nakayama et al., 1985; Park et al., 1991). It is widely distributed in human tissues (Wirth and Wermuth, 1992), and typically occurs in multiple molecular forms that differ in size and charge (Wermuth, 1981). Recently, autocatalytic reductive alkylation by 2-oxocarboxylic acids, e.g. pyruvate and 2-oxoglutarate, was suggested to be the cause of the heterogeneity (Krook et al., 1993; Wermuth et al., 1993). The primary structure (Wermuth et al., 1988; Forrest et al., 1990) and gene organization (Forrest et al., 1991) of human carbonyl reductase have been determined, and the enzyme was identified as a member of the short-chain dehydrogenase superfamily (Wermuth, 1992) which includes a number of animal and bacterial dehydrogenases with specificities for steroids, prostaglandins, pterins and other alcohol and carbonyl compounds with special functions in bacteria (Persson et al., 1991). More recently, the cDNA encoding carbonyl reductase from rat testis was cloned, revealing 84 % identity with the human cDNA and 86 % homology between the two gene products (Wermuth et al.). Although the rat and human enzymes show many similarities with respect to physicochemical, immunochemical and enzymatic properties they differ significantly in their tissue distribution. In contrast to the human enzyme which is ubiquitously distributed in the body (Wirth and Wermuth, 1992), the rat enzyme is only expressed in reproductive tissues and the adrenals (Iwata et al., 1989; 1990). Moreover, the rat enzyme appears to catalyze the reduction of steroids with greater specificity and more efficiently than its human counterpart, suggesting a specific role of the rat enzyme in steroid metabolism.

To better understand the structural and functional relationships between the two reductases the cDNAs coding for carbonyl reductase from human placenta and rat testis, respectively, were cloned into an expression vector and the two proteins overexpressed in E. coli. I here report on the autocatalytic modification and the enzymatic properties of the two recombinant reductases.

Enzymology and Molecular Biology of Carbonyl Metabolism 5
Edited by H. Weiner *et al.*, Plenum Press, New York, 1995

EXPERIMENTAL PROCEDURES

Materials

The materials and their sources were as follows: enzymes and nucleotides from Boehringer Mannheim; oligonucleotides from Microsynth (Windisch, Switzerland); agar and media for bacterial cultures from GIBCO-BRL; radioactive compounds from Amersham Corp.; all other chemicals not specified from Sigma, Fluka or Merck.

Isolation of Carbonyl Reductase cDNAS and Expression of the Encoded Proteins in E. Coli

The cDNA complementary to carbonyl reductase from human placenta was isolated from a cDNA library in phage gt11 and transferred into a pGEM plasmid vector as described previously (Wermuth et al., 1988). The human cDNA was used to screen a rat testis cDNA library in gt11 (Clontech) under reduced stringency (1 x SSC, 0.1% SDS, 60°C). Clones hybridizing with the [^{32}P]deoxycytidine-labeled human probe were isolated, and the insert cDNA was cloned into phage M13mp18.

The polymerase chain reaction was used to create NdeI restriction endonuclease sites on both 5'- and 3'-ends of the coding region of human and rat carbonyl reductase cDNA, respectively, as well as to eliminate an internal NdeI site in the rat cDNA. Primers were designed to contain at least 5 nucleotides between the 5'-end and the NdeI endonuclease recognition site. Amplifications were performed in a total volume of 100 µl reaction buffer, containing 100 ng template DNA, 15 pmol of each primer, 50 µM each of the four dNTPs and 2 units Taq DNA polymerase. The DNA was denatured at 94° for 7 min followed by 30 cycles at 94° for 30 s, 52° for 90 s, and 72° for 60 s, and a final extension step at 72° for 6 min. The amplification products were digested with NdeI, ligated into the NdeI cloning site of the expression vector pET11a (Novagen) and introduced into competent E. coli BL21(DE3) by electroporation. Screening for recombinant clones was by immunochemical detection of expressed proteins as described by Studier et al. (1990) using anti human carbonyl reductase antibodies (Wirth and Wermuth, 1985). Bound IgGs were visualized using protein A-peroxidase conjugate (Dr. Bommeli, Berne, Switzerland) with 3,3'-diaminobenzidine and H_2O_2 as substrates (Wirth and Wermuth, 1992).

Clones expressing carbonyl reductase were grown in 100 ml medium (1 % casein hydrolysate, 0.5 % NaCl, 20 µg/ml ampicillin) to an absorbance at 620 nm of 0.6-0.7. IPTG (0.4 mM) was added, and after 4-6 h the cells were sedimented by centrifugation, washed once with 10 mM Tris-Cl (pH 8.6), containing 0.1 mM EDTA, and disrupted by freeze-thawing and sonification. Cell debris were removed by centrifugation, and recombinant carbonyl reductase was purified by DEAE-cellulose and affinity chromatography on 8-(6-amino-hexyl)-amino-2'-phosphoadenosine diphosphoribose Sepharose as established for the purification of the enzyme from human brain (Wermuth, 1981). Plasmid DNA was isolated by standard methods (Sambrook et al., 1989) and sequenced using synthetic oligonucleotide primers designed from the derived sequences.

Enzyme Assay

Carbonyl reductase activity was determined spectrophotometrically by recording the change of NADPH absorbance at 340 nm. The standard assay mixture consisted of 0.1 M sodium phosphate buffer (pH 7.0), 0.05 mM NADPH, 0.2 mM menadione or substrates as indicated in the text, and 5-100 µl enzyme solution in a total volume of 1 ml. Reactions were initiated by the addition of enzyme, and blanks without substrates or enzyme were routinely included. One enzyme unit is defined as the change in absorbance at 340 nm corresponding to the oxidation of 1 μmol coenzyme per min.

In addition to following the oxidation of NADPH, the reduction of steroids was monitored by analysis of the alcohol products. Steroids (0.2 mM), NADPH (0.1 mM) and carbonyl reductase were incubated in 20 mM sodium phosphate (pH 7.0) at 30°C for 4 h. Steroids were extracted with chloroform / methanol (2 : 1), dried under a stream of nitrogen and treated with N,O-bis(trimethylsilyl)trifluoroacetamide. The trimethylsilyl derivatives were analyzed by gas chromatography-mass spectrometry (GC 5890-MSD 5970B, Hewlett-Packard) using a HP SE54 capillary column.

RESULTS AND DISCUSSION

The introduction of the coding regions of the human and rat carbonyl reductase cDNAs into the NdeI cloning site of the expression vector pET11a yielded the constructs pET11a-hCR [1] [*] and pET11a-rCR, respectively. Sequence analysis of the insert DNAs showed that they were identical to the original cDNAs with the exception of the T582C mutation in pET11a-rCR introduced to eliminate the internal NdeI site. Addition of IPTG to cultures of E. coli BL21(DE3) harboring pET11a-hCR or pET11a-rCR was followed by the appearance of a protein on Western blots which migrated together with carbonyl reductase from human and rat tissues, respectively. Concomitantly, an increase in menadione reductase activity became detectable. Both enzyme protein and activity reached a plateau after 4-6 h. Passage of bacterial extracts over DEAE-cellulose yielded preparations of the recombinant enzymes that were essentially free of bacterial proteins (Fig. 1A). Significant amounts of nucleic acids that were still present after DEAE-cellulose chromatography were completely removed by the subsequent affinity chromatography. Typical 1-litre cultures yielded 10-15 mg of homogeneous recombinant carbonyl reductase after purification.

In contrast to the tissue enzymes which occur in at least three charge and size-heterogeneous forms, the enzymes expressed in bacteria were homogeneous. We recently showed that incubation of the smallest, most basic tissue form from human brain with pyruvate and 2-oxoglutarate, respectively, in the presence of NADPH resulted in the formation of enzyme species with the same electrophoretic mobility as the two larger, more acidic enzyme forms (Wermuth et al., 1993). Similarly in this study, incubation of the recombinant human and rat enzymes with the two oxocarboxylic acids resulted in the formation of enzyme forms with slightly higher apparent molecular weights and lower isoelectric points. The isolation of each form was achieved by DEAE-cellulose chromatography (Fig. 1B).

The recombinant enzymes readily catalyzed the reduction of a number of characteristic substrates of carbonyl reductase from human and rat tissues (Table 1). Similar to the tissue enzymes, the recombinant enzymes most efficiently catalyzed the reduction of quinones, e.g. menadione, followed by xenobiotic aromatic aldehydes and ketones, whereas endogenous carbonyl compounds, such as dihydrotestosterone, were poor substrates. Previous studies of the substrate specificity of rat testis carbonyl reductase had indicated marked

---

[*] A similar construct was originally obtained by K.M. Bohren and K.H. Gabbay (Department of Pediatrics, Baylor College of Medicine, Houston, USA) using PCR primers with only 3 nucleotides between the 5'-end and the NdeI endonuclease recognition site. It differs from the present construct in that it misses 4 bases of the Shine Dalgarno sequence of the pET vector and yields about half the amount of recombinant protein.

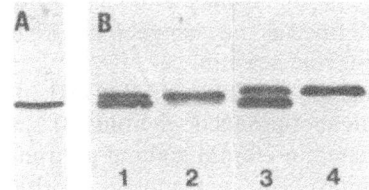

Figure 1. SDS-PAGE of recombinant carbonyl reductase. (A) Recombinant rat testis enzyme (silver stain). (B) Recombinant human placenta enzyme after autocatalytic reductive carboxyalkylation. Lane 1, incubation with 10 mM pyruvate; lane 2, isolated pyruvate-modified form; lane 3, incubation with 10 mM 2-oxoglutarate; lane 4, isolated 2-oxoglutarate-modified form (immunostaining).

differences between the three molecular forms in both Km and V for various substrates (Iwata et al., 1990). In contrast, analogous studies with the human enzyme had yielded similar kinetic constants for all molecular forms (Wermuth, 1981; Bohren et al., 1987; Inazu et al., 1992a). In the present study, all three the native recombinant and the pyruvate- and 2-oxoglutarate-modified forms of both human and rat carbonyl reductase exhibited similar relative velocities for all substrates (Table 2).

The reason for the discrepancy between the enzyme forms from rat testis and those derived from the recombinant enzyme is not known. Different procedures of enzyme purification and assay conditions may be responsible. Alternatively, the differences may be related to structural differences, e.g. different N-termini - methionine or free serine in the recombinant enzyme versus N-acetylserine in the tissue forms - or ligands other than pyruvate and 2-oxoglutarate in the higher-molecular weight enzyme forms from testis.

The physiological role of carbonyl reductase is not clear. Reflecting the wide substrate specificity of the enzyme specific roles in the metabolism of prostaglandins (Lee and Levine, 1974), anthracycline antibiotics (Ahmed et al., 1978) and tetrahydrobiopterin (Park et al., 1991), as well as a general role in the detoxification of foreign and endogenous reactive carbonyl compounds (Wermuth et al., 1986) have been suggested. The finding by Satoh and coworkers (Iwata et al.,1989; 1990) of the exclusive expression of carbonyl reductase in rat reproductive tissues and adrenals suggested a new role for the enzyme in steroid metabolism. In line with this idea, certain 5-reduced steroids, e.g. dihydrotestosterone, are among the best substrates of carbonyl reductase from rat ovary, whereas 5ß-reduced steroids are not reduced. Moreover, the expression of the enzyme in the rat ovary is under the control of gonadotropins and estrogens (Inazu et al., 1992b). Although carbonyl reductase from rat testis has been shown to catalyze the reduction of dihydrotestosterone (Iwata et al., 1990) its specificity for steroids has not been

Table 1. Substrate specificity of recombinant human and rat carbonyl reductase

Substrate	Human rCR		Rat rCR	
	Km	V[1]	Km	V[1]
	mM	*%*	*mM*	*%*
Menadione	0.02	100	0.10	100
4-Benzoylpyridine	0.28	104	1.4	103
Pyridine-3-carbaldehyde	6.5	79	7.4	80
Pyridine-4-carbaldehyde	1.2	122	2.5	109
2-Nitrobenzaldehyde	0.8	116	1.2	98
4-Nitrobenzaldehyde	1.2	92	1.7	71
Phenylglyoxal	1.3	45	2.8	46
Dihydrotestosterone (100 μM)		1.2		2

[1] Expressed as percentage of the activity with menadione The absolute specific activities of the human and rat enzyme using menadione as substrate were 12.6 and 6.2 U/mg protein, respectively.

Table 2. Substrate specificity of the native and carboxyalkylated forms of recombinant human and rat carbonyl reductase

| Substrate | Conc. mM | Relative activity (%) | | | | | |
| | | Human rCR | | | Rat rCR | | |
		native	pyr.[1]	2-OG[2]	native	pyr[1]	2-OG[2]
Menadione	0.25	100	100	100	100	100	100
4-Benzoylpyridine	1.0	80	69	86	65	67	68
4-Nitrobenzaldehyde	1.0	48	27	42	37	30	38
Pyridine-3-aldehyde	10.0	56	37	52	69	65	73
Pyridine-4-aldehyde	5.0	116	68	93	109	98	108
Dihydrotestosterone	0.1	<5	<5	<5	≈2	≈2	≈2

[1] Pyruvate-modified form

[2] 2-Oxoglutarate-modified form

investigated in detail. Hence, the recombinant enzyme was incubated with various 3- and 17-oxo steroids, and the reaction products were analyzed by gas chromatography-mass spectrometry (Table 3). In contrast to the ovarian enzyme, the recombinant testis enzyme did not distinguish between the 5- and 5ß-configuration, and the rates of reduction for the best steroid substrates were less than 5% of the rate obtained with menadione. In addition to the 3-oxo group, the recombinant enzyme also catalyzed the reduction of 17-oxosteroids although less 17-hydroxy than 3-hydroxy product was formed from 5- and 5ß-androstane-3,17-dione, respectively. These results are essentially identical to the findings of a previous study with carbonyl reductase from human tissue (Inazu et al., 1992a).

The present data suggest the occurrence of sex-specific isoenzymes of carbonyl reductase in rat tissues, one with high activity and selectivity for steroids in females and the other with low activity and broad specificity in males. While this work was in progress, Tanaka et al. (1992) reported the sequence of neonatal pig testicular 20ß-hydroxysteroid dehydrogenase and showed that it is 84.5 % homologous to human carbonyl reductase. In addition to quinones and aromatic aldehydes and ketones, the pig enzyme efficiently catalyzed the reduction of dihydrotestosterone and other 3-oxosteroids. More recently, Nakajin et al. (1994) reported the purification and characterization of the corresponding enzyme from the testis of adult pigs. This enzyme also efficiently catalyzed the reduction of quinones and aromatic aldehydes and ketones but showed very low activity towards steroids. The enzyme from adult pigs thus resembles carbonyl reductase from human and rat testis, whereas the neonatal form of the enzyme shows more similarity with the rat ovarian enzyme although it does not distinguish between 5- and 5ß-reduced steroids.

The reduction of the oxo group at C_3 of dihydrotestosterone is thought to inactivate this male hormone. The fact that both rat female reproductive tissues and the testes of immature pigs contain isoenzymes with high dihydrotestosterone reductase activity suggests that this enzyme may regulate androgen action in the tissues concerned. The availability of recombinant carbonyl reductase in abundant quantities combined with the ability to probe its function with appropriate mutations will facilitate the study of this enzyme in man and animals.

ACKNOWLEDGEMENTS

I thank Mmes. E. Ernst and G. Mäder-Heinemann for skillful technical assistance. This work was supported by a grant from the Swiss National Science Foundation.

Table 3. Reduction of steroids by human brain and recombinant rat testis
carbonyl reductase

Substrate	Product	Human[1]	Rat
		%[2]	%[2]
5α-Androstan-17ß-ol-3-one	unreacted substrate	89	56
	5α-androstane-3,17ß-diol	11	34
	5α-androstane-3ß,17ß-diol	trace	10
5ß-Androstan-17ß-ol-3-one	unreacted substrate	71	46
	5ß-androstane-3α,17ß-diol	27	48
	5ß-androstane-3ß,17ß-diol	2	6
5α-Androstan-3α-ol-17-one	unreacted substrate	100	100
5α-Androstan-3ß-ol-17-one	unreacted substrate	96	78
	5α-androstane-3ß,17ß-diol	4	22
5ß-Androstan-3α-ol-17-one	unreacted substrate	91	85
	5ß-androstane-3,α17α-diol	9	15
5ß-Androstan-3ß-ol-17-one	unreacted substrate	100	95
	5ß-androstane-3ß,17α-diol		3
	5ß-androstane-3ß,17ß-diol		2
5α-Androstane-3,17-dione	unreacted substrate	54	43
	5α-Androstan-3α-ol-17-one	32	37
	5α-Androstan-3ß-ol-17-one	5	11
	5α-Androstan-17ß-ol-3-one	9	6
	5α-androstane-3α,17ß-diol	trace	2
	5α-androstane-3ß,17ß-diol		1
5ß-Androstane-3,17-dione	unreacted substrate	62	39
	5ß-Androstan-3α-ol-17-one	37	57
	5ß-Androstan-3ß-ol-17-one	1	3
	5ß-androstane-3α,17α-diol		1

[1] Inazu et al. (1992a)
[2] Ratio of peak areas (total ion) of product and the sum of all products plus substrate
expressed as percentage

REFERENCES

Ahmed, N.K., Felsted, R.L., and Bachur, N.R., 1978, Heterogeneity of anthracycline antibiotic carbonyl reductases in mammalian livers. *Biochem. Pharmacol.* 27:2713.

Bohren, K.M., von Wartburg, J.P., and Wermuth, B., 1987, Kinetics of carbonyl reductase from human brain. *Biochem. J.* 244:65.

Forrest, G.L., Akman, S., Krutzik, S. Paxton, R.J., Sparkes, R.S., Doroshow, J., Felsted, R.L., Glover, C.J., Mohandas, T., and Bachur, N.R., 1990, Induction of human carbonyl reductase gene located on chromosome 21. *Biochim. Biophys. Acta* 1048:149.

Forrest, G.L., Akman, S., Doroshow, J., Rivera, H., and Kaplan, W.D., 1991, Genomic sequence and expression of cloned human carbonyl reductase gene with daunorubicin reductase activity. *Mol. Pharmacol.* 40:502.

Inazu, N., Ruepp, B., Wirth, H., and Wermuth, B., 1992a, Carbonyl reductase from human testis: Purification and comparison with carbonyl reductase from human brain and rat testis. *Biochim. Biophys. Acta* 1116:50.

Inazu, N., Inaba, N, Satoh, T., and Fujii, T., 1992b, Human chorionic gonadotropin causes an estrogen-mediated induction of rat ovarian carbonyl reductase. *Life Sciences* 51:817.

Iwata, N., Inazu, N., and Satoh, T., 1989, The purification and properties of NADPH-dependent carbonyl reductases from rat ovary. *J. Biochem.* 105:556.

Iwata, N., Inazu, N. Takeo, S., and Satoh, T., 1990, Carbonyl reductases from rat testis and vas deferens. *Eur. J. Biochem.* 193:75.

Jarabak, J., Luncsford, A., and Berkowitz, D., 1983, Substrate specificity of three prostaglandin dehydrogenases. *Prostaglandins* 26:849.

Krook, M., Ghosh, D., Stömberg, R., Carlquist, M., and Jörnvall, H., 1993, Carboxyethyllysine in a protein: Native carbonyl reductase/NADP⁺-dependent prostaglandin dehydrogenase. *Proc. Natl. Acad. Sci. USA* 90:502.

Lee, S.-C., and Levine, L., 1974, Prostaglandin metabolism: Cytoplasmic NADPH-dependent and microsomal NADH-dependent prostaglandin E 9-ketoreductase activities in monkey and pigeon tissues. *J. Biol. Chem.* 249:1369.

Nakajin, S., Fujita, Y., Ohno, S., Uchida, M., Aoki, M., and Shinoda, M., 1994, Purification and characterization of 3/ß-hydroxysteroid dehydrogenase from mature porcine testicular cytosol. *J. Steroid Biochem. Molec. Biol.* 48:249.

Nakayama, T., Hara, A., Yashiro, K., and Sawada, H., 1985, Reductases for carbonyl compounds in human liver. *Biochem. Pharmacol.* 34:107.

Park, Y.S., Heizmann, C.W., Wermuth, B., Levine, R.A., Steinerstauch, P., Guzman, J., and Blau, N., 1991, Human carbonyl and aldose reductases: New catalytic functions in tetrahydrobiopterin biosynthesis. *Biochem. Biophys. Res. Commun.* 175:738.

Persson, B., Krook, M., and Jörnvall, H., 1991, Characteristics of short-chain alcohol dehydrogenases and related enzymes. *Eur. J. Biochem.* 200:537.

Sambrook, J., Fritsch, E.F., and Maniatis, T., 1989, *Molecular cloning, a laboratory manual*, 2nd Ed., Cold Spring Harbor Laboratory, Cold Spring Harbor, NY.

Studier, F.W., Rosenberg, A.H., Dunn, J.J., and Bubendorff, J.W., 1990, Use of T7 polymerase to direct expression of cloned genes. *Meth. Enzymol.* 185:60.

Tanaka, M., Ohno, S., Adachi, S., Nakajin, S., Shinoda, M., and Nagahama, Y., 1992, Pig testicular 20ß-hydroxysteroid dehydrogenase exhibits carbonyl reductase-like structure and activity. *J. Biol. Chem.* 267:13451.

Wermuth, B., 1981, Purification and properties of an NADPH-dependent carbonyl reductase from human brain: Relationship to prostaglandin 9-ketoreductase and xenobiotic ketone reductase. *J. Biol. Chem.* 256:1206.

Wermuth, B., 1992, NADP-dependent 15-hydroxyprostaglandin dehydrogenase is homologous to NAD-dependent 15-hydroxyprostaglandin dehydrogenase and other short-chain alcohol dehydrogenases. *Prostaglandins* 44:5.

Wermuth, B., Platt, K.L., Seidel, A., and Oesch, F., 1986, Carbonyl reductase provides the enzymatic basis of quinone detoxication in man. *Biochem. Pharmacol.* 35:1277.

Wermuth, B., Bohren, K.M., Heinemann, G., von Wartburg, J.P., and Gabbay, K.H., 1988, Human carbonyl reductase: Nucleotide sequence analysis of a cDNA and amino acid sequence of the encoded protein. *J. Biol. Chem.* 263:16185.

Wermuth, B., Bohren, K.M., and Ernst, E., 1993, Autocatalytic modification of human carbonyl reductase by 2-oxocarboxylic acids. *FEBS Lett.* 335:151.

Wermuth, B., Mäder-Heinemann, G., and Ernst, E., Cloning and expression of carbonyl reductase from rat testis. Submitted.

Wirth, H.P., and Wermuth, B., 1985, Immunochemical characterization of aldo-keto reductases from human tissues. *FEBS Lett.* 187:280.

Wirth, H., and Wermuth, B., 1992, Immunohistochemical localization of carbonyl reductase in human tissues. *J. Histochem. Cytochem.* 40:1857.

27

MOLECULAR CLONING AND SEQUENCING OF MOUSE HEPATIC 11ß-HYDROXYSTEROID DEHYDROGENASE/CARBONYL REDUCTASE

A MEMBER OF THE SHORT CHAIN DEHYDROGENASE SUPERFAMILY

Edmund Maser and Udo C.T. Oppermann

Department of Pharmacology and Toxicology
School of Medicine
Philipps-University of Marburg
Karl-von-Frisch-Strasse, 35033 Marburg, Germany

INTRODUCTION

Extensive studies on carbonyl reducing enzymes in recent years have led to the establishment of the aldo-keto reductase family, members of which are aldehyde reductase (EC 1.1.1.2) and aldose reductase (EC 1.1.1.21) (Bohren et al., 1989). In addition to these "classic" aldo-keto reductases, enzymes like carbonyl reductase (EC 1.1.1.184), dihydrodiol dehydrogenase (EC 1.3.1.20), NAD(P)H:quinone-oxidoreductase (EC 1.6.99.2) and several hydroxysteroid dehydrogenases (3α-, 17ß-, 20α-, 3α/20ß-) were also shown to be involved in the reductive metabolism of xenobiotic carbonyl compounds (for Review see Maser, 1994). The involvement of hydroxysteroid dehydrogenases in the reductive metabolism of xenobiotic carbonyl compounds and in the inactivation of proximal carcinogens, derived from polycyclic aromatic hydrocarbons, points to other important roles of these proteins besides their normal endocrinological function.

One of the most characteristic features of these carbonyl reducing enzymes is their subcellular localization in the cytosolic fraction of the cell. Nevertheless, for a variety of pharmacologically and toxicologically relevant substances, such as acetohexamide (Imamura et al., 1993), metyrapone (Maser, 1989), warfarin (Hermans and Thijssen, 1989) and haloperidol (Inaba and Kovacs, 1989), enzymatic conversion via carbonyl reduction within the microsomal fraction is known. However, the lack of exact data disables an exact assignment to a specific enzyme unless the respective activity has been proved with homogenous enzyme preparations, or, using molecular biology methods, respective cDNAs have been expressed and characterized in cell culture systems.

Enzymology and Molecular Biology of Carbonyl Metabolism 5
Edited by H. Weiner *et al.*, Plenum Press, New York, 1995

In previous investigations it was found that the cytochrome P450 inhibitor metyrapone undergoes carbonyl reduction in liver microsomes (Kahl, 1970; Maser, 1989; Maser
et al., 1991; Oppermann et al., 1991), in addition to the cytosol. Whereas the participation
of a quercitrin-sensitive carbonyl reductase in cytosolic metyrapone reduction was well
documented (Maser and Netter, 1991), little was known on the membrane-associated
metyrapone reducing enzyme, although it has been purified recently (Maser and Netter,
1989). With regard to its subcellular localization in the endoplasmic reticulum and its
insensitivity towards the diagnostic carbonyl reductase inhibitor quercitrin, the microsomal
enzyme differed in its characteristics from those, common to the cytosolic carbonyl reductases (Maser, 1989).

Surprisingly, N-terminal amino acid sequence analysis of the mouse liver microsomal
metyrapone reductase revealed sequence homology to rat liver 11ß-hydroxysteroid dehydrogenase (11ß-HSD) of around 54%. It then could be shown that the homogeneously
purified enzyme indeed is able to catalyze the reversible 11ß-oxidoreduction of glucocorticoids, thus providing evidence that the microsomal metyrapone reductase is identical to
11ß-HSD (Maser and Bannenberg, 1994a).

The enzyme 11ß-HSD (EC 1.1.1.146) is considered to confer mineralocorticoid
specificity on the non-selective type I adrenocorticoid receptor by converting the receptor-
active 11-hydroxy glucocorticoids cortisol and corticosterone to their receptor-inactive
11-oxo metabolites cortisone and dehydrocorticosterone in mineralocorticoid target tissues
like the kidney (Edwards et al., 1988; Funder et al., 1988). Deficiency of 11ß-HSD, either
congenital (Stewart et al., 1988) or when the enzyme is inhibited by liquorice (Stewart et
al., 1987), results in the activation of renal mineralocorticoid receptors by cortisol or
corticosterone with resultant sodium retention and hypertension.

11ß-HSD is also found in glucocorticoid target tissues, notably the liver (Bush et al.,
1968; Keorner, 1969) and preliminary studies indicate that it may regulate exposure of active
glucocorticoids to the classical glucocorticoid (Typ II) receptor. Molecular cloning of rat,
human, sheep and monkey homologous 11ß-HSD forms (Agarwal et al., 1989; Tannin et al.,
1991; Yang et al., 1992; Moore et al., 1993) established this enzyme to belong to the short
chain dehydrogenase superfamily (Persson et al., 1991; Tannin et al., 1991; Krozowski,
1992). Kinetic, histochemical and genetic analyses, however, suggest that this particular
isoform (termed 11ß-HSD 1A) and its assumed truncated, catalytically inactive form (termed
11ß-HSD 1B) are not responsible for the above described receptor protection mechanism in
mineralocorticoid target cells (Mercer et al., 1993; Krozowski et al., 1992; Moisan et al.,
1992), which rather seems to be achieved by the structurally and functionally distinct forms
11ß-HSD 2 and/or 11ß-HSD 3 (Seckl, 1993; Naray-Fejes-Toth et al., 1993). A possible
physiological function of the 11ß-HSD 1A isoform might, therefore, be the regulation of
access of glucocorticoid hormones to the type II glucocorticoid receptor (Stewart and
Whorwood, 1994), thus modulating for example the vascular tone in resistance vessels
(Walker et al., 1992) or mediating testicular Leydig cell androgen hormone secretion
(Phillips et al., 1989; Monder et al., 1994).

Because of the much higher activities of hepatic 11ß-HSD 1A compared to that of
11ß-HSD 2 and 11ß-HSD 3 in other tissues, the liver isoform possibly has additional
functions besides the metabolism of glucocorticoids, which, according to previous investigations (Maser and Bannenberg, 1994a) may be the reduction of non-steroidal xenobiotic
carbonyl compounds. Therefore, 11ß-HSD may play an important role in the phase I drug
metabolism of pharmacologically relevant carbonyl compounds as well as protecting organisms against toxic carbonyl compounds by converting them to less lipophilic and more
soluble and conjugatable metabolites.

In this study we report on the isolation of a full length 11ß-HSD cDNA (11ß-HSD
1A) from a lamda gt 11 mouse liver library, the elucidation of its primary structure and its

relationship to the previously described microsomal carbonyl reductase by immunoprecipitation after coupled *in vitro* translation/transcription. Structural elements, which on the one hand, are obtained from the deduced amino acid sequence of the 11ß-HSD 1 A cDNA, and which on the other hand, represent common features for the mechanistic function of other hydroxysteroid dehydrogenases of the short chain dehydrogenase protein superfamily are discussed.

MATERIALS AND METHODS

Screening of a Lambda Gt 11 cDNA Library and 11ß-HSDm cDNA Isolation

A lambda gt 11 mouse liver cDNA library (Balb/c strain; Clontech, Heidelberg, Germany) was screened with a rat 11ß-HSD 1A probe amplified by PCR (30 cycles; annealing: 1 min at 55 °C, extension: 2 min at 72 °C, melting: 1 min at 94 °C) from a lambda zap II rat liver cDNA library (Stratagene, Heidelberg, Germany) using 21 mer sense and antisense oligonucleotides spanning the entire coding region of rat liver 11ß-HSD 1A cDNA (Agarwal et al., 1989). The probe was labeled by use of digoxigenin dUTP (DIG DNA labeling and detection kit, Boehringer Mannheim, Germany). Screening, isolation and purification of lambda plaques and DNA preparation was carried out as described in Sambrook et al. (1982) using moderate stringency hybridization and washing conditions (hybridization in 5x SSC, 1 % blocking reagent, 0.1 % Na-laurylsarcosine, 0.02 % SDS, at 45 °C overnight; stringency washing at 0.5x SSC, 0.1 % SDS at 60 °C for 1 hr). Detection of positive plaques was achieved by chemoluminescence detection using AMPPD (Boehringer Mannheim, Germany) as substrate for alkaline phosphatase coupled anti-DIG antibodies. Initial screening of 2.5 x 10^5 independent plaques with the rat probe resulted in the identification of 8 positive clones, none of them containing any full length cDNA. Rescreening of 3 x 10^5 lambda plaques under more stringent conditions (hybridization as above, stringency washing at 0.1x SSC, 0.1 % SDS at 68 °C for 1 hr) with a mouse probe (nucleotides 420-1200 in the mouse sequence), isolated from the first screening, resulted in the isolation of 3 full length cDNAs, which were supposed to be identical based on restriction enzyme fragmentation pattern and DNA sequencing.

After subcloning of full length cDNAs into SK Bluescript vector (Stratagene, Heidelberg, Germany) the primary structure of the cDNA was determined by sequencing both strands using the Sequenase II kit (USB/Amersham, Braunschweig, Germany). Sequencing strategy was as depicted in Figure 1.

Coupled *in Vitro* Transcription/Translation/Immunoprecipitation of 11ß-HSDm 1A

In vitro transcription of the linearized 11ß-HSDm 1A/SK Bluescript plasmid was carried out with the Promega Riboprobe Kit (Heidelberg, Germany) using T3 and T7 RNA polymerase, resulting in sense and antisense riboprobes which served as templates for *in vitro* translation. *In vitro* translation was conducted using a rabbit reticulocyte system (Promega, Heidelberg, Germany) and ^{35}S labeled methionine (NEN-DuPont, Dreieich, Germany) using the manufacturer's protocol. Immunoprecipitation of translated polypeptides was achieved with antibodies against the mouse liver enzyme and Protein A Sepharose (Pharmacia, Freiburg, Germany). Identification of precipitated 11ß-HSD polypeptide suc-

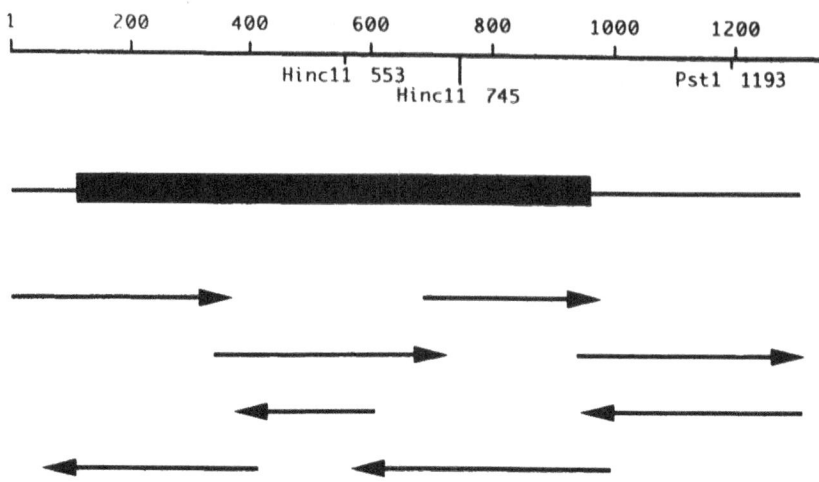

Figure 1. Restriction map and sequencing strategy of full length 11ß-HSDm 1A cDNA clones. The heavy bar represents the coding region, the light bars indicate the 5'- and 3'-untranslated regions. The selected restriction sites are HincII and PstI. Sequencing strategy was as shown by arrows.

ceeded after separation on SDS-PAGE and subsequent signal enhancement by fluorography (Amplify, Amersham, Braunschweig, Germany) and exposure of the dried gel for 48 hrs.

N-Terminal Sequencing of Purified 11ß-HSD 1A

11ß-HSD 1A protein was purified as described elsewhere (Maser and Netter, 1989; Maser and Bannenberg, 1994b). When subjected to Edman degradation in a gas-phase sequencer (Applied Biosystems, Weiterstadt, Germany) a N-terminal amino acid sequence was obtained, which exhibits high homologies to rat liver and human testicular 11ß-HSD (underlined in Figure 3).

RESULTS

cDNA Structure of 11ß-HSDm 1A and Deduced Protein Sequence

A mouse liver lambda gt 11 cDNA library was initially screened with a 11ß-HSDr 1A rat liver cDNA probe and subsequently rescreened with an isolated 11ß-HSDm 1A mouse liver oligonucleotide probe. Positive plaques were isolated and sequenced, resulting in the identification of three full length cDNAs, which were judged to be identical based on restriction enzyme fragmentation pattern and on DNA sequencing. The cDNA contains an open reading frame of 876 bases, predicting a protein of 292 amino acid residues with a calculated molecular weight of 32 kDa (Figures 2 and 3). Potential glycosylation sites are located at Asn 162 and Asn 207, being in agreement with previous findings on glycosylation of the purified active mouse liver enzyme, after specific carbohydrate staining within a polyacrylamide gel (Maser and Bannenberg, 1994b). This explains the difference in the molecular mass of the *in vitro* translated protein of 32 kDa (Figure 4) to that of the native protein of 34 kDa (Maser and Netter, 1989; Maser and Bannenberg, 1994b). This kind of post-translational modification seems to be a determining factor concerning the direction of enzymatic catalysis, favouring either 11ß-dehydrogenation or 11-oxo reduction (Agarwal et al., 1990).

```
  1        GGATGAGACAGAAGGATAGAGAGGAGGAGAGAGAGAGAGAGAAGAGAAGCAACCAGAAAT
 61 AGGCAGCCAATAAAAAGGAGCCGCACTTATCTGAAGCCTCAAGGGGCCTGAGCCAGGTCCCTGTTTG
    Met Ala Val Met Lys Asn Tyr Leu Leu Pro Ile Leu Val Leu Ser Leu Ala      17
128 ATG GCA GTT ATG AAA AAT TAC CTC CTC CCG ATC CTG GTG CTC TCC CTG GCC
    Tyr Tyr Tyr Tyr Ser Thr Asn Glu Glu Phe Arg Pro Glu Met Leu Gln Gly      34
180 TAC TAC TAC TAT TCT ACA AAT GAA GAG TTC AGA CCA GAA ATG CTC CAG GGA
    Lys Lys Val Ile Val Thr Gly Ala Ser Lys Gly Ile Gly Arg Glu Met Ala      51
231 AAG AAA GTG ATT GTC ACT GGG GCC AGC AAA GGG ATT GGA AGA GAA ATG GCA
    Tyr His Leu Ser Lys Met Gly Ala His Val Val Leu Thr Ala Arg Ser Glu      68
282 TAT CAT CTG TCA AAA ATG GGA GCC CAT GTG GTA TTG ACT GCC AGG TCG GAG
    Glu Gly Leu Gln Lys Val Val Ser Arg Cys Leu Glu Leu Gly Ala Ala Ser      85
333 GAA GGT CTC CAG AAG GTA GTG TCT CGC TGC CTT GAA CTC GGA GCA GCC TCT
    Ala His Tyr Ile Ala Gly Thr Met Glu Asp Met Thr Phe Ala Glu Gln Phe     102
384 GCT CAC TAC ATT GCT GGC ACT ATG GAA GAC ATG ACA TTT GCG GAG CAA TTT
    Ile Val Lys Ala Gly Lys Leu Met Gly Gly Leu Asp Met Leu Ile Leu Asn     119
435 ATT GTC AAG GCG GGA AAG CTC ATG GGC GGA CTG GAC ATG CTT ATT CTA AAC
    His Ile Thr Gln Thr Ser Leu Ser Leu Phe His Asp Asp Ile His Ser Val     136
486 CAC ATC ACT CAG ACC TCG CTG TCT CTC TTC CAT GAC GAC ATC CAC TCT GTG
    Arg Arg Val Met Glu Val Asn Phe Leu Ser Tyr Val Val Met Ser Thr Ala     153
537 CGA AGA GTC ATG GAG GTC AAC TTC CTC AGC TAC GTG GTC ATG AGC ACA GCC
    Ala Leu Pro Met Leu Lys Gln Ser Asn Gly Ser Ile Ala Val Ile Ser Ser     170
588 GCC TTG CCC ATG CTG AAG CAG AGC AAT GGC AGC ATT GCC GTC ATC TCC TCC
    Leu Ala Gly Lys Met Thr Gln Pro Met Ile Ala Pro Tyr Ser Ala Ser Lys     187
639 TTG GCT GGG AAA ATG ACC CAG CCT ATG ATT GCT CCC TAC TCT GCA AGC AAG
    Phe Ala Leu Asp Gly Phe Phe Ser Thr Ile Arg Thr Glu Leu Tyr Ile Thr     204
690 TTT GCT CTG GAT GGG TTC TTT TCC ACC ATT AGA ACA GAA CTC TAC ATA ACC
    Lys Val Asn Val Ser Ile Thr Leu Cys Val Leu Gly Leu Ile Asp Thr Glu     221
741 AAG GTC AAC GTG TCC ATC ACT CTC TGT GTC CTT GGC CTC ATA GAC ACA GAA
    Thr Ala Met Lys Met Ile Ser Gly Ile Ile Asp Ala Gln Ala Ser Pro Lys     238
792 ACA GCT ATG AAG GAA ATC TCT GGG ATA ATT GAC GCC CAA GCT TCT CCC AAG
    Glu Glu Cys Ala Leu Glu Ile Ile Lys Gly Thr Ala Leu Arg Lys Ser Glu     255
843 GAG GAG TGC GCC CTG GAG ATC ATC AAA GGC ACA GCT CTA CGC AAA AGC GAG
    Val Tyr Tyr Asp Lys Leu Pro Leu Thr Pro Ile Leu Leu Gly Asn Pro Gly     272
894 GTG TAC TAT GAC AAA TTG CCT TTG ACT CCA ATC CTG CTT GGG AAC CCA GGA
    Arg Lys Ile Met Glu Phe Phe Ser Leu Arg Tyr Tyr Asn Lys Asp Met Phe     289
945 AGG AAG ATC ATG GAA TTT TTT TCA TTA CGA TAT TAT AAT AAG GAC ATG TTT
    Val Ser Asn Stop                                                        292
996 GTA AGT AAC TAG GAACTCCTGAGCCCTGGTGAGTGGTCTTAGAACAGTCCTGCCTCATACTTC
1059 AGTAAGCCCTACCCACAAAAGTATCTTTCCAGAGATACACAAATTTTGGGGTACACCTCATCATGAG
1126 AAATTCTTGCAACACTTGCACAGTGAAAATGTAATTGTAATAAATGTCACAAACCACTTTGGGCCTG
1193 CAGTTGTGAACTTGATTGTAACTATGGATATAAACACATAGTGGTTGTATCGGCTTTACCTCACACT
1260 GAATGAAACAATGATAACTAATGTAACATTAAATATAATAAAGGTAATATCAACTTCGTAAATGCAA
1327 AAAAAAAAAAAAAAAAAAAAAAAAAA
```

Figure 2. cDNA and deduced amino acid sequence of mouse liver 11ß-HSDm 1A. Nucleotide sequence of the murine 11ß-HSD 1A cDNA and the deduced amino acid sequence of the protein are shown. Nucleotides are numbered at the left with position 1 being assigned to the first nucleotide in the cDNA. The amino acids are numbered at the right with position 1 assigned to the first methionine encoded by the nucleotide sequence. Potential N-linked glycosylation sites are underlined.

The derived structural data suggest that the encoded protein belongs to the short chain dehydrogenase superfamily (Persson et al., 1991). Moreover, the isolated mouse liver 11ß-HSDm 1A cDNA shows high homologies to human testicular (Tannin et al., 1991) and rat hepatic (Agarwal et al., 1989) 11ß-HSD 1A, displaying 78 % and 86 % residue identity, respectively, on the protein level (Figure 3). Comparison of the deduced protein sequence to the N-terminal sequence of the purified 11ß-HSD/carbonyl reductase after Edman degradation revealed Met 4 of the deduced protein sequence as the first amino acid in the native polypeptide chain, due either to posttranslational cleavage or to starting of translation at AUG of Met 4 instead of Met 1 (Figure 3).

The alignment and comparison to other hydroxysteroid dehydrogenases of the short chain dehydrogenase superfamily allows the identification of important residues in the 11ß-HSD primary structure. All proteins in this alignment show typical primary structure elements of the short chain dehydrogenase superfamily (Persson et al., 1991; Krozowski, 1992). For example, compared to the soluble 3α/20ß-HSD from *Streptomyces hydrogenans* (Marekov et al., 1990), the membrane associated 11ß-HSDs from rat (Agarwal et al., 1989), human (Tannin et al., 1991), squirrel monkey (Moore et al., 1993) and sheep (Yang et al., 1992) extend the N-terminal part of the protein by a hydrophobic segment, likely to anchor in or span the membrane of the endoplasmic reticulum, and possibly containing a cleavable signal peptide sequence. High homology regions are restricted to certain segments, which are essential for the mechanistic function of the short chain dehydrogenases. A conserved region near the N-terminus is part of the coenzyme binding site (Persson et al., 1991). In addition, three highly conserved glycine residues (Gly-41, Gly-45, Gly-47) are found close to the N-terminus (Figure 3) and are predicted to be involved in dinucleotide binding. The C-terminal part of the short chain dehydrogenases is considered to contain the substrate binding site (Persson et al., 1991). This part exhibits much variations in terms of the primary structure which might explain the broad substrate spectrum among the short chain hydroxysteroid dehydrogenases.

Immunoprecipitation with antibodies against the mouse liver enzyme

The identity of the cDNA coded protein with the purified mouse liver 11ß-HSD/carbonyl reductase was confirmed by immunoprecipitation (Figure 4) after *in vitro* transcription and translation, as well as by N-terminal amino acid sequencing of the purified protein (underlined in Figure 3). The identification of the *in vitro* transcribed peptide was achieved by precipitation with anti liver 11ß-HSDm 1A antibodies coupled to protein A sepharose and

```
          1                                                          47
HSD11r         MKKY LLPVLVLCLG  YYYSTNEEF RPEMLQGKKV IVTGASKGIG
HSD11m      MAVMKNY LLPILVLSLA YYYYSTNEEF RPEMLQGKKV IVTGASKGIG
HSD11h      MAFMKKY LLPILGLFMA YYYYSANEEF RPEMLQGKKV IVTGASKGIG

          48                                                         97
HSD11r    REMAYHLSKM GAHVVLTARS EEGLQKVVSR CLELGAASAH YIAGTMEDMA
HSD11m    REMAYHLSKM GAHVVLTARS EEGLQKVVSR CLELGAASAH YIAGTMEDMT
HSD11h    REMAYHLAKM GAHVVVTARS KETLQKVVSH CLELGAASAH YIAGTMEDMT

          98                                                        147
HSD11r    FAERFVVEAG KLLGGLDMLI LNHITQTTMS LFHDDIHSVR RSMEVNFLSY
HSD11m    FAEQFIVKAG KLMGGLDMLI LNHITQTSLS LFHDDIHSVR RVMEVNFLSY
HSD11h    FAEQFVAQAG KLMGGLDMLI LNHITNTSLN LFHDDIHHVR KSMEVNFLSY

          148                                                       197
HSD11r    VVLSTAALPM LKQSNGSIAI ISSMAGKMTQ PLIASYSASK FALDGFFSTI
HSD11m    VVMSTAALPM LKQSNGSIAV ISSLAGKMTQ PMIAPYSASK FALDGFFSTI
HSD11h    VVLTVAALPM LKQSNGSIVV VSSLAGKVAY PMVAAYSASK FALDGFFSSI

          198                                                       247
HSD11r    RKEHLMTKVN VSITLCVLGF IDTETALKET SGIILSQAAP KQECALE IK
HSD11m    RTELYITKVN VSITLCVLGL IDTETAMKEI SGIIDAQASP KEECALEIIK
HSD11h    RKEYSVSRVN VSITLCVLGL IDTETAMKAV SGIVHMQAAP KEECALEIIK

          248                                                       297
HSD11r    GTVLRKDEVY YDKSSWTPLL LGNPGRRIME FLSLRSYNRD LFVSN
HSD11m    GTALRKSEVY YDKLPLTPIL LGNPGRKIME FFSLRYYNKD MFVSN
HSD11h    GGALRQEEVY YDSSLWTTLL IRNPCRKILE FLYSTSYNMD RFINK
```

Figure 3. Comparison of the predicted amino acid sequences for 11ß-HSD 1A from rat (HSD11r), mouse (HSD11m) and human (HSD11h). Single-letter codes are used. The amino acid residues are numbered in the amino-terminus to carboxy-terminus direction. The N-terminal peptide sequence, as previously determined by Edman degradation, as well as the three highly conserved glycine residues which are predicted to be involved in cofactor binding, are underlined.

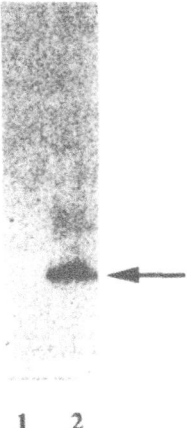

1　　2

Figure 4. Immunoprecipitation of *in vitro* transcribed/translated 11ß-HSDm 1A cDNA. 11ß-HSD 1A cDNA was used as template for *in vitro* transcription using T7 and T3 RNA polymerase. 1 µg of *in vitro* transcribed RNA from each reaction was used for *in vitro* translation and subsequent immunoprecipitation with antibodies against the mouse liver 11ß-HSD 1A (Maser and Bannenberg, 1994a). Lane 1: control (T3 polymerase reaction); lane 2: 11ß-HSD 1A immunoprecipitation (T7 reaction). A single band with a molecular mass of 32 kDa is detectable indicating an *in vitro* translated and precipitated 11ß-HSD 1A cDNA product. Control precipitation with pre-immune serum was also performed, which resulted in blank films (data not shown).

subsequent SDS PAGE and autoradiography. A single polypeptide band was detected with a molecular mass of 32 kDa, being in agreement with the expected cDNA deduced protein.

DISCUSSION

The reduction of xenobiotic and endogenous carbonyl compounds to their respective alcohol metabolites is mediated by structurally and functionally diverse enzymes (reviewed in Maser, 1994). Cytosolic carbonyl reducing enzymes include species and tissue specific isoforms of prostaglandin dehydrogenases and various hydroxysteroid dehydrogenases (3α-, 17ß-, 20α-, 3α/20ß-) and are up to now grouped into two protein superfamilies: the aldo-keto reductases (Bohren et al., 1989) and the short chain dehydrogenases (Krook et al., 1993; Wermuth et al., 1988). The oxidation of trans-dihydrodiols of ultimate carcinogenic aromatic compounds to their non-carcinogenic catechol metabolites is also mediated by enzymes which belong to the above mentioned groups, and which are often identical to each other, like for example, cytosolic rat liver 3α-HSD/dihydrodiol dehydrogenase/bile acid binding protein (Stolz et al., 1987; Smithgall and Penning, 1988; Stolz et al., this volume).

Microsomal linked carbonyl reducing enzymes are associated with 3α-, 17ß- (Hara et al., 1987; Usui et al., 1984) or 11ß-HSD activities (Maser and Bannenberg, 1994a), which might also belong to the aldo-keto reductase or short chain dehydrogenase branch. Considering that most carbonyl compounds are lipid soluble and are expected to distribute in membranes rich in lipids, microsomal hydroxysteroid dehydrogenases may play a more significant role in the reductive metabolism of xenobiotic carbonyl compounds than the cytoplasmic reductases.

The number of distinct reductases in a tissue or organism and whether or not they are the same as similar enzymes in other tissues or organisms is of particular relevance with regard to the search for the physiological role of these enzymes. A physiological function of

11ß-HSD has recently been demonstrated in the kidney, where it protects mineralocorticoid receptors from exposure to the active glucocorticoids cortisol and corticosterone by converting them to the receptor-inactive metabolites cortisone or dehydrocorticosterone, respectively (Edwards et al., 1988; Funder et al., 1988). By this action 11ß-HSD ensures Type I (mineralocorticoid) receptor specificity for the physiological ligand aldosterone which, after binding, mediates target gene transcription.

The isolation and cloning of the rat liver form 11ß-HSD 1A (Agarwal et al., 1989) and subsequent studies on its expression in mineralocorticoid target tissues put doubts on the role of 11ß-HSD 1A as a protector of type I receptors, a role which seems to be fulfilled by the yet only partially purified and characterized distinct isoforms 11ß-HSD 2 and 11ß-HSD 3 (Naray-Fejes-Toth et al., 1993; Seckl, 1993; Brown et al., 1993). A defined role for 11ß-HSD 1A as a modulator of glucocorticoid hormone action in cell lines containing type II receptors has been shown (Stewart and Whorwood, 1994; Duperrex et al., 1993). The 11ß-HSD 1A isoform also plays a fundamental role in glucocorticoid dependent androgen secretion of Leydig cells or could mediate cortisol dependent regulation of vascular tone in resistance vessels (Phillips et al., 1989; Monder et al., 1994; Walker et al., 1992).

The structure determination of murine 11ß-HSD/carbonyl reductase, as presented in this report, reveals that it belongs to the short chain dehydrogenase protein superfamily and shows that it possesses high homologies to other vertebrate 11ß-HSDs. The presence of primary structure data of several 11ß-HSDs and their comparison to other hydroxysteroid dehydrogenases of the short chain dehydrogenase protein superfamily allows the identification of important residues for the catalytic function of these enzymes.

Strictly conserved elements of the short chain dehydrogenases constitute parts of the cosubstrate and substrate binding sites (Persson et al., 1991; Krook et al., 1993; Krozowski, 1992; Baker, 1994) and are essential for catalysis (Chen et al., 1993; Obeid and White, 1992). The N-terminal hydrophobic segment seems to represent the membrane spanning domain, which is required for catalytic activity. Expression of a N-terminal truncated (lacking the hydrophobic N-terminal part) 11ß-HSD 1A cDNA (termed 11ß-HSD 1B) resulted in a non-glycosylated, catalytically inactive protein species, suggesting membrane attachment and surrounding as well as glycosylation as important structural features for the mechanistic function of this class of hydroxysteroid dehydrogenases (Obeid et al., 1993). This is further underlined by the finding that tunicamycin mediated inhibition of glycosylation resulted in a 50% decrease of the 11ß-dehydrogenation reaction, whereas the 11-oxo reductase activity of heterologous expressed 11ß-HSD 1A rat cDNA was unaffected (Agarwal et al., 1990). The attachment of N-linked sugar residues could probably provide optimal three-dimensional stabilization and inter/intrasubunit aggregation via H-bonding or proper orientation of the holoenzyme within the membrane environment.

A conserved region near the N-terminus of the short chain dehydrogenases is considered to be part of the coenzyme binding site (Persson et al., 1991), containing three highly conserved glycine residues, which have been also found in the N-terminus of the murine 11ß-HSD 1A as Gly-41, Gly-45 and Gly-47. The C-terminal part of the short chain dehydrogenases is considered to contain the substrate binding site (Persson et al., 1991). This part exhibits much variations in terms of the primary structure, which might explain the broad substrate spectrum among the short chain dehydrogenases.

From the three-dimensional structures of 3α,20ß-HSD (Ghosh et al., 1991) and dihydropteridine reductase (Varughese et al., 1992), as well as from the computer modelling of NAD$^+$dependent prostaglandin dehydrogenase/carbonyl reductase (Krook et al., 1993) the topologies of the active site with coenzyme bound are known and some homologies are found. They include a well conserved pentapeptide between Tyr-152 and Lys-156 (Tyr-184 and Lys-187 in the murine 11ß-HSD 1A sequence) together with adjacent conserved residues, which are supposed to have a general role in catalysis. Although exhibiting residue identities only at the 25% level, 3α/20ß-HSD, dihydropteridine reductase and NAD^+-dependent prostaglandin dehydrogenase/carbonyl reductase (Krook et al., 1993) are clearly related in terms of their tertiary structure. Accordingly, the 11ß-HSD protein may also share a similar three-dimensional architecture.

In conclusion, structure determinations of 11ß-HSD isoforms will become an essential part in the development of specific effectors that could be used in glucocorticoid hormone based therapy as well as in the prediction of side effects or interactions with non-steroidal xenobiotic carbonyl compounds. The present study was intended to solve the primary structure of the microsomal bound 11ß-HSD/carbonyl reductase as a prerequisite for further investigations, such as for example, heterologous expression in respective cell lines for clarifying its role in detoxification processes.

REFERENCES

Agarwal, A.K., Monder, C., Eckstein, B., and White, P.C., 1989, Cloning and expression of rat cDNA encoding corticosteroid 11 beta-dehydrogenase, *J. Biol. Chem.*, 264: 18939-18943.

Agarwal, A.K., Tusie Luna, M.T., Monder, C., and White, P.C., 1990, Expression of 11 beta-hydroxysteroid dehydrogenase using recombinant vaccinia virus, *Mol. Endocrinol.*, 4: 1827-1832.

Baker, M.E., 1994, Sequence analysis of steroid- and prostaglandin-metabolizing enzymes: Application to understanding catalysis, *Steroids*, 59: 248-258.

Bohren, K.M., Bullock, B., Wermuth, B., and Gabbay, K.H., 1989, The aldo-keto reductase superfamily. cDNAs and deduced amino acid sequences of human aldehyde and aldose reductases, *J. Biol. Chem.*, 264: 9547-9551.

Brown, R.W., Chapman, K.E., Edwards, C.R., and Seckl, J.R., 1993, Human placental 11 beta-hydroxysteroid dehydrogenase: evidence for and partial purification of a distinct NAD-dependent isoform, *Endocrinology*, 132: 2614-2621.

Bush, I.E., Hunter, S.A., and Meigs, R.A., 1968, Metabolism of 11-oxygenated steroids, *Biochem. J.*, 107: 239-257.

Chen, Z., Jiang, J.C., Lin, G.Z., Lee, W.R., Baker, M.E., and Chang, S.H., 1993, Site-specific mutagenesis of *Drosophila* alcohol dehydrogenase: Evidence for involvement of tyrosine-152 and lysine-156 in catalysis, *Biochemistry*, 32: 3342-3346.

Duperrex, H., Kenouch, S., Gaeggeler, H.P., Seckl, J.R., Edwards, C.R., Farman, N., and Rossier, B.C., 1993, Rat liver 11 beta-hydroxysteroid dehydrogenase complementary deoxyribonucleic acid encodes oxoreductase activity in a mineralocorticoid-responsive toad bladder cell line, *Endocrinology*, 132: 612-619.

Edwards, C.R., Stewart, P.M., Burt, D., Brett, L., McIntyre, M.A., Sutanto, W.S., de Kloet, E.R., and Monder, C., 1988, Localisation of 11 beta-hydroxysteroid dehydrogenase--tissue specific protector of the mineralocorticoid receptor, *Lancet*, 2: 986-989.

Funder, J.W., Pearce, P.T., Smith, R., and Smith, A.I., 1988, Mineralocorticoid action: target specificity is enzyme, not receptor mediated, *Science*, 242: 583-585.

Ghosh, D., Weeks, C.M., Grochulski, P., Duax, W.L., Erman, M., Rimsay, R.L., and Orr, J.C., 1991, Three-dimensional structure of holo 3α,20ß-hydroxysteroid dehydrogenase: A member of a short-chain dehydrogenase family, *Proc. Natl. Acad. Sci. U S A*, 88: 10064-10068.

Hara, A., Usui, S., Hayashibara, M., Horiuchi, T., Nakayama, T., and Sawada, H. (1987). Microsomal carbonyl reductase in rat liver. Sex difference, hormonal regulation, and characterization. In: H. Weiner and T.G. Flynn (Eds.), *Enzymology and Molecular Biology of Carbonyl Metabolism: Aldehyde Dehydrogenase, Aldo-Keto Reductase, and Alcohol Dehydrogenase*, pp. 401-414, Alan R. Liss, New York.

Hermans, J.J.R. and Thijssen, H.H.W., 1989, The *in vitro* ketone reduction of warfarin and analogues. Substrate stereoselectivity, product stereoselectivity and species differences, *Biochem. Pharmacol.*, 38: 3365-3370.

Imamura, Y., Iwamoto, K., Yanachi, Y., Higuchi, T., and Otagiri, M., 1993, Postnatal development, sex-related difference and hormonal regulation of acetohexamide reductase activities in rat liver kidney, *J. Pharmacol. Exp. Ther.*, 264: 166-171.

Inaba, T. and Kovacs, J., 1989, Haloperidol reductase in human and guinea pig livers, *Drug Metab. Dispos.*, 17: 330-333.

Kahl, G.F., 1970, Experiments on the metyrapone reducing microsomal enzyme system, *Naunyn-Schmiede-bergs Arch. Pharmacol.*, 266: 61-74.

Keorner, D.R., 1969, Assay and substrate specificity of liver 11ß-hydroxysteroid dehydrogenase, *Biochim. Biophys. Acta*, 179: 377-382.

Krook, M., Ghosh, D., Duax, W., and Jornvall, H., 1993, Three-dimensional model of NAD(+)-dependent 15-hydroxyprostaglandin dehydrogenase and relationships to the NADP(+)-dependent enzyme (carbonyl reductase), *FEBS Lett.*, 322: 139-142.

Krozowski, Z., 1992, 11 beta-hydroxysteroid dehydrogenase and the short-chain alcohol dehydrogenase (SCAD) superfamily, *Mol. Cell Endocrinol.*, 84: C25-C31.

Krozowski, Z., Obeyesekere, V., Smith, R., and Mercer, W., 1992, Tissue-specific expression of an 11 beta-hydroxysteroid dehydrogenase with a truncated N-terminal domain. A potential mechanism for differential intracellular localization within mineralocorticoid target cells, *J. Biol. Chem.*, 267: 2569-2574.

Marekov, L., Krook, M., and Jornvall, H., 1990, Prokaryotic 20 beta-hydroxysteroid dehydrogenase is an enzyme of the 'short-chain, non-metalloenzyme' alcohol dehydrogenase type, *FEBS Lett.*, 266: 51-54.

Maser, E., 1989, Characterization of microsomal and cytoplasmic metyrapone reducing enzymes from mouse liver, *Arch. Toxicol. Suppl.*, 13: 271-274.

Maser, E., 1994, Xenobiotic carbonyl reduction and physiological steroid oxidoreduction - the pluripotency of several hydroxysteroid dehydrogenases, *Biochem. Pharmacol.* (in press).

Maser, E. and Bannenberg, G., 1994a, 11ß-Hydroxysteroid dehydrogenase mediates reductive metabolism of xenobiotic carbonyl compounds, *Biochem. Pharmacol.*, 47: 1805-1812.

Maser, E. and Bannenberg, G., 1994b, The purification of 11ß-hydroxysteroid dehydrogenase from mouse liver microsomes, *J. Steroid Biochem. Molec. Biol.*, 48: 257-263.

Maser, E., Gebel, T., and Netter, K.J., 1991, Carbonyl reduction of metyrapone in human liver, *Biochem. Pharmacol*, 42 Suppl: S93-S98.

Maser, E. and Netter, K.J., 1989, Purification and properties of a metyrapone-reducing enzyme from mouse liver microsomes--this ketone is reduced by an aldehyde reductase, *Biochem. Pharmacol*, 38: 3049-3054.

Maser, E. and Netter, K.J., 1991, Reductive metabolism of metyrapone by a quercitrin-sensitive ketone reductase in mouse liver cytosol, *Biochem. Pharmacol.*, 41: 1595-1599.

Mercer, W., Obeyesekere, V., Smith, R., and Krozowski, Z., 1993, Characterization of 11 beta-HSD1B gene expression and enzymatic activity, *Mol. Cell Endocrinol.*, 92: 247-251.

Moisan, M.P., Edwards, C.R., and Seckl, J.R., 1992, Differential promoter usage by the rat 11 beta-hydroxysteroid dehydrogenase gene, *Mol. Endocrinol.*, 6: 1082-1087.

Monder, C., Hardy, M.P., Blanchard, R.J., and Blanchard, D.C., 1994, Comparative aspects of 11ß-hydroxysteroid dehydrogenase. Testicular 11ß-hydroxysteroid dehydrogenase: Development of a model for the mediation of Leydig cell function by corticosteroids, *Steroids*, 59: 69-73.

Moore, C.C., Mellon, S.H., Murai, J., Siiteri, P.K., and Miller, W.L., 1993, Structure and function of the hepatic form of 11 beta-hydroxysteroid dehydrogenase in the squirrel monkey, an animal model of glucocorticoid resistance, *Endocrinology*, 133: 368-375.

Naray-Fejes-Toth, A., Rusvai, E., Denault, D.L., St.Germain, D.L., and Fejes-Toth, G., 1993, *Am. J. Physiol.*, 265: F896-F900.

Obeid, J., Curnow, K.M., Aisenberg, J., and White, P.C., 1993, Transcripts originating in intron 1 of the HSD11 (11 beta-hydroxysteroid dehydrogenase) gene encode a truncated polypeptide that is enzymatically inactive, *Mol. Endocrinol.*, 7: 154-160.

Obeid, J. and White, P.C., 1992, Tyr-179 and Lys-183 are essential for enzymatic activitiy of 11ß-hydroxysteroid dehydrogenase, *Biochem. Biophys. Res. Com.*, 188: 222-227.

Oppermann, U.C., Maser, E., Mangoura, S.A., and Netter, K.J., 1991, Heterogeneity of carbonyl reduction in subcellular fractions and different organs in rodents, *Biochem. Pharmacol.*, 42 Suppl: S189-S195.

Persson, B., Krook, M., and Jornvall, H., 1991, Characteristics of short-chain alcohol dehydrogenases and related enzymes, *Eur. J. Biochem.*, 200: 537-543.

Phillips, D.M., Lakshmi, V., and Monder, C., 1989, Corticosteroid 11 beta-dehydrogenase in rat testis, *Endocrinology*, 125: 209-216.

Sambrook, J., Maniatis, T., and Fritsch, E.F., 1982, Molecular cloning: a laboratory manual, Cold Spring Harbor Laboratory, New York.

Seckl, J.R., 1993, 11ß-Hydroxysteroid dehydrogenase isoforms and their implications for blood pressure regulation., *Eur. J. Clin. Invest.*, 23: 589-601.

Smithgall, T.E. and Penning, T.M., 1988, Electrophoretic and immunochemical characterization of 3 alpha-hydroxysteroid/dihydrodiol dehydrogenases of rat tissues, *Biochem. J.*, 254: 715-721.

Stewart, P.M., Corrie, J.E., Shackleton, C.H., and Edwards, C.R., 1988, Syndrome of apparent mineralocorticoid excess. A defect in the cortisol-cortisone shuttle, *J. Clin. Invest.*, 82: 340-349.

Stewart, P.M., Valentino, R., Wallace, A.M., Burt, D., Shackleton, C.H.L., and Edwards, C.R.W., 1987, Mineralocorticoid activity of liquorice: 11 beta-hydroxysteroid dehydrogenase deficiency comes of age., *Lancet*, ii: 821-824.

Stewart, P.M. and Whorwood, C.B., 1994, 11ß-Hydroxysteroid dehydrogenase activity and corticoid hormone action, *Steroids*, 59: 90-95.

Stolz, A., Takikawa, H., Sugiyama, Y., Kuhlekamp, J., and Kaplowitz, N., 1987, 3α-hydroxysteroid dehydrogenase activity of the Y'-bile acid binders in rat liver cytosol, *J. Clin. Invest.*, 79: 427-434.

Tannin, G.M., Agarwal, A.K., Monder, C., New, M.I., and White, P.C., 1991, The human gene for 11 beta-hydroxysteroid dehydrogenase. Structure, tissue distribution, and chromosomal localization, *J. Biol. Chem.*, 266: 16653-16658.

Usui, S., Hara, A., Nakayama, T., and Sawada, H., 1984, Purification and characterization of two forms of microsomal carbonyl reductase in guinea pig liver, *Biochem. J.*, 223: 679-705.

Varughese, K.I., Skinner, M.M., Whiteley, J.M., Matthews, D.A., and Xuong, N.H., 1992, Crystal structure of rat liver dihydropteridine reductase, *Proc. Natl. Acad. Sci. U S A*, 89: 6080-6084.

Walker, B.R., Connacher, A.A., Webb, D.J., and Edwards, C.R.W., 1992, Glucocorticoids and blood pressure: a role for the cortisol/cortisone shuttle in the control of vascular tone in man, *Clin. Sci.*, 83: 171-178.

Wermuth, B., Bohren, K.M., Heinemann, G., von Wartburg, J.P., and Gabbay, K.H., 1988, Human carbonyl reductase. Nucleotide sequence analysis of a cDNA and amino acid sequence of the encoded protein, *J. Biol. Chem.*, 263: 16185-16188.

Yang, K., Smith, C.L., Dales, D., Hammond, G.L., and Challis, J.R., 1992, Cloning of an ovine 11 beta-hydroxysteroid dehydrogenase complementary deoxyribonucleic acid: tissue and temporal distribution of its messenger ribonucleic acid during fetal and neonatal development, *Endocrinology*, 131: 2120-2126.

28

MOLECULAR MODELLING CALCULATIONS ON THE BINDING OF D- AND L-XYLOSE TO WILD-TYPE ALDOSE REDUCTASE AND ITS H110Q AND H110A MUTANTS

Hans L. De Winter and Mark von Itzstein

School of Pharmaceutical Chemistry
Victorian College of Pharmacy
Monash University
381 Royal Parade
Parkville, 3052, Victoria, Australia

INTRODUCTION

Human aldose reductase (hAR), an NADPH-dependent enzyme, catalyzes the reversible reduction of a wide variety of carbonyl-containing compounds to their respective alcohol counterparts (Flynn and Green, 1993). Although a physiological role for this enzyme has not yet been established, inhibitors of hAR appear to be effective in the treatment of diabetic neuropathy (Masson and Boulton, 1990).

The active site pocket of hAR is lined by seven aromatic and four apolar residues, and is very hydrophobic. The only polar side chains within the pocket are Gln49, Cys298, Tyr48 and His110. The nicotinamide ring of the NADPH cofactor is located at the bottom of the pocket, with the hydroxyl group of Tyr48 and the imidazole ring of His110 positioned above it.

The catalytic mechanism of hAR appears to be relatively simple, involving the protonation of the substrate by an acid-base catalyst and a stereospecific transfer of the 4-*pro-R* hydrogen, presumably in the form of a hydride ion, from the C4 of the nicotinamide ring to the carbonyl carbon of the substrate. An important step in the mechanism is the transfer of a proton from a proton donor group in the enzyme to the carbonyl oxygen of the substrate. Cys298, Tyr48 and His110 are all proton donor candidates, however because Cys298 is not conserved in the aldo-keto superfamily, its role as a putative proton donor group in hAR is considered less likely (Bohren et al., 1989). In order for His110 to act as a proton donor, it is necessary that its Nε atom is protonated when the carbonyl substrate is bound. Unfortunately, X-ray diffraction cannot locate hydrogen atoms as a result of their weak electron density. Thus, any experimental information concerning the protonation state of His110 in hAR is not available. Additionally, the preferred binding conformation of substrate in the active site pocket of hAR has not yet become available, presumably because of the difficulties in either crystallising a suitable hAR-substrate complex or soaking substrate into crystals. Therefore, molecular modelling calculations on the binding of two substrates, **D**- and **L**-xylose, in the active site of wild-type hAR and two of its site-directed mutants (H110A and H110Q) have been performed in order to elucidate a picture of the protonation state of His110 with these substrates bound (De Winter and von Itzstein, 1994). In this study we have built the eight different enzyme-substrate (ES) complexes for which the interaction enthalpies between enzyme (E) and substrate (S) were calculated in the following manner. Firstly, the H110A and H110Q mutants of the wild-type enzyme were constructed by replacing His110 with the appropriate residue. In addition, two wild-type configurations were considered, one in which His110 was configured with Nε protonated (hereafter referred to as the HisI configuration), and one in which Nδ was protonated (HisII configuration). Thus, in total, four different E systems were investigated. Secondly, either **D**- or **L**-xylose was manually docked into the active site of each enzyme, yielding eight different ES complexes. Finally, conformational space of the ES complexes within the active site was explored using a slightly adapted procedure of Taylor and von Itzstein (1994), which gave eleven interaction enthalpies [ΔH_{inter}(E-S)] for each complex. These interaction enthalpies have then been used to deduce an estimation of the corresponding K_M constants, and the differences between the predicted and experimental K_M's have been used to obtain a picture of the protonation state of His110.

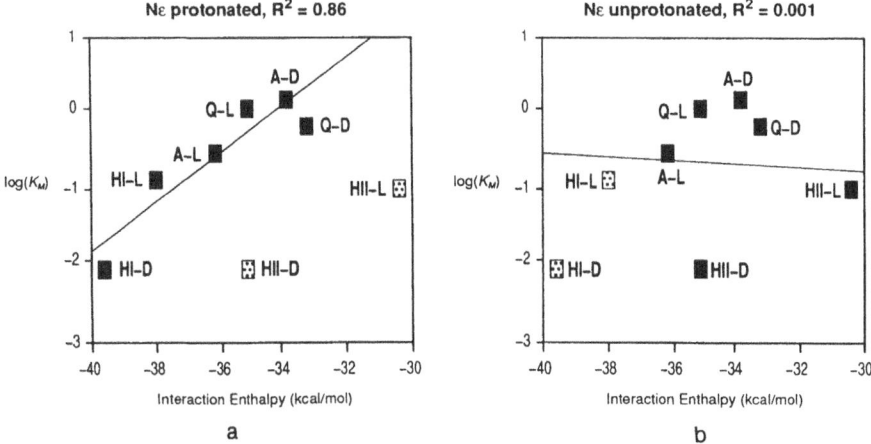

Figure 1. (a) A plot of the average interaction enthalpies against the experimental log(K_M) values for the eight enzyme-substrate complexes. The six points which were used for the calculation of the linear regression line are shown by black squares. The equation of the fitted line through these points is log(K_M) = 11.2 + 0.33ΔH_{inter}. The abbreviations used are: **HI-D** (HisI/**D**-xylose); **HII-D** (HisII/**D**-xylose); **A-D** (Ala/**D**-xylose); **Q-D** (Gln/**D**-xylose); **HI-L** (HisI/**L**-xylose); **HII-L** (HisII/**L**-xylose); **A-L** (Ala/**L**-xylose); **Q-L** (Gln/**L**-xylose). Dotted squares show the corresponding values of the HisII configuration complexes with either **D**-xylose (**HII-D**) or **L**-xylose (**HII-L**). (b) As for (a) except that the HisII values instead of HisI values were used for the calculation of the regression line. Dotted squares show the values of the HisI configurations with either **D**-xylose (**HI-D**) or **L**-xylose (**HI-L**).

THE TAUTOMERIC STATE OF THE ACTIVE SITE RESIDUE HIS110

A picture of the His110 binding configuration can be obtained from a correlation plot of the calculated interaction enthalpies with the experimental Michaelis-Menten (K_M) constants which is shown in Figure 1.

A good correlation exists between the $\log(K_M)$ values and the interaction enthalpies calculated for the binding between substrate and the enzyme in which His110 is protonated at Nε (HisI configuration, Figure 1a), while no significant correlation is apparent for the other tautomer (HisII, Figure 1b). We believe that an explanation for this trend becomes clear when one considers that the interaction enthalpies calculated for the HisII enzyme are much too small (too unfavourable) to account for the small K_M values for the wild-type enzyme (compared with the H110A and H110Q mutants). For example, the average interaction enthalpy calculated for the interaction between **D**-xylose and hAR in the HisII configuration, is comparable with the corresponding values for the binding of **L**-xylose with Gln or Ala enzyme, although the corresponding K_M values differ by almost two orders of magnitude.

The large difference in binding enthalpies between HisI and HisII can be explained by the difference in the way that the substrate is bound. The obvious difference between HisI and HisII is the existence (HisI) or absence (HisII) of a proton at position Nε of residue His110. While in HisI this proton is strongly involved in substrate binding through interactions with both the carbonyl- and 2'-hydroxyl group of the substrate, such interactions are obviously not feasible in HisII. Consequently, the average total interaction energy between **D**-xylose and HisII is significantly reduced compared to HisI. In nine of the eleven **D**-xylose/HisII complexes, a 180° rotation around the C1'-C2' bond has occurred so that the carbonyl oxygen points towards the six-membered ring of Trp20 and a hydrogen bond between the aldehydic proton and Nε of His110 is formed. This orientation leads to more favourable interaction enthalpies because the unfavourable electrostatic repulsion between the carbonyl oxygen and Nε of His110 is significantly reduced. However, such a substrate orientation is not consistent with experiment, as this would lead to the incorrect enantiomeric product (vander Jagt et al., 1992). In contrast to this, the **D**-xylose orientation in HisI (Nε protonated) is in good agreement with experiment, having its *re* face oriented towards the nicotinamide ring in all of the eleven structures. In addition, in this orientation the reactive carbonyl carbon is firmly locked in its position and at a distance close enough to the 4-*pro-R* hydrogen of NADPH to enable its transfer to the carbonyl carbon. The corresponding distance in the HisII tautomer is about 0.3 Å longer, which further supports the HisI configuration upon substrate binding.

BINDING CONFORMATION OF D-XYLOSE TO THE ACTIVE SITE OF ALDOSE REDUCTASE

The conformations adopted by **D**-xylose upon binding hAR in the HisI configuration can be classified into two conformationally distinct clusters. The main structural difference is formed by a torsional rotation of approximately 90° around the C2'-C3' and C3'-C4' bonds which only affects the conformation near the C5' end of the substrate. The conformational freedom in this part of the substrate is due to the absence of any suitable interactions of the hydroxyl groups at positions C3', C4' and C5' of the polar substrate with the hydrophobic active site pocket of the enzyme. The only active site residues in HisI which contribute significantly to the binding of **D**-xylose are Tyr48, His110, Trp111, and the nicotinamide ring of NADPH. The outcome of this is that a tight binding along the carbonyl end of the

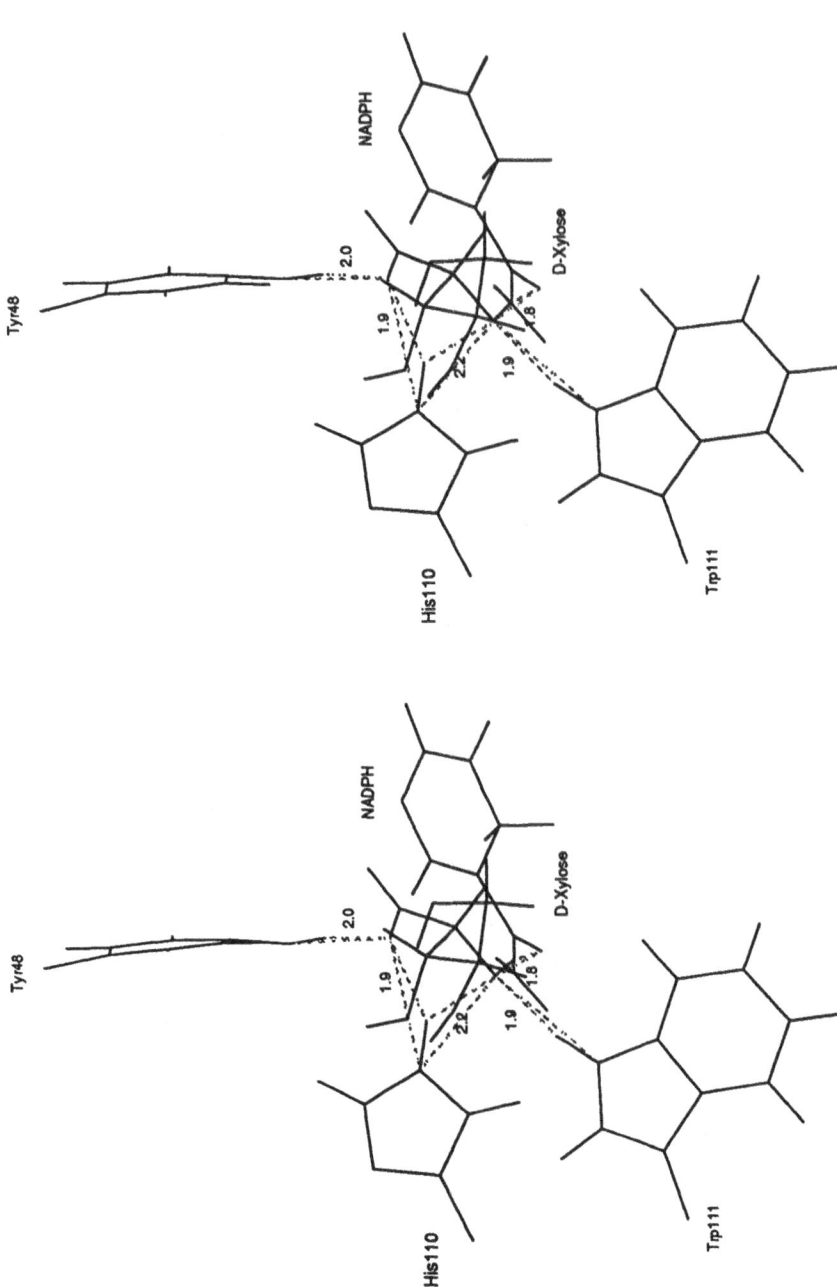

Figure 2. A stereo plot of the predicted binding geometry of **D**-xylose to the active site of wild-type human aldose reductase. Shown are the three active site residues which contribute to most of the total binding enthalpy, and the nicotinamide ring of NADPH. Distances, which are averaged values over the eleven minimized structures, are in Å.

substrate is observed, favouring, in contrast to the C5'-tail, a single binding conformation which is shown schematically in Figure 2.

Although the contribution of Trp20 is not significant for the total binding energy, its close proximity to the substrate carbonyl group might be important for the substrate stereospecifity of hAR. Chemical groups larger than hydrogen are likely to be sterically hindered by Trp20.

IMPLICATIONS FOR THE CATALYTIC MECHANISM OF ALDOSE REDUCTASE

By correlating experimental K_M values to calculated interaction enthalpies, we have been able to propose that the Nϵ atom of His110 appears to be protonated when substrate is bound. This finding might have important implications for the catalytic mechanism of aldose reductase. The availability of a proton at Nϵ implies that His110 is indeed a possible candidate to be the proton donor group in the reduction reaction catalysed by hAR. However, the present calculations do not unequivocally distinguish between His110 and Tyr48 as the actual proton donor group in this reduction reaction. To this end, other molecular modelling calculations, using a combined quantum mechanical/molecular mechanical method (Bash et al., 1991), are currently underway in our laboratory and these calculations could provide further useful information.

ACKNOWLEDGEMENT

HLDW is a post-doctoral fellowship recipient of the D. Collen Research Foundation, Leuven, Belgium.

REFERENCES

Bash,P.A., Field,M.J., Davenport,R.C., Petsko,G.A., Ringe,D., and Karplus,M., 1991, Computer Simulation and Analysis of the Reaction Pathway of Triosephosphate Isomerase, *Biochemistry* 30:5826-5832.

Bohren,K.M., Bullock,B., Wermuth,B., and Gabbay,K.H., 1989, The Aldo-Keto Reductase Superfamily. cDNAs and Deduced Amino Acid Sequences of Human Aldehyde and Aldose Reductases, *J. Biol. Chem.* 264:9547-9551.

De Winter,H.L., and von Itzstein,M., 1994, submitted (*Biochemistry*).

Flynn,T.G., and Green,N.C., 1993, The aldo-keto reductases: an overview, *in:* Enzymology and Molecular Biology of Carbonyl Metabolism 4:251-257. Edited by H. Weiner, Plenum Press, New York.

Masson,E.A., and Boulton,A.J.M., 1990, Aldose Reductase Inhibitors in the Treatment of Diabetic Neuropathy. A Review of the Rationale and Clinical Evidence, *Drugs* 39:190-202.

Taylor,N.R., and von Itzstein,M., 1994, Molecular Modeling Studies on Ligand Binding to Sialidase from Influenza Virus and the Mechanism of Catalysis, *J. Med. Chem.* 37:616-624.

Vander Jagt,D.L., Robinson,B., Taylor,K.K., and Hunsaker,L.A., 1992, Reduction of Trioses by NADPH-dependent Aldo-Keto Reductases, *J. Biol. Chem.* 267:4364-4369.

29

STOPPED-FLOW STUDIES OF HUMAN ALDOSE REDUCTASE REVEAL WHICH ENZYME FORM PREDOMINATES DURING STEADY-STATE TURNOVER IN EITHER REACTION DIRECTION

Charles E. Grimshaw and Chung-Jeng Lai

Lutcher Brown Department of Biochemistry
The Whittier Institute for Diabetes and Endocrinology
La Jolla, CA 92037

Progress in the aldo-keto reductase field has been quite rapid since the solution of the 3-dimensional structure of aldose reductase (ALR2) by the French group (Rondeau et al., 1992) and the Baylor group (Wilson et al., 1992), with additional contributions from Washington University School of Medicine with BioCryst Pharmaceuticals (Borhani et al., 1992) and with Dr. Quiocho's laboratory (Wilson et al., 1993). Most recently, a definitive assignment of the active site constellation of amino acid residues and their likely roles in the catalytic mechanism was established by the collaborative efforts of researchers at Baylor, Brandeis and The Whittier Institute (Harrison et al., 1994; Bohren et al., 1994).

Based on these results, we now understand the gross structural changes that must occur during the binding of NADP(H) to the apoenzyme. In brief, the sequence of residues (213-227) which comprise loop 7, between the end of β-sheet 7 and the start of α-helix 8 (numbering scheme of Wilson et al., 1992), shows the greatest change, folding down over the pyrophosphate group and apparently locking the dinucleotide cofactor in place. The resulting "tunnel" through which the pyrophosphate moiety connecting the 2'-monophosphoadenosine and nicotinamide ribose ends of the dinucleotide cofactor must snake, is lined on the bottom by a portion of loop 8 (transition from β-sheet 8 to α-helix 8) which contains Lys 262 and Arg 268, and by residues from loop 1 (transition from β-sheet 1 to α-helix 1) which contains Lys 21 and Trp 20. Thus, loop 7 contains Asp 216 which forms a salt bridge with both Lys 21 and Lys 262. Arg 268 in loop 8 lays over the adenine ring and, along with Lys 262, appears to form a strong interaction with the adenine ribose 2'-phosphate group. Trp 20 in loop 1 forms part of the hydrophobic substrate binding pocket, at the bottom of which lies the active site constellation including the OH of Tyr 48, the C4N of nicotinamide and the Nε of His 110. This overall structural reorganization can be best appreciated by comparing the ribbon structures for the binary E:2'-monophospho-adenosine-5'-diphosphoribose complex of pig ALR2 (Rondeau et al., 1992) with the ternary E:NADP⁺:citrate complex of human ALR2 (Wilson et al., 1992; Harrison et al., 1994). In view of the pronounced loop movement(s), this structural transition can be likened to the action of a "Chinese finger trap," in which the interwoven strands (i.e., the 8-stranded $\alpha\beta$-barrel) clamp down on the annulus following nucleotide binding.

Much of this crystallographic progress has been made despite significant questions remaining unanswered regarding the detailed kinetic properties of this enzyme. Previous studies established that aldose reductase follows a compulsory ordered mechanism with NADPH and NADP$^+$ binding first and leaving last from the free enzyme (Grimshaw et al., 1990; Kubiseski et al., 1992), although there has been some disagreement in the literature (Bhatnagar et al., 1988; 1994). Steady-state kinetic methods further suggested, on the basis of kinetic isotope effects and a comparison of the calculated value of the rate constant for NADP$^+$ release with the turnover number in the direction of aldehyde reduction, that NADP$^+$release, or the conformational change that precedes release, limits the overall rate (Grimshaw et al., 1989, 1990). Kubiseski et al. (1992) provided direct evidence from stopped-flow measurements of nucleotide binding to pig ALR2 that this conformational change was the slow step in the direction of aldehyde reduction but not alcohol oxidation. Unfortunately, interpretation of their stopped-flow data relied on analysis of the nucleotide concentration dependence of a rather poorly defined "second phase" in a two-phase exponential transient. Furthermore, these studies did not address rate constants for the formation and interconversion of central complexes where the actual chemistry takes place. For this reason we have undertaken an in-depth analysis of the kinetic mechanism of aldose reductase using recombinant human enzyme.

In the present study we have analyzed, using rapid reaction kinetic methods, the binding of nucleotides to the free enzyme, single turnover and multiple turnover transients for aldehyde reduction using NADPH and NADPD, and multiple turnover progress curves for alcohol oxidation catalyzed by recombinant human aldose reductase (hALR2). An appropriate substrate, namely D-xylose and D-xylitol, and the natural cofactors, NADPH and NADP$^+$, have been used to study the reaction at pH 8.0 in both directions. The results described clearly define each of the rate constants for the overall reaction mechanism using both protio- and deuterio-substrates, demonstrating for the first time burst kinetics for aldehyde reduction, and significantly revising the rate constant values for nucleotide binding reported previously (Kubiseski et al., 1992). Hopefully, the availability of a complete kinetic model as described herein will aid in the interpretation of past and future site-directed mutagenesis studies of this intriguing class of enzymic redox catalysts.

MATERIALS AND METHODS

Recombinant human aldose reductase prepared as described (Bohren et al., 1991) was the generous gift of Drs. Kurt M. Bohren and Kenneth H. Gabbay, Baylor College of Medicine. Enzyme was stripped of all residual nucleotide and the protein concentration was determined using $\varepsilon^{280\,nm} = 50$ mM^{-1}cm^{-1} as described (Ehrig et al., 1994). Stereospecifically labeled [4R-^{2}H]-NADPH (NADPD) was synthesized as described (Bohren et al., 1992). All other chemicals were from Sigma.

A Bio-Logic MPS-51 stopped-flow instrument with MOS-1000 modular optical system (absorbance and fluorescence detection), a path length of 1 cm and a dead time of 2 ms, along with Bio-Kine rapid kinetics data analysis software package were used to obtain transient data and to calculate apparent first-order and zero-order rate constants. Assays were conducted at 25 °C using 33 mM Na-phosphate buffer (pH 8) containing 0.5 mM EDTA and 0.1 mM DTT.

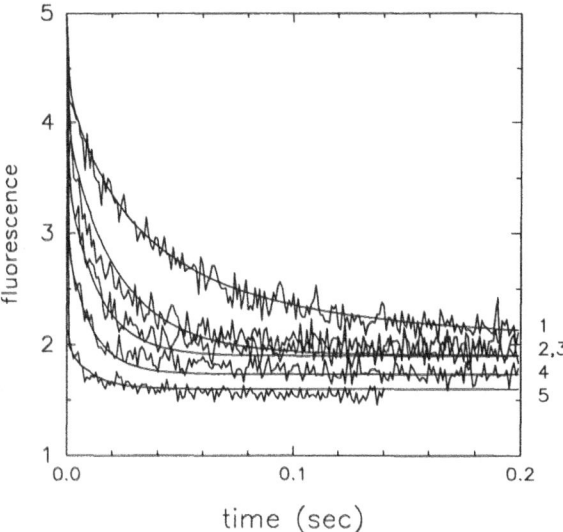

Figure 1. Stopped-flow progress curves for $NADP^+$ binding to hALR2 monitored as quenching of enzyme fluorescence ($_{294\ nm}F_{>>315\ nm}$). Reactions were conducted in 33 mM Na-phosphate buffer (pH 8.0) containing 0.5 mM EDTA and 0.1 mM DTT, with 1.5 μM final enzyme concentration. Final $NADP^+$ concentrations are: 1) 2; 2) 4; 3) 6; 4) 10; and 5) 45 μM. Actual progress curve data are shown; the solid lines were calculated using KINSIM and the kinetic constants listed in Table I.

Values for microscopic rate constants were estimated by progress curve analysis using the manual kinetic simulation program KINSIM (Barshop et al., 1983) and the automatic fitting routine FITSIM (Zimmerle et al., 1987) as modified for use under Windows 3.1 by Drs. Bryce V. Plapp and Gary X. Hua. Maximum values for apparent rate constants for steady-state and pre-steady-state xylose reduction and for steady-state xylitol oxidation were also obtained by fitting the observed rate constants as a function of substrate concentration to the expression $k_{obs} = k_{max}[S]/(K_{1/2} + [S])$ using the HYPER program of Cleland (1979). SigmaPlot ver. 5.0 was used for other general nonlinear regression analysis.

RESULTS AND DISCUSSION

Nucleotide binding

The overall reaction mechanism is shown in Scheme I:

$$E + NH \underset{k_2}{\overset{k_1}{\rightleftharpoons}} E{:}NH \underset{k_4}{\overset{k_3}{\rightleftharpoons}} {}^*E{:}NH \underset{k_6}{\overset{k_5}{\rightleftharpoons}} {}^*E{:}NH{:}RO$$

$$k_8 \uparrow\downarrow k_7$$

$$E + N \underset{k_{13}}{\overset{k_{14}}{\rightleftharpoons}} E{:}N \underset{k_{11}}{\overset{k_{12}}{\rightleftharpoons}} {}^*E{:}N \underset{k_9}{\overset{k_{10}}{\rightleftharpoons}} {}^*E{:}N{:}ROH$$

Scheme I

where E is hALR2, NH is NADPH, N is $NADP^+$, and RO and ROH are D-xylose and D-xylitol, respectively. Initially, we studied the rate of nucleotide binding to the enzyme,

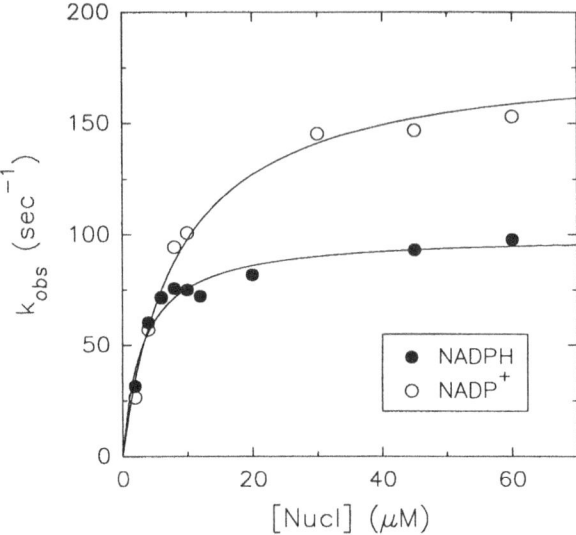

Figure 2. Plot of k_{obs} versus nucleotide concentration for rate constant values determined from Bio-Kine fits to a single exponential of stopped-flow progress curves similar to those shown in Figure 1. Actual data points are shown for NADPH (●) and NADP$^+$ (○); the solid lines were calculated from fits of the data to eq 4.

which comprises the first part of the reaction in either direction. As shown in Figure 1, progress curves determined for binding of NADP$^+$ to hALR2 were well fitted using a single exponential. This is in marked contrast to the report by Kubiseski et al. (1992) of biphasic progress curves with a fast exponential phase followed by a second slower phase. In some cases we did observe a slow rate after completion of the exponential fluorescence decay. However, careful analysis revealed no systematic variation of this slow phase with either enzyme or nucleotide concentration. Furthermore, the extent of this slow phase was a function of the particular enzyme sample used, suggesting that it might have been due to a small amount of contamination, perhaps by the so-called "activated" or "oxidized" enzyme form that has often plagued kinetic studies of ALR2 (Srivastava et al., 1985; Grimshaw et al., 1989; Del Corso et al., 1989; Vander Jagt and Hunsaker, 1993).

As shown in Figure 2, the k_{obs} values determined from this series of progress curves displayed saturation behavior. Again, this result is in contrast to the results of Kubiseski et al. (1992), but here the difference may be simply that the previous studies were limited to a maximum nucleotide concentration of 8 µM, while our results cover a much broader range (0 - 60 µM). Inspection of Figure 2 clearly shows that over the 0 - 8 µM concentration range our data are quite similar to those reported by Kubiseski et al. (1992). Experiments conducted using NADPH revealed a similar family of progress curves (data not shown), and the plot of k_{obs} versus [NADPH] is shown in Figure 2 for comparison. As is apparent, the limiting rate at saturating nucleotide is slightly faster for NADP$^+$ than for NADPH.

Based on the KINSIM simulations, there are three possible mechanisms that could account for the saturation behavior observed for nucleotide binding to hALR2:

(1)
$$E + Nucl \underset{k_2}{\overset{k_1}{\rightleftarrows}} E:Nucl \underset{k_4}{\overset{k_3}{\rightleftarrows}} {}^*E:Nucl$$

(2)
$$E \underset{k_2}{\overset{k_1}{\rightleftarrows}} {}^*E \underset{k_4}{\overset{k_3}{\rightleftarrows}} {}^*E:Nucl$$

(3)
$$\text{E:Nucl} \underset{k_2}{\overset{k_1}{\rightleftarrows}} \text{E} \underset{k_4}{\overset{k_3}{\rightleftarrows}} {}^*\text{E:Nucl}$$

where Nucl is either NADPH or NADP$^+$. The mechanism shown in eq 1, which gives the best overall fits to the experimental data and which is most consistent with the results reported here and elsewhere, is discussed in more detail below. Eq 2, which involves isomerization of the free apoenzyme, was eliminated from further consideration for several reasons. First, eq 2 would predict noncompetitive inhibition by NADP$^+$ versus NADPH, when in fact the inhibition is competitive (Grimshaw et al., 1990; Kubiseski et al., 1992). Second, eq 2 predicts that k_{obs} should decrease with increasing nucleotide concentration, because as [Nucl] increases, the rate of equilibration of E and *E must decrease from the sum $(k_1 + k_2)$ to k_1 at infinite [Nucl]. Third, there is no evidence from crystallographic studies for more than one form of the apoenzyme as eq 2 would require. Isomerization to *E only occurs when the full dinucleotide is present; removal of nicotinamide ring alone is sufficient to prevent the structural shift from occurring (Rondeau et al., 1992). Eq 3 was ruled out because it predicts a decrease in the amplitude of the fluorescence decay with increasing [Nucl]. Thus, at high [Nucl] eq 3 requires that E partition between E:Nucl and *E:Nucl with the relative amounts determined by the values of k_1 and k_2.

The nucleotide binding kinetic data were analyzed in two ways. Initially, the k_{obs} data in Figure 2 were fitted to eq 4, which was derived for the mechanism shown in eq 1:

(4)
$$k_{obs} = k_4 + k_3/(1.0 + K_d/[\text{Nucl}])$$

where $K_d = (k_2/k_1)$. The results are shown in Table I. Because k_4 is not well-determined by these fits, we used as an initial estimate a value of $k_4 = V/E_t$ (i.e., k_{cat}) for reaction in the opposite direction, since each of the first order rate constants for reaction in this direction must be at least as fast as the overall turnover number. Subsequently, the entire set of progress curves for either NADPH or NADP$^+$ were also analyzed using the KINSIM/FITSIM methodology, and the results are listed for comparison in Table I. As one can see, the agreement is quite good between the values determined using these two independent methods of data analysis.

At first glance, our results appear to support the conclusions reached by Kubiseski et al. (1992) that the isomerization step for NADP$^+$ (*E:NADP$^+$ $\leftrightarrow$ E:NADP$^+$) is overall rate-determining for reaction in the aldehyde reduction direction, but the analogous step for NADPH (*E:NADPH $\leftrightarrow$ E:NADPH), while indeed quite slow, does not limit the rate of reaction in the direction of alcohol oxidation. Thus, the k_4 values determined here with hALR2 are essentially the same as those reported for the pig enzyme (Kubiseski et al., 1992). However, that is where the similarities between the two studies end. Because we were able to directly show saturation behavior for

Table 1. Rate constants for nucleotide binding to hALR2

Nucleotide	Fitting Method	K_d (μM)	k_1 (M^{-1}s^{-1})	k_2 (s^{-1})	k_3 (s^{-1})	$k_4{}^a$ (s^{-1})	$K_{ia}{}^b$ (nM)
NADP$^+$	SigmaPlot	8 ± 3	---	---	180 ± 12	$< 0.20 >$	11
	Fitsim	3.1	2.0×10^8	550 ± 40	130 ± 6	0.5 ± 0.1	6
NADPH	SigmaPlot	3 ± 2	---	---	100 ± 30	$< 0.05 >$	26
	Fitsim	1.5	1.2×10^8	170 ± 13	90 ± 3	0.8 ± 0.1	13

[a] Brackets contain steady-state $<k_{cat}>$ values for reaction in the opposite direction.
[b] $K_{ia} = k_2/(1.0 + (k_3/k_4))$ is the dissociation constant for the tight complex *E:Nucl.

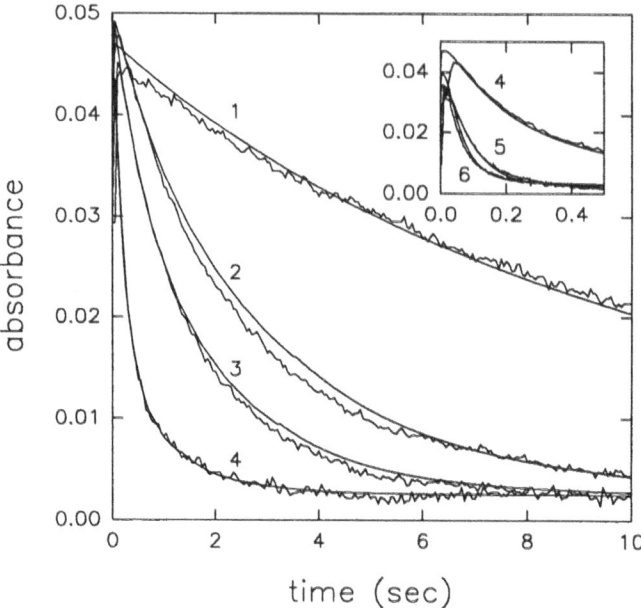

Figure 3. Single-turnover stopped-flow progress curves for hALR2-catalyzed D-xylose reduction using NADPH. Reaction progress monitored as 363 nm absorbance decrease. Reactions conducted in 33 mM Na-phosphate buffer containing 0.5 mM EDTA and 0.1 mM DTT, with 12.5 μM enzyme and 11.5 μM NADPH. Final D-xylose concentrations are: 1) 1 ; 2) 3; 3) 6; 4) 30; 5) 75; and 6) 150 mM. Actual progress curve data are shown; the solid lines were calculated using KINSIM and the kinetic constants shown in Scheme II.

k_{obs} versus [Nucl], we were able to establish the value of k_3 for each nucleotide to within narrow error limits. Values for k_1 and k_2 likewise were obtained from the KINSIM/FITSIM analysis. As a result, the values listed in Table I for k_1, k_2, and k_3 are on the order of 100-fold larger than those estimated by Kubiseski et al. (1992). Yet, as one can see from the solid curves in Figure 1, the agreement is excellent between the calculated fits and the experimental data. The fortuitous agreement in k_4 values must then be due to the fact that k_4 is by far the slowest step in eq 1, and therefore controls the overall kinetic behavior of the system.

Based on the results shown in Table I, hALR2 displays an exceptional degree of "clamping" between the initial weak E:Nucl complex ($K_d \cong 2$ -3 μM) and the final tight *E:Nucl complex ($K_{ia} \cong 5$ - 15 nM)! Ehrig et al. (1994) have recently reported similar K_{ia} values for NADPH and NADP$^+$ binding to hALR2 determined by fluorescence titration. These values correspond to a 100- to 500-fold increase in binding affinity for NADPH and NADP$^+$, respectively. Again, this represents the kinetic manifestation of the "Chinese finger trap" structural change detected by crystallographic studies. For comparison, a similar conformational change documented for equine liver alcohol dehydrogenase (Sekhar and Plapp, 1988; 1990) shows only a 4-fold tightening of binding (230 μM → 56 μM), while formation of the tight ternary complex E:NADPH:MTX between human dihydrofolate reductase, NADPH, and methotrexate (Appleman et al., 1988) displays a 60-fold effect (200 pM → 3 pM). Formation of the competent *E:NADH binary complex in the reverse reaction catalyzed by sheep liver aldehyde dehydrogenase in the presence of an appropriate anhydride shows a 120-fold effect (160 μM → 1.2 μM) (L. Blackwell, personal communication).

Turnover Studies

Having established the rate constants for nucleotide binding (k_1 - k_4 and k_{11} - k_{14} in Scheme I), we next set about looking at the sequence of steps for substrate binding and catalysis (k_5 - k_{10}). First we determined single turnover progress curves using 0.9 equivalents of NADPH (relative to enzyme) and varying the initial concentration of D-xylose. Figure 3 shows a representative family of progress curves. As with the nucleotide binding transients, each progress curve was best fitted by a single exponential, and the k_{obs} values obtained displayed saturation behavior when plotted versus [D-xylose] (data not shown). Further- more, when the same analysis was conducted using stereospecifically labeled NADPD, the resulting k_{max} obtained from saturation curve analysis displayed a significant primary deuterium isotope effect of 3.6 ± 2.9. (The large standard error arises from the extrapolation of k_{obs} data for [D-xylose] = 1 - 150 mM where the apparent $K_{1/2} \cong 300$ mM.) The detection of a large $^Dk_{max}$ (read: deuterium isotope effect on k_{max}) confirms that the hydride transfer step (k_7) is a large component of the rate limitation for these single turnover transients.

Next, the same experimental protocol was employed except that now 10 equivalents of NADPH were used in order to monitor both the pre-steady-state transient and the steady-state turnover as well. As shown in Figure 4, we observed a rapid pre-steady-state "burst" of NADPH disappearance followed by a much slower steady-state rate over a wide range of D-xylose concentrations. A similar family of progress curves was determined using NADPD (data not shown). As for the single turnover transients, the initial burst phase was best fitted in all cases using a single exponential. Both the first-order rate constant (k_{burst}) for the burst phase and the zero-order rate constant (k_{ss}) for the steady-state phase of the progress curves, determined using the Bio-Kine analysis software package, displayed

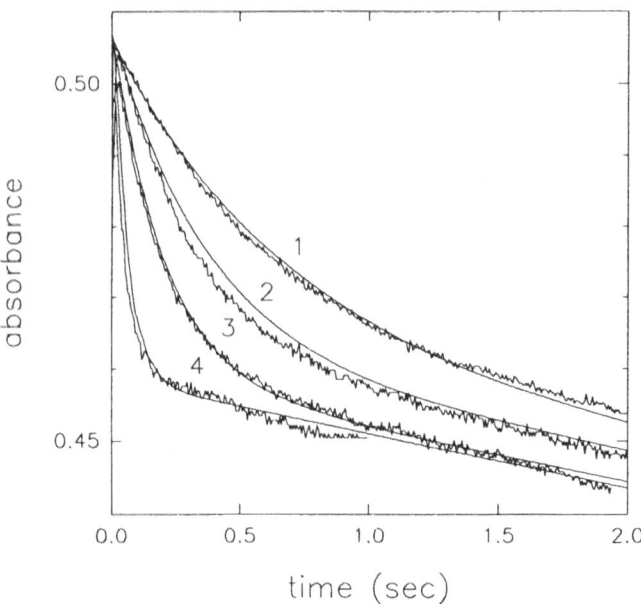

Figure 4. Multiple-turnover stopped-flow progress curves for hALR2-catalyzed D-xylose reduction using NADPH. Experimental conditions similar to those for Figure 3, except that 150 µM NADPH final concentra- tion was used in all assays. Actual progress curve data are shown; the solid lines were calculated using KINSIM and the kinetic constants shown in Scheme II. The burst magnitude (extrapolated to zero time) preceding the steady-state reaction corresponds to 0.88 enzyme equivalents.

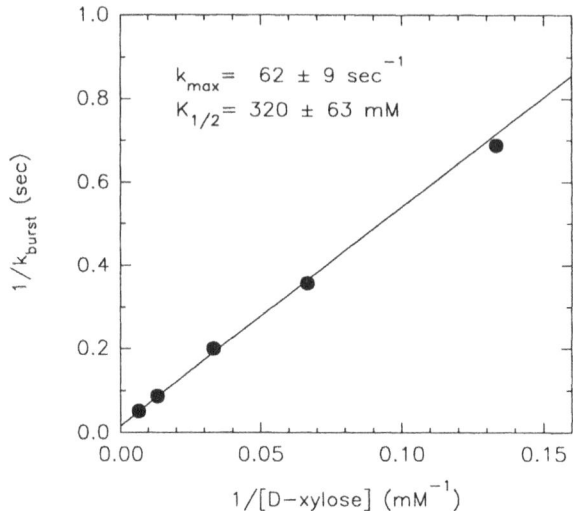

Figure 5. Double reciprocal plot of $1/k_{burst}$ versus $1/[D\text{-xylose}]$ for the k_{burst} values determined from Bio-Kine analysis of the pre-steady-state burst phase of progress curves for multiple-turnover stopped-flow progress curves such as those shown in Figure 4. Actual data points are shown; the straight line is the best fit of the data to a saturation function using HYPER. By contrast, no burst phase was detected for D-xylitol oxidation in the reverse reaction (data not shown). As a result, only values for the zero-order rate constant (k_{ss}) for steady-state appearance of NADPH could be obtained from the linear progress curves. These were analyzed as a function of [D-xylitol] using HYPER, and the resulting kinetic parameter values determined using stopped-flow (k_{max} $_{(ss)} = 0.07 \pm 0.02$ s^{-1} ($= V_2/E_t$, or k_{cat} for D-xylitol oxidation) and $K_{1/2} = 200 \pm 30$ mM ($= K_p$, or K_m for D-xylitol)) were again in reasonable agreement with the values for these same kinetic parameters ($V_2/E_t = 0.05$ s^{-1}; $K_p = 150 \pm 20$ mM) measured using conventional steady-state kinetic methods.

saturation behavior as a function of [D-xylose]. Figure 5 shows a representative plot of k_{ss} versus [D-xylose] measured using NADPH. As expected based on the single turnover studies, the burst phase k_{max} derived from saturation curve analysis of data for NADPD displayed a large primary deuterium isotope effect ($^D k_{max\ (burst)} = 4.1 \pm 0.6$). Analysis of the k_{ss} data as a function of [D-xylose] using NADPH gave values for $k_{max\ (ss)} = 0.22 \pm 0.02$ s^{-1} ($= V_1/E_t$, or k_{cat} for D-xylose reduction) and $K_{1/2} = 5 \pm 2$ mM ($= K_b$, or K_m for D-xylose), in reasonable agreement with the values for these same kinetic parameters ($V_1/E_t = 0.20$ s^{-1}; $K_b = 1.5 \pm 0.2$ mM) measured using conventional steady-state kinetic methods.

Kinetic Model

The resulting kinetic model derived from the entire set of stopped-flow data is shown in Scheme II including each of the individual rate constants:

$$E+NH \underset{k_2\ (174\ s^{-1})}{\overset{k_1\ (1.2\ \times\ 10^8 M^{-1}s^{-1})}{\rightleftarrows}} E{:}NH \underset{k_4\ (0.83\ s^{-1})}{\overset{k_3\ (89\ s^{-1})}{\rightleftarrows}} {}^*E{:}NH \underset{k_6\ (25\ s^{-1})}{\overset{k_5\ (220\ M^{-1}s^{-1})}{\rightleftarrows}} {}^*E{:}NH{:}RO$$

$$k_8\ (0.60\ s^{-1}) \uparrow \downarrow k_7\ (130\ s^{-1})$$

$$E+N \underset{k_{13}\ (623\ s^{-1})}{\overset{k_{14}\ (2.0\ \times\ 10^8 M^{-1}s^{-1})}{\rightleftarrows}} E{:}N \underset{k_{11}\ (0.23\ s^{-1})}{\overset{k_{12}\ (130\ s^{-1})}{\rightleftarrows}} {}^*E{:}N \underset{k_9\ (1\ \times\ 10^6 s^{-1})}{\overset{k_{10}\ (5\ \times\ 10^6 M^{-1}s^{-1})}{\rightleftarrows}} {}^*E{:}N{:}ROH$$

Scheme II

Using this model we can now rationalize a great many details of hALR2 catalysis. First, we can quantitate precisely which enzyme form(s) are present during steady-state turnover in each direction. For reduction of D-xylose, the overall slow step corresponds to k_{11} for the isomerization: $*E:NADP^+ \rightarrow E:NADP^+$. From a structural standpoint, we would ascribe this step to reversal of the conformational change demonstrated to occur upon nucleotide binding, as described in detail in the introductory section. Since k_{11} (or more precisely the "net rate constant" (Cleland, 1975) for this step: $k_{11}' = k_{11}k_{13}/(k_{12} + k_{13})$) contributes more than 99% to the rate limitation under conditions of saturating NADPH and D-xylose, this requires that $*E:NADP^+$ comprise more than 99% of the enzyme present during steady-state turnover. In the back reaction of D-xylitol oxidation, both k_8' (hydride transfer from alcohol to $NADP^+$) and k_4' are slow. However, because of the unfavorable partitioning of E:NH:RO following reverse hydride transfer ($k_7/k_6 \cong 5$), k_8' will be only 0.097 s^{-1} and thus will be roughly 85% rate-limiting under conditions of saturating $NADP^+$ and D-xylitol. As such, $*E:NADP^+:ROH$ will comprise

Table 2. Comparison of calculated and measured kinetic parameters for hALR2

Parameter[a]	Calculated	Measured[b]
V_1/E_t (s^{-1})	0.19	0.20
K_a (μM)	0.005	< 0.10
K_b (mM)	1.0	1.5
K_{ia} (μM)	0.013	0.010[c]
K_{da} (μM)	1.5	----
V_2/E_t (s^{-1})	0.08	0.05
K_q (μM)	0.003	< 0.10
K_p (mM)	160	150
K_{iq} (μM)	0.005	0.006[c]
K_{dq} (μM)	3.1	----
$^D V_1$	1.01	1.00
$^D V_1/K_b$	1.89	1.82
$^D V_2/K_p$	2.04	----
$^D V_2$	2.23	----

[a] A = NADPH, B = D-xylose, P = D-xylitol, Q = $NADP^+$, V_1/E_t = k_{cat} for xylose reduction, V_2/E_t = k_{cat} for xylitol oxidation, K_{ia} = dissociation constant for tight $*E:NADPH$ complex, K_{da} = dissociation constant for weak E:NADPH complex, $^D V_1$ = primary deuterium isotope effect on k_{cat} for xylose reduction measured by direct comparison using NADPD versus NADPH, $^D V_1/K_b$ = primary deuterium isotope effect on $k_{cat}/K_{m\ xylose}$.

[b] Determined from initial velocity studies in 33 mM Na-phosphate buffer (pH 8.0) at 25 °C.

[c] Determined by fluorescence titration in 5 mM phosphate buffer (pH 7.0) at 20 °C (Ehrig et al., 1994).

about 85% of the total enzyme present during steady-state turnover in the reverse direction, with *E:NADPH comprising the remaining 15% or so. In view of the poor binding affinity of alcohols for *E:NADP$^+$, however, at any reasonable level of D-xylitol (e.g., 20 mM) the equilibrium between *E:NADP$^+$:ROH and *E:NADP$^+$ will lie mostly towards the latter. Thus, at reasonable levels of D-xylitol, most of the enzyme (e.g., > 70%) will again be present as the *E:NADP$^+$ complex!

These calculations finally provide a quantitative rationale for the anomalous inhibition patterns described for various aldose reductase inhibitors (ARIs). Thus, several laboratories have now shown that compounds that are un- or noncompetitive versus the aldehyde substrate are competitive inhibitors when tested versus the alcohol substrate in the back reaction for either ALR2 or the closely related aldehyde reductase (Griffin & McNatt, 1986; Wermuth, 1990; Sato & Kador, 1990; Liu et al., 1992; Harrison et al., 1994; Ehrig et al., 1994). Walter Ward of ICI probably came closest to the truth during the previous meeting of this symposium in Dublin (Ward et al., 1993), when he ascribed all ARI inhibition to binding to the "E:Q" enzyme form (equivalent to our *E:NADP$^+$). The answer is now clear. Because the predominant enzyme form present under *both* sets of conditions is *E:NADP$^+$ (competitive inhibition versus D-xylitol represents an extrapolation to zero alcohol concentration), ARI binding can occur at a single site to give both types of inhibition. In fact, these results predict that the K_{ii} value for uncompetitive inhibition versus aldehyde and the K_{is} value for competitive inhibition versus alcohol should be nearly equal, since both reflect binding of the ARI to the same enzyme form. This has indeed been demonstrated recently for Alrestatin (Ehrig et al., 1994), and in our laboratory for a series of ARIs (Grimshaw & Lai, unpublished results).

Using the kinetic model shown in Scheme I, we have derived expressions for each of the observable kinetic parameters (e.g., V_1/E_t, K_a, K_b, etc.). Substituting the rate constant values shown in Scheme II into these expressions yields calculated values for these kinetic parameters that are in excellent agreement with those determined using conventional steady-state kinetic methods, as shown in Table II. Also included are calculated values for $^DV_1$ and $^DV_1/K_b$ which are sensitive indicators of reaction mechanism, and these too are in quite good agreement with the independently measured steady-state values. Note that the values for $^DV_2$ and $^DV_2/K_p$ are predicted to be large and roughly equal. Although these values have not yet been determined using [1-^{2}H]-D-xylitol, both Liu et al. (1993) and our laboratory (Grimshaw & Lai, unpublished results) have observed large primary deuterium isotope effects on the order of 2.5 - 4.0 on both $^DV_2$ and $^DV_2/K_p$ using [1-^{2}H$_2$]-benzyl alcohol as the alcohol substrate in the reverse reaction.

Finally, a practical note on the use of K_m values to denote "binding affinity" of a given substrate. Using the mechanism shown in Scheme I, the following expression (eq 5) was derived for K_b, the Michaelis constant for D-xylose:

$$(5) \quad K_b = \frac{k_3 k_{11} k_{13} (k_6 k_8 + k_6 k_9 + k_7 k_9)(1.0 + (k_4/k_3))}{[k_3 k_5 k_7 k_9(k_{11} + k_{12} + k_{13}) + k_3 k_5 k_{11} k_{13}(k_7 + k_8 + k_9) + k_5 k_7 k_9 k_{11} k_{13}]}$$

As one can readily see, this expression bears little resemblance to the ratio k_6/k_5 which reflects the true "binding affinity" of D-xylose to the *E:NADPH complex. In fact, K_b calculated using eq 5 is roughly 1.0 mM, while $k_6/k_5 = 110$ mM! On the other hand, because of the magnitude of the various rate constants, the calculated K_m value for D-xylitol ($K_p = 169$ mM) is quite close to the ratio $k_9/k_{10} = 200$ mM.

By using the actual rate constant values from Scheme II we can simplify eq 5. Thus, the expression for V_1/K_b is:

$$(6) \quad V_1/K_b = \frac{k_5 k_7 k_9}{[k_6 k_8 + k_6 k_9 + k_7 k_9](1.0 + k_4/k_3)}$$

Since $k_8 \ll k_9$ and $k_4 \ll k_3$, we have:

(7)
$$V_1 / K_b = \frac{k_5 k_7}{(k_6 + k_7)} = \frac{k_5}{(1.0 + k_6 / k_7)}$$

We also know that $k_6/k_7 \cong 0.2$, so that $V_1/K_b \cong k_5$. In addition, we have already established earlier that $V_1/E_t \cong k_{11}$, which means that:

(8)
$$K_b = \frac{V_1}{(V_1 / K_b)} \cong \frac{k_{11}}{k_5}$$

In other words K_b, the Michaelis constant for D-xylose, which represents the *steady-state* dissociation constant, is equal to the ratio of the forward isomerization rate constant (k_{11}) divided by the rate constant for D-xylose binding (k_5). Thus, if k_{11} can be shown to be essentially equal for a wide range of substrates (as is the case for hALR2 (Grimshaw et al., 1989)), a comparison of apparent K_b values, or better yet V_1/K_b values, can be used as an estimate of relative k_5 values for different substrates. However, this comparison only reflects changes in the on-rate, k_5, and should not be confused with "binding affinity" which must, of course, include contributions from both k_5 and k_6. The simplification described above follows directly from the derived expressions using the individual rate constant values. However, this is not a general case, and investigators are therefore cautioned to avoid equating K_m, the Michaelis constant, with K_i, the dissociation constant, unless there is good evidence to support such a contention.

ACKNOWLEDGMENTS

This work was supported by R01-DK43595 and the Charles E. Culpeper Foundation. The authors thank Dr. Bryce V. Plapp and the members of his laboratory at the University of Iowa, in particular Dr. Keehyuk Kim, for use of the stopped-flow instrument and for helpful discussions.

REFERENCES

Appleman, J.R., Prendergast, N., Delcamp, T.J., Freisheim, J.H., and Blakely, R.L. (1988) Kinetics of formation and isomerization of Methotrexate complexes of recombinant human dihydrofolate reductase. *J. Biol. Chem.* **263**:103004-10313

Barshop, B.A., Wrenn, R.F., & Frieden, C. (1983) Analysis of numerical methods for computer simulation of kinetic processes: Development of KINSIM -- A flexible, portable system. *Anal. Biochem.* **130**:134-145

Bhatnagar, A., Das, B., Gavva, S.R., Cook, P.F., and Srivastava, S.K. (1988) The kinetic mechanism of human placental aldose reductase and aldehyde reductase II. *Arch. Biochem. Biophys.* **261**:264-274

Bhatnagar, A., Liu, S.-Q., Ueno, N., Chakrabarti, B., and Srivastava, S.K. (1994) Human placental aldose reductase: Role of Cys-298 in substrate and inhibitor binding. *Biochim. Biophys. Acta* **1205**:207-214

Bohren, K.M., Page, J.L., Shankar, R., Henry, S.P., and Gabbay, K.H. (1991) Expression of human aldose and aldehyde reductase. Site-directed mutagenesis of a critical lysine 262. *J. Biol. Chem.* **266**:24031-24037

Bohren, K.M., Grimshaw, C.E., and Gabbay, K.H. (1992) Catalytic effectiveness of human aldose reductase. Critical role of C-terminal domain. *J. Biol. Chem.* **267**:20965-20970

Bohren, K.M., Grimshaw, C.E., Lai, C.-J., Harrison, D.A., Ringe, D., Petsko, G.A., and Gabbay, K.H. (1994) Tyrosine-48 is the proton donor and histidine-110 directs substrate stereochemical selectivity in the reduction reaction of human aldose reductase. Enzyme kinetics and crystal structure of the Y48H mutant enzyme. *Biochemistry* **33**:2021-2032

Borhani, D.W., Harter, T.M., and Petrash, J.M. (1992) The crystal structure of the aldose reductase-NADPH binary complex. *J. Biol. Chem.* **267**:24841-24847

Cleland, W.W. (1975) Partition analysis and the concept of net rate constants as tools in enzyme kinetics. *Biochemistry* **14**:3220-3224

Cleland, W.W. (1979) Statistical analysis of enzyme kinetic data. *Methods Enzymology* **63**:103-138

Del Corso, A., Barsacchi, D., Giannessi, M., Tozzi, M.G., Camici, M., and Mura, U. (1989) Change in stereospecificity of bovine lens aldose reductase modified by oxidative stress. *J. Biol. Chem.* **264**:17653-17655

Ehrig, T., Bohren, K.M., Prendergast, F.G., and Gabbay, K.H. (1994) Mechanism of aldose reductase inhibition: Binding of NADP+/NADPH and Alrestatin-like inhibitors. *Biochemistry* **33**:7157-7165

Griffin, B.W, and McNatt, L.G. (1986) Characterization of the reduction of 3-acetylpyridine adenine dinucleotide phosphate by benzyl alcohol catalyzed by aldose reductase. *Arch. Biochem. Biophys.* **246**:75-81

Grimshaw, C.E., Shahbaz, M., Jahangiri, G., Putney, C.G., McKercher, S.R., and Mathur, E.J. (1989) Kinetic and structural effects of activation of bovine kidney aldose reductase. *Biochemistry* **28**:5343-5353

Grimshaw, C.E., Shahbaz, M., and Putney, C.G. (1990) Mechanistic basis for nonlinear kinetics of aldehyde reduction catalyzed by aldose reductase. *Biochemistry* **29**:9947-9955

Harrison, D.H., Bohren, K.M., Ringe, D., Petsko, G.A., and Gabbay, K.H. (1994) An anion binding site in human aldose reductase: Mechanistic implications for the binding of citrate, cacodylate and glucose 6-phosphate. *Biochemistry* **33**:2011-2020

Kubiseski, T.J., Hyndman, D.J., Morjana, N.A., and Flynn, T.G. (1992) Studies on pig muscle aldose reductase. Kinetic mechanism and evidence for a slow conformational change upon nucleotide binding. *J. Biol. Chem.* **267**:6510-6517

Liu, S.-Q., Bhatnagar, A., and Srivastava, S.K. (1992) Does Sorbinil bind to the substrate binding site of aldose reductase? *Biochem. Pharmacol.* **44**:2427-2429

Liu, S.-Q., Bhatnagar, A., and Srivastava, S.K. (1993) Bovine lens aldose reductase. pH-Dependence of steady-state kinetic parameters and nucleotide binding. *J. Biol. Chem.* **268**:25494-25499

Rondeau, J.-M., Tête-Favier, F., Podjarny, A., Reymann, J.-M., Barth, P., Biellmann, J.-F., and Moras, D. (1992) Novel NADPH-binding domain revealed by the crystal structure of aldose reductase. *Nature* **355**:469-472

Sato, S., and Kador, P.F. (1990) Inhibition of aldehyde reductase by aldose reductase inhibitors. *Biochem. Pharmacol.* **40**:1033-1042

Sekhar, V.C., and Plapp, B.V. (1988) Mechanism of binding of horse liver alcohol dehydrogenase and nicotinamide adenine dinucleotide. *Biochemistry* **27**:5082-5088

Sekhar, V.C., and Plapp, B.V. (1990) Rate constants for a mechanism including intermediates in the interconversion of ternary complexes for horse liver alcohol dehydrogenase. *Biochemistry* **29**:4289-4295

Srivastava, S.K., Hair, G.A., and Das, B. (1985) Activated and unactivated forms of human erythrocyte aldose reductase. *Proc. Natl. Acad. Sci. USA* **82**:7222-7226

Vander Jagt, D.L., and Hunsaker, L.A. (1993) Substrate specificity of reduced and oxidized forms of human aldose reductase. *Adv. Exp. Med. Biol.* **328**:279-288

Ward, W.H.J., Cook, P.N., Mirrlees, D.J., Brittain, D.R., Preston, J., Carey, F., Tuffin, D.P., and Howe, R. (1993) Inhibition of aldose reductase by (2,6-dimethylphenylsulphonyl)nitromethane: Possible implications for the nature of an inhibitor binding site and a cause of biphasic kinetics. *Adv. Exp. Med. Biol.* **328**:301-311

Wermuth, B. (1990) Inhibition of aldehyde reductase by carboxylic acids. *Adv. Exp. Med. Biol.* **284**:197-204

Wilson, D.K., Bohren, K.M., Gabbay, K.H., and Quiocho, F.A. (1992) An unlikely sugar substrate binding site in the 1.65 Å structure of human aldose reductase holoenzyme implicated in diabetic complications. *Science* **257**:81-84

Wilson, D.K., Tarle, I., Petrash, J.M., and Quiocho, F.A. (1993) Refined 1.8 Å structure of human aldose reductase complexed with the potent inhibitor Zopolrestat. *Proc. Natl. Acad. Sci. USA* **90**:9847-9851

Zimmerle, C.T., Patane, K., & Frieden, C. (1987) Analysis of progress curves by simulations generated by numerical integration. *Biochemistry* **26**:6545-6552

30

LYSINE RESIDUES IN THE COENZYME-BINDING REGION OF MOUSE LUNG CARBONYL REDUCTASE

Yoshihiro Deyashiki, Masayuki Nakanishi, Masaki Sakai, and Akira Hara

Laboratory of Biochemistry
Gifu Pharmaceutical University
Mitahora-higashi, Gifu 502, Japan

INTRODUCTION

Tetrameric carbonyl reductase (CR, EC 1.1.1.184) of guinea-pig, mouse and pig lung differs from CRs of other mammalian tissues in subunit structure, broad substrate specificity for aromatic and aliphatic carbonyl compounds, reversibility of the reaction and sensitivity to pyrazole (Nakayama et al., 1982, 1986; Oritani et al., 1992). It is also uniquely activated by fatty acids and dipyridyl compounds (Hara et al., 1992a, 1993). The cDNA for pig lung has been cloned (Nakanishi et al., 1993). The enzyme is composed of 244 amino acids, and is structurally related to members of the short-chain alcohol dehydrogenase (SCAD) family, which includes eucaryotic and procaryotic enzymes with different substrate specificity (Persson et al., 1990; Neidle et al., 1992; Krozowski, 1992).

Sequence comparisons among the members of the SCAD family have pointed out at least two highly conserved sequences (Persson et al., 1990; Neidle et al., 1992; Krozowski, 1992). One sequence is Thr-Gly-Xaa-Xaa-Xaa-Gly-Xaa-Gly (where Xaa is any amino acid) near the N-terminal regions of the SCAD family proteins and the lung CR, and another sequence of Tyr-Xaa-Xaa-Xaa-Lys is distal to this area; this is observed at residues 149-153 in the lung CR sequence (Nakanishi et al., 1993). The importance of the tyrosyl and lysyl residues in the latter sequence for catalytic function has been demonstrated by site-directed mutagenesis and chemical modification studies of some dehydrogenases of this family (Ensor and Tai, 1992; Obeid and White, 1992; Prozorovski et al., 1992; Chen et al., 1993) and by an X-ray crystallographic study of 3α,20β-hydroxysteroid dehydrogenase from Streptomyces hydrogenans (Ghosh et al., 1991). On the other hand, the former sequence has been proposed to constitute a part of the binding site for the coenzyme on the basis of sequence and secondary-structure comparisons of the SCADs against the coenzyme-binding domains of other oxidoreductases (Persson et al., 1990; Neidle et al., 1992; Krozowski, 1992). A site-directed mutagenesis of the second glycine in the putative coenzyme-binding sequence of one SCAD, Drosophila alcohol dehydrogenase, shows that the residue plays a role in maintaining the coenzyme-binding fold (Chen et al., 1990), and the crystallographic

Enzymology and Molecular Biology of Carbonyl Metabolism 5
Edited by H. Weiner *et al.*, Plenum Press, New York, 1995

study of S. hydrogenans 3α,20β-hydroxysteroid dehydrogenase-NADH complex indicates the importance of the threonine in the sequence for the binding to the coenzyme (Ghosh et al., 1991). However, there has been no information on the roles of the other residues in or near the putative coenzyme-binding sequence.

We previously showed that a lysine-modifying reagent, 2,4,6-trinitrobenzene-1-sulfonic acid (TNBS), inactivated guinea-pig lung CR and NADPH protected against the inactivation (Hara et al., 1992b). Similar inactivation by TNBS and protective effects of NADPH and its analogs against inactivation were also observed with lung CRs of mouse and pig, which suggested that lysyl residue(s) lie at or near the coenzyme-binding site of lung CR. Recently, we have cloned a cDNA for mouse lung CR and showed that the mouse enzyme is also composed of 244 amino acids, sharing a 85% sequence identity with pig lung CR (Nakanishi et al., 1994). In this study, we chemically modified mouse lung CR by TNBS in the absence and presence of the protecting agents and identified the location of the lysyl residues.

MATERIALS AND METHODS

Chemical Modification

Homogeneous mouse lung CR with a specific activity of 9.3 mmol/min/mg was purified as described (Nakayama et al., 1986). The enzyme (1 mg) was dialyzed for 7 h at 4°C against 50 mM N-2-hydroxyethylpiperazine-N'-2-ethanesulfonic acid (HEPES)-KOH, pH 8.0, containing 0.15 M KCl before the treatment by TNBS, and diluted with the buffer without KCl. TNBS inactivation reactions were carried out in the dark at 25°C in 0.5-ml reaction mixtures containing 50 mM HEPES buffer, pH 8.0, the enzyme (0.05-0.1 mg/ml) and the modifier. At time intervals, aliquots (25 μl) were removed from the reaction mixtures, and added to 0.1 ml of the HEPES buffer containing 20 mM lysine to terminate the modification. Protection experiments were performed by pre-incubating the enzyme with various ligands at 4°C for 5 min before adding TNBS. A control containing no modifying reagent was routinely included in a set of modification experiments and the activity of the modified enzyme at any given time was calculated relative to the control. The enzyme activity was assayed at 25°C with 0.1 mM NADPH and 1 mM 4-nitroacetophenone as the coenzyme and substrate, respectively (Nakayama et al., 1986).

The absorption and difference spectra of the modified enzymes by TNBS were observed with a Hitachi UV-3000 spectrophotometer. The number of 2,4,6-trinitrophenyl (TNP)-lysyl residues was estimated from the absorbance at 346 nm as described (Hollenberg et al., 1971).

Peptide Mapping and Sequencing

CR solution (0.2 mg in 0.16 ml of the HEPES buffer) was incubated with 1 mM TNBS at 25°C for 30 min in the absence or presence of protecting compound. After terminating the reaction by adding 40 μl of 0.1 M lysine, the mixture was dialyzed for 24 h against 500 ml of 0.1 M ammonium bicarbonate which was changed every 8 h. The dialyzed sample was digested at 37°C for 24 h with L-1-tosylamido-2-phenylethyl chloromethyl ketone (TPCK)-treated trypsin (Sigma Chemicals Co., 4 μg). Peptide mapping was performed by reverse-phase high pressure liquid chromatography (RP-HPLC) as described (Nakanishi et al., 1993). The effluent was pooled according to UV absorbance at 225 nm or 346 nm (for detection of peptides containing TNP-amino acids) and lyophilized. Some of the pools were separately rechromatographed on RP-HPLC with a linear gradient 0-60%

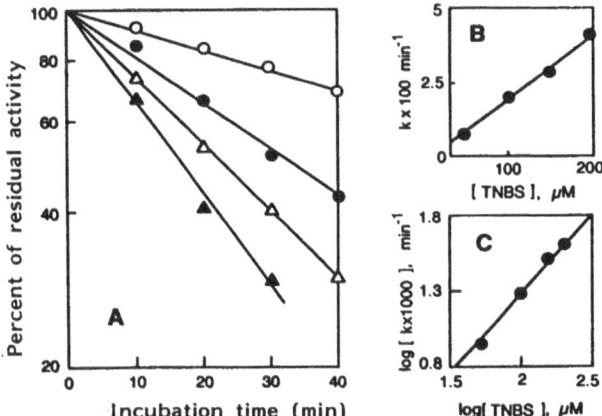

Figure 1. Inactivation of mouse lung CR by TNBS. Panel A is the time-dependent inactivation of the reductase activity. The TNBS concentrations were 50 μM (O), 100 μM (●), 150 μM (Δ) and 200 μM (▲). Panels, B and C, are the replots of the data to obtain the second-order rate constant and apparent order of reaction, respectively.

(v/v) acetonitrile/10 mM ammonium formate, pH 6.3. The purified peptides were sequenced by automated Edman degradation on a 473A protein sequencer (Applied Biosystems).

RESULTS

Modification by TNBS

Mouse lung CR was rapidly and irreversibly inactivated at pH 8.0 by TNBS. The inactivation reaction showed time-dependent pseudo-first-order kinetics with a kinetic stoichiometry of 1.05 and a second-order rate constant of 205 M-1 min-1 (Fig. 1).

The absorption spectrum of the TNBS-modified enzyme showed a peak with a broad maximum around 346 nm and a shoulder at about 420 nm, which is indicative of TNP-amino acid derivative (Hollenberg et al., 1971). The equivalents of the amino acid modified per molecule of the enzyme as a function of time were determined (Fig. 2). Although a linear correlation between the rate of loss of enzymatic activity and the lysines modified per mole of CR subunit was not obtained, extrapolation of the enzyme activity in the earlier period of the modification to zero suggests that there is a single essential lysine for the enzyme subunit.

The inactivation of the enzyme by TNBS was inhibited by the addition of NADPH and its analogs (Table 1), of which NADPH and 2'-AMP gave the greatest protective effect, and completely protected against the inactivation at their concentrations of more than 10 mM. It should be noted that the substrate, 1 mM 4-nitroacetopheone, and an activator, 1 mM 1,10-phenanthroline, did not show the protective effect.

Identification of TNBS Modified Peptides

CR was completely inactivated by incubation with 1 mM TNBS for 30 min, and then digested with TPCK-treated trypsin. The peptide mixture was separated by RP-HPLC. As shown in Fig. 3, of the major 18 peaks detected by monitoring at 225 nm, six peaks (B6-B9, B11, B12) exhibited absorbance at 346 nm, which suggests that the six peptides contain TNP-lysine residues. When CR was similarly treated with TNBS in the presence of 0.1 mM

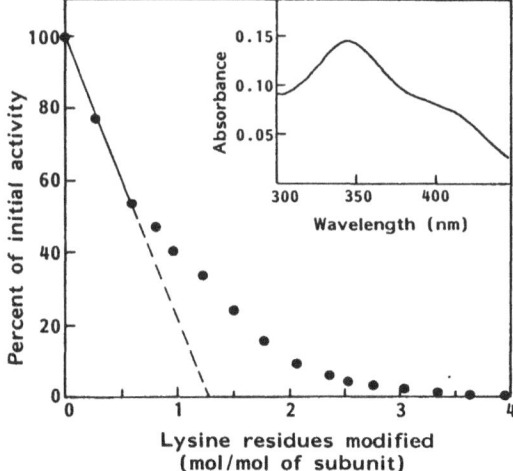

Figure 2. Effect of TNBS on the enzymatic activity and the lysine content of mouse lung CR. The enzyme (0.21 mg) was incubated with 0.6 mM TNBS in 0.4 ml of 50 mM HEPES buffer, pH 8.0. Aliquots (10 μl) of the reaction mixture were then removed at different times for catalytic activity. To another portion (10 μl), 20 μl of 10% SDS and 20 μl of 1.2 M HCl were added, and the absorbance due to TNP-lysine was determined at 346 nm. The difference spectrum of the enzyme modified for 2 h is shown in the inset. The reference cuvette contained the enzyme which was carried out through the same treatment without TNBS.

Table 1. Effects of coenzyme analogs on the inactivation of mouse lung CR by TNBS.

Coenzyme analog	$k \times 10^3$	Protection
	min^{-1}	%
Control	38.1	--
2.0 μM NADPH	8.8	77
2.0 μM 2'-AMP	9.7	75
2.0 μM ATP-ribose	25.5	33
2.0 μM 2',5'-ADP	26.8	30
0.3 mM 5'-ATP	12.4	67
0.3 mM 5'-ADP	21.4	44
2.0 mM NADH	15.8	59
2.0 mM 5'-AMP	25.7	33

The enzyme (0.05 mg) was incubated with 0.1 mM TNBS in the absence (control) or the presence of one of the coenzyme analogs, and then the residual activity in the incubation mixture was assayed. The protection percentage is expressed as $(k_c - k_p)/k_c \times 100$, in which k_c and k_p are pseudo-first order rate constant of the control and that with modifier plus protecting agent, respectively.

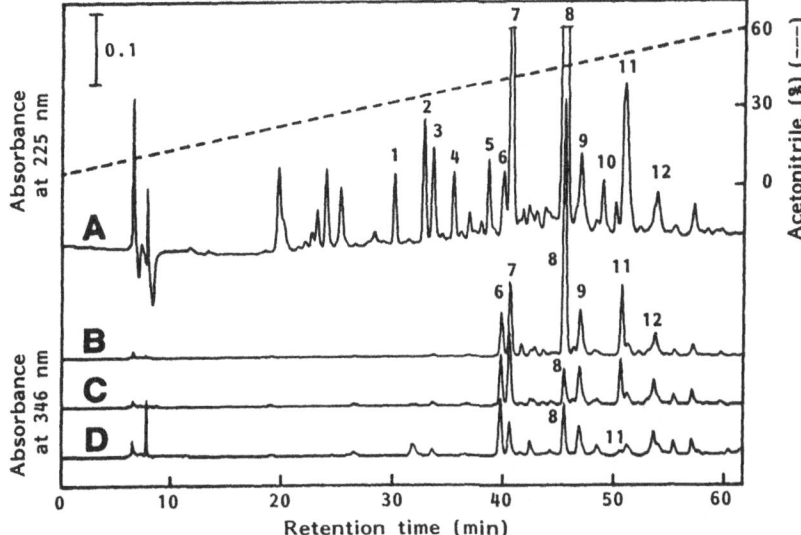

Figure 3. Comparison of the elution profiles on RP-HPLC of tryptic peptides of TNBS-treated mouse lung CRs in the absence and presence of 2'-AMP or NADPH. The enzyme was incubated with 1 mM TNBS for 30 min in the absence (A and B) or the presence of 0.1 mM 2'-AMP (C) or 0.1 mM NADPH (D). After dialysis, the samples were digested with trypsin, and then subjected to RP-HPLC. Top chromatogram was monitored at 225 nm, and bottom three chromatograms were at 346 nm to detect peptides containing TNP-lysine. The same sample without the protecting agent was analyzed at both 225 nm (A) and 346 nm (B). Peptides analyzed are numbered corresponding to the identifications given in Table 2.

Table 2. Summary of sequence data of tryptic peptides derived from TNBS-treated mouse lung CR.

Peptide	Sequence	Amino acid residue position	TNP-lysyl residue position	Repetitive yield (%)
B8-2	ALVTGAGX(K)GIGR	10-21	17	62
B12	TNSDLVSLAX(K)EX(C)PGIEPVX(C)VDLGDWDA	40-66	49	95
B9	AALVIMQPFLEVTX(K)EAFDR	84-102	97	76
B11-2	VNSVNPTVVLTDMGX(K)X(K)VSADPEFAR	174-198	188, 189	99
B6	KLX(K)ER	199-203	201	42
B7	LX(K)ER	200-203	201	56

The peptides shown in Fig. 3 were purified by RP-HPLC rechromatography, and then sequenced by Edman degradation. The positions of the peptides and lysine or cysteine residue are assigned by alignment of the sequences with the amino acid sequence of the enzyme deduced from its cDNA.

X in the sequence is an unidentifiable PTH-amino acid derivative. The amino acid assigned by the alignment is indicated in a parenthesis behind X.

NADPH or 2'-AMP, and the elution patterns (at 346 nm) of the tryptic peptides from the samples were compared with that of the unprotected enzyme, only peak B8 was greatly decreased in the case of the 2'-AMP-protected enzyme, whereas, in addition to B8, B11 appeared considerably smaller in the NADPH-protected enzyme than in the unprotected enzyme. On rechromatography of the peaks on RP-HPLC, monitoring at 225 nm, the peak B8 from the unprotected enzyme was separated into two peptide peaks (B8-1 and B8-2), whereas both peak C8 from the 2'-AMP-protected enzyme and peak D8 from the NADPH-protected enzyme yielded only one peak corresponding to B8-1. The peak B11 from the unprotected enzyme also divided into three peptides (B11-1, B11-2 and B11-3), of which only B11-2 was not seen in the rechromatography of the peak D11 from the NADPH-protected enzyme. The two peptides, B8-2 and B11-2, had absorbance at 346 nm, but the other peptides did not. The results indicate that the two peptides contain lysyl residues, the modification of which is protected by 2'-AMP and/or NADPH.

The amino acid sequences of peptides B8-2 and B11-2 contained unidentifiable phenylthiohydantoin (PTH)-amino acid derivatives (X, in Table 2). Sequence alignment of the two peptides with the primary structure deduced from the cDNA for mouse lung CR (Nakanishi et al, 1994) assigned the peptides B8-2 and B11-2 to positions 10-21 and 174-198, respectively, of the amino acid sequence. Thus, the unidentified amino acid of peptide B8-2 was found to be Lys-17, and those of peptide B11-2 were Lys-188 and Lys-189.

To identify the modified amino acids in the other peptide fractions (B6, B7, B9 and B12) of the TNBS-treated CR, the peptides with absorbance at 346 nm were further purified from these fractions by RP-HPLC rechromatography. The amino acid sequences of the peptides are also listed in Table 2, and were aligned with the primary structure of mouse lung CR deduced from its cDNA. The purified peptides contained unidentifiable PTH-amino acid derivatives which corresponded to Lys-49, Lys-97 and Lys-201, although Lys-199 in the peptide B6 was not modified by TNBS.

DISCUSSION

The kinetic and stoichiometric analyses of the inactivation of mouse lung CR by TNBS suggested the presence of at least one rapidly reacting lysyl residue per active site. Since the reaction catalyzed by lung CR has been kinetically shown to follow a di-iso ordered bi bi mechanism with the coenzyme as the first substrate (Matsuura et al., 1988), the complete protective effects of NADPH and 2'-AMP against the inactivation support the conclusion that the modified lysyl residue is present at the coenzyme-binding site of lung CR.

The sequence analysis of the tryptic peptides of the TNBS-treated enzymes revealed that the modification of Lys-17 was protected by 2'-AMP and NADPH, and those of Lys-188 and Lys-189 were only by NADPH. This suggests that three lysyl residues exist near the coenzyme-binding site of mouse lung CR, and that Lys-17 is the rapidly reacting residue in the TNBS modification. Although Lys-188 and Lys-189 of mouse CR (Nakanishi et al., 1994) are substituted for arginine and serine, respectively, in pig lung CR (Nakanishi et al., 1993), Lys-17 is conserved in lung CRs of mouse, pig and guinea-pig. Thus, Lys-17, rather than the other two residues, may be important for the coenzyme-binding. As recent studies on several oxidoreductases have established a role of lysyl residue in the binding of NADPH (Mas and Colman, 1984; Pai, 1988; Huang et al., 1990; Hurley et al., 1991; Haeffner-Gormley et al., 1992), it appears likely that the lysyl residue in the coenzyme-binding site of mouse lung CR interacts with the 2'-phosphate group of the coenzyme. Alternatively, the inactivation of the enzyme by TNBS may have resulted from steric hindrance by the TNP moiety on the lysyl residue on the binding of NADPH to the binding site, even if this residue would not interact directly with the coenzyme.

The sequence of Thr-Gly-Xaa-Xaa-Xaa-Gly-Xaa-Gly in the N-terminal regions of the SCAD family proteins has been thought to be a part of the coenzyme-binding site (Persson et al., 1990; Ghosh et al., 1991; Neidle et al., 1992; Krozowski, 1992). Mouse and pig lung CRs conserve the putative coenzyme-binding sequence (Nakanishi et al, 1993, 1994), in which Lys-17 corresponds to the third Xaa before the second Gly. The present data demonstrated the significance of the sequence around Lys-17 as the coenzyme-binding site, although the role of this lysine should be elucidated by site-directed mutagenesis.

REFERENCES

Chen, Z., Jiang, J.C., Lin, Z.-G., Lee, W.R., Baker, M.E., & Chang, S.H., 1993, Site-specific mutagenesis of Drosophila alcohol dehydrogenase: Evidence for involvement of tyrosine-152 and lysine-156 in catalysis, Biochemistry, 32:3342.

Chen, Z., Lu, L., Shieley, M., Lee, W.R., & Chang, S.H., 1990, Site-directed mutagenesis of glycine-14 and two "critical" cysteinyl residues in Drosophila alcohol dehydrogenase, Biochemistry, 29:1112.

Ensor, C.M., & Tai, H.-H., 1992, Site-directed mutagenesis of the conserved tyrosine 151 of human placental NAD+-dependent 15-hydroxyprostaglandin dehydrogenase yields a catalytically inactive enzyme, Biochem. Biophys. Res. Commun., 176:840.

Ghosh, D., Weeks, C.M., Grochulski, P., Duax, W.L., Erman, M., Rimsay, R.L., & Orr, J.C., 1991, Three-dimensional structure of holo 3a,20b-hydroxysteroid dehydrogenase: A member of short-chain dehydrogenase family, Proc. Natl. Acad. Sci. U.S.A., 88:10064.

Haeffner-Gormley, L., Chen, Z., Zalkin, H., & Colman, R.F., 1992, Importance of lysine-286 at the NADP site of glutamate dehydrogenase from Salmonella typhimurium, Biochemistry, 31:7807.

Hara, A., Oritani, H., Deyashiki, Y., Nakayama, T., & Sawada, H., 1992a, Activation of carbonyl reductase from pig lung by fatty acids, Arch. Biochem. Biophys. 292:548.

Hara, A., Sakai, M., Nakayama, T., Deyashiki, Y., & Sawada, H., 1993, Activation of pulmonary carbonyl reductase by aromatic amines and pyridine ring-containing compounds, in Enzymology and Molecular Biology of Carbonyl Metabolism 4, Weiner, H., ed., Plenum Press, New York, p. 361.

Hara, A., Yamamoto, H., Deyashiki, Y., Nakayama, T., Oritani, H., & Sawada, H., 1992b, Aldehyde dismutation catalyzed by pulmonary carbonyl reductase: Kinetic studies of chloral hydrate metabolism to trichloroacetic acid and trichloroethanol, Biochim. Biophys. Acta, 1075:61.

Hollenberg, P.F., Flashner, M., & Coon, M.J., 1971, Role of lysyl e-amino groups in adenosine diphosphate binding and catalytic activity of pyruvate kinase, J. Biol. Chem., 246:946.

Huang, S., Appleman, J.R., Tan, X., Thompson, P.D., Blakley, R.L., Sheridan, R.P., Venkataraghanvan, R., & Freisheim, J.H., 1990, Role of lysine-54 in determining cofactor specificity and binding in human dihydrofolate reductase, Biochemistry, 29:8063.

Hurley, J.H., Dean, A.M., Koshland, D.E.Jr., & Stroud, R.M., 1991, Catalytic mechanism of NADP+-dependent isocitrate dehydrogenase: Implications from the structures of magnesium-isocitrate and NADP+ complex, Biochemistry, 30:8671.

Krozowski, Z., 1992, 11b-Hydroxysteroid dehydrogenase and the short-chain alcohol dehydrogenase (SCAD) superfamily, Mol. Cell. Endocrinol., 84:C25.

Mas, M.T., & Colman, R.F., 1984, Phosphorus-31 Nuclear Magnetic Resonance studies of the binding of nucleotides to NADP+-specific isocitrate dehydrogenase, Biochemistry, 23:1675.

Matsuura, K., Nakayama, T., Nakagawa, M., Hara, A., & Sawada, H., 1988, Kinetic mechanism of pulmonary carbonyl reductase, Biochem. J., 252:17.

Nakanishi, M., Deyashiki, Y., Nakayama, T., Sato, K., & Hara, A., 1993, Cloning and sequence analysis of a cDNA encording tetrameric carbonyl reductase of pig lung, Biochem. Biophys. Res. Commun. 194:1311.

Nakanishi, M., Deyashiki, Y., Oshima, K., & Hara, A., 1994, Cloning and expression of mouse lung carbonyl reductase, (in preparation, the sequence of the enzyme has been submitted to the GSDB/DDBJ/EMBL/NCBI Data Bank with accession number D26123).

Nakayama, T., Hara, A., & Sawada, H., 1982, Purification and characterization of a novel pyrazole-sensitive carbonyl reductase in guinea pig lung, Arch. Biochem. Biophys. 217:564.

Nakayama, T., Yashiro, K., Inoue, Y., Matsuura, K., Ichikawa, H., Hara, A., & Sawada, H., 1986, Characterization of pulmonary carbonyl reductases of mouse and guinea pig, Biochim. Biophys. Acta 882:220.

Neidle, E., Hartnett, C., Ornston, L.N., Bairoch, A., Rekiki, M., & Harayama, S., 1992, Cis-diol dehydrogenases encoded by the TOL pWWO plasmid xylLgene and the Acinetobacter calcoaceticus chromosomal benD gene are members of the short-chain alcohol dehydrogenase superfamily, Eur. J.
Biochem., 204:113.

Obeid, J., & White, P.C., 1992, Tyr-179 and Lys-183 are essential for enzymatic activity of 11b-hydroxysteroid
dehydrogenase, Biochem. Biophys. Res. Commun., 188:222.

Oritani, H., Deyashiki, Y., Nakayama, T., Hara, A., Sawada, H., Matsuura, K., Bunai, Y., & Ohya, I., 1992,
Purification and characterization of pig lung carbonyl reductase, Arch. Biochem. Biophys. 292:539.

Pai, E.F., 1988, Crystallographic analysis of the binding of NADPH, NADPH fragments, and NADPH
analogues to glutathione reductase, Biochemistry, 27:4465.

Persson, B., Krook, M., & Jornvall, H., 1990, Characterization of short-chain alcohol dehydrogenases and
related enzymes, Eur. J. Biochem. 200:537.

Prozorovski, V., Krook, M., Atrian, S., Gonzalez-Duarte, R., & Jornvall, H., 1992, Identification of reactive
tyrosine residues in cysteine-reactive dehydrogenases: Differences between liver sorbitol, liver
alcohol and Drosophila alcohol dehydrogenases, FEBS Lett., 304:1.

31

SUBSTRATE SPECIFICITY AND KINETIC MECHANISM OF *TETRAHYMENA* 20α-HYDROXYSTEROID DEHYDROGENASE

Akira Hara*, Ayako Inazu*, Yoshihiro Deyashiki*, and Yoshinori Nozawa**

*Biochemistry Laboratory
Gifu Pharmaceutical University
Gifu 502, Japan
**Department of Biochemistry
Gifu University School of Medicine
Gifu 500, Japan

INTRODUCTION

20α-Hydroxysteroid dehydrogenase (20HSD) is distributed in mammalian tissues (Gower, 1984) and micro-organisms (Dorfman and Ungar, 1965). 20HSDs purified from mammalian tissues are NADP+-dependent monomeric proteins with Mr values of 35,000-40,000 (Shikita et al., 1967; Sato et al., 1972; Nakajin et al., 1989; Noda et al., 1991), whereas the bacteral enzymes are NAD+-dependent tetramers of Mr 162,000 (Krafft and Hylemon, 1989) and monomers of Mr 48,000 (Rimsay et al., 1988). Recently, the cDNAs encoding the enzymes of bovine testis (Warren et al., 1993), rabbit ovary (Lacy et al., 1993) and rat ovary (Miura et al., 1994) have been cloned, and the enzymes have been shown to be members of the aldo-keto reductase superfamily which includes monomeric NADPH-dependent oxidoreductases (Flynn and Green, 1993; bifunctional enzymes: Aldehyde and aldose reductases exhibit dihydrodiol dehydrogenase activity (Matsuura et al., 1987; Hara et al., 1985, 1991) rat and human liver 3a-hydroxysteroid dehydrogenases associate with both carbonyl reductase and dihydrodiol dehydrogenase activities (Penning et al., 1986; Pawlowski et al., 1991; Deyashiki et al., 1992, 1994), bovine liver prostaglandin F synthase shows carbonyl reductase activity (Chen et al., 1992), and bovine testicular 20HSD possesses aldose reductase activity (Warren et al., 1993).

We previously purified an NADP+-dependent dimeric 20HSD with an Mr of 68,000 from *Tetrahymena pyriformis* (Inazu et al., 1994). Although this enzyme is distinct from mammalian 20HSDs in its dimeric structure, its N-terminal sequence shows a low degree of similarity to those of the aldo-keto reductase superfamily proteins. In addition, the enzyme exhibits reductase activity for several nonsteroidal carbonyl compounds and dihydrodiol dehydrogenase activity, and is inhibited by inhibitors of aldehyde and aldose reductases. In

Enzymology and Molecular Biology of Carbonyl Metabolism 5
Edited by H. Weiner *et al.*, Plenum Press, New York, 1995

this study, we co-purified 20HSD with aldehyde reductase and dihydrodiol dehydrogenase from *T. pyriformis* and *T. thermophila*, and examined its peptide sequence, substrate specificity for various carbonyl compounds and kinetic mechanisms of the reaction and inhibition, in order to elucidate the relationship of *Tetrahymena* 20HSD with known aldo-keto reductases.

MATERIALS AND METHODS

Cells of *Tetrahymena thermophila* and *T. pyriformis* were grown at 28°C in a medium composed of 2% proteose-peptone, 0.5% glucose and 0.2% yeast extract. The cells were harvested by centrifugation and washed with 50 mM glycine-NaOH buffer, pH 8.6. 20HSD was purified from the cells as described previously (Inazu et al., 1994).

The activity of 20HSD was determined using 0.1 mM NADPH and 50 μM 17α-hydroxyprogesterone as the coenzyme and substrate, respectively, at pH 6.5, as described (Inazu et al., 1994). Reductase activity for other substrates was assayed in a similar manner, and dehydrogenase activity was determined in 0.1 M glycine-NaOH buffer, pH 10.0 containing 0.25 mM NADP+. One unit (U) of the enzyme activity was defined as the amount that catalyzes the oxidation or formation of 1 μmol of NADPH at 25 °C. Protein concentration was determined using bovine serum albumin as the standard by the method of Bradford (1976).

The kinetic mechanism and parameters of the enzyme in both reductase and dehydrogenase reactions were analyzed, at pH 7.0 and 25 °C, according to Cleland (1963a, 1963b). The assay system, in a final volume of 2.0 ml, contained 0.1 M potassium phosphate, pH 7.0, coenzyme, substrate and enzyme.

The purified enzyme was digested with lysylendopeptidase after reductive pyridylethylation of the protein. The peptides were purified by reversed-phase high performance liquid chromatography as described (Deyashiki et al., 1994). The sequence of the peptide was determined by automated Edman degradation using an Applied Biosystems 473A gas-phase sequencer.

RESULTS AND DISCUSSION

Co-Purification of 20HSD with Aldehyde Reductase and Dihydrodiol Dehydrogenase

Since the purified 20HSD of *T. pyriformis* exhibits NADPH-linked reductase activity for pyridine-3-aldehyde and NADP+-linked dehydrogenase activity for *trans*-benzene dihydrodiol (Inazu et al., 1994), the three enzyme activities were assayed in the 105,000xg supernatants of *T. thermophila* and *T. pyriformis* lysates. The specific activities of 20HSD, pyridine-3-aldehyde reductase and benzene dihydrodiol dehydrogenase were 24 ± 6, 86 ± 28 and 41 ± 20 mU/mg (n = 4), respectively, for *T. thermophila*, and the respective values were 12 ± 5, 48 ± 13, and 31 ± 17 mU/mg for *T. pyriformis*. To clarify whether the three enzyme activities are due to an identical protein, 20HSD activity was co-purified with pyridine-3-aldehyde reductase and benzene dihydrodiol dehydrogenase activities from the two *Tetrahymena* species. A typical result of purification of the enzyme from *T. thermophila* is shown in Table 1, indicating that the three enzyme activities were not separated by the five steps of purification, during which the activity ratios of 20HSD to the aldehyde reductase or to the dihydrodiol dehydrogenase were essentially consistent. Similar result was also

Table 1. Co-purification of 20HSD, aldehyde reductase (ALR) and dihydrodiol dehydrogenase (DD) from *T. thermophila*

Step	Protein (mg)	20HSD activity Total (U)	Specific (U/mg)	Total ALR activity (U)	Ratio of ALR/20HSD	Total DD activity (U)	Ratio of DD/20HSD
Extract	531	24.1	0.045	87.9	3.6	33.8	1.4
40-75% (NH4)2SO4	392	22.7	0.058	83.3	3.7	31.1	1.4
Sephadex G-100	116	15.4	0.155	56.1	3.6	22.5	1.5
Matrex Red A	5.8	11.4	1.97	42.0	3.7	18.5	1.6
Chromatofocusing	2.2	8.6	3.95	31.7	3.7	13.8	1.6
HA-Ultrogel	1.3	6.5	5.00	23.5	3.6	10.7	1.6

ALR and DD activities were assayed with 1 mM pyridine-3-aldehyde and 1.8 mM *trans*-benzene dihydrodiol, respectively, as substrate.

obtained for 20HSD from *T. pyriformis*. Since the enzyme from either *Tetrahymena* species showed a single band on polyacrylamide gel electrophoresis, the 20HSD may be a major sepcies of the aromatic aldehyde reductase and dihydrodiol dehydrogenase in the cells.

Molecular Mass and Peptide Sequence

The molecular masses of the native and denatured enzymes from *T. thermophila* were 68kDa and 33kDa respectively, which are the same as those of the *T. pyriformis* enzyme (Inazu et al., 1994). Edman degradation of the *T. thermophila* enzyme did not yield any phenylthiohydantoin-amino acid derivative. This was different from the *T. pyriformis* enzyme, where the N-terminal sequence was determined (Inazu et al., 1994). Three peptides derived from the enzymatic digestion of the *T. thermophila* enzyme were sequenced, and their sequences were YVREDLYIVSK for peptide 4, AIGVSNFNVQSLLDLCSYA for peptide 8, and TVPLNDGTNFPIFGLG for peptide 10. Of these peptides, the sequence of peptide 10 showed a high homology with the N-terminal sequence of the *T. pyriformis* enzyme (identical amino acids are underlined in the peptide 10 sequence) which has been described to show a low degree of similarity to those of the aldo-keto reductase superfamily proteins (Inazu et al., 1994). In addition, the sequences of peptides 4 and 8 were similar to two regions, starting from the 70th and 156th residues respectively, in the amino acid sequences of human aldehyde and aldose reductases (Bohren et al., 1989). These results suggest that dimeric 20HSDs of *T. pyriformis* and *T. thermophila*, despite slight sequence difference between them, may belong to the aldo-keto reductase superfamily.

Substrate Specificity

20HSDs from *T. thermophila* and *T. pyriformis* showed essentially the same substrate specificity for carbonyl compounds (Table 2). The enzymes reduced a limited number of compounds with a carbonyl group: These substrates were aromatic aldehydes such as pyridine aldehydes and 4-nitrobenzaldehyde, aliphatic hydroxyaldehydes such as glyceraldehyde and lactaldehyde, and aliphatic ketones with a hydroxy group adjacent to the carbonyl group (acetoin and 17α-hydroxypregnenes). No significant activity was observed when other aromatic aldehydes, aldoses, aliphatic aldehydes, aliphatic ketones and aromatic ketones were used as the substrates.

The enzymes exhibited greater reactivity to α-dicarbonyl compounds and β-ketoacid esters than the above simple carbonyl compounds. For α-dicarbonyl compounds (R_1-CO-

Table 2. Substrate specificity for carbonyl compounds

Substrate	*T. pyriformis* 20HSD			*T. thermophila* 20HSD		
	Km	Vmax	Vmax/Km	Km	Vmax	Vmax/Km
	(μM)	(%)		(μM)	(%)	
Compounds with a carbonyl group						
Pyridine-3-aldehyde	51	100	1.96	49	100	2.04
Pyridine-4-aldehyde	1060	95	0.09	945	51	0.05
4-Nitrobenzaldehyde	254	36	0.14	221	44	0.20
D-Glyceraldehyde	300	24	0.08	464	51	0.11
DL-Glyceraldehyde-3-phosphate	1300	89	0.07	852	12	0.01
D-Lactaldehyde	587	81	0.14	534	82	0.15
Acetoin	186	90	0.48	272	104	0.39
17α-Hydroxyprogesterone	3	38	12.7	3	35	11.7
17α-Hydroxypregnenolone	3	19	6.3	2	36	18.0
α-Dicarbonyl compounds						
Methylglyoxal	548	114	0.21	356	96	0.27
Phenylglyoxal	591	115	0.20	447	95	0.21
Ethyl pyruvate	25	112	4.48	25	115	4.60
2,3-Butanedione	15	84	5.60	18	100	5.56
2,3-Pentanedione	18	107	5.94	21	127	6.05
2,3-Hexanedione	16	97	6.06	10	75	7.50
2,3-Heptanedione	8	179	22.4	5	118	23.6
1-Phenyl-1,2-propanedione	6	75	12.5	6	82	14.2
3,4-Hexanedione	570	111	0.20	230	68	0.30
1,2-Cyclohexanedione	129	66	0.51	92	50	0.54
Isatin	45	81	1.80	39	74	1.90
2,4-Pentanedione	243	131	0.54	131	98	0.75
β-Ketoacid esters						
Methyl acetoacetate	123	96	0.78	117	92	0.79
Ethyl acetoacetate	32	106	3.31	20	104	5.20
Ethyl 2-methylacetoacetate	15	97	6.47	12	86	7.17
Ethyl 4-chloroacetoacetate	27	141	5.22	27	122	4.52
n-Butyl acetoacetate	4	105	26.3	4	105	26.3
n-Hexyl acetoacetate	1.5	95	66.8	2.4	127	52.9
n-Octyl acetoacetate	1.3	135	108	1.6	119	74.5
Benzyl acetoacetate	8	106	13.3	8	101	12.6
Acetoacetyl CoA	20	87	4.35	22	93	4.23
Ethyl 3-oxovalerate	447	88	0.20	222	60	0.27
Ethyl benzoylacetate	654	112	0.17	303	77	0.25

Vmax value is relative to those for pyridine-3-aldehyde which were 19.7 and 18.8 U/mg for *T. pyriformis* and *T. thermophila* enzymes, respectively.

The following compounds were not reduced: ethyl 3-oxohexonate, ethyl 3-oxoheptanate, pyruvic acid, phenylpyruvic acid, acetoacetic acid, oxaloacetic acid and 2-ketoglutaric acid.

CO-R_2), the enzymes showed high Vmax/Km values for the compounds with a methyl group as the R_1 group and long alkyl chains as the R_2 group. The enzymes also reduced 2,4-pentanedione and acetoacetic acid esters including acetoacetyl CoA. Even in the reduction of these compounds (R_1-CO-CH_2-CO-R_2), the structural requisites were the same as those for the α-dicarbonyl substrates: (1) Ethyl acetoacetate with a methyl group as the R_1 group showed the highest Vmax/Km value of β-ketoacid ethyl esters, and (2) the catalytic

Table 3. Substrate specificity for alcohols

Substrate	T. pyriformis 20HSD			T. thermophila 20HSD		
	Km	Vmax	Vmax/Km	Km	Vmax	Vmax/Km
	(μM)	(%)		(μM)	(%)	
17α,20α-Dihydroxyprogesterone	13	19	1.5	14	25	1.8
5β-Pregnan-20α-ol-3-one	73	26	0.4		(21)	
17α,20β-Dihydroxyprogesterone		(0)			(0)	
trans-Benzene-1S,2S-dihydrodiol	1600	107	0.07	1600	85	0.05
cis-Benzene-1,2-dihydrodiol*	2020	84	0.04	2900	87	0.03
trans-1S,2S-Cyclohexanediol	50900	125	0.003	79700	173	0.002
2,3-Butanediol	66300	125	0.002	34800	64	0.002
(S)-Ethyl 3-hydroxybutyrate	1000	82	0.08	1300	87	0.07
(S)-Methyl 3-hydroxybutyrate	5000	67	0.01	4800	60	0.01
3-Hydroxybutyryl CoA*		(15)			(15)	
(R)-Ethyl 3-hydroxybutyrate		(0)			(0)	

Vmax value is relative to that of the respective enzyme for pyridine-3-aldehyde. Value in parenthesis is the relative activity with 0.1 mM steroid or 1 mM 3-hydroxybutyric acid esters. *Racemic compound was used as the substrate.

efficiency for the acetoacetic acid esters was dramatically augmented as the chain length of the R_2 group was increased up to C:8. It should be noted that ethyl levulinate with two carbon units between the two carbonyl groups was not reduced by the enzymes. Thus, the R_1 group of the good substrates must be a methyl group, which suggests that the enzyme may reduce the carbonyl group adjacent to the R_1 group of the dicarbonyl compounds. In addition, the high catalytic efficiency for the substrates with long alkyl chains at R_2 suggests the presence of a hydrophobic site in the active center of the enzyme.

In the reverse reaction with NADP+ as the coenzyme, both T. pyriformis and T. thermophila 20HSDs efficiently oxidized 20α-hydroxysteroids, but not 20β-hydroxysteroids. The enzymes slowly oxidized several nonsteroidal alcohols listed in Table 3, and showed strict stereospecificity for (S)-isomers of 3-hydroxybutyric acid esters and for (1S,2S)-isomers of trans-cyclohexanediol and trans-benzene dihydrodiol. These results suggest that the enzymes catalyze the stereoselective reduction of both 20-ketosteroids and β-ketoacid esters, and also support the above idea that the carbonyl group adjacent to the R1 group at least on the acetoacetic acid esters is reduced by the enzymes.

The substrate specificity of Tetrahymena 20HSD in both directions is clearly distinct from those of aldehyde reductase, carbonyl reductase and diacetyl reductase from mammalian tissues (Hara et al., 1985; Wermuth, 1985; and references cited therein) and fungal carbonyl reductase (Shimizu et al., 1988). The specificity for α-dicarbonyl compounds and β-ketoacid esters is similar to those of yeast aldehyde reductase (Kataoka et al., 1992) and dog adrenal aldose reductase (Hara et al., 1994), although the Tetrahymena enzyme is different from the aldehyde and aldose reductases in its inability to reduce aldoses and various aldehydes. Such specificity for the nonsteroidal compounds has not been examined for mammalian and bacterial 20HSDs. There have been reports on two reductases of baker's yeast, β-keto reductase (Shieh et al., 1985) and (S)-diacetyl reductase (Heidlas and Tressl, 1990), which show molecular masses and specificity for β-ketoacid esters similar to those of Tetrahymena 20HSD. Thus, reductases with such unique substrate specificity may be

Table 4. Kinetic parameters for *T. thermophila* 20HSD
determined from the initial rate measurements at pH 7.0

Parameter	Value
K_i^A: inhibition constant for NADPH	$3.2 \pm 0.2\ \mu M$
K_i^Q: inhibition constant for NADP$^+$	$11 \pm 1\ \mu M$
K_m^A: Km for NADPH	$2.0 \pm 0.1\ \mu M$
K_m^Q: Km for NADP$^+$	$12 \pm 1\ \mu M$
K_m^B: Km for ethyl acetoacetate	$10 \pm 1\ mM$
K_m^P: Km for (S)-ethyl 3-hydroxybutyrate	$1870 \pm 60\ mM$
V_f: Vmax for the reductase reaction	$18.4 \pm 0.7\ U/mg$
V_r: Vmax for the dehydrogenase reaction	$6.9 \pm 0.4\ U/mg$
$V_f K_i^A / V_r K_m^A$	4.3
$V_r K_i^Q / V_f K_m^Q$	0.34

distributed in some eucaryotic unicellular organisms. While the physiological role of the yeast enzymes remains unknown, the enzymes have been studied from a view of microbial asymmetric catalysis on organic synthesis. Since the Km values of the *Tetrahymena* enzyme for the carbonyl compounds are lower than those of the yeast reductases, the *Tetrahymena* enzyme may be useful for the production of chiral starting materials to synthesize enantiomeric pure compounds.

Kinetic Mechanism

Since *Tetrahymena* 20HSD is inhibited by inhibitors of aldehyde and aldose reductases (Inazu et al., 1994), we examined the kinetic mechanism of the inhibition with the *T. thermophila* enzyme which showed inhibitor sensitivity similar to that of the *T. pyriformis* enzyme. Steady-state kinetic mechanism of the reaction catalyzed by the enzyme was first analyzed by initial velocity patterns of the reduction of ethyl acetoacetate and the oxidation of (*S*)-ethyl 3-hydroxybutyrate. The double-reciprocal plots of initial rate versus concentration of the coenzyme at fixed levels of the carbonyl or alcohol substrate, yielded a series of intersecting lines. In the forward reaction, NADP$^+$ inhibited competitively with respect to NADPH (Kis = 3.3 μM) and noncompetitively with respect to ethyl acetoacetate (Kis = 12 μM; Kii = 198 μM). In the backward reaction, the inhibition by ethyl acetoacetate was uncompetitive versus NADP$^+$ (Kii = 118 μM) and was noncompetitive versus (*S*)-ethyl 3-hydroxybutyrate (Kis = 12 μM; Kii = 88 μM). Product inhibition by (*S*)-ethyl 3-hydroxybutyrate (in the forward reaction) or NADPH (in the backward reaction) could not be accurately determined, because the enzyme slowly reduced ethyl 3-hydroxybutyrate in the presence of NADPH. Another method for establishing the kinetic mechanism of enzyme catalysis, entailing the use of alternative substrates, has been put forward by Radhika and Northrop (1984). When the concentration of NADPH was varied in the presence of 50 μM 17α-hydroxyprogesterone, 0.1 mM n-octyl acetoacetate, 1 mM pyridine-4-aldehyde or 0.1 mM 2,3-heptanedione, the double-reciprocal plots of initial rate versus NADPH concentration gave a set of parallel lines. These results indicate that the reaction catalyzed by *Tetrahymena* 20HSD follows an ordered bi bi mechanism with the coenzyme binding to the free enzyme. The kinetic parameters are summarized in Table 4. It has been described that two equations, VfKiA/VrKmA≥1 and VrKiQ/VfKmQ≥1, should hold for the enzyme reaction which follows an ordered bi-bi mechanism with no isomerizations (Cornish-Bowden,

Table 5. Inhibition patterns and constants of different
inhibitors in ethyl acetoacetate (EA) reduction and (S)-ethyl
3-hydroxybutyrate (EHB) oxidation at pH 7.0

Inhibitor	Varied substrate			
	NADPH	EA	NADP$^+$	EHB
Quercitrin	UC (36)	UC (36)	UC (37)	C (22)
Phenobarbital	UC (7.1)	UC (11)	UC (5.6)	C (6.2)
Tolrestat	UC (6.8)	UC (6.2)	UC (6.0)	C (4.3)
Sorbinil	UC (8.6)	UC (9.4)	UC (5.9)	C (7.7)
Dienstrol	UC (7.0)	UC (5.5)	UC (6.0)	C (4.0)
Lithocholic acid	UC (5.4)	UC (4.1)	UC (3.0)	C (3.0)
Myristic acid	UC (7.4)	UC (8.9)	UC (7.6)	C (6.0)

The activity was assayed in the presence of 0.1 mM NADPH, 0.1 mM EA,
0.25 mM NADP$^+$ or 4 mM EHB as a fixed substrate. UC, uncompetitive
inhibition; C, competitive inhibition. The value in parenthesis is the Ki value
(μM) for the inhibitor.

1976). In inserting the kinetic values into the equations, the relationship of the latter equation did not hold. This suggests that the kinetic mechanism for *Tetrahymena* 20HSD includes a NADP+-induced isomerization step.

In addition to known inhibitors such as quercitrin, sorbinil, tolrestat and phenobarbital (Inazu et al., 1994), bile acids and fatty acids were found to inhibit *Tetrahymena* 20HSD. Of the bile acids, lithocholic acid showed the highest inhibition, and the inhibitory potency of fatty acids increased with elongation in carbon chain from caprylic acid to myristic acid. The inhibition patterns and constants of the representative inhibitors with different structures were compared. As shown in Table 5, all the inhibitors showed the same inhibition patterns in both directions of the reaction. The inhibitors showed uncompetitive inhibitions with respect to NADPH, ethyl acetoacetate and NADP$^+$, but acted as competitive inhibitors with respect to (S)-ethyl 3-hydroxybutyrate. Cibacron blue, used as a coenzyme analog, inhibited competitively with respect to the two coenzymes (Kis = 0.6 μM) under the same conditions. The results indicate that the inhibitors, except Cibacron blue, selectively bind to the enzyme-NADP$^+$ binary complex. Since the enzyme reaction follows an iso-ordered bi bi mechanism, the selective binding of the inhibitors might be caused by the isomerization of the enzyme-NADP$^+$ binary complex. To further test whether the different inhibitors bind to the same site on the enzyme, the kinetics of inhibition by a mixture of two different inhibitors were performed as described by Siegel (1976). As illustrated in the Dixon plots with the representative inhibitors (Fig. 1), the set of parallel lines suggests that all the inhibitors bind to overlapping sites on the enzyme. Since these inhibitors are hydrophobic, they might bind to the hydrophobic site of the active center of the enzyme which was suggested by the data of substrate specificity (Table 2).

The proposed kinetic mechanism of the reaction catalyzed by *Tetrahymena* 20HSD resembles those for mammalian aldose reductase (Kubiseski et al., 1992) and aldehyde reductase (Daly and Mantle, 1982), although the kinetic mechanisms of aldose and aldehyde reductases include two isomerization steps which are induced by the binding of NADPH and NADP$^+$. Recently, Ward et al. (1993) have reported that different aldose reductase inhibitors bind to the enzyme-NADPH and enzyme-NADP$^+$ binary complexes of aldose reductase after isomerization, and suggested that the inhibitors have overlapping sites. The kinetic inhibition mechanism of *Tetrahymena* 20HSD is also similar to that of aldose reductase.

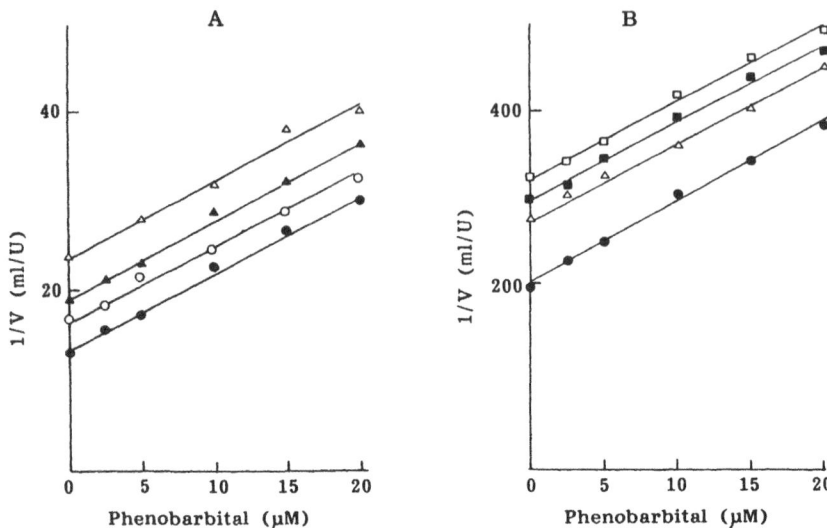

Figure 1. Dixon plots of the combined inhibition of oxidoreductase activities of *Tetrahymena* 20HSD by the representative inhibitors. The activity was assayed in the presence of phenobarbital at the indicated concentrations as well as one of the following inhibitors. The control activity was assayed only with phenobarbital (●). The other inhibitors are 5 μM sorbinil (○), 5 μM myristic acid (▲), 6 μM dienstrol (△), 2 μM lithocholic acid (■) and 10 μM Tolrestat (□). A, ethyl acetoacetate reductase activity. B, (*S*)-ethyl 3-hydroxybutyrate dehydrogenase activity.

CONCLUSION

We have demonstrated that both *T. pyriformis* and *T. thermophila* contain dimeric 20HSDs with almost the same molecular and enzymatical properties. This enzyme, capable of reducing aromatic aldehydes, aliphatic hydroxyaldehydes, α-dicarbonyl compounds and β-ketoacid esters, is a major species of dihydrodiol dehydrogenase and pyridine-3-aldehyde reductase in these protozoan cells. Although the enzyme was inactive towards other aldoses, aldehydes and aromatic ketones, its peptide sequence, inhibitor sensitivity and kinetic mechanism were similar to those of aldehyde and aldose reductases. Thus, *Tetrahymena* 20HSD may be related or belong to the aldo-keto reductase superfamily, in which this enzyme is the first dimeric protein. The broad substrate specificity of the *Tetrahymena* enzyme suggests its role in the detoxication or metabolism of carbonyl compounds taken up by the protozoan cells. The precise roles of the enzyme, however, await further investigation.

REFERENCES

Bohren, K.M., Bullock, B., Wermuth, B. & Gabbay, K.H., 1989, The aldo-keto reductase superfamily. cDNAs and deduced amino acid sequences of human aldehyde and aldose reductases, *J. Biol. Chem.*, 264:9547.

Bradford, M.M., 1976, A rapid and sensitive method for the quantitation of microgram quantities of protein utilizing the principle of protein-dye binding, *Anal. Biochem.*, 72:248.

Chen, L.-Y., Watanabe, K. & Hayaishi, O., 1992, Purification and characterization of prostaglandin F synthase from bovine liver, *Arch. Biochem. Biophys.*, 296:17.

Cleland, W.W., 1963a, The kinetics of enzyme-catalyzed reactions with two or more substrates or products. I. Nomenclature and equations, *Biochim. Biophys. Acta*, 67:104.

Cleland, W.W., 1963b, The kinetics of enzyme-catalyzed reactions with two or more substrates or products. III. Prediction of initial velocity and inhibition patterns by inspection, *Biochim. Biophys. Acta*, 67:188.

Cornish-Bowden, A., 1976, Principles of Enzyme Kinetics, Butterworths, London, p. 90.

Daly, A.K. & Mantle, T.J., 1982, The kinetic mechanism of the major form of ox kidney aldehyde reductase with D-glucuronic acid, *Biochem. J.*, 205:381.

Deyashiki, Y., Ogasawara, H., Nakayama, T., Nakanishi, M., Miyabe, Y. & Hara, A., 1994, Molecular cloning of two human liver 3α-hydroxysteroid/dihydrodiol dehydrogenase isozymes that are identical with chlordecone reductase and bile-acid binder, *Biochem. J.*, 299:545.

Deyashiki, Y., Taniguchi, H., Amano, T., Nakayama, T., Hara, A. & Sawada, H., 1992, Structural and functional comparison of two human liver dihydrodiol dehydrogenase associated with 3a-hydroxysteroid dehydrogenase activity, *Biochem. J.*, 282:741.

Dorfman, R.I. & Ungar, F., 1965, The role of cytochrome P-450 in steroidogenesis and properties of some of the steroid-transforming enzymes, in Metabolism of Steroid Hormones, Academic Press, New York, p 224.

Flynn, T.G. & Green, N.C., 1993, The aldo-keto reductases: An overview, in Enzymology and Molecular Biology of Carbonyl Metabolism 4, Weiner, H., ed., Plenum Press, New York, p. 251.

Gower, D.B., 1984, Steroid transformations by microorganisms, in Biochemistry of Steroid Hormones, 2nd edn., Makin, H.L.J., ed., Backwell Scientific Publications, Oxford, p. 230.

Hara, A., Hayashibara, M., Nakayama, T., Hasebe, K., Usui, S. & Sawada, H., 1985, Guinea-pig liver testosterone 17b-dehydrogenase (NADP+) and aldehyde reductase exhibit benzene dihydrodiol dehydrogenase activity, *Biochem. J.*, 225:177.

Hara, A., Matsuura, K., Sato, K., Deayashiki, Y., Miyabe, Y., Bunai, Y. & Ohya, I., 1994, Adrenal aldose reductase: Its characterization, localization and role, (in preparation).

Hara, A., Nakayama, T., Harada, T., Kanazu, T., Shinoda, M., Deayashiki, Y. & Sawada, H., 1991, Distribution and characterization of dihydrodiol dehydrogenases in mammalian ocular tissues, *Biochem. J.*, 275:113.

Hara, A., Seiriki, K., Nakayama, T. & Sawada, H., 1985, Discrimination of multiforms of diacetyl reductase in hamster liver, in Enzymology of Carbonyl Metabolism 2, Flynn, T.G. & Weiner, H., eds., Alan R. Liss, New York, p. 291.

Heidlas, J. & Tressl, R., 1990, Purification and properties of two oxidoreductases catalyzing the enantioselective reduction of diacetyl and other diketones from baker's yeast, *Eur. J. Biochem.*, 188:165.

Inazu, A., Sato, K., Nakayama, T., Deyashiki, Y., Hara, A. & Nozawa, Y., 1994, Purification and characterization of a novel dimeric 20α-hydroxysteroid dehydrogenase from *Tetrahymena pyriformis*. *Biochem. J.*, 297:195.

Kataoka, M., Sakai, H., Morikawa, T., Katho, M., Miyoshi, T., Shimizu, S. & Yamada, H., 1992, Characterization of aldehyde reductase of *Sporobolomyces salmonicolor*, *Biochim. Biophys. Acta*, 1122:57.

Krafft, A.E. & Hylemon, P.B., 1989, Purification and characterization of a novel form of 20α-hydroxysteroid dehydrogenase from *Clostridium scindens*, *J. Bacteriol.* 171:2925.

Kubiseski, T.J., Hyndman, D.J., Morjana, N.A. & Flynn, T.G., 1992, Studies on pig muscle aldose reductase. Kinetic mechanism and evidence for a slow conformation change on coenzyme binding, *J. Biol. Chem.*, 267:6510.

Lacy, W.R., Washenik, K.J., Cook, R.G. & Dunbar, B.S., 1993, Molecular cloning and expression of an abundant rabbit ovarian protein with 20α-hydroxysteroid dehydrogenase activity, *Mol. Endocrinol.*, 7:58.

Matsuura, K., Hara, A., Nakayama, T., Nakagawa, M. & Sawada, H., 1987, Purification and characterization of two multiple forms of dihydrodiol dehydrogenase from guinea-pig testis, *Biochim. Biophys. Acta*, 912:270.

Miura, R., Shito, K., Noda, K., Yagi, S., Ogawa, T. & Takahashi, M., 1994, Molecular cloning of cDNA for rat ovarian 20α-hydroxysteroid dehydrogenase (HSD1), *Biochem. J.* 299:561.

Nakajin, S., Kawai, Y., Ohno, S. & Shinoda, M., 1989, Purification and characterization of pig adrenal 20α-hydroxysteroid dehydrogenase, *J. Steroid Biochem.*, 33:1181.

Noda, K., Shota, K. & Takahashi, M., 1992, Purification and characterization of rat ovarian 20α-hydroxysteroid dehydrogenase, *Biochim. Biophys. Acta*, 1079:112.

Pawlowski, J.E., Huizinga, M. & Penning T.M., 1991, Cloning and sequencing of the cDNA for rat liver 3a-hydroxysteroid dehydrogenase, *J. Biol. Chem.*, 266:8820.

Penning, T.M., Smithgall, T.E., Askonas, L.J. & Sharp, B., 1986, Rat liver 3α-hydroxysteroid dehydrogenase, *Steroids*, 47:221.

Radhika, K. & Northrop, D., 1984, A new kinetic diagnostic for enzymatic mechanisms using alternative substrate, *Anal. Biochem.*, 141:413.

Rimsay, R.L., Murphy, G.W., Martin, C.J. & Orr, J.C., 1988, The 20α-hydroxysteroid dehydrogenase of *Streptomyces hydrogenans, Eur. J. Biochem.* 174:437.

Sato, F., Takagi, Y. & Shiota, M., 1972, 20α-Hydroxysteroid dehydrogenase of procine testes. Purification and properties, *J. Biol. Chem.*, 147:815.

Shieh, W.-R., Gopalan, A.S. & Sin, C.J., 1985, Stereochemical control of yeast reduction 5. Characterization of the oxidoreductases involved in the reduction of β-keto esters, *J. Am. Chem. Soc.*, 107:2993.

Shikita, M., Inano, H. & Tamaoki, B., 1967, Further studies on 20α-hydroxysteroid dehydrogenase of rat testes, *Biochemistry*, 6:1760.

Shimizu, S., Hattori, S., Hata, H. & Yamada, H., 1988, A novel fungal enzyme, NADPH-dependent carbonyl reductase, showing high specificity to conjugated polyketones. Purification and characterization, *Eur. J. Biochem.*, 174:37.

Siegel, I. H., 1975, Enzyme Kinetics, John Wiley & Sons, New York, p. 465.

Warren, J.C., Murdock, G.L., Ma, Y., Goodman, S.G. & Zimmer, W.E., 1993, Molecular cloning of testicular 20α-hydroxysteroid dehydrogenase: Identity with aldose reductase, *Biochemistry*, 32:1401.

Ward, W.H.J., Cook, P.N., Mirrlees, D.J., Brittain, D.R., Preston, J., Carey, F., Tuffin, D.P. & Howe, R., 1993, Inhibition of aldose reductase by (2,6-dimethylphenylsulphonyl)-nitromethane: Possible implications for the nature of an inhibitor binding site and a cause of biphasic kinetics, in Enzymology and Molecular Biology of Carbonyl Metabolism 4, Weiner, H., ed., Plenum Press, New York, p. 301.

Wermuth, B., 1985, Aldo-keto reductases, in Enzymology of Carbonyl Metabolism 2, Flynn, T.G. & Weiner, H., eds., Alan R. Liss, New York, p. 209.

32

PURIFICATION AND CHARACTERIZATION OF RECOMBINANT HUMAN PLACENTAL AND RAT LENS ALDOSE REDUCTASES EXPRESSED IN ESCHERICHIA COLI[*]

Sanai Sato, Susan Old*, Deborah Carper*, and Peter F. Kador

Laboratory of Ocular Therapeutics and
*Laboratory of Mechanism of Ocular Diseases
National Eye Institute
National Institutes of Health
Bethesda, MD 20892

INTRODUCTION

Aldose reductase (aditol NADP$^+$ oxidoreductase, EC 1.1.1.21) with sorbitol dehydrogenase (1-iditol dehydrogenase, EC 1.1.1.4) together form the polyol pathway where glucose is converted to fructose through the sugar alcohol sorbitol. In diabetes the increased flux of glucose through the polyol pathway results in the accumulation of sorbitol which is linked to the onset of various diabetic complications such as cataract formation, retinopathy, neuropathy and nephropathy (Kinoshita, 1974; Kador, 1988; Kinoshita et al., 1990; Kador et al., 1990). This has spurred the world-wide interest in the development of aldose reductase inhibitors as a new approach toward the treatment of diabetic complications.

Human aldose reductases have been sequenced using cDNA libraries from placenta (Bohren et al., 1989; Chung and LaMendola, 1989; Grundmann et al., 1990), liver (Graham et al., 1989), retina and muscle (Nishimura et al., 1990). These sequences from human tissues are virtually identical except for one amino acid difference in some clones (Nishimura et al., 1990). In addition, the sequences of aldose reductase from rat lens (Carper et al., 1987) and rabbit kidney medulla (Garcia-Perez et al., 1989) and bovine lens (Schade et al., 1990) have been determined. Comparisons of these sequences indicate more than 80% homology between the animal and human aldose reductases.

Gaining an understanding of the enzyme structure and the relationship between this enzyme and its inhibitor(s) is the first step in the rational development of more potent and specific aldose reductase inhibitors. Both pig lens aldose reductase (Rondeau et al., 1992)

---

[*] The material in this chapter was not presented at the meeting, but is based upon an extension of work presented at the 1990 workshop (Carper et al., 1990).

and recombinant human placental aldose reductase (Wilson et al., 1992; Borhani et al., 1992; Wilson et al., 1993) have been crystallized and the nucleotide binding site and substrate/inhibitor site described.

Recombinant protein technology represents a valuable tool for the large scale *in vitro* production of aldose reductase required for crystallographic, inhibitor, and nucleotide binding site studies (Bohren et al., 1991; Petrash et al., 1992; Bohren et al., 1992; Yamaoka et al., 1992; Tarle et al., 1993; Bohren et al., 1994; Harrison et al., 1994; Kubiseski et al., 1994). It is, however, important to establish the specific similarities and/or differences of the secondary and tertiary structures around both active and inhibitor sites of the recombinant proteins and native tissue enzymes. We have reported that the recombinant proteins from both human placental and rat lens aldose reductase clones expressed in *E. coli* possess aldose reductase activity (Carper et al., 1991; Old et al., 1990). In this experiment the difference and/or similarity between tissue enzymes and recombinant proteins expressed in *E. coli* have been investigated.

MATERIALS AND METHODS

Materials

Unless otherwise stated, all chemicals employed were of reagent grade. Mono P (HR 5/20), Polybuffer 74 and all reagents for PhastSystem were obtained from Pharmacia Fine Chemicals (Piscataway, NJ). Aldose reductase inhibitors Al 1576, Ponalrestat, FK 366, sorbinil and tolrestat were gifts from Alcon Laboratories (Fort Worth, TX), ICI Americas (Wilmington, DE), Fujisawa Pharmaceutical Co. Ltd. (Osaka, Japan), Pfizer Central Research (Groton, CT) and Wyeth-Ayerst Research (Princeton, NJ), respectively.

Frozen rat lenses were obtained commercially from Biotrol (Indianapolis, IN). Human placentas, frozen immediately after delivery, were fractionated by ammonium sulfate as previously described (Kador et al. 1981a) and stored at -70°C until subsequent purification.

Enzyme Assay and Kinetic Study

Reductase activity was spectrophotometrically assayed as previously described (Kador et al., 1981a). One enzyme unit was defined as the amount of enzyme consuming one micromole of NADPH per min under the assay condition. Kinetic calculations were conducted on the NIH PROPHET computer system, and IC_{50} calculation was performed as previously described (Kador et al., 1981b).

Protein concentrations were measured according to Bradford (1976) using bovine serum albumin as standard protein.

Enzyme Purification

Aldose reductases from both rat lens and human placenta were purified by the series of chromatographic procedures as previously described (Sato, 1992).

Either the homogenate from 100 rat lenses or 5 ml of crude human placental solution (30 to 50% saturated fraction) were centrifuged at 10,000 x *g* for 10 min. The obtained supernatant was applied to a Sephadex G-75 column (2.5 x 90 cm) equilibrated with 10 mM imidazole-HCl buffer, pH 7.5 containing 7 mM 2-mercaptoethanol and eluted with the same imidazole buffer. The eluent was collected in 220-drop aliquots (*ca.* 10 ml) and fractions containing the aldose reductase activity were applied to a Matrex Gel Orange A column (2.5

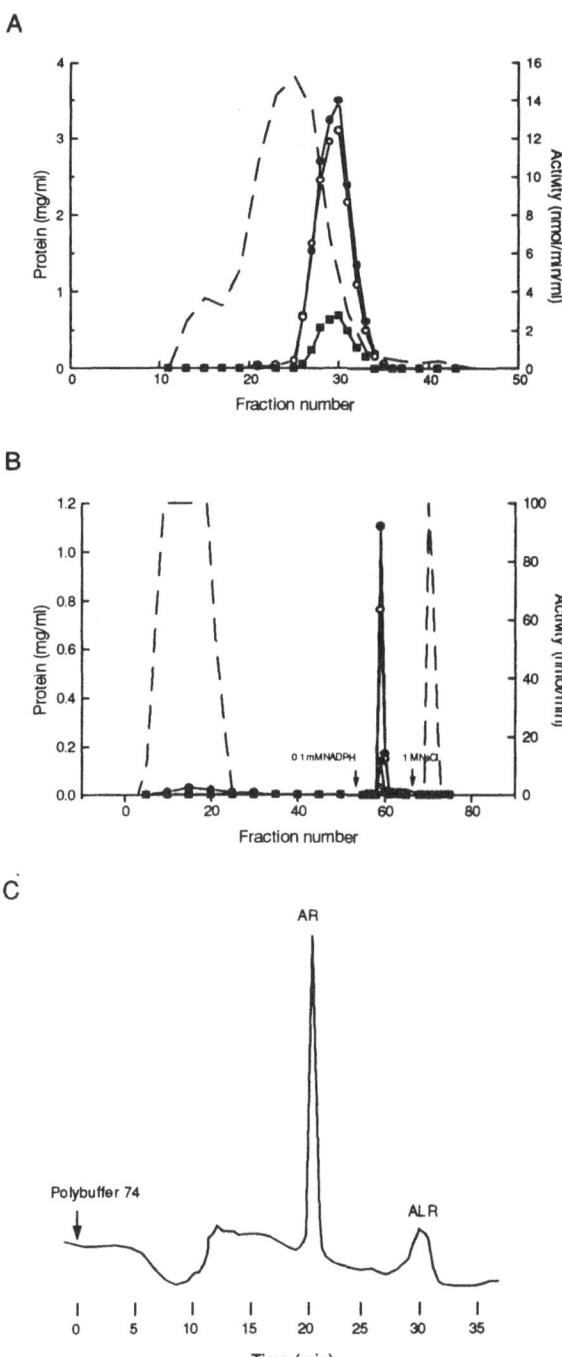

Figure 1. Chromatographic patterns observed in the purification of human placental aldose reductase by gel filtration on Sephadex G-75 (A), affinity chromatography on Matrex Gel Orange A (B) and chromatofocusing on Mono P (C). The solid circle, open circle and solid square indicate the activities with DL-glyceraldehyde, D-glucuronate and D-xylose (10 mM) as substrates, respectively. The broken lines in A and B indicate the protein concentrations assayed by the method of Bradford. The line in C indicates the protein concentration monitored by absorbance at 280 nm. AR; aldose reductase. ALR; aldehyde reductase.

x 15 cm). The column was washed with the imidazole buffer (*ca.* 500 ml), and then the enzyme was eluted with the same imidazole buffer containing 0.1 mM NADPH. Fractions eluted with NADPH were then chromatofocused on a Mono P (HR 5/20) column developed at the flow rate of 1 ml/min with Polybuffer 74 which was diluted 10 times with 7 mM 2-mercaptoethanol. The protein concentration of the eluent was monitored at 280 nm and the peaks containing aldose reductase activity were collected.

Purification of Recombinant Proteins from Rat Lens and Human Placental Aldose Reductase Clones

After release from *E. coli* harboring either rat lens aldose reductase (Old et al., 1990) or human placental aldose reductase constructs (Carper et al., 1991) by sonication, the recombinant rat lens and human placental aldose reductase proteins were purified by the same chromatographic procedures as described above.

Sodium Dodecyl Sulfate-Polyacrylamide Gel Electrophoresis (SDS-PAGE) and Isoelectric Focusing

SDS-PAGE and isoelectric focusing were conducted with the PhastSystem (Pharmacia-LKB) on acrylamide gels PhastGel gradient 8-25 (Pharmacia-LKB) and PhastGel IF 3-9 (Pharmacia-LKB), respectively. Proteins were visualized with Coomassie blue stain.

Double Immunodiffusion

Double immunodiffusion was performed on Ouchterlony plates (ICN ImmunoBiologicals, Lisle, IL) using antibodies against human placental aldose reductase (Kador et al., 1981a) and rat lens aldose reductase (Shiono et al., 1987) prepared in goats.

RESULTS

Aldose reductase was purified from human placenta by the combination of chromatographic procedures which included gel filtration, affinity chromatography and chromatofocusing. Figure 1 illustrates the typical chromatographic profiles of human placenta. A single peak of NADPH-dependent reductase activity with either DL-glyceraldehyde, D-glucuronate or D-xylose as substrate was observed with either gel filtration on Sephadex G-75

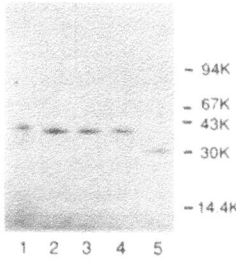

Figure 2. SDS-PAGE of the purified human placental aldehyde reductase (lane 1), wild type human placental aldose reductase (lane 2), recombinant human placental aldose reductase (lane 3 and 4) and glyceraldehyde reductase of *E. coli* (lane 5). Proteins are stained with Coomassie brilliant blue.

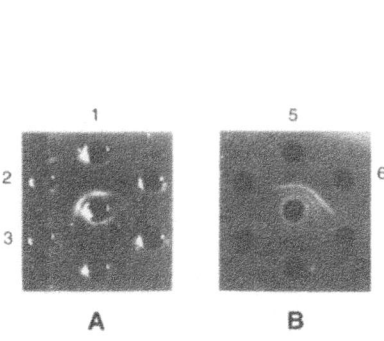

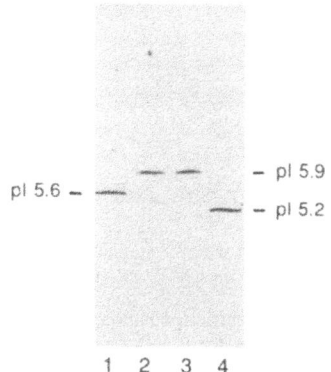

Figure 3. Double immunodiffusion of the recombinant human placental aldose reductase (well 1), wild type human placental aldose reductase (well 2), human placental aldehyde reductase (well 3), *E. coli* glyceraldehyde reductase (well 4), recombinant rat lens aldose reductase (well 5), and wild type rat lens aldose reductase (well 6). The center well contains antibody against either human placental aldose reductase (A) or rat lens aldose reductase (B).

Figure 4. Isoelectric focusing of purified human placental aldose reductase (lane 1), recombinant human placental aldose reductase (lane 2 and 3) and *E. coli* glyceraldehyde reductase (lane 4).

(Figure 1A) or affinity chromatography on Matrex Gel Orange A (Figure 1B). Aldose reductase (EC 1.1.1.21) was separated from aldehyde reductase (EC 1.1.1.2) in the final step of chromatofocusing (Figure 1C). Aldose reductase from rat lens was similarly purified with aldose reductase separated from the small amount of aldehyde reductase present in the rat lens by chromatofocusing (Sato and Kador, 1989).

Recombinant rat lens and human placental aldose reductase, expressed in *E. coli,* were purified by the same chromatographic procedures (Old et al., 1990; Carper et al., 1991). Although *E. coli* do not contain aldehyde reductase, the third step of chromatofocusing was utilized to separate the recombinant aldose reductases from a DL-glyceraldehyde-specific reductase present in *E. coli*.

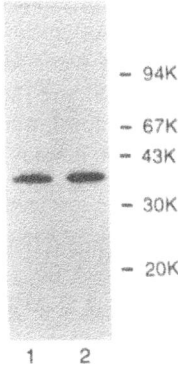

Figure 5. SDS-PAGE of the purified rat lens aldose reductase (lane 1) and recombinant enzyme from rat lens aldose reductase clone (lane 2).

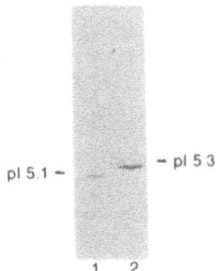

Figure 6. Isoelectric focusing of the purified rat lens aldose reductase (lane 1) and recombinant enzyme from rat lens aldose reductase clone (lane 2).

Both purified recombinant and wild type human placental aldose reductases comigrated as single bands on SDS-PAGE with molecular weights corresponding to 38 kDa (Figure 2). In addition, both were immunologically distinct from either human placental aldehyde reductase or *E coli* glyceraldehyde reductase. On double immunodiffusion, the recombinant human enzyme displayed a line of identity with wild type human aldose reductase with antibodies prepared against human placental aldose reductase. This antibody did not cross-react with either human placental aldehyde reductase or *E. coli* glyceraldehyde reductase (Figure 3A). Despite these similarities the recombinant and wild type human placental aldose reductases differed on isoelectric focusing with the pI value of the recombinant enzyme (pI 5.9) 0.3 pH units higher than that of wild type enzyme (pI 5.6) (Figure 4). The pI values of both the recombinant and wild type human placental aldose reductase differed from the pI value of *E. coli* glyceraldehyde reductase (pI 5.2).

Similarly, the purified recombinant and wild type rat lens aldose reductases comigrated on SDS-PAGE with similar molecular weights (Figure 5) and displayed immunological identity (Figure 3B). On isoelectric focusing, the recombinant aldose reductase (pI 5.3) was 0.2 pH units higher than the wild type aldose reductase (pI 5.1) (Figure 6).

Sequence analysis of the first 20 amino acids from the N-terminal of purified recombinant human placental and rat lens aldose reductases indicated that the amino acid sequences of the recombinant proteins were identical to the deduced amino acid sequences (Table 1). Resequencing of the human placental cDNA (Chung and LaMendola, 1989) showed no differences from a previously published human placental cDNA clone by Bohren et al. (1989). Methionine, the first amino acid in the cDNA sequences of both human placental and rat lens aldose reductase, was missing in both purified recombinant proteins. Sequences 2-21 (alanine to tryptophan) in recombinant human placental aldose reductase and recombinant rat lens aldose reductase were identical to the amino acid sequences obtained from the cDNAs of human placental and rat lens aldose reductases, respectively.

The kinetic properties of purified recombinant and wild type enzymes are summarized in Table 2. Both recombinant enzymes displayed kinetic properties that are virtually

Table 1. Amino acid sequences of human placental (HPAR) and rat lens aldose reductase (RLAR)

HPAR[1]:   Met-Ala-Ser-Arg-Leu-Leu-Leu-Asn-Asn-Gly-Ala-Lys-Met-Pro-Ile-Leu-Gly-Leu-Gly-Thr-Trp-Lys-Ser-

RLAR[2]:   Met-Ala-Ser-His-Leu-Glu-Leu-Asn-Asn-Gly-Thr-Lys-Met-Pro-Thr-Leu-Gly-Leu-Gly-Thr-Trp-Lys-Ser-

---

[1]Bohren et al. 1989. [2]Carper et al. 1987. The sequences indicated by underlines were identified by amino acid analysis.

Table 2. Kinetic properties of human placental (HPAR) and rat lens aldose reductases (RLAR)

	HPAR				RLAR			
	Recombinant		Wild type		Recombinant		Wild type[3]	
Substrate	K_m[1]	$\%V_{max}$[2]	K_m	$\%V_{max}$	K_m	$\%V_{max}$	K_m	$\%V_{max}$
DL-Glyceraldehyde	0.047	100	0.046	100	0.12	100	0.08	100
D-Glucuronate	6.4	37.8	4.8	35.1	13.9	21	19.2	25
D-Xylose	8.7	49.4	8.2	44.7	16.8	23	13.5	26
D-Glucose	98.1	4.7	86.8	4.3	281	1.7	204	2.1
D-Galactose	42.9	12.0	46.7	10.6	78	4.7	83	4.6

[1]mM [2]$\%V_{max}$ was expressed by the relative activity at 10 mM concentration to that with DL-glyceraldehyde as substrate.
[3]Sato et al. 1989.

Table 3. Effects of sulfate and chloride ions on the activity of human placental (HPAR) and rat lens aldose reductases (RLAR)

		Relative activity (%)			
Compound		HPAR		RLAR	
		Recombinant	Wild type	Recombinant	Wild type[1]
None		100	100	100	100
$(NH_4)_2SO_4$	(0.1 M)	193	171	122	180
Na_2SO_4	(0.1 M)	174	159	185	195
Li_2SO_4	(0.1 M)	193	166	182	197
NaCl	(0.2 M)	75	75	95	75
LiCl	(0.2 M)	85	85	93	95

[1] Sato et al. 1989

identical to those of the corresponding wild type enzymes. With the substrates utilized, both recombinant and wild type human placental aldose reductases displayed greater activity with DL-glyceraldehyde (K_m ca. 0.05 mM) than with the aldose sugars D-xylose, D-glucose and D-galactose (K_m's of 10-100 mM). Both the recombinant and wild type enzymes were activated by sulfate ions and inhibited by chloride ions (Table 3).

The inhibitions of purified recombinant and wild type enzymes by aldose reductase inhibitors are summarized in Table 4. The recombinant and wild type human placental aldose reductases and rat lens aldose reductase were similarly susceptible to inhibition by five commonly used aldose reductase inhibitors. With the exception of sorbinil, all inhibitors inhibited both recombinant and wild type human placental aldose reductases with IC_{50}'s in the 30-50 nanomolar range. IC_{50}'s of the rat lens enzymes were in the 10-20 nanomolar range.

Table 4. Inhibition of human placental (HPAR) and rat lens aldose reductases (RLAR)

Inhibitor	IC$_{50}$ (µM)			
	HPAR		RLAR[1]	
	Recombinant	Wild type	Recombinant	Wild type
Sorbinil	0.64	0.40	0.165	0.122
Tolrestat	0.051	0.036	0.018	0.023
Ponalrestat	0.048	0.049	0.017	0.010
Al 1576	0.040	0.041	0.016	0.015
FK 366	0.035	0.025	0.022	0.011

[1]Old et al. 1990 (except for data with FK 366).

DISCUSSION

The development of recombinant protein technology and site-directed mutagenesis have become valuable tools for gaining insight into structure-function relationships and mechanisms of substrate, nucleotide and inhibitor binding of aldose reductase. These studies, however, are based on the premise that the overall structure and function of recombinant enzymes are identical to those of wild type enzymes. Aldose reductase has been expressed in two different cells, *Spodoptera frugiperda* insect cells and *Escherichia coli* bacterial cells. The *S. frugiperda* system has been reported to be especially beneficial for producing large amounts of recombinant human protein with purification of recombinant protein achieved in a single step by affinity chromatography (Nishimura et al., 1990; Nishimura et al., 1991). However, Bohren et al. (1991) suggest that such one-step purification can not be achieved since the insect cell also contains an aldehyde reductase which should co-elute from an affinity column. Instead, they propose the use of *E. coli* since this cell does not contain a DL-glyceraldehyde reducing reductase. However, we have found that DH5α does contain a small amount of DL-glyceraldehyde reducing reductase that can be separated from recombinant aldose reductase by chromatofocusing (Carper et al., 1991). This DL-glyceraldehyde reducing reductase differs from both aldose reductase and aldehyde reductase in molecular weight, isoelectric point, substrate specificities, and immunological cross-reactivities.

The present studies, which compare the enzymatic properties and susceptibility to inhibition by aldose reductase inhibitors of purified recombinant rat lens and human placental aldose reductase expressed in *E. coli* with those of purified wild type enzymes, indicate that the recombinant enzymes possess structural and enzymatic properties that are similar to those of native human placental and rat lens aldose reductases. Recombinant enzyme expressed in *Spodoptera frugiperda* cells has also been reported to possesses similar kinetic properties and susceptibility to inhibition by aldose reductase inhibitors to those of native aldose reductase (Nishimura et al., 1991). Nevertheless, recombinant human placental and rat lens aldose reductases expressed in *E. coli* possess isoelectric points that are 0.2-0.3 pH units more basic in their isoelectric points (pI's) compared to the wild type enzymes. In contrast, the pI of the recombinant enzyme expressed in *S. frugiperda* cells has been reported to be identical to that of native enzyme (Nishimura et al., 1991). The difference in pI of

recombinant proteins expressed in *E. coli* most likely results from the lack of post-translational modifications in the N-terminal amino acid. While the N-terminal end of native
mammalian aldose reductases are blocked with an acetyl group, the N-terminal amino acids
of both recombinant human placental and rat lens aldose reductases expressed in *E. coli* are
not. In contrast, the N-terminal amino acid of the recombinant human muscle aldose
reductase enzyme expressed in *S. frugiperda* also contains a blocked N-terminal amino acid
which has been identified as acetylalanine (Nishimura et al., 1991). This suggests that
expression in *S. frugiperda* is closer to the tissue counterpart than expression in *E. coli* since
it yields enzyme with blocked ends identical to the native form. However, the similarities in
enzyme function and inhibition observed in the present studies indicate that the recombinant
enzymes possess secondary and tertiary structures similar to those of the native enzymes
despite the presence of a free N-terminal end. Therefore, expression of aldose reductase in
E. coli provides an excellent enzyme source for the further structural study of aldose
reductase.

ACKNOWLEDGEMENTS

The authors thank Dr. Hao-Chia Chen, National Institute of Child Health and Human
Development, National Institutes of Health for analyzing the amino acid sequence of
recombinant proteins.

REFERENCES

Bohren, K.M., Bullock, B., Wermuth, B., and Gabby, K.H., 1989, The aldo-keto reductase superfamily. cDNAs
    and deduced amino acid sequences of human aldehyde and aldose reductases, *J. Biol. Chem.*
    265:9547.
Bohren, K.M., Page, J.L., Shankar, R., Henry, S.P., and Gabbay, K.H., 1991, Expression of human aldose and
    aldehyde reductases. Site-directed mutagenesis of a critical lysine 262, *J. Biol. Chem.* 266:24031.
Bohren, K.M., Grimshaw C.E., and Gabbay, K.H., 1992, Catalytic effectiveness of human aldose reductase.
    Critical role of C-terminal domain, *J. Biol. Chem.* 267:20965.
Bohren, K.M., Grimshaw, C.E., Lai, C.-J., Harrison, D.H., Ringe, D., Petsko, G.A., and Gabbay, K.H., 1994,
    Tyrosine-48 is the proton donor and histidine-110 directs substrate stereochemical selectivity in the
    reduction reaction of human aldose reductase: Enzyme kinetics and crystal structure of the Y48H
    mutant enzyme, *Biochemistry* 33:2021.
Borhani, D.W., Harter, T.M., and Petrash, J.M., 1992, The crystal structure of the aldose reductase-NADPH
    binary complex, *J.Biol. Chem.* 267:24841.
Bradford, M.M., 1976, A rapid and sensitive method for the quantitation of microgram quantities of protein
    utilizing the principle of protein-dye binding, *Anal. Biochem.* 72:248.
Carper, D., Nishimura, C., Shinohara, T., Dietzchold, B., Wistow, G., Craft, C., Kador, P., and Kinoshita, J.H.,
    1987, Aldose reductase and rho-crystallin belong to the same protein superfamily as aldehyde
    reductase, *FEBS Letters* 220:209.
Carper, D., Sato, S., Old, S., Chung, S., and Kador, P.F., 1991, *In vitro* expression of human placental aldose
    reductase in *Escherichia coli*, *Adv. Exp. Med. Biol.* 284:129.
Chung, S., and LaMendola, J., 1989, Cloning and sequence determination of human placental aldose reductase
    gene, *J. Biol. Chem.* 264:14775.
Garcia-Perez, A., Martin, B., Murphy, H.R., Uchida, S., Murer, H., Cowley, B.D., Jr., Handler, J.S., and Burg,
    M.B., 1989, Molecular Cloning of cDNA coding for kidney aldose reductase. Regulation of specific
    mRNA accumulation by NaCl-mediated osmotic stress, *J. Biol. Chem.* 264:16815.
Graham, A., Hedge, P.J., Powell, S.J., Riley, J., Brown, L., Gammack, A., Carely, F., and Markham, A.F., 1989,
    Nucleotide sequence of cDNA for human aldose reductase, *Nucleic Acid Res.* 17:8368.
Grundmann, U., Bohn, H., Obermeier, R., and Amann, E., 1990, Cloning and prokaryotic expression of a
    biologically active human placental aldose reductase, *DNA Cell Biol.* 9:149.

Harrison, D.H., Bohren, K.M., Ringe, D., Petsko, G.A., and Gabbay, K.H., 1994, An anion binding site in human aldose reductase: Mechanistic implications for the binding of citrate, cacodylate, and glucose 6-phosphate, *Biochemistry* 33:2021.

Kador, P.F.,1988, The role of aldose reductase in the development of diabetic complications, *Med. Res. Rev.* 8:325.

Kador, P.F., Akagi, Y., Takahashi, Y., Ikebe, H., Wyman, M., and Kinoshita, J.H., 1990, Prevention of retinal vessel changes associated with diabetic retinopathy in galactose-fed dogs by aldose reductase inhibitors, *Arch. Ophthalmol.* 108:1301.

Kador, P.F., Carper, D., and Kinoshita, J.H., 1981a, Rapid purification of human placental aldose reductase, *Anal. Biochem.* 114:53.

Kador, P.F., Goosey, J.D., Sharpless, N.E., Kolish, J., and Miller, D.D., 1981b, Stereoscopic inhibition of aldose reductase, *Eur. J. Med. Chem.* 16:293.

Kinoshita, J.H., 1974, Mechanism initiating cataract formation, *Invest. Ophthalmol.* 13:713.

Kinoshita, J.H., Datiles, M.B., Kador, P.K., and Robison, W.G., 1990, Aldose reductase and diabetic eye complications. *in* Diabetes Mellitus: Theory and Practice, H. Rifkin and D. Porte, Jr., eds., Elsevier, New York.

Kubiseski, T.J., Green, N.C., Borhani, D.W., and Flynn, T.G., 1990, Studies on pig aldose reductase. Identification of an essential arginine in the primary and tertiary structure of the enzyme, *J. Biol. Chem.* 269:2183.

Nishimura, C., Matsuura, Y., Kokai, Y., Akera, T., Carper, D., Morjana, N., Lyons, C., and Flynn, T.G., 1990, Cloning and expression of human aldose reductase, *J. Biol. Chem.* 265:9788.

Nishimura, C., Yamaoka, T., Mizutani, M., Yamashita, K., Akera, T., and Tanimoto, T., 1991, Purification and characterization of the recombinant human aldose reductase expressed in baculovirus system, *Biochem Biophys Acta* 1078:171.

Old, S.E., Sato, S., Kador, P.F., and Carper, D.A., 1990, *In vitro* expression of rat lens aldose reductase in *Escherichia coli, Proc. Natl. Acad. Sci. USA* 87:4942.

Petrash, J.M., Harter, T.M., Devine, C.S., Olins, P.O., Bhatnagar, A., Liu, S.-Q., and Srivastava, S.K., 1992, Involvement of cysteine residues in catalysis and inhibition of human aldose reductase. Site-directed mutagenesis of Cys-80, 298, and -303, *J. Biol. Chem.* 267:24833.

Rondeau, J.-M., Tete-Favier, F., Podjarny, A, Reymann, J.-M., Barth, P., Biellmann, J.-F., and Moras, D., 1992, Novel NADPH-binding domain revealed by the crystal structure of aldose reductase, *Nature* 355:469.

Sato, S., 1992, Rat kidney aldose reductase and aldehyde reductase and polyol production in rat kidney, *Am. J. Physiol.* 263:F799.

Sato, S., and Kador, P.F., 1989, Rat lens aldehyde reductase, *Invest. Ophthalmol. Vis. Sci.* 30:1618.

Sato, S., and Kador, P.F., 1990, NADPH-dependent reductases in the dog lens, *Exp. Eye Res.* 50:629.

Shiono, T., Sato, S., Reddy, V.N., Kador, P.F., and Kinoshita, J.H., 1987, Rapid purification of rat lens aldose reductase, *Prog Clin Biol Res* 232:317.

Schade, S.Z., Early, S.L., Williams, T.R., Kezdy, F.J., Heinrikson, R.L., Grimshaw, C.E., and Doughty, C.C., 1990, Sequence analysis of bovine lens aldose reductase, *J. Biol. Chem.* 265:3628.

Tarle, I., Borhani, D.W., Wilson, D.K., Quiocho, F.A., and Petrash, J.M., 1993, Probing the active site of human aldose reductase. Site-directed mutagenesis of Asp-43, Tyr-48, Lys-77, and His-110, *J. Biol. Chem.* 268:25687.

Wilson, D.K., Bohren, K.M., Gabbay, K.H., and Quiocho, F.A., 1992, An unlikely sugar substrate site in the 1.65 Å structure of the human aldose reductase holoenzyme implicated in diabetic complications, *Science* 257:81.

Wilson, D.K., Tarle, I., Petrash, J.M., and Quicho, F.A., 1993, Refined 1.8 Å structure of human aldose reductase complexed with the potent inhibitor zopolrestat, *Proc. Natl. Acad. Sci. USA* 90:9847.

Yamaoka, T., Matsuura, Y., Yamashita, K., Tanimoto, T., and Nishimura, C., 1992, Site-directed mutagenesis of His-42, His-188 and Lys-263 of human aldose reductase, *Biochem. Biophys. Res. Commun.* 183:327.

33

RAT AND HUMAN BILE ACID BINDERS ARE MEMBERS OF THE MONOMERIC REDUCTASE GENE FAMILY

A. Stolz, L. Hammond and H. Lou

University of Southern California
LAC-USC 11-221
2025 Zonal Avenue
Los Angeles, CA 90033

Major advances have provided important new information on the fundamental mechanism of bile acid transport and metabolism in the liver. Previous studies of bile acid transport in enriched hepatocyte plasma membrane domains, isolated cells, couplets and whole organ perfusion are now complemented by the recent molecular identification of membrane spanning bile acid transporters and specific cytosolic binding proteins. With these new tools, the interrelationship of bile acid uptake, efflux and transport within the cell can be examined. We have focused on the molecular characterization of both human and rat cytosolic bile acid binding proteins in order to define their physiological function. This chapter will review the physiological role and biochemical characteristic of specific cytosolic bile acid binding proteins in human and rat liver. Molecular cloning of these genes reveals them to be members of a newly emerging family of monomeric reductase suggesting multifunctional role for this class of proteins.

BILE FORMATION AND ROLE OF CYTOSOLIC BILE ACID BINDING PROTEINS

Bile acids play a crucial role in many different physiological functions that require their efficient transcellular hepatic transport (Meier, 1988; Nathanson and Boyer, 1991; Coleman, 1987). In addition to their roles in bile formation and intestinal fatty acid absorption, bile acids are critical components in total body cholesterol homeostasis because they provide a major driving force for bile formation which contains significant amounts of cholesterol. In addition, replacement of bile acids lost during their enterohepatic circulation can also lower the intrahepatic cholesterol pool that has profound global effects on lipid metabolism by the liver.

In the liver, bile acids are transported into the hepatocyte by sinusoidal (basolateral) bile acid transporters from the portal vascular system rich in bile acids returning to the liver as part of their enteroheptic circulation. These bile acids present in the portal system are absorbed from the intestinal lumen by bile acid transporters located at the apical surface of enterocytes. Multiple sinusoidal bile acid transporters exist based on their sodium dependence and sensitivity to inhibition by other bile acids, cyclic peptides and the organic anion transport inhibitor, DIDS (Bellentani et al., 1987; Van Dyke et al., 1982; Meier, 1988; Frimmer and Ziegler, 1988; Zimmerli et al., 1989). Passive, non-saturating diffusion across sinusoidal plasma membrane by the more hydrophobic monohydroxyl and non-amidated bile acids has also been demonstrated (Van Dyke et al., 1982). Using a Xenopus Oocyte expression cloning strategy, Meier cloned a hepatic 1.7 kB sodium-dependent taurocholate transporter (Ntcp) encoding for a 39 kD protein that contains seven potential transmembrane spanning regions (Hagenbuch et al., 1991; Hagenbuch et al., 1990). In addition, Levy demonstrated that microsomal epoxide hydrolase (mEH), which reduces the mutagenic potential of polyaromatic hydrocarbon epoxide metabolites was also identical to a sinusoidal bile acid transporter identified by a photoaffinity bile acid probe (von Dippe and Levy, 1990; von Dippe et al., 1993).

Identification of Intracellular Bile Acid Binding Proteins

Identification of a specific cytosolic bile acid binding protein has been a major goal of our laboratory. We have purified two 37 kD monomeric bile acid binding proteins in human and rat hepatic cytosol by monitoring binding affinities using equilibrium dialysis during protein purification (Stolz et al., 1984; Stolz et al., 1989; Sugiyama et al., 1983). In rat, this 37 kD protein migrates with an apparent molecular weight of 33 kD on SDS-PAGE. Other investigators have utilized chemical and photoaffinity radio-labeled bile acid probes to covalently bind and selectively identify cytosolic and membrane proteins. Using intact, isolated rat hepatocytes with different radio-labeled bile acid photoprobes, Kurz preferentially labeled a 33 kD protein in SDS-PAGE from rat liver cytosol whereas a different pattern of labeling was observed when cytosol and bile acid were mixed and irradiated in vitro (Abberger et al., 1981; Abberger et al., 1983). A similar 33 kD bile acid binding protein was identified by Ziegler using a different, radio-labeled bile acid probe (Ziegler, 1985). Identification of similar size bile acid binding protein by different methodology substantiates our approach of identifying specific binding proteins by their in vivo binding affinities. In the rat, we also demonstrated that the 37 kD bile acid binding protein co-purified with 3a hydroxysteroid dehydrogenase activity (3a-HSD) (Stolz et al., 1987).

Canalicular Transport of Bile Acids

Enriched hepatic canalicular membrane fractions have been extensively used to define the biochemical properties of bile acid transporter present in this fraction. Excretion of bile acid across the canalicular (apical) membrane of the hepatocyte is the rate limiting component in net transcellular transport of bile acid by the hepatocyte (Coleman, 1987; Nathanson and Boyer, 1991). Characterization of bile acid transport activity in sealed canalicular enriched plasma membrane reveals increase in transport activity with hydrolysis of ATP on the outside of the enriched membrane fractions (Arias et al., 1993; Nishida et al., 1991). In addition, an electrogenic mediated bile acid transport activity has been identified and appears to be separate from the ATP inducible transport activity. In contrast to the sinusoidal membrane, no sodium dependent bile acid transport activity was identified. No specific bile acid canalicular transporter or gene has been identified, but a candidate gene Ecto-ATP'ase has been shown to mediate bile acid efflux in transfected Cos 7 cells (Sippel et al., 1993; Sippel et al., 1994). Characterization of this candidate bile acid transporter has demonstrated the requirement of a cytoplasmic domain for bile acid transport and increase in bile acid transport capacity by ATP hydrolysis as observed in enriched canalicular membrane fractions

Mechanism of Intracellular Bile Acid Transport

The physiological role of the bile acid binding protein in net transport of bile acid across the rat liver has been extensively evaluated by our group. Photoaffinity studies with bile acid probes can demonstrate only a single time point during bile acid transport within the cell and can not assess the dynamics of this rapid transcellular event. We have directly examined this issue in both isolated cells and in a single pass rat liver perfusion model by defining the transport process of unique C24 [^{14}C] and 3b [^{3}H] dual labeled bile acid probes which would provide a crude estimation of interacting of bile acids with cytosolic proteins, including the bile acid binder, 3a-HSD. We reasoned that these dual labeled bile acids, when binding (interacting) with cytosolic 3a-HSD would undergo transient oxidation resulting in transfer of ^{3}H b hydrogen to the NADP+ cofactor with subsequent reduction to reform the native bile acid. We confirmed that these two reactions can only be mediated by the cytosolic 3a-HSD and thus the ^{3}H/^{14}C ratio provides a simple but crude estimation of bile acid binding to the cytosolic compartment. Significant decrease in this ratio for di and tri hydroxyl bile acids occurred in isolated hepatocytes, whereas the monohydroxy bile acid, lithocholate underwent little change in its ratio (Takikawa et al., 1987a). This result suggested that hydrophobic bile acids such as lithocholate portioned into a non-cytosolic compartment consistent with its lipid partition coefficient. Similar alterations were also found in a rat liver perfusion model (Takikawa et al., 1987b).

We then assessed the role of bile acid interactions with the cytosol by using indomethacin, a non substrate competitive inhibitor of the substrate binding site of 3a-HSD (Penning et al., 1984; Penning and Talalay, 1983; Takikawa et al., 1987a; Takikawa et al., 1987b). Co-incubation of bile acids and indomethacin with isolated hepatocytes led to displacement of the bile acids into the media. Indomethacin inhibited the tritium loss of the di- and tri-hydroxy metabolites of dual labeled lithocholate and chenodeoxycholate; this confirmed the displacement of bile acids out of the cytosolic compartment within the hepatocyte. These studies clearly demonstrate that bile acids are capable of entering into the cytosolic compartment and that indomethacin is capable of inhibiting the interaction with the cytosolic 3a-HSD.

In order to determine the physiological role of the cytosolic bile acid binders in the intact organ, we evaluated the effect of indomethacin on transcellular transport of glycocholate. We monitored the biliary excretion of glycocholate to eliminate the potential cofounding effect of bile acid biotransformation on net transport of bile acid coinfused with and without the 3a-HSD inhibitor, indomethacin (50 uM). A brief summary follows. Steady state infusion of glycocholate with indomethacin caused a significant delay in biliary excretion with reduced bile acid extraction by the liver. No toxicity was observed during the 20 minute indomethacin co-infusion experiments. Reduced extraction in these coinfusion studies is in part due to inhibition of uptake by indomethacin, whereas delayed excretion is not due to inhibition of canalicular transport. Based on all these results, we hypothesize that bile acids retained in the cytosolic compartment can be rapidly excreted by the canalicular transport system at low bile acid concentrations. When the intracellular protein binding sites are occupied by indomethacin, bile acids are diverted into a non-cytosolic compartment from which they may eventually enter into bile, re-enter into the cytosolic compartment, or are regurgitated out of the cell and into the plasma

Other Intracellular Bile Acid Transport System

Multiple sources suggest the existence of an organelle bile acid transport system which would be a logical candidate for the site to which hydrophobic bile acids and those displaced by indomethacin in our experiments are sequestered (Crawford and Gollan, 1988; Crawford et al., 1988; Coleman, 1987). Histological studies revealed the accumulation of pericanalicular organelles during large bile acid flux suggesting their delivery into bile as well as immunohistochemical identification of bile acid in organelles (Suchy et al., 1983; Jones et al., 1979; Erlinger, 1990). A series of carefully performed studies by Crawford demonstrated a specific delay in bile acid biliary excretion by colchicine when large bile acid loads were infused in a bile acid depleted rat suggesting the existence of microtubule dependent vesicular transport system for bile acids (Crawford et al., 1988). Alternatively, colchicine has been suggested to reduce the excretion of taurocholate from hepatocytes by inhibiting insertion of additional canalicular bile acid transporter thereby causing a reduction in hepatic bile acid excretory capacity (Haussinger et al., 1993).

Potential Function of Cytosolic Bile Acid Binder on Canalicular Bile Acid Transport

Bile acid binders may influence canalicular transport in several ways. They may serve to retain bile acid within the cytosol which may be required for their subsequent elimination by the bile acid canalicular transporter. If bile acids were not retained by these proteins, they may bind non-specifically to intracellular membranes or re-distribute to an intracellular compartment from which they may ultimately be excreted into bile. Alternatively, they may regurgitate back into the plasma. The canalicular transporter most likely transports free and not bound bile acids. The intracellular bile acid binders may be critical for providing the free bile acids and therefore may serve to accumulate a bile acid pool within the cytosol. This pool of bound bile acids would be in equilibrium with the free bile acids providing a reservoir of bile acids which could then be eliminated by the canalicular transporter. Thus, accumulation of bile acids within the cytosol may be the mechanism by which bile acid binding proteins influence net transport. Figure 1 summarizes our proposed model for the role of these bile acid binding proteins.

BIOCHEMICAL CHARACTERISTICS AND cDNA CLONING OF THE RAT AND HUMAN BILE ACID BINDERS

In addition to binding bile acids, human and rat bile acid binders are oxidoreductases that preferentially utilize NADP(H) as a cofactor. In the rat, 3a-HSD metabolizes endogenous steroids as well as xenobiotic compounds. Both primary bile acid precursors and other

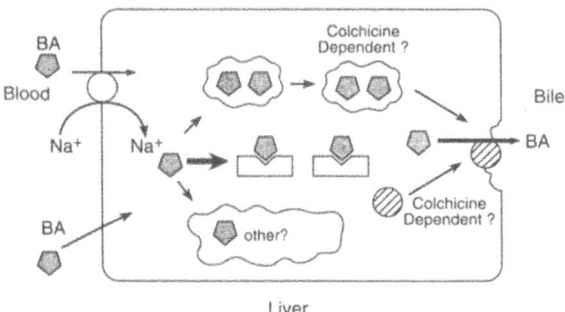

Figure 1. Proposed Function of Bile Acid Binding Proteins in Intracellular Transport of Bile Acids in the Liver.

steroids such as progesterone and androstenedione are stereospecifically reduced by 3a-HSD (Takikawa et al., 1990; Stolz et al., 1987). In the rat liver, 3a-HSD also co-elutes with dihydrodiol dehydrogenase (DDH) activity, which has been shown to reduce the mutagenic potential of benzopyrene metabolites in a bacterial mutagenicity test system (Vogel et al., 1980; Glatt et al., 1982; Woerner and Oesch, 1984). DDH prevents the formation of the genotoxic dihydrodiol epoxide metabolites through the elimination of the proximal carcinogen by catechol formation. DDH activity is expressed by many enzymes in different tissues (Woerner and Oesch, 1984). Sensitivity to inhibitors has been used to classify the various subtypes of DDH present in specific organs (Smithgall and Penning, 1985; Smithgall and Penning, 1988). As more of these proteins are cloned and sequenced, the relationship between these various enzymes can be directly compared amongst them selves and the human and rat bile acid binders.

cDNA Cloning of the Rat Bile Acid Binder/3a-HSD

We were the first group to report the cDNA cloning of the entire rat hepatic 3a-HSD gene and to characterize its tissue distribution (Stolz et al., 1991). A 2.4 kB cDNA clone was isolated and found to encode for 322 amino acids with a predicted weight of 37,022 daltons. The deduced amino acid sequence completely agreed with primary amino acid sequence determined by microsequence analysis. A unique feature of this gene as compared to other gene family members is a large, 1.3 kB 3' untranslated region. Northern blot analysis of Sprague-Dawley male and female rat organs correlated well with protein distribution determined by a radioimmunoassay (Stolz et al., 1986). The 3a-HSD mRNA is predominantly expressed in the liver and intestine, with respectively decreasing expression in the stomach, colon, and lung. Other tissues containing 3a-HSD activity, such as the kidney, brain, and testis, failed to hybridize with the rat hepatic 3a-HSD cDNA, indicating that this activity is encoded by non-related gene(s).

Regulation of Rat Hepatic 3a-HSD We examined changes in steady state 3a-HSD gene expression by both primary rat hepatocytes and the bile fistulae rat model in response to manipulation of bile acid synthesis, bile acid flux and hormone manipulation. Because 3a-HSD is required for bile acid synthesis, we postulated that coordinate gene expression with other bile acid synthetic enzymes may be occurring as well as to examine its regulation by bile acid flux. We determined the effects of a combination of bile acid replacement therapy with and without HMG-CoA reductase inhibitors in a bile acid depleted rat and steady state mRNA levels were determined and compared with a noninducible gene, cyclophilin (Pandak et al., 1991; Stravitz et al., 1994). No increase in steady state 3a-HSD mRNA was found in bile fistulae rats treated with cholesterol, cholestyramine or mevinolin, which all transcriptionally induced expression of the rate limiting enzyme in bile acid synthesis, cholesterol 7a hydroxylase. Bile acid replacement treatment with chenodeoxycholate or cholate in a depleted rat inhibited cholesterol 7a hydroxylase expression but did not alter 3a-HSD gene expression suggesting non-coordinate regulation We also determined hormonal requirements for maintenance of 3a-HSD gene expression in primary adult rat hepatocyte tissue culture as well as to define the site of regulation by these agents. A combination of Thyroxine and Dexamethasone was able to maintain 3a-HSD expression after 72 hours, whereas cells plated without supplemental hormones had minimally detectable mRNA. Cells plated with a single hormone had increased steady state mRNA levels and protein (by western blot), but this effect was significantly potentiated when hormones were combined. Thyroxine caused an increase in transcriptional activity whereas Dexamethasone functioned by stabilization of mRNA half life (Stravitz et al., 1994).

Table 1. Biochemical features of Human and Rat Bile Acid Binding Proteins

	Monomer	MW	3α-HSD	DDH	Lithocholate	Chenodeoxycholate	Cholate
Rat	+	37,022	0.55	1.15	1-2	1-2	100
Human	+	37,325	<0.03	2.3	0.05	0.1	1

Biochemical Characterization of Human Bile Acid Binder

We purified a 36 kD human bile acid binding protein (HBAB) that has a remarkably high binding affinity using a similar purification protocol as we used in the purification of the rat protein (Takikawa et al., 1986; Stolz et al., 1984). No information is available on other potential human cytosolic bile acid binding protein using the previous cited bile acid affinity probes. Table 1 list the biochemical features of these two similar proteins. Thus, in contrast to the rat, the HBAB appears unmatched in its ability to bind bile acids and demonstrates 1 to 2 orders of magnitude higher affinities for bile acids then the rat.

We sought oxidoreductase activity for the HBAB because these activities eluted with the rat protein. In rat liver, only one DDH isoform is present whereas, a complex elution pattern of DDH and 3a-HSD activities were found in human liver cytosol, as reported by us and Drs. Hara and Sawada (Hara et al., 1990; Takikawa et al., 1990). At least six fractions of DDH activity were identified in the chromatofocusing chromatogram of the 30 to 40 kD human liver cytosol fraction, using the single hydroxyl substrate, 1-acenaphtenol. Only one of these fractions demonstrated high affinity bile acid binding. The 3a-HSD activity co-eluted with some other fractions of DDH activity but was clearly distinct from the high affinity bile acid binding activity. Thus, unlike the rat, the human protein is unable to utilize bile acid as a substrate for enzymatic activity (Takikawa et al., 1990).

cDNA Cloning and Expression of HBAB

Human bile acid binder was cloned by probing a human Hep G2 cDNA library with the 5' Eco R1 proximal fragment of the rat hepatic 3a-HSD cDNA (Stolz et al., 1993). A 1252 bp cDNA clone was sequenced and its deduced peptide sequence compared with primary sequence data. In order to confirm the specificity of the clone and to begin to characterize its structural domains, the human and rat bile acid binders were recombinantly expressed as either a single protein (Pharmacia pKK 233-2) or as part of a GST fusion protein, using a Pharmacia pGEX-2T modified plasmid.

Table 1 lists the Bile acid binding and enzyme kinetic features of the rat and human bile acid binding protein. Both recombinant rat and human protein expressed the same catalytic rates and substrate specificities as the native protein suggesting that recombinantly expressed protein are properly folded. We previously demonstrated in the native 3a-HSD that Kd for bile acids are the same as the Km, indicating that binding occurs at the substrate site. Km for bile acids for recombinant protein are the same as the native protein, yet we

Figure 2. Multiple Sequence Alignment for Members of the Aldo-Keto Reductase Gene Family. Sequences compared are: Human bile acid binder (HBAB) (Stolz et al., 1993), ORF Genbank Accession number D17793, rabbit 20a hydroxysteroid dehydrogenase (20 HSD) (Lacy et al., 1993), human chloredecone reductase (HUMCR) (Winters et al., 1990), bovine prostaglandin f synthetase (BOVPGF) (Watanabe et al., 1988), rat hepatic 3a hydroxysteroid dehydrogense (RH3HSD) (Stolz et al., 1991), major mouse vas deferens proteins (MUSMVD) (Pailhoux et al., 1990), human aldose reductase (HUMALR) (Bohren et al., 1989). Similar amino acids are identified by a dash and gaps inserted by Bestfit program by an asterisk.

```
              10        20        30        40        50        60
HBAB    MDSKYQCVKLNDGHFMPVLGFGTYAPAEVPKSKALEATKLAIEAGFRHIDSAHLYNNEEQ
ORF     ----Q-----------------P---R-----V-----------------------
20 HSD  --P-F-R-A-S----I----------E-------M----I--D---------YF-K--KE
HUMCR   -----------P---RNR-V-V---------------Y------
BOVPGF  --P-S-R---------I----------E-----E------F---V----V------Q----
RH3HSD  ---ISLR-A----N-I-------TV-EK-A-DEVIK---I--DN----F---Y--EV--E
              10        20        30        40        50
MUSMVD  MATF-E-STKAK--PL-L--WKS***SPGQVKEAV-A--D--Y----C-YV-H--NE
HUMALR  MASRLL--N-AK--I--L--WKS***-PGQVTE-V-V--DV-Y----C--V-Q--NE


              70        80        90        100       110       120
HBAB    VGLAIRSKIADGSVKREDIFYTSKLWCNSHRPELVRPALERSLKNLQLDYVDLYLIHFPV
ORF     ------------------------STF-----------N---KA----------HS-M
20 HSD  ------------------------TF--------S--D-------------I----T
HUMCR   ------------------------TFFQ-QM-Q----S---K----------L---M
BOVPGF  --Q---------T----------------LQ---------K--Q---------I--S--
RH3HSD  --Q-----E--T----------STF-------TC--KT--ST-------I----M
              60        70        80        90        100       110
MUSMVD  --E--QE--KENA-----L-IV----ATFFEKS--KK-FDNT-SD-K---L----V-W-Q
HUMALR  --V--QE--LREQV----EL-IV-----TY-EKG--KG-CQKT-SD-K---L------W-T


              130       140       150       160       170       180
HBAB    SVKPGEEVIPKDENGKILFDTVDLCATWEAVEKCKDAGLAKSIGVSNFNRRQLEMILNKP
ORF     -L-----LS-T-----VI--I----T----M----------------------------
20 HSD  AL---V-I--T--H--AI---------M----------------------------
HUMCR   AL----TPL------VI------S----VM----------------C----------
BOVPGF  -L---GNKF----E--KL--D----C----L---------TK--------HK---K-----
RH3HSD  ALQ--DIFF-R--H--L--E---I-D----M----------------C----R-----
              120       130       140       150       160       170
MUSMVD  GFQA-NALL---NK--V-LSKSTFLDA---M-ELV-Q--V-AL-I----HF-I-RL----
HUMALR  GF---K-FF-L--S-NVVPSDTNILD--A-M-ELV-E-LV-A--I----HL-V-------


              190       200       210       220       230       240
HBAB    GLKYKPVCNQVECHPYFNQRKLLDFCKSKDIVLVAYSALGSHREEPWVDPNSPVLLEDPV
ORF     -----------------RS-----------------------Q-DKR-------------
20 HSD  ---------------L--G---E-----G--------------PE---QSA-------L
HUMCR   ---------------L--S----------------H----TQ-HKL-------------
BOVPGF  ---------------L--S---E----H-------A---AQLLSE--NS-N--------
RH3HSD  -------------L-L--S-M--Y------I--S-CT---S-DKT---QK-----D---
              180       190       200       210       220       230
MUSMVD  ---H---T--I-S---LT-E--IQY-Q--G-AVT---PLGS*PDR-YAK-ED--VM-I-K
HUMALR  ------AV--I-----LT-E--IQY-Q--G--VT---P----*PDR--AK-ED-S-----R


              250       260       270       280       290       300
HBAB    LCALAKKHKRTPALIALRYQLQRGVVVLAKSYNEQRIRQNVQVFEFQLTSEEMKAIDGLN
ORF     --------------------------------------------------A-D------
20 HSD  IG-------QQ-------------I------FT-K--KE-I--------P-DMKV--S--
HUMCR   ------------------------------------E-I----------D--VL----
BOVPGF  ---I-----Q----V-----V---------F-KK--KE-M---D-E--P-D--------
RH3HSD  ---I---Y-Q----V----------P-IR-F-AK--KELT-------A--D---L----
              240       250       260       270       280       290
MUSMVD  IKEIAA---K-V-QVLIRFHV--N---IP--VTPS--QE-L---D---SE-D-A--LSF-
HUMALR  IK-I-A--NK-T-QVLI-FPM--NL--IP--VTPE--AE-FK--D-E-S-QD-TTLLSYN


              310       320
HBAB    RNVRYLTLDIFAGPPNYPFSDEY
ORF     --LH-FNS-S--SH----Y----
20 HSD  --F--V-A-FAI-H---------
HUMCR   --Y--VVM-FLMDH-D-------
BOVPGF  --I--YDFQKGI-H-E----E--
RH3HSD  --F--NNAKY-DDH--H--TD-
              300       310
MUSMVD  --W-ACD-LDARTEED---HE--
HUMALR  --W-VCA-LSCTSHKD---HE-F
```

were unable to determine bile acid binding by equilibrium dialysis because of protein precipitation.

Rat and Human Bile Acid Binding Proteins Are Members of a Novel Monomeric Reductase Gene Family

Comparison of the deduced cDNA sequence for our rat and human proteins identified a new family of highly related oxidoreductases (>50% amino acid homology). These genes share common features of functioning as reductases in vivo, preferential utilization of NADP(H), and are composed of monomers. These genes share 50% sequence homology with the evolutionary distant gene, gamma lens crystalline suggesting evolution from a common ancestral gene with potential conservation of catalytic and nucleotide cofactor binding sites (Baker, 1990; Baker, 1989; Borras et al., 1989; Wistow and Piatigorsky, 1987; Jörnvall et al., 1981).

The aldose reductase and other recently cloned NADP(H) oxidoreductases represent a major group of proteins that share 50% to 80% sequence identity, suggesting that these proteins are members of the newly emerging aldo-keto reductase gene family (Lacy et al., 1993; Watanabe et al., 1988; Bohren et al., 1989; Winters et al., 1990; Pailhoux et al., 1990; Stolz et al., 1991). Of these proteins, the aldose reductase has been extensively analyzed because of its hypothesized role in mediating the sorbitol toxicity of diabetes. Recently, the X-ray crystal coordinates for aldose reductase with the $NADP^+$ cofactor was reported by two different groups (Wilson et al., 1992; Rondeau et al., 1992). The aldose reductase protein is a parallel alpha/ beta barrel structure that establishes a new motif for the $NADP^+$ binding domain of an oxidoreductase. This crystallographic data provides exquisitely detailed information about the cofactor binding site, the candidate residues involved in the transfer of hydrogen from cofactor to the substrate and the dimensions of the catalytic pocket. Figure 2 list the amino acid sequence of some members of this emerging gene family. This is a rapidly progressing field because a majority of these proteins were cloned within the last two years.

The greatest sequence divergence amongst all these genes is at the carboxyl end, which suggests the possibility that catalytic specificity or substrate binding site may reside within this region. Deletion of the 13 residue of the carboxyl terminal end of aldose reductase markedly reduced catalytic rates for uncharged residues suggesting a critical role of the carboxyl terminal end in orienting substrate to the active site (Bohren et al., 1992). We have recently discovered by determining HBAB's genomic organization that the carboxyl terminal end of the protein beginning at position 310 and the entire 3' untranslated region of the gene is contained within the last exon, which is also true for other members of this gene family (Pailhoux et al., 1992; Graham et al., 1991; Lou et al., 1994). Conservation of exon domains is a feature of gene family members. As an initial hypothesis, the shuffling of the last exon, which may potentially be critical for substrate specificity, would provide a mechanism for the generation of highly related proteins with different substrate specificity. We have identified at least six isoforms of DDH activity in the 30 to 40 kD in the human liver and other related genes during genomic cloning of the human bile acid binder suggesting a large number of highly related genes. We will require additional information about the genomic organization of these other potential family members to substantiate this hypothesis.

Preliminary Studies to Identify Substrate Specificity of the Rat and Human Bile Acid Binding Proteins

The availability of both the human and rat proteins which are highly related, yet having distinct catalytic activity, will be beneficial for defining the role of specific residues

Table 2. Catalytic Activity of Recombinant Wildtype and Chimeric Bile Acid Binding Proteins

Enzyme activity	Rat	H 1-63 / R64-322	Human	H 1-285 / R 286-322
3α-HSD	0.61	1.2	<.03	0.3
DDH	1.3	0.6	2.3	0.6

Enzyme activity determined as in Table 1.

to determine substrate specificity and binding properties of these enzymes. Comparing these similar yet different proteins allows us to potentially define the protein domains responsible for these differences. In addition, the availability of the crystal coordinates for the human aldose reductase provides a paradigm for other members of this gene family. Our preliminary strategy for defining substrate and bile acid binding sites on these proteins has been to first generate and express chimeric human-rat proteins and compare catalytic active to wildtype protein.

We first generated and expressed chimeric proteins composed of human and rat bile acid binding proteins to dissect the contributions of adjoining peptide regions. We recognized the limits of this approach because functional protein domains may be constructed from non-contiguous amino acids. Amino terminal (1 to 63) and carboxyl terminal (285-322(3)) peptide regions of the human and rat bile acid binders were exchanged at mutual endonuclease restriction sites, which did not disrupt the open reading frame. Constructs were confirmed by restriction mapping and DNA sequence analysis. Only 2 of 9 chimeric and truncated proteins were expressed either because of precipitation or inability to be cleaved by thrombin from the GST fusion protein. Table 2 lists preliminary features of these two chimeric proteins that were expressed. The addition of the rat carboxyl terminal end to the human protein resulted in 3a-HSD activity, although with a lower Vmax then the rat protein. These preliminary data support the concept that the carboxyl terminal is a major determinant of substrate specificity. The relationship of binding to substrate site will require additional studies.

CONCLUSION

In conclusion, human and rat bile acid binding proteins serve an important role in maintaining bile acids within the cytosolic compartment of the hepatocytes so that they may be rapidly and efficiently transported out of the cell. In addition to binding, these homologous proteins are members of the newly emerging monomeric aldo-keto reductase gene family. Preliminary studies with chimeric protein demonstrate potential significance of the carboxyl terminal end in dictating substrate specificity. The ability of these proteins to both bind bile acids and metabolize hydrophobic molecules suggest a new, multifunctinal role for these monomeric reductases.

REFERENCES

Abberger, H., Bickel, U., Buscher, H-P., Fuchte, K., Gerok, W., Kramer, W. and Kurz, G., 1981, Transport of bile acids: lipoproteins, membrane polypeptides, and cytosolic protein carriers, in: Bile Acids and Lipids. Edited by Paumgartner, G., Stiehl, A. and Gerok, W. MTP Press, Lancaster, England.

Abberger, H., Buscher, H-P., Fuchte, K., Gerok, W., Giese, U., Kramer, W., Kurz, G. and Zanger, U., 1983, Compartmentation of bile salt synthesis and transport revealed by photoaffinity labelling of isolated hepatocytes, in: Bile Acids and Cholesterol in Health and Disease. Edited by Paumgartner, G., Stiehl, A. and Gerok, W. MTP Press, Lancaster, England.

Arias, I.M., Che, M., Gatmaitan, Z., Leveille, C., Nishida, T. and St Pierre, M., 1993, The biology of the bile canaliculus, 1993, *Hepatology* 17:318-329.

Baker, M.E., 1989, Human placental 17 beta-hydroxysteroid dehydrogenase is homologous to NodG protein of Rhizobium meliloti, *Mol.Endocrinol.* 3:881-884.

Baker, M.E., 1990, A common ancestor for human placental 17 beta-hydroxysteroid dehydrogenase, Streptomyces coelicolor actIII protein, and Drosophila melanogaster alcohol dehydrogenase, *FASEB.J.* 4:222-226.

Bellentani, S., Hardison, W.G., Marchegiano, P., Zanasi, G. and Manenti, F., 1987, Bile acid inhibition of taurocholate uptake by rat hepatocytes: role of OH groups, *Am.J.Physiol.* 252:G339-G344.

Bohren, K.M., Bullock, B., Wermuth, B. and Gabbay, K.H., 1989, The aldo-keto reductase superfamily. cDNAs and deduced amino acid sequences of human aldehyde and aldose reductases, *J.Biol.Chem.* 264:9547-9551.

Bohren, K.M., Grimshaw, C.E. and Gabbay, K.H., 1992, Catalytic effectiveness of human aldose reductase. Critical role of C-terminal domain, *J.Biol.Chem.* 267:20965-20970.

Borras, T., Persson, B. and Jörnvall, H., 1989, Eye lens zeta-crystallin relationships to the family of "long-chain" alcohol/polyol dehydrogenases. Protein trimming and conservation of stable parts, *Biochemistry* 28:6133-6139.

Coleman, R., 1987, Biochemistry of bile secretion, *Biochem J.* 244:249-261.

Crawford, J.M., Berken, C.A. and Gollan, J.L., 1988, Role of the hepatocyte microtubular system in the excretion of bile salts and biliary lipid: implications for intracellular vesicular transport, *J.Lipid.Res.* 29:144.

Crawford, J.M. and Gollan, J.L., 1988, Hepatocyte cotransport of taurocholate and bilirubin glucuronides: role of microtubules, *Am.J.Physiol.* 255:G121-G131.

Erlinger, S., 1990, Role of intracellular organelles in the hepatic transport of bile acids, *Biomed.Pharmacother.* 44:409-416.

Frimmer, M. and Ziegler, K., 1988, The transport of bile acids in liver cell, *Biochim.Biophys.Acta* 947:75.-99

Glatt, H.R., Cooper, C.S., Grover, P.L., Sims, P., Bentley, P., Merdes, M., Waechter, F., Vogel, K., Guenthner, T.M. and Oesch, F., 1982, Inactivation of a diol epoxide by dihydrodiol dehydrogenase but not by two epoxide hydrolases, *Science* 215:1507-1509.

Graham, A., Brown, L., Hedge, P.J., Gammack, A.J. and Markham, A.F., 1991, Structure of the human aldose reductase gene, *J.Biol.Chem.* 266:6872-6877.

Hagenbuch, B., Luebbert, H., Stieger, B. and Meier, P.J., 1990, Expression of the hepatocyte Na+/bile acid cotransporter in Xenopus laevis oocytes, *J.Biol.Chem.* 265:5357.

Hagenbuch, B., Stieger, B., Foguet, M., Luebbert, H. and Meier, P.J., 1991, Functional expression cloning and characterization of the hepatocyte Na+/bile acid cotransport system, *Proc.Natl.Acad.Sci.U.S.A.* 88:10629-10633.

Hara, A., Taniguchi, H., Nakayama, T. and Sawada, H., 1990, Purification and properties of multiple forms of dihydrodiol dehydrogenase from human liver, *J.Biochem.Tokyo.* 108:250-254.

Haussinger, D., Saha, N., Hallbrucker, C., Lang, F. and Gerok, W., 1993, Involvement of microtubules in the swelling-induced stimulation of transcellular taurocholate transport in perfused rat liver, *Biochem J.* 291:355-360.

Jörnvall, H., Persson, M. and Jeffery, J., 1981, Alcohol and polyol dehydrogenases are both divided into two protein types, and structural properties cross-relate the different enzyme activities within each type, *Proc.Natl.Acad.Sci.U.S.A.* 78:4226-4230.

Jones, A.L., Schmucker, D.L., Mooney, J.S., Ockner, R.K. and Adler, R.D., 1979, Alterations in hepatic pericanalicular cytoplasm during enhanced bile secretory activity, *Lab.Invest.* 40:512-517.

Lacy, W.R., Washenick, K.J., Cook, R.G. and Dunbar, B.S., 1993, Molecular cloning and expression of an abundant rabbit ovarian protein with 20 alpha-hydroxysteroid dehydrogenase activity, *Mol.Endocrinol.* 7:58-66.

Lou, H., Hammond, L., Sharma, V., Sparkes, R.S., Lusis, A.J. and Stolz, A., 1994, Genomic organization and chromosomal localization of a novel human hepatic dihydrodiol dehydrogenase with high affinity bile acid binding, *J.Biol.Chem.* 269:8416-8422.

Meier, P.J., 1988, Transport polarity of hepatocytes, *Semin.Liver.Dis.* 8:293-307.

Nathanson, M.H. and Boyer, J.L., 1991, Mechanisms and regulation of bile secretion, *Hepatology* 14:551-566.

Nishida, T., Gatmaitan, Z., Che, M. and Arias, I.M., 1991, Rat liver canalicular membrane vesicles contain an ATP-dependent bile acid transport system, *Proc.Natl.Acad.Sci.U.S.A.* 88:6590-6594.

Pailhoux, E., Veyssiere, G., Fabre, S., Tournaire, C. and Jean, C., 1992, The genomic organization and DNA sequence of the mouse vas deferens androgen-regulated protein gene, *J.Steroid.Biochem Mol.Biol.* 42:561-568.

Pailhoux, E.A., Martinez, A., Veyssiere, G.M. and Jean, C.G., 1990, Androgen-dependent protein from mouse vas deferens. cDNA cloning and protein homology with the aldo-keto reductase superfamily, *J.Biol.Chem.* 265:19932-19936.

Pandak, W.M., Li, Y.C., Chiang, J.Y., Studer, E.J., Gurley, E.C., Heuman, D.M., Vlahcevic, Z.R. and Hylemon, P.B., 1991, Regulation of cholesterol 7 alpha-hydroxylase mRNA and transcriptional activity by taurocholate and cholesterol in the chronic biliary diverted rat, *J.Biol.Chem.* 266:3416-3421.

Penning, T.M., Mukharji, I., Barrows, S. and Talalay, P., 1984, Purification and properties of a 3 alpha-hydroxysteroid dehydrogenase of rat liver cytosol and its inhibition by anti-inflammatory drugs, *Biochem.J.* 222:601-611.

Penning, T.M. and Talalay, P., 1983, Inhibition of a major NAD(P)-linked oxidoreductase from rat liver cytosol by steroidal and nonsteroidal anti-inflammatory agents and by prostaglandins, *Proc.Natl.Acad.Sci.U.S.A.* 80:4504-4508.

Rondeau, J.M., Tete Favier, F., Podjarny, A., Reymann, J.M., Barth, P., Biellmann, J.F. and Moras, D., 1992, Novel NADPH-binding domain revealed by the crystal structure of aldose reductase, *Nature* 355:469-472.

Sippel, C.J., McCollum, M.J. and Perlmutter, D.H., 1994, Bile acid transport by the rat liver canalicular bile acid transport/ecto-ATPase protein is dependent on ATP but not on its own ecto-ATPase activity, *J.Biol.Chem.* 269:2820-2826.

Sippel, C.J., Suchy, F.J., Ananthanarayanan, M. and Perlmutter, D.H., 1993, The rat liver ecto-ATPase is also a canalicular bile acid transport protein, *J.Biol.Chem.* 268:2083-2091.

Smithgall, T.E. and Penning, T.M., 1985, Indomethacin-sensitive 3 alpha-hydroxysteroid dehydrogenase in rat tissues, *Biochem.Pharmacol.* 34:831-835.

Smithgall, T.E. and Penning, T.M., 1988, Electrophoretic and immunochemical characterization of 3 alpha-hydroxysteroid/dihydrodiol dehydrogenases of rat tissues, *Biochem J.* 254:715-721.

Stolz, A., Hammond, L., Lou, H., Takikawa, H., Ronk, M. and Shively, J.E., 1993, cDNA cloning and expression of the human hepatic bile acid-binding protein. A member of the monomeric reductase gene family, *J.Biol.Chem.* 268:10448-10457.

Stolz, A., Rahimi Kiani, M., Ameis, D., Chan, E., Ronk, M. and Shively, J.E., 1991, Molecular structure of rat hepatic 3 alpha-hydroxysteroid dehydrogenase. A member of the oxidoreductase gene family, *J.Biol.Chem.* 266:15253-15257.

Stolz, A., Sugiyama, Y., Kuhlenkamp, J. and Kaplowitz, N., 1984, Identification and purification of a 36 kDa bile acid binder in human hepatic cytosol, *FEBS Lett.* 177:31-35.

Stolz, A., Sugiyama, Y., Kuhlenkamp, J., Osadchey, B., Yamada, T., Belknap, W., Balistreri, W. and Kaplowitz, N., 1986, Cytosolic bile acid binding protein in rat liver: radioimmunoassay, molecular forms, developmental characteristics and organ distribution, *Hepatology* 6:433-439.

Stolz, A., Takikawa, H., Ookhtens, M. and Kaplowitz, N., 1989, The role of cytoplasmic proteins in hepatic bile acid transport, *in:* Annual Review Physiolology. Edited by G. Sachs, Annual Review, Palo Alto, California.

Stolz, A., Takikawa, H., Sugiyama, Y., Kuhlenkamp, J. and Kaplowitz, N., 1987, 3 alpha-hydroxysteroid dehydrogenase activity of the Y' bile acid binders in rat liver cytosol. Identification, kinetics, and physiologic significance, *J.Clin.Invest.* 79:427-434.

Stravitz, R.T., Vlahcevic, Z.R., Pandak, W.M., Stolz, A. and Hylemon, P.B., 1994, Regulation of rat hepatic 3 alpha-hydroxysteroid dehydrogenase in vivo and in primary cultures of rat hepatocytes, *J.Lipid.Res.* 35:239-247.

Suchy, F.J., Balistreri, W.F., Hung, F., Miller, P. and Garfield, S.A., 1983, Intracellular bile acid transport in rat liver as visualized by electron microscope autoradiography using a bile acid analogue, *Am.J.Physiol.* 245:G681-G689.

Sugiyama, Y., Yamada, T. and Kaplowitz, N., 1983, Newly identified bile acid binders in rat liver cytosol. Purification and comparison with glutathione S-transferases, *J.Biol.Chem.* 258:3602-3607.

Takikawa, H., Ookhtens, M., Stolz, A. and Kaplowitz, N., 1987a, Cyclical oxidation-reduction of the C3 position on bile acids catalyzed by 3 alpha-hydroxysteroid dehydrogenase. II. Studies in the prograde and retrograde single-pass, perfused rat liver and inhibition by indomethacin, *J.Clin.Invest.* 80:861.

Takikawa, H., Stolz, A. and Kaplowitz, N., 1987b, Cyclical oxidation-reduction of the C3 position on bile acids catalyzed by rat hepatic 3 alpha-hydroxysteroid dehydrogenase. I. Studies with the purified enzyme, isolated rat hepatocytes, and inhibition by indomethacin, *J.Clin.Invest.* 80:852.

Takikawa, H., Stolz, A., Kuroki, S. and Kaplowitz, N., 1990, Oxidation and reduction of bile acid precursors by rat hepatic 3 alpha-hydroxysteroid dehydrogenase and inhibition by bile acids and indomethacin, *Biochim.Biophys.Acta* 1043:153-156.

Takikawa, H., Stolz, A., Sugimoto, M., Sugiyama, Y. and Kaplowitz, N., 1986, Comparison of the affinities of newly identified human bile acid binder and cationic glutathione S-transferase for bile acids, *J.Lipid.Res.* 27:652-657.

Takikawa, H., Stolz, A., Sugiyama, Y., Yoshida, H., Yamanaka, M. and Kaplowitz, N., 1990, Relationship between the newly identified bile acid binder and bile acid oxidoreductases in human liver, *J.Biol.Chem.* 265:2132-2136.

Van Dyke, R.W., Stephens, J.E. and Scharschmidt, B.F., 1982, Bile acid transport in cultured rat hepatocytes, *Am.J.Physiol.* 243:G484-G492.

Vogel, K., Bentley, P., Platt, K.L. and Oesch, F., 1980, Rat liver cytoplasmic dihydrodiol dehydrogenase. Purification to apparent homogeneity and properties, *J.Biol.Chem.* 255:9621-9625.

von Dippe, P., Amoui, M., Alves, C. and Levy, D., 1993, Na(+)-dependent bile acid transport by hepatocytes is mediated by a protein similar to microsomal epoxide hydrolase, *Am.J.Physiol.* 264:G528-G534.

von Dippe, P. and Levy, D., 1990, Reconstitution of the immunopurified 49-kDa sodium-dependent bile acid transport protein derived from hepatocyte sinusoidal plasma membranes, *J.Biol.Chem.* 265:14812-14816.

Watanabe, K., Fujii, Y., Nakayama, K., Ohkubo, H., Kuramitsu, S., Kagamiyama, H., Nakanishi, S. and Hayaishi, O., 1988, Structural similarity of bovine lung prostaglandin F synthase to lens epsilon-crystallin of the European common frog, *Proc.Natl.Acad.Sci.U.S.A.* 85:11-15.

Wilson, D.K., Bohren, K.M., Gabbay, K.H. and Quiocho, F.A., 1992, An unlikely sugar substrate site in the 1.65 A structure of the human aldose reductase holoenzyme implicated in diabetic complications, *Science* 257:81-84.

Winters, C.J., Molowa, D.T. and Guzelian, P.S., 1990, Isolation and characterization of cloned cDNAs encoding human liver chlordecone reductase, *Biochemistry* 29:1080-1087.

Wistow, G. and Piatigorsky, J., 1987, Recruitment of enzymes as lens structural proteins, *Science* 236:1554-1556.

Woerner, W. and Oesch, F., 1984, Identity of dihydrodiol dehydrogenase and 3 alpha-hydroxysteroid dehydrogenase in rat but not in rabbit liver cytosol, *FEBS Lett.* 170:263-267.

Ziegler, K., 1985, Further characterization of 3'-isothiocyanatobenamido[3H]cholate binding to hepatocytes. Correlation with bile acid transprot inhibition and protection by substrates and inhibitors, *Biochim.Biophys.Acta* 819:37-44.

Zimmerli, B., Valantinas, J. and Meier, P.J., 1989, Multispecificity of Na^+-dependent taurocholate uptake in basolateral(sinusoidal) rat liver plasma membrane vesicles, *J.Pharm.Exp.Therap.* 250:301-308.

34

THE ALCOHOL DEHYDROGENASE SYSTEM

Hans Jörnvall, Olle Danielsson, Lars Hjelmqvist, Bengt Persson, and
Jawed Shafqat

Department of Medical Biochemistry and Biophysics
Karolinska Institutet
S-171 77 Stockholm, Sweden

INTRODUCTION

Alcohol dehydrogenases of different types are common enzymes in nature. Two of these families, the *m*edium-chain *d*ehydrogenase/*r*eductase family, MDR, and the *s*hort-chain *d*ehydrogenase/*r*eductase family, SDR, are well studied and known since long, but have experienced a recent "explosion" of new knowledge, extension and importance. The MDR family includes the classical zinc-containing liver alcohol dehydrogenases encompassing the classes of human liver alcohol dehydrogenase, while the SDR family includes the *Drosophila* alcohol dehydrogenase, which has shorter subunits, no similar metal requirements, other sub-domain arrangements with different structural relationships, and other subunit interactions.

In addition to the MDR and SDR families, further families with alcohol dehydrogenases exist, reflecting iron-dependent enzymes, long-chain enzymes, and several types of mainly prokaryotic enzymes with other factor requirements (Scopes, 1983; Williamson and Paquin, 1987; Inoue et al., 1989; van Ophem et al., 1992). Some of these sub-groups may in fact turn out to belong to the MDR or SDR families, but are hitherto little studied. Notably, still further NAD(P)-dependent dehydrogenase families of course exist, including for example the glycolytic dehydrogenases, aldehyde dehydrogenases, and glutamate dehydrogenases (all separate families), just to name a few others.

The MDR family has received much attention. It was the first mammalian alcohol dehydrogenase type purified (Bonnichsen and Wassén, 1948), the first one analyzed in primary and tertiary structure (cf Brändén et al., 1975) and was reviewed at the previous volume in this series on carbonyl metabolizing enzymes (Jörnvall et al., 1993a). The mammalian forms of these enzymes were early distinguished into three classes (Vallee and Bazzone, 1983), had been extended to five at the previous meeting in 1992, and now appear to be six (cf Jörnvall and Höög, 1994) by the inclusion of a type characterized from a strain of deer mice lacking the gene for the classical enzyme (Zheng et al., 1993). In addition to alcohol dehydrogenases, the MDR family was already at the previous meeting known to contain sorbitol and threonine dehydrogenases, as well as reductases, of which ζ-crystallin with quinone

Enzymology and Molecular Biology of Carbonyl Metabolism 5
Edited by H. Weiner *et al.*, Plenum Press, New York, 1995

reductase activity (Rao and Zigler, 1991; Rao et al., 1992) and structural properties typical of enzymes (Jörnvall et al., 1993b) was the first reductase characterized in this family (Borrás et al., 1989).

Insight into the SDR family started with the distinction of *Drosophila* alcohol dehydrogenase as related to prokaryotic ribitol dehydrogenase (Jörnvall et al., 1981). Interest in this family was greatly increased when steroid (Marekov et al., 1990; Peltoketo et al., 1988) and prostaglandin (Krook et al., 1990) dehydrogenases were added, and a previous review summarized data on 20 different enzymes of the SDR family, showing them to have a functionally important Tyr-X-X-X-Lys structure, presumably at the active site (Persson et al., 1991) in catalytic functions (McKinley-McKee et al., 1991). Addition of carbonyl reductase and a posttranslationally modified form of that enzyme containing carboxyethyllysine (Krook et al., 1993) and other substituted amino acids (Wermuth et al., 1993) added further interest into the family.

This was roughly the situation at the previous meeting. Since then, progress has been rapid on the knowledge of both the MDR and SDR families. Both have increased in size, now containing hundreds of enzymes, including species variants (or each tens to fifties of enzymes, excluding species variants), but also with further characterized tertiary structures, additional forms, known evolutionary properties, novel isozyme/family relationships, and illustrations of general principles of protein build-up. These aspects are updated below.

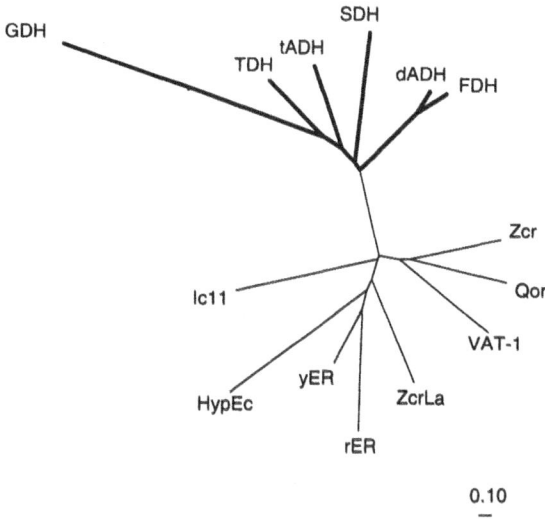

0.10
—

Figure 1. Unrooted phylogenetic tree illustrating relationships among different types of MDR proteins. The top part (bold) shows the dehydrogenases, the bottom part (thin) the reductases. Within both these separate parts, branchings are close and therefore still tentative, while these two main parts are more distantly separated constituting clearly different branchings. ADH, alcohol dehydrogenase (t, tetrameric, like the yeast enzyme, d, dimeric like the liver enzyme); FDH, formaldehyde dehydrogenase; SDH, sorbitol dehydrogenase; TDH, threonine dehydrogenase; GDH, *Thermoplasma acidophilum* glucose dehydrogenase; Zcr, ζ-crystallin; Qor, *E. coli* quinone reductase; VAT-1, *Torpedo californica* VAT-1 synaptic vesicle protein; ZcrLa, *Leishmania amazonensis* ζ-crystallin homolog; rER and yER, enoyl reductase of rat fatty acid synthase, and *Saccharopolyspora erythraea* erythronolid synthase, respectively; HypEc and Ic11, hypothetical *E. coli* protein, and *Trichoderma harzianum* protein of unknown function, respectively. Bottom bar indicates the distance for 10% apparent sequence divergence.

THE MDR FAMILY

Additional Enzymes

The family has broadened considerably, by inclusion of both dehydrogenases and reductases (cf Persson et al., 1994b). In particular, enoyl reductases of fatty acid synthesis are clearly homologous with the MDR dehydrogenases and now combined with them to constitute the MDR family. Minimally seven different activities are known for members of the MDR family, and additional activities seem likely, since some of the proteins structurally ascribed to this family, like the VAT-1 protein of synaptic vesicles, are little known in functional terms and may well represent additional activities. All these enzymes are fairly distantly related, hence reflecting distant duplicatory origins. Nevertheless, the dehydrogenases and the reductases are distinguishable as two sub-lines. A schematic representation, showing this division and the structural relationships among the separate lines, is given in Fig. 1.

In functional terms, this extension of the family brings additional importance to the whole group of enzymes by establishing the role of MDR proteins into further metabolic routes. It also completes a symmetry with the SDR family, by showing these large families both to involve a number of dehydrogenases and reductases. In addition, the extension shows that the MDR enzymes are involved in both degradative and synthetic pathways and that detoxications, suggested to be the function of the glutathione-dependent formaldehyde activity of class III alcohol dehydrogenase, may not be the sole or even major function of the whole enzyme family.

Furthermore, an alcohol dehydrogenase of the MDR family has finally been found also in insects (Danielsson et al., 1994a; Luque et al., 1994), where previously just the SDR alcohol dehydrogenase was known. This new finding of a class III alcohol dehydrogenase completes the pattern by showing the ubiquitous presence of MDR class III alcohol dehydrogenase in all organisms. Even more important, this new enzyme allows extensive comparisons between all the enzymes of class I and III type, distinguishing fundamental differences in molecular variability of these two related proteins (Danielsson et al., 1994a). Thus, three variable segments outline the class differences (Persson et al., 1993b; Danielsson et al., 1994a). They furthermore coincide with the loop differences that distinguish the class transitions as deduced by recent crystallography on the cod class I/III hybrid enzyme (El-Ahmad et al., 1994), and with differences among isozymes (Hjelmqvist et al., 1994), establishing special importance to the molecular segments defined from the structural comparisons. A figure showing the locations of these segments in the protein is given in relation to the isozyme discussion (Fig. 4, below).

Additional Classes

Gene duplications more recent than those giving rise to the separate enzymes have occurred and have produced a set of different classes for some of the enzymes, in particular the alcohol dehydrogenases (Table 1). Recently, a novel alcohol dehydrogenase was detected in mammals (Zheng et al., 1993), interpreted to represent a sixth class, and additional classes appear likely. Furthermore, studies on species variants within each class has given information during the last few years as to the origin and datings of the classes and their corresponding gene duplications. Available data show that class III appears to be present in all life forms. It has been detected in animals, plants, yeasts, and prokaryotes, representing a distant origin and an original and long-lasting function for this enzyme as a glutathione-dependent formaldehyde dehydrogenase.

Table 1. Nomenclature of characterized forms of the human alcohol dehydrogenase classes, genes, and subunits

Class	Human gene	Human subunit	Characteristics
I	ADH1-3	α, β, γ	Classical liver enzyme
II	ADH4	π	
III	ADH5 ·	χ	FDH; Ubiquitous
IV	ADH7	σ	Stomach/Epithelial
V	ADH6		
VI			

Class origins appear to be fairly distant, before the mammalian radiation, and classes are therefore likely to be similar in at least all or most mammals. Six classes have been suggested to occur in mammals, but classes V and VI have thus far been isolated from different species and should perhaps not yet be regarded as definitely separate. For five classes the corresponding human genes have been analyzed, and for four the protein subunits have been named. Empty spaces, not characterized. Three forms have special distributions or substrates, as shown. FDH, glutathione-dependent formaldehyde dehydrogenase.

Class I has been traced to a duplicatory origin from class III at about early vertebrate times. Significantly, three different methods arrive at similar time estimates for this duplication and the class I origin. Thus, extrapolation backwards from known species variants of class I gave a presumed origin at about 450 MYA (million years ago), corresponding to early vertebrate times (Cederlund et al., 1991). Second, purification of ADH from a large number of animals, gave recovery of class I alcohol dehydrogenase activity from all vertebrates down to bony fish (Danielsson et al., 1992), while cartilaginous fish (unpublished) and cyclostomes (Danielsson et al., 1994b) appear to give just class III enzymes, again compatible with an origin of class I at early vertebrate times (after separation of bony fish from cartilaginous fish) as schematically shown in Fig 2. Finally, the most distant of the characterized class I enzymes (from cod liver) has some class-hybrid structures, with overall residue identities larger to class III than to class I, but functional properties typical of class I, presumably reflecting the origin ("enzymogenesis") of class I from class III in the fish line (Danielsson and Jörnvall, 1992). It appears that early changes were rapid, before functional restrictions got fixed and later produced a slower rate of evolutionary changes. This change of acceptance of mutational differences probably explains why programs for construction of phylogenetic trees have difficulties assigning the fish enzyme, activity-wise of class I, in the tree of the class I vertebrate enzymes.

Of the recently characterized classes, class IV appears to be of particular interest. It has a special distribution in epithelial tissues, including in particular the stomach (Moreno and Parés, 1991; Farrés et al., 1994a,b; Satre et al., 1994), thus reflecting special functional needs. It has the highest ethanol activity of all the classes, an extensive difference in activity between mammals, and a retinol dehydrogenase activity of interest in retinoic acid formation and thus in regulation of differentiation (Parés et al., 1994; Farrés et al., 1994b; Duester, 1994a). It can further oxidize products of lipid peroxidation, and in this context its presence in epithelial tissues may be significant. These aspects are specially treated in other chapters of this volume (Farrés et al., 1994b; Duester, 1994b).

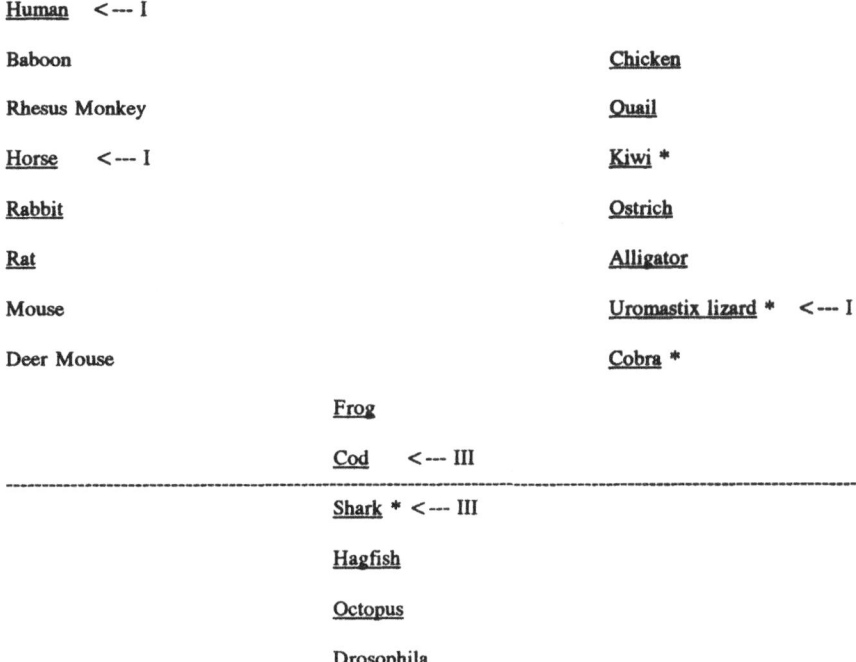

Figure 2. Animal sources of alcohol dehydrogenases characterized, showing mammals (top left), birds and reptiles (top right) and lower vertebrates or invertebrates (bottom center). Dotted line indicates the border above which both class I and class III have been found, but below which routine purificatons thus far have yielded only class III enzymes. Hence, the duplicatory origin of class I from class III as apparent from routine purifications coincides with the estimate from extrapolation of accumulation of mutational differences, both suggesting a time at roughly the fish divergence. Arrows marked I or III indicate those forms where isozymes have thus far been characterized in class I or III, respectively, showing the isozyme duplications to be multiple, taxon-specific, and apparently more frequent in class I than class III. The positions of the isozyme duplications in the phylogenetic tree of class I are further detailed in Fig. 3. Underlining indicate those species for which alcohol dehydrogenase structures have been reported from our laboratory (references in Persson et al., 1993b; Danielsson et al., 1994a,b), while asterisks indicate structures still not reported, and remaining structures were either available at the previous meeting (then referenced in Jörnvall et al., 1993a) or have been reported by others since then (deer mouse, Zheng et al., 1993).

Class IV appears to have a duplicatory origin from class I, much like class I appears to have so from class III, but the class IV/I duplication is of later date than the class I/III duplication. Still, it is early and apparently before the mammalian radiation, suggesting class IV to be present in all mammals, compatible with its possible role in differentiation inmammals (Duester, 1994a,b). Finally, class IV exhibits an evolutionary pattern of inter-mediate variability (Farrés et al., 1994a), in between that typical of the "variable" class I ethanol dehydrogenases and the "constant" class III formaldehyde dehydrogenases (Yin et al., 1991).

Isozyme Formation

Isozymes have been characterized thus far in three different lines of the class I enzyme, i.e. in the human line (cf Jörnvall et al., 1993a), the horse line (cf Brändén et al., 1975), and the lizard *Uromastix* line (Hjelmqvist et al., 1994). In all cases, the corresponding

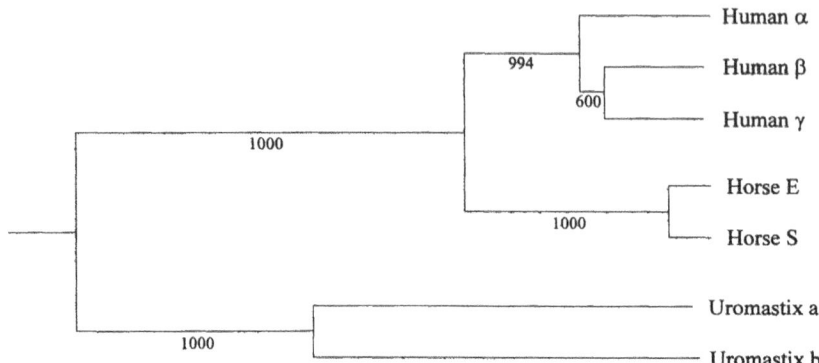

Figure 3. Phylogenetic tree of characterized forms of class I alcohol dehydrogenases, showing the human, horse and *Uromastix* lizard lines to represent separate and taxon-specific duplications. Numbers indicate confidence by giving the results from analysis of 1000 bootstrap replicates (Felsenstein, 1985).

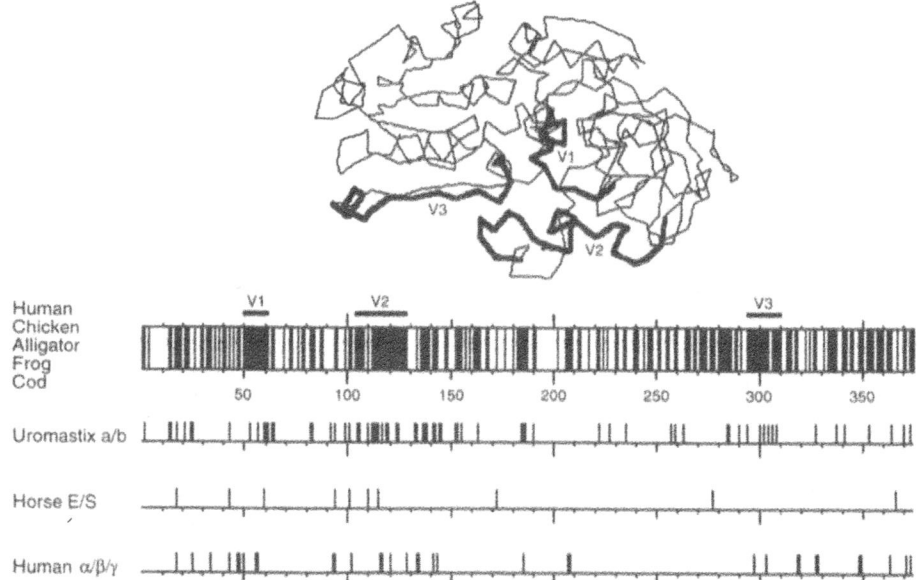

Figure 4. Positions of class and isozyme variabilities in the MDR alcohol dehydrogenases. Top panel indicates three segments (V_1, V_2, V_3, bold lines, in the conformation of the class I enzyme) that are discerned as variable between the classes of the enzyme (Persson et al., 1993b; Danielsson et al., 1994a) and which also appear to differ in the class hybrid analyzed crystallographically (El-Ahmad et al., 1994). Middle bar panel indicates the V_{1-3} regions in the linear sequence of the protein, at the same time as it shows the positions of species variability among the five differnt class I forms. The bottom three panels similarly indicates the positions of isozyme variabilities in the three species known to have class I isozymes. As shown, the three class-distinguishing regions (V_{1-3}) closely coincide with the species variability within class I (middle) and largely so also with the isozyme variabilities (bottom), hence indicating three regions of general divergence in the MDR dehydrogenases.

gene duplications represent separate events, and thus separate origins for the isozymes in a taxon-specific manner. The isozyme occurrence is shown in Fig 2, and their separate origins in Fig. 3.

In contrast, for the class III enzyme, isozymes have thus far been characterized only in one case, the class III enzyme from cod (Danielsson and Jörnvall, 1992; and Danielsson et al., unpublished), where the two forms differ widely, corresponding to a distant origin. Although further isozymes and duplications may well exist for both class I and III forms, the available data suggest that isozyme formation is more common in class I than in class III, at least in the more recent vertebrate classes (Fig. 2). In spite of the fact that each of the isozyme duplications traced represent separate events, several of the positions affected are identical (Fig. 4), suggesting, like for the class divergence, that certain positions are more prone to accept mutations than others. This difference among segments appears to correlate fairly well with the segment division of molecular building units that has been defined from studies on the variability in structure between the enzymes from different species (Danielsson et al., 1994a). Hence, isozyme and class formations appear to follow similar rules, since segments affected are at least partly similar (cf Fig. 4).

Distinguishing Properties

The crucial class distinction appears to be enzymatic activity, not structure. Thus, single residue replacements at the active site have been interpreted from computer graphics modelling to explain the separate substrate specificities of the enzyme classes (Eklund et al, 1990), and some of these conclusions have been directly confirmed by experimental proof from site-directed mutagenesis (Höög et al., 1992; Engeland et al., 1993; Estonius et al., 1994) as dependent on just a few replacements. Similarly, single active site exchanges between species variants of class IV explain large catalytic differences between these human and rat enzymes (Farrés et al., 1994a). In contrast to these large changes from just limited replacements, the overall class assignments of structures change much more slowly, explaining why the fish class I enzyme, in spite of its activity, still has an overall residue relationship to class III, reflecting its origin, as early noted (Danielsson et al., 1992).

In conclusion, activity measurements are essential for correct understanding of all the enzymes of different classes. Structural relationships reflect origins, but not necessarily functions, and the restrictions on the variability work on the functional activity, not on remaining, structural residues per se. Hence, enzymatic measurements are essential for correct class assignments and functional conclusions of these enzymes. This general fact does not prevent overall structural relationships to give a hint to the assignments, classes typically having residue identities at the 60-70% level, while mammalian species variabilities usually exhibit higher conservation.

Conformation-wise, the importance of Gly residues and their positions at reverse turns have been previously reported for the alcohol dehydrogenase families (Jörnvall et al., 1978; 1993a; Persson et al., 1993a). When the number of known enzymes and variants within each family now is large, this fundamental rule is even more pronounced: when all members of the MDR family are compared, 3 Gly residues only are found to constitute the strictly conserved residues (Table 2). Hence, they constitute the limits that actually define the family, reflecting the true importance of the conservation of turns and conformation at large. In contrast, catalysis and functional properties can change and have done so repeatedly, reflected already by the fact that some of the enzymes are metalloenzymes (the alcohol dehydrogenases, with catalytic zinc), whereas other relatives (the reductases) are not. Significantly, they share no catalytic residues, and no polar residues are strictly conserved, whereas the structurally important Gly residues are the only residues conserved when the presently wide spread of all the enzymes are scrutinized (Table 2). A similar tendency at Gly

Table 2. Conserved residues in the enzymes of the MDR and SDR families

Residue	MDR			SDR		
	4DH	4DH/Red	All	4DH	Most DH/Red	All
Ala	2	2	-	3	-	-
Cys	1	-	-	-	-	-
Asp	3	-	-	3	-	-
Glu	2	-	-	1	-	-
Gly	15	9	3	10	4	-
His	2	1	-	-	-	-
Ile	1	-	-	1	-	-
Lys	3	-	-	1	1	-
Leu	-	2	-	5	-	-
Asn	-	1	-	4	-	-
Pro	1	1	-	1	1	-
Arg	-	-	-	1	-	-
Ser	1	-	-	2	1	-
Thr	1	1	-	2	-	-
Val	4	1	-	3	-	-
Tyr	-	-	-	1	1	1
Sum	36	18	3	38	8	1

The columns 4DH refer to the residues conserved when four different dehydrogenase lines are compared within each family, as reported (Persson et al., 1993a). Columns 4DH/Red and All for the MDR family refer to the conservations observed when four major lines of dehydrogenases/reductases are included in the comparison (Persson et al., 1994b), and when all known MDR proteins are included, respectively. For the SDR family, columns Most DH/Red and All similarly refer to the values obtained when a conserved residue is found in 90 % and 100% (Persson et al., 1994a), respectively, of all SDR proteins characterized.

conservation is observed also in the SDR family (Table 2), but in this case, the constant functional pattern with a catalytic Tyr residue is more conserved, and when all the 50-odd different short-chain enzymes are included in the comparisons, only that Tyr residue is conserved (Table 2), while exclusion of just a few enzymes (to include 90% of all the enzymes) highlights the conservation of Gly also in the SDR family (Table 2, penultimate column).

Three-Dimensional Structures

Progress has been great also in establishment of novel three-dimensional structures for the MDR enzymes. For a long time, only the horse class I enzyme tertiary structure was available (Eklund et al., 1976) and used repeatedly for modelling and comparisons. In the early 90's a second structure was established in the form of the corresponding human enzyme (Hurley et al., 1991), still a class I form, though. Recently, however, class III forms have also been crystallized, as well as the hybrid class I enzyme of cod liver withoverall class III residue relationships (Ramaswamy et al., 1994). Data are presently under consideration although conformational differences between the classes appear to be localized, giving extensive but limited differences in three loops and a domain movement (El-Ahmad et al., 1994), but keeping the overall conformational relationships.

Interestingly, this pattern is the one predicted from evolutionary conclusions based on analyses of species variants (Danielsson et al., 1994a). Thus, inter-class differences involve three segments (Persson et al., 1993b), and they are where class I in general differs from class III in general. These three segments are a part of the entrance to the active site, a part of the major subunit interacting surface, and the loop around the second zinc atom, as shown in Fig. 4. When crystallographic data now start to become available, exactly these loop segments are those found to differ between the traditional class I enzymes and the class hybrid form in fish (El-Ahmad et al., 1994). Similarly, another class distinction deduced from species variabilities is a difference in the domain-connecting segment of class III (Danielsson et al., 1994a), suggesting that classes may differ in domain movements upon catalysis. Again, from the preliminary crystallographic data available, such movement differences indeed appear to exist between the traditional class I enzymes and the hybrid cod enzyme (El-Ahmad et al., 1994).

Overall, these agreements between conclusions obtained from observations of species variants, and those from crystallography of novel classes, highlight the power of both techniques, and the value of analyses of natural variants from a wide range of sources, as those indicated in Fig 2. Notably, this spread of species analyzed forms the basis for the construction of the variability regions in Fig 4, now supported by crystallographic data.

CONCLUSION

Knowledge about the structure, function and evolution of the MDR family has increased rapidly. A series of gene duplications gave rise early to many enzymes (minimally seven activities presently distinguished), intermediately to many classes of some of these enzymes (presently six classes of the mammalian alcohol dehydrogenases appear discernible), and recently to isozymes within some of these classes (presently taxon-specific isozymes in three class I lines and two class III lines; cf Fig. 2). Investigations of all major animal lines have defined molecular building units and segments of variability (Fig. 4), and they appear to correlate with crystal structures starting to become available. Finally, the wide analysis of all major animal lines have established enzymogenesis of new activities, tracing of gene duplications, and help to make alcohol dehydrogenases of the MDR type a well-studied model of protein evolution in general. Few other proteins have been analyzed from so many divergent sources (Fig. 2), illustrating gradual changes with divergence as well as convergence.

THE SDR FAMILY

Additional Enzymes: Presently 50-Odd Members Within the Family

Progress has been similarly rapid in the SDR family. Formally, the number of enzymes characterized within SDR is even larger than that within MDR, and presently at least 57 different enzymes of the SDR type have been characterized (Persson et al., 1994a). Like for the MDR family, they represent both dehydrogenase and reductase sub-lines, but in addition also completely different enzymes of other classes, including transferases, hydrolases and aldolases. In effect, the SDR family is remarkable in exhibiting a large spread of forms with a correspondingly extensive divergence in function.

In spite of this wide range of enzymes and extensive divergence, the family also exhibits the usual pattern of some strict conservation. Disregarding the extremes by including only 90 % of all the enzymes known, the pattern of Gly conservation is observed (Table

2), and considering all enzymes, a catalytic Tyr residue appears to be the most crucial characteristic of the SDR family, much as a catalytic Ser residue is a hallmark of serine proteases.

Chemical Modifications

Many aspects of this divergent family of SDR enzymes can be noticed, and several are treated in a special chapter in this volume (Persson et al., 1994a). Protein-wise, a recent finding of general interest is the fact that a new type of amino acid has been found in mammalian SDR enzymes, carboxyethyllysine. This residue was first defined in carbonyl reductase (identical to NADP-dependent 15-hydroxyprostaglandin dehydrogenase) and there found to represent a specific modification of one Lys residue, Lys-238 (Krook et al., 1993), concluded to be derived from autocatalytic reductase activity on a Schiff base formed betwen pyruvate and that Lys residue side-chain. Later, this pattern was confirmed by further definitions of additionally substituted forms of this enzyme, where other 2-oxo-metabolites, like α–ketoglutarate and oxaloacetate, were also found to form adducts in this enzyme (Wermuth et al. , 1993). Combined, all these forms explain the multiplicity known since long for the enzyme.

Three-dimensional structures

Also for the SDR enzymes, novel three-dimensional structures have recently become available. Three such enzymes are now known in tertiary structure, a steroid dehydrogenase (Ghosh et al, 1991), a pteridine reductase (Varughese et al., 1992), and a sugar epimerase (Bauer et al., 1992). These structures are related, but distantly so, and involve frequent insertions/deletions of superficial loop segments (Holm et al., 1994) as summarized in the SDR chapter of this volume (Persson et al., 1994a).

Interestingly, mapping of conserved residues in the natural variants of these enzymes into the conformational properties highlight the importance of the segment around the active site pocket (cf Persson et al., 1994a). Like for the MDR family, overall conservation is extensive, but single replacements appear to have critical influences on details of enzymatic activities, and natural variants help to define the residues of interest that can then be further studied by site directed mutagenesis.

OVERALL CONCLUSIONS AND FUTURE ASPECTS

Alcohol dehydrogenases and related enzymes now constitute a true enzyme system, affecting parts of several protein families. Two of these, the MDR and SDR families, are well-known and have experienced a recent rapid progress regarding characterized enzyme multiplicities, natural variants, and tertiary structures (Fig. 5). Combined, the data give new insight into mechanisms of catalysis, and even define the SDR enzymes as Tyr-enzymes, where a single Tyr residue appears to be the only feature absolutely conserved (Table 2). In contrast, the MDR enzymes vary at the active site, including dependence or not on metals, and therefore in metalloenzyme properties. Overall both families illustrate patterns of protein evolution in general, at the same time as they show the value of extensive studies of species and enzyme variants in nature.

It will be of interest now to complement this new information by solid data on further tertiary structures, on enzymatic properties, and on effects of site-directed mutants. It may also be anticipated that further analysis of additional native enzymes may show both the

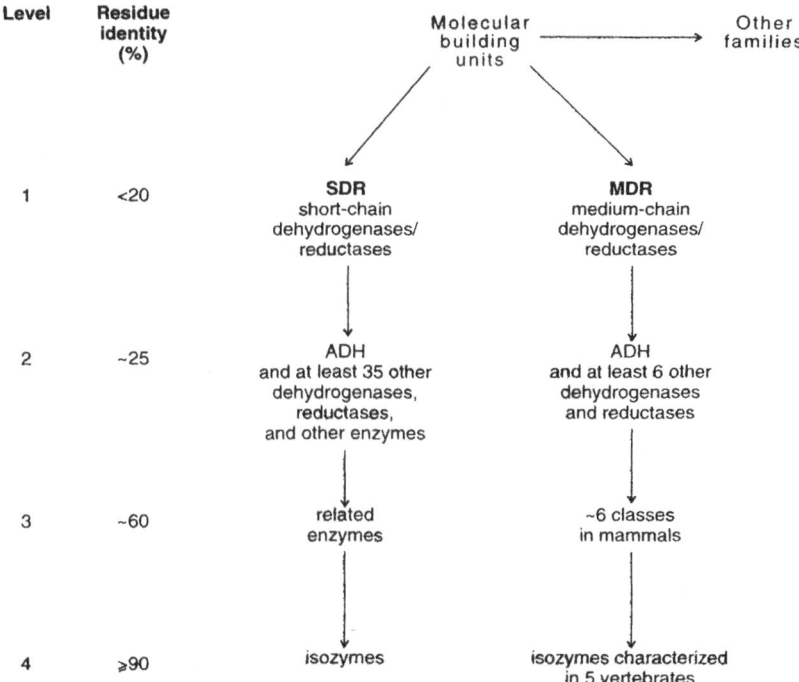

Figure 5. Schematic summary of relationships within the SDR and MDR families. In both lines, at least four levels of duplications, ranging from very distant to more recent, have produced different families, different enzymes, different classes, and different isozymes.

SDR and MDR families to be still more extensive, perhaps making the SDR family be one of the largest protein families known, including enzymes of all major classes of activity.

ACKNOWLEDGMENTS

Studies quoted from this laboratory were supported by grants from the Swedish MRC (project 13X-3532), the Swedish Alcohol Research Fund, the Endowment for Research in Human Biology (Boston, US), and Peptech Ltd (Sydney, Australia).

REFERENCES

Bauer, A.J., Rayment, I., Frey, P.A. and Holden, H.M., 1992, The molecular structure of UDP-galactose 4-epimerase from *Escherichia coli* determined at 2.5 Å resolution. *Proteins* 12:372-381.

Bonnichsen, R.K. and Wassén, A.M., 1948, Crystalline alcohol dehydrogenase from horse liver. *Arch.. Biochem. Biophys.* 18:361-363.

Borrás, T., Persson, B. and Jörnvall, H., 1989, Eye lens ζ-crystallin relationships to the family of "long-chain" alcohol/polyol dehydrogenases. Protein trimming and conservation of stable parts. *Biochemistry* 28:6133-6139.

Brändén, C.-I., Jörnvall, H., Eklund, H. and Furugren, B., 1975, Alcohol dehydrogenases. In *The Enzymes*, 3rd ed., Vol. 11, Academic Press, pp. 103-190.

Cederlund, E., Peralba, J.M., Parés, X. and Jörnvall, H., 1991, Amphibian alcohol dehydrogenase, the major frog liver enzyme. Relationships to other forms and assessment of an early gene duplication separating vertebrate class I and class III alcohol dehydrogenases. *Biochemistry* 30:2811-2816.

Danielsson, O. and Jörnvall, H., 1992, "Enzymogenesis": Classical liver alcohol dehydrogenase origin from the glutathione-dependent formaldehyde dehydrogenase line. *Proc. Natl. Acad. Sci. USA* 89:9247-9251.

Danielsson, O., Eklund, H. and Jörnvall, H., 1992, The major piscine liver alcohol dehydrogenase hasclass-mixed properties in relation to mammalian alcohol dehydrogenases of classes I and III. *Biochemistry* 31:3751-3759.

Danielsson, O., Atrian, S., Luque, T., Hjelmqvist, L., Gonzàlez-Duarte, R. and Jörnvall, H., 1994a, Fundamental molecular differences between alcohol dehydrogenase classes. *Proc. Natl. Acad. Sci. USA* 91:4980-4984.

Danielsson, O., Shafqat, J., Estonius, M. and Jörnvall, H., 1994b, Alcohol dehydrogenase, class III contrasted to class I: characterization of the Cyclostome enzyme, multiplicity like for the human form, and distant cross-species hybridization. *Eur. J. Biochem.*, in press.

Duester, G., 1994a, Retinoids and the alcohol dehydrogenase gene family. In *Toward a Molecular Basis of Alcohol Use and Abuse* (Jansson, B., Jörnvall, H., Rydberg, U., Terenius, L., and Vallee, B.L., eds.) Birkhäuser Verlag, Basel, pp. 279-290.

Duester, G., 1994b, Class I and Class IV alcohol dehydrogenase (retinol dehydrogenase) gene expression in mouse embryos. This vol., in press

Eklund, H., Nordström, B., Zeppezauer, E., Söderlund, G., Ohlsson, I., Boiwe, T., Söderberg, B.-O., Tapia, O., Brändén, C.-I. and Åkeson, Å., 1976, Three-dimensional structure of horse liver alcohol dehydrogenase at 2.4 Å resolution. *J. Mol. Biol.* 102:27-59.

Eklund, H., Müller-Wille, P., Horjales, E., Futer, O., Holmquist, B., Vallee, B.L., Höög, J.-O., Kaiser, R. and Jörnvall, H., 1990, Comparison of three classes of human liver alcohol dehydrogenase. Emphasis on different substrate-binding pockets. *Eur. J. Biochem.* 193:303-310.

El-Ahmad, M., Ramaswamy, S., Danielsson, O., Karlsson, C., Estonius, M., Höög, J.-O., Eklund, H. and Jörnvall, H., 1994, Crystallizations of novel forms of alcohol dehydrogenase. This vol., in press.

Engeland, K., Höög, J.-O., Holmquist, B., Estonius, M., Jörnvall, H. and Vallee, B.L., 1993, Mutation of Arg-115 of human class III alcohol dehydrogenase: a binding site required for formaldehyde dehydrogenase activity and fatty acid activation. *Proc. Natl. Acad. Sci. USA* 90:2491-2494.

Estonius, M., Höög, J.-O., Danielsson, O. and Jörnvall, H., 1994, Residues specific for class III alcohol dehydrogenase. *Biochemistry*, submitted.

Farrés, J., Moreno, A., Crosas, B., Peralba, J.M., Allali-Hassani, A., Hjelmqvist, L., Jörnvall, H. andParés, X., 1994a, Human stomach class IV alcohol dehydrogenase: cDNA sequence and primary structure correlations with enzymatic properties. *Eur. J. Biochem.* in press.

Farrés, J., Moreno, A., Crosas, B., Cederlund, E., Allali-Hassani, A., Peralba, J.M., Hjelmqvist, L., Jörnvall, H. and Parés, X., 1994b, Human and rat class IV alcohol dehydrogenases. Correlations of primary structures with enzymatic properties. This vol., in press.

Felsenstein, J., 1985, Confidence limits on phylogenies: an approach using the bootstrap. *Evolution* 39:783-791.

Ghosh, D., Weeks, C.M., Grochulski, P., Duax, W.L., Erman, M., Rimsay, R.L. and Orr, J.C., 1991, Three-dimensional structure of holo 3α,20β-hydroxysteroid dehydrogenase: A member of a short-chain dehydrogenase family. *Proc. Natl. Acad. Sci. USA.* 88:10064-10068.

Hjelmqvist, L., Shafqat, J., Siddiqi, A.R. and Jörnvall, H., 1994, The vertebrate alcohol dehydrogenasesystem. Gene duplications traced by isozyme heterogeneity of the reptilian class I enzyme from *Uromastix hardwickii*. *Eur. J. Biochem.*, submitted.

Holm, L., Sander, C. and Murzin, A., 1994, Three sisters, different names. *Nature Structural Biology* 1:146-147.

Höög, J.-O., Eklund, H. and Jörnvall, H., 1992, A single-residue exchange gives human recombinant ββ alcohol dehydrogenase γγ isozyme properties. *Eur. J. Biochem.* 213:31-38

Hurley, T.D., Bosron, W.F., Hamilton, J.A. and Amzel, L.M., 1991, Structure of $\beta_1\beta_1$ alcoholdehydrogenase: catalytic effects of non-active site substitutions. *Proc. Natl. Acad. Sci. USA* 88, 8149-8153.

Inoue, T., Sunagawa, M., Mori, A., Imai, C., Fukuda, M., Takagi, M. and Yano., 1989, Cloning and sequencing of the gene encoding the 72-kilodalton dehydrogenase subunit of alcohol dehydrogenase from *Acetobacter aceti*. *J. Bacteriol.* 171:3115-3122.

Jörnvall, H. and Höög, J.-O., 1994, Nomenclature of alcohol dehydrogenases. *Alcohol and Alcoholism*, in press.

Jörnvall, H., Eklund, H. and Brändén, C.-I., 1978, Subunit conformation of yeast alcohol dehydrogenase. *J. Biol. Chem.* 253:8414-8419.

Jörnvall, H., Persson, M. and Jeffery, J., 1981, Alcohol and polyol dehydrogenases are both divided into two protein types, and structural properties cross-relate the different enzyme activities within each type. *Proc. Natl. Acad. Sci. USA* 78:4226-4230.

Jörnvall, H., Danielsson, O., Eklund, H., Hjelmqvist, L., Höög, J.-O., Parés, X. and Shafqat, J., 1993a, Enzyme and isozyme developments within the medium-chain alcohol dehydrogenase family. In *Enzymology and Molecular Biology of Carbonyl Metabolism 4* (Weiner, H., ed.), Plenum, New York, pp. 533-544.

Jörnvall, H., Persson, B., Du Bois, G.C., Lavers, G.C., Chen, J.H., Gonzalez, P., Rao, P.V. and Zigler, Jr., J.S., 1993b, ζ-crystallin versus other members of the alcohol dehydrogenase super-family. Variability as a functional characteristic. *FEBS Lett.* 322:240-244.

Krook, M., Marekov, L. and Jörnvall, H., 1990, Purification and structural characterization of placental NAD⁺-linked 15-hydroxyprostaglandin dehydrogenase. The primary structure reveals the enzyme to belong to the short-chain alcohol dehydrogenase family. *Biochemistry* 29:738-743.

Krook, M., Ghosh, D., Strömberg, R., Carlquist, M. and Jörnvall, H., 1993, Carboxyethyllysine in a protein: Native carbonyl reductase/NADP⁺-dependent prostaglandin dehydrogenase. *Proc. Natl. Acad. Sci. USA* 90:502-506.

Luque, T., Atrian, S., Danielsson, O., Jörnvall, H. and Gonzàlez-Duarte, R., 1994, Structure of the *Drosophila melanogaster* glutathione-dependent formaldehyde dehydrogenase/octanol dehydrogenase gene (class III alcohol dehydrogenase). Evolutionary pathway of the alcohol dehydrogenase genes. *Eur. J. Biochem.*, in press.

Marekov, L., Krook, M. and Jörnvall, H., 1990, Prokaryotic 20β-hydroxysteroid dehydrogenase is an enzyme of the "short-chain, non-metalloenzyme" alcohol dehydrogenase type. *FEBS Lett.* 266:51-54.

McKinley-McKee, J.S., Winberg, J.-O. and Pettersson, G., 1991, Mechanism of action of *Drosophila melanogaster* alcohol dehydrogenase. *Biochem. Internat.* 25:879-885.

Moreno, A. and Parés, X., 1991, Purification and characterization of a new alcohol dehydrogenase from human stomach. *J. Biol. Chem.* 266:1128-1133.

Parés, X., Cederlund, E., Moreno, A., Hjelmqvist, L., Farrés, J. and Jörnvall, H., 1994, Mammalian class IV alcohol dehydrogenase (stomach ADH): structure, origin and correlation with enzymology. *Proc. Natl. Acad. Sci. USA* 91:1893-1897.

Peltoketo, H., Isomaa, V., Mäentausta, O. and Vihko, R., 1988, Complete amino acid sequence of human placental 17β-hydroxysteroid dehydrogenase deduced from cDNA. *FEBS Lett.* 239:73-77.

Persson, B., Krook, M. and Jörnvall, H., 1991, Characteristics of short-chain alcohol dehydrogenses and related enzymes. *Eur. J. Biochem.* 200:537-543.

Persson, B., Hallborn, J., Walfridsson, M., Hahn-Hägerdal, B., Keränen, S., Penttilä, M. and Jörnvall, H., 1993a, Dual relationships of xylitol and alcohol dehydrogenases in families of two protein types. *FEBS Lett.* 324:9-14.

Persson, B., Bergman, T., Keung, W.M., Waldenström, U., Holmquist, B., Vallee, B.L. and Jörnvall, H., 1993b, Basic features of class-I alcohol dehydrogenase: variable and constant segments coordinated by inter-class and intra-class variability. Conclusions from characterization of the alligator enzyme. *Eur. J. Biochem.* 216:49-56.

Persson, B., Krook, M. and Jörnvall, H., 1994a, Short-chain dehydrogenases/reductases. This vol., in press.

Persson, B., Zigler, Jr., J.S. and Jörnvall, H., 1994b, A super-family of medium-chain dehydrogenases/-reductases (MDR): sub-lines including ζ-crystallin, alcohol and polyol dehydrogenases, quinone oxidoreductases, enoyl reductases, VAT-1 and further proteins. *Eur. J. Biochem.*, in press.

Ramaswamy, S., El-Ahmad, M., Danielsson, O., Jörnvall, H. and Eklund, H., 1994, Crystallization and crystallographic investigations of cod alcohol dehydrogenase class I and class III enzymes. *FEBS Lett.* 350:122-124.

Rao, P.V. and Zigler, J.S., Jr, 1991, ζ-crystallin from guinea pig lens is capable of functioning catalytically as an oxidoreductase. *Arch. Biochem. Biophys.* 284:181-185.

Rao, P.V., Krishna, C.M. and Zigler, J.S., Jr, 1992, Identification and characterization of the enzymatic activity of zeta-crystallin from guinea-pig lens. A novel NADPH:quinone oxidoreductase. *J. Biol. Chem.* 267:96-102.

Satre, M.A., îgombi...-Knight, M. and Duester, G., 1994, The complete structure of human class IV alcohol dehydrogenase (retinol dehydrogenase) determined from the *ADH7* gene. *J. Biol. Chem.* 269:15606-15612.

Scopes, R.K., 1983, An iron-activated alcohol dehydrogenase. *FEBS Lett.* 156:303-306.

Vallee, B.L. and Bazzone, T.J., 1983, Isozymes of human liver alcohol dehydrogenases. *Isozymes: Curr. Top. Biol. Med. Res.* 8:219-244.

van Ophem, P.W., Van Beeumen, J and Duine, J.A., 1992, NAD-linked, factor-dependent formaldehy dedehy-drogenase or trimeric, zinc-containing, long-chain alcohol dehydrogenase from *Amycolatopsis methanolica. Eur. J. Biochem.*, 206:511-518.

Varughese, K.I., Skinner, M.M., Whiteley, J.M., Matthews, D.A. and Xuong, N.H., 1992, Crystal structure of rat liver dihydropteridine reductase. *Proc. Natl. Acad. Sci. USA.* 89:6080-6084.

Wermuth, B., Bohren, K.M. and Ernst, E., 1993, Autocatalytic modification of human carbonyl reductase by 2-oxocarboxylic acids. *FEBS Lett.* 335, 151-154.

Williamson, V.M. and Paquin, C.E., 1987, Homology of *Saccharmomyces cerevisiae* ADH4 to an iron-activated alcohol dehydrogenase from *Zymomonas mobilis. Mol. Gen. Genet.* 209:374-381.

Yin, S.-J., Vagelopoulos, N., Wang, S.-L. and Jörnvall, H., 1991, Structural features of stomach aldehyde dehydrogenase distinguish dimeric aldehyde dehydrogenase as a "variable" enzyme. "Variable" and "constant" enzymes within the alcohol and aldehyde dehydrogenase families. *FEBS Lett.* 283:85-88.

Zheng, Y.-W., Bey, M., Liu, H. and Felder, M.R., 1993, Molecular basis of the alcohol dehydrogenase-negative deer mouse. *J. Biol. Chem.* 268:24933-24939.

35

PROMOTERS OF THE MAMMALIAN CLASS III ALCOHOL DEHYDROGENASE GENES

Howard J. Edenberg, Wei-Hsien Ho and Man-Wook Hur

Department of Biochemistry and Molecular Biology
Indiana University School of Medicine
635 Barnhill Drive, MS418
Indianapolis, IN 46202-5122

INTRODUCTION

Class III ADHs are not saturable with ethanol as substrate; long chain alcohols and ω–hydroxyfatty acids are better substrates (Pares and Vallee, 1981; Wagner et al., 1984; Kaiser et al., 1988; Giri et al., 1989b). These ADHs also function as glutathione-dependent formaldehyde dehydrogenases (FDH), as shown by Koivusalo et. al. (1989). Class III ADH, in contrast to the other ADH classes, is expressed in virtually all tissues (Adinolfi et al., 1984; Seeley et al., 1984; Duley et al., 1985; Holmes et al., 1986b; Holmes et al., 1986a; Julia et al., 1987; Edenberg, 1991). The amount of expression is higher in liver and kidney than in other tissues (Adinolfi et al., 1984; Seeley et al., 1984; Duley et al., 1985; Holmes et al., 1986b; Holmes et al., 1986a; Julia et al., 1987; Hur et al., 1992). The class III alcohol dehydrogenases (ADH) have been highly conserved during evolution, suggesting important roles in cellular metabolism (Sharma et al., 1989; Giri et al., 1989a; Edenberg et al., 1991; Hur et al., 1992; Gutheil et al., 1992).

The human and mouse class III ADH cDNAs have been cloned (Sharma et al., 1989; Giri et al., 1989a; Hur et al., 1992). The class III ADH cDNAs have diverged less from their common ancestral gene than have the class I ADH cDNAs (Hur et al., 1992). A high degree of sequence conservation generally reflects strong functional constraints upon the protein. We have demonstrated that silent (synonymous) substitutions in the coding region, that do not alter the amino acids encoded by the gene, are also less frequent than would be expected (Hur et al., 1992). This was not explained by differences in codon bias. Thus the entire nucleotide sequence of the class III cDNAs appeared to be evolving slower than that of the other ADHs.

The human χ–ADH is the prototype of class III; it is encoded by the *ADH5* gene, which we have cloned and analyzed (Hur and Edenberg, 1992). We have also cloned the closely-related mouse *Adh-2* gene (Ho, 1991, and in preparation). Both genes have 8 introns, located in the same places as in other *ADH* genes. To understand the expression of these genes, we are characterizing their 5' upstream regions.

Enzymology and Molecular Biology of Carbonyl Metabolism 5
Edited by H. Weiner *et al.*, Plenum Press, New York, 1995

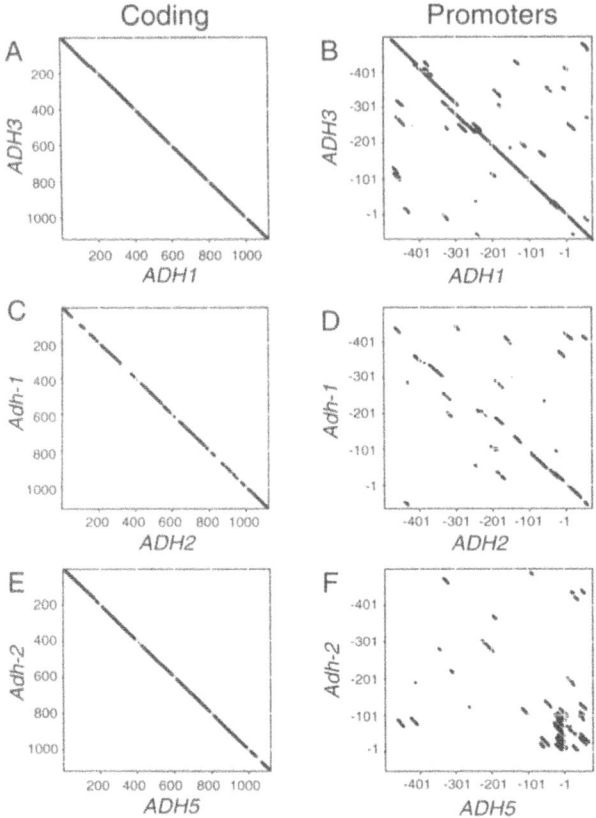

Figure 1. Alignments of human and mouse alcohol dehydrogenase genes. A. *ADH3 vs. ADH1* coding sequences. B. *ADH3 vs. ADH1* promoters. C. *Adh-1 vs. ADH2* coding sequences. D. *Adh-1 vs. ADH2* promoters. E. *Adh-2 vs. ADH5* coding sequences. F. *Adh-2 vs. ADH5* promoters. Dot plots of Pustell Matrix comparisons done using the MacVector sequence analysis program (IBI/Kodak). For all coding comparisons (A, C, E) the regions from the initial ATG to the last codon were compared with the stringency set high: 80% identity in a window of 30 bp, with a match of at least 6 consecutive bp within the window. For all promoter comparisons (B, D, F) the regions between -500 bp and the translation start site were compared as above, but at lower stringency: 55% identity in a window of 30 bp, with a match of at least 4 consecutive bp within the window. Promoter comparisons at 50% identity look similar along the diagonal, but show more very short patches of homology scattered off the diagonal. Sequences were from Ikuta et al. (1986), Edenberg et al. (1985), Stewart et al. (1990a), Carr et al. (1989a), Ho (1991), Hur and Edenberg (1992), and unpublished data.

LOW SEQUENCE CONSERVATION IN THE 5' REGION

Although the coding sequences of the class III *ADH* genes have been highly conserved, the promoter regions have diverged widely. This is true despite the overall similarity in base composition of the promoters, discussed below. The difference between high sequence conservation in the coding region and low conservation in the promoters can be illustrated by comparing the alignments of three pairs of genes and cDNAs (Figure 1).

The three human class I genes are very closely related. *ADH1* and *ADH3* are compared as examples: their coding regions are approximately 94% identical for both amino acid and nucleotide sequences. Fig. 1A shows the alignment of coding sequences, which is excellent. Fig. 1B shows the alignment of the promoter region extending from -500 bp to

+71 bp (the nucleotide just before the ATG that initiates translation). [The stringency of matching was reduced for all promoter comparisons.] There is obviously an excellent overall alignment of these closely-related promoters.

Adh-1 is the mouse homolog of the human class I genes; their coding regions are approximately 83% identical in amino acid sequence (Edenberg, 1991) and 82% identical in nucleotide sequence to *ADH2*. The alignment of coding sequences is good (Fig. 1C), but not as good as that between *ADH1* and *ADH3*. The alignment of promoters is good up to about bp -100 and then falls off, although there are still patches of homology along the diagonal further upstream (Fig. 1D). The first 100 bp upstream of the transcription start site contain several cis-acting sequences known to be important in transcription of these genes (Carr et al., 1989b; Carr and Edenberg, 1990; Stewart et al., 1990b; Brown et al., 1992; Lin et al., 1993; Edenberg and Brown, 1992; van Ooij et al., 1992; Brown et al., 1994).

Adh-2 is the mouse homolog of the human class III gene, *ADH5*. These genes diverged at the same time as the rodent and primate class I genes, from the last common ancestor of mouse and human, but have evolved slower. The similarity between mouse and human class III genes (Hur et al., 1992) in amino acid sequence (93% identity) is very high,

3' Nontranslated

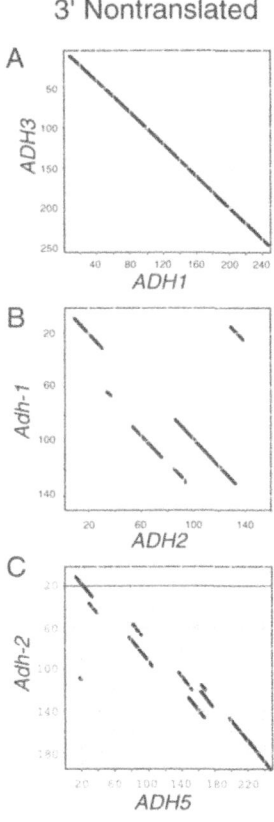

Figure 2. Alignments of human and mouse alcohol dehydrogenase genes in the 3' nontranslated regions. A. *ADH3 vs. ADH1*. B. *Adh-1 vs. ADH2*. C. *Adh-2 vs. ADH5*. The sequences beyond the translation termination site were compared as above. For A, 65% identity in a window of 30 bp, with a match of at least 4 consecutive bp within the window was used. For B and C, the stringency was reduced to 50%. Sequences were from Edenberg et al. (1985), Ikuta et al. (1986), Carr et al. (1989a), Edenberg et al. (1989), Ho (1991) and Hur et al. (1992).

comparable to that among the very recently diverged human class I genes, despite the longer period since their last common ancestor. Their similarity in nucleotide sequence (88% identity) is midway between the 93-94% identity among the human class I coding regions, and the 82% identity between the mouse and human class I coding regions. Fig. 1E shows the excellent alignment of *Adh-2* and *ADH5* coding regions (compare to 1A,C). In contrast, Fig. 1F shows the very poor alignment of the 5' regions of *Adh-2* and *ADH5*: there are only short patches of homology, mainly in a very G+C-rich region in which matches are expected based simply on the base composition.

The 3' nontranslated regions of the genes are compared in Figure 2. There is excellent conservation between the human class I genes *ADH1* and *ADH3* (Fig. 2A), but weaker alignments between human and mouse class I and class III genes (Fig. 2B,C). The 3' nontranslated regions of the class III genes are more similar to each other than their promoters, but much less similar than their coding regions.

These comparisons present something of a paradox: the class III genes have coding regions (including synonymous positions) that have been better conserved than those of the class I genes, attached to promoters that have been less well conserved. This suggests that there are much weaker constraints upon the promoter sequences of these genes than upon their coding sequences. Few changes in the amino acid sequence are compatible with maintenance of the physiologically important enzymatic functions of the class III ADHs. In contrast, the appropriate expression of these physiologically important genes can be maintained despite more changes in the promoter sequence. Thus preservation of a particular set of *cis*-acting promoter elements is not as important as preservation of the amino acid sequences. Rather, a general type of promoter region which can accommodate variations in the exact sequences and their placement may provide sufficient specificity for proper expression. This does not, of course, mean that alterations in the sequence of either of the promoters will not have effects, but rather that compensating changes have been selected to provide appropriate function.

The conservation of the 3' nontranslated regions is intermediate between that of the promoters and of the coding sequences. This indicates that the 3' non-translated regions play important roles in the expression of these genes.

G+C-RICH PROMOTERS

The promoters of both *ADH5* and *Adh-2*, in contrast to the promoters of all other cloned mammalian *ADH* genes, lack a TATA box (Hur and Edenberg, 1992; Ho, 1991; and unpublished data). The 5' regions of these class III *ADH* genes are very G+C rich. Although the dinucleotide sequence CpG is greatly underrepresented in mammalian DNA (relative to that predicted from the base composition or to the number of GpC sequences), these promoters contain as many CpG doublets as expected based upon the G+C content, approximately as many as GpC doublets (Hur and Edenberg, 1992; Ho, 1991). Thus both promoters can be classified as CpG islands (Gardiner-Garden and Frommer, 1987; Bird, 1986). The high G+C content and lack of a TATA box are characteristic of the promoters of housekeeping genes, genes expressed in all tissues. The class III ADH genes are expressed in all tissues, although to different levels, as noted above. Despite their overall similarity in base composition, the promoters of the *ADH5* and *Adh-2* genes are surprisingly different, as discussed above. They only align in short patches of 8-12 bp.

Both promoters contain sequences that resemble the consensus sequence motif recognized by the transcription factor Sp1, as would be expected for a G+C-rich promoter. There are also sequences resembling consensus motifs for several tissue-specific transcription factors, including AP-2, C/EBP, HNF-5, E-boxes, and heat shock elements (Hur and

Edenberg, 1992; and unpublished data). These latter elements may be responsible for the differences in the amount of expression in some tissues, particularly the higher expression in liver and kidney.

We are characterizing the promoter regions of *ADH5* by DNase I footprinting assays and transient transfection assays in several cell lines. We have previously shown that a fragment extending approximately 1.5 kb upstream of the transcriptional start site can function as a promoter in CV-1 cells (Hur and Edenberg, 1992). Preliminary data indicate the presence of multiple *cis*-acting elements, and that the promoter can function in other cell types.

ADH5 also has a feature uncommon in eukaryotic mRNAs, two ATGs in the 5' nontranslated region of the message (Hur and Edenberg, 1992). Neither of these is in a good context for initiation of translation (Kozak, 1991), however. They would encode overlapping peptides of 20 and 10 amino acids, and terminate several nucleotides before the ATG that initiates translation of χ–ADH. We have speculated that these could affect the regulation of χ–ADH expression (Hur and Edenberg, 1992).

ACKNOWLEDGMENTS

This research was supported by PHS grant AA06460 from the National Institute of Alcohol Abuse and Alcoholism.

REFERENCES

Adinolfi, A., Adinolfi, M. and Hopkinson, D.A.: Immunological and biochemical characterization of the human alcohol dehydrogenase χ-ADH isozyme. Ann.Hum.Genet. 48 (1984) 1-10.

Bird, A.P.: CpG-rich islands and the function of DNA methylation. Nature 321 (1986) 209-213.

Brown, C.J., Baltz, K.A. and Edenberg, H.J.: Expression of the human *ADH2* gene: an unusual Sp1-binding site in the promoter of a gene expressed at high levels in liver. Gene 121 (1992) 313-320.

Brown, C.J., Zhang, L. and Edenberg, H.J.: Tissue-specific differences in the expression of the human *ADH2* alcohol dehydrogenase gene and in binding of factors to *cis*-acting elements in its promoter. DNA Cell Biol. 13 (1994) 235-247.

Carr, L.G., Xu, Y., Ho, W.-H. and Edenberg, H.J.: Nucleotide Sequence of the *ADH2*3* Gene Encoding the Human Alcohol Dehydrogenase β3 Subunit. Alcohol.Clin.Exp.Res. 13 (1989a) 594-586.

Carr, L.G., Zhang, K. and Edenberg, H.J.: Protein-DNA interactions in the 5' region of the mouse alcohol dehydrogenase gene *Adh-1*. Gene 78 (1989b) 277-285.

Carr, L.G. and Edenberg, H.J.: Cis-acting sequences involved in protein binding and in vitro transcription of the human alcohol dehydrogenase gene *ADH2*. J.Biol.Chem. 265 (1990) 1658-1664.

Duley, J.A., Harris, O. and Holmes, R.S.: Analysis of human alcohol- and aldehyde-metabolizing isozymes by electrophoresis and isoelectric focusing. Alcoholism (NY) 9 (1985) 263-271.

Edenberg, H.J., Zhang, K., Fong, K., Bosron, W.F. and Li, T.-K.: Cloning and sequencing of cDNA encoding the complete mouse liver alcohol dehydrogenase. Proc.Natl.Acad.Sci.USA. 82 (1985) 2262-2266.

Edenberg, H.J., Dailey, T.L. and Zhang, K.: Human alcohol dehydrogenase cDNAs: structure and expression. Prog.Clin.Biol.Res. 290 (1989) 181-192.

Edenberg, H.J.: Molecular biological approaches to studies of alcohol-metabolizing enzymes. In Crabbe, J.C. and Harris, R.A. (Eds.), The Genetic Basis of Alcohol and Drug Actions. Plenum Press, N.Y., 1991, pp.165-223.

Edenberg, H.J., Brown, C.J., Carr, L.G., Ho, W.H. and Hur, M.W.: Alcohol dehydrogenase gene expression and cloning of the mouse chi-like ADH. Adv.Exp.Med.Biol. 284 (1991) 253-262.

Edenberg, H.J. and Brown, C.J.: Regulation of Human Alcohol Dehydrogenase Genes. Pharmacogenet. 2 (1992) 185-196.

Gardiner-Garden, M. and Frommer, M.: CpG islands in vertebrate genomes. J.Mol.Biol. 196 (1987) 261-282.

Giri, P.R., Krug, J.F., Kozak, C., Moretti, T., O'Brien, S.J., Seuanez, H.N. and Goldman, D.: Cloning and comparative mapping of a human class III (chi) alcohol dehydrogenase cDNA. Biochem.Biophys.Res.Commun. 164 (1989a) 453-460.

Giri, P.R., Linnoila, M., O'Neill, J.B. and Goldman, D.: Distribution and possible metabolic role of class III alcohol dehydrogenase in the human brain. Brain Res. 481 (1989b) 131-141.

Gutheil, W.G., Holmquist, B. and Vallee, B.L.: Purification, characterization, and partial sequence of the glutathione-dependent formaldehyde dehydrogenase from Escherichia coli: a class III alcohol dehydrogenase. Biochemistry. 31 (1992) 475-481.

Ho, W.-H.: Molecular cloning of the mouse Adh-2 gene and Adh-B2 cDNA. M.S. Thesis, Indiana University Graduate School, Indianapolis, IN, 1991.

Holmes, R.S., Courtney, Y.R. and VandeBerg, J.L.: Alcohol dehydrogenase isozymes in baboons: tissue distribution, catalytic properties, and variant phenotypes in liver, kidney, stomach, and testis. Alcoholism (NY) 10 (1986a) 623-630.

Holmes, R.S., Duley, J.A., Algar, E.M., Mather, P.B. and Rout, U.K.: Biochemical and genetic studies on enzymes of alcohol metabolism: the mouse as a model organism for human studies. Alcohol.Alcohol. 21 (1986b) 41-56.

Hur, M.-W. and Edenberg, H.J.: Cloning and characterization of the *ADH5* gene encoding human alcohol dehydrogenase 5, formaldehyde dehydrogenase. Gene 121 (1992) 305-311.

Hur, M.-W., Ho, W.-H.,,, Brown, C.J., Goldman, D. and Edenberg, H.J.: Molecular cloning of mouse alcohol dehydrogenase-B2 cDNA: nucleotide sequences of the class III ADH genes evolve slowly even for silent substitutions. DNA SEQ.-J.DNA Sequencing and Mapping 3 (1992) 167-175.

Ikuta, T., Szeto, S. and Yoshida, A.: Three human alcohol dehydrogenase subunits: cDNA structure and molecular and evolutionary divergence. Proc.Natl.Acad.Sci.USA. 83 (1986) 634-638.

Julia, P., Farres, J. and Pares, X.: Characterization of three isoenzymes of rat alcohol dehydrogenase. Tissue distribution and physical and enzymatic properties. Eur.J.Biochem. 162 (1987) 179-189.

Kaiser, R., Holmquist, B., Hempel, J., Vallee, B.L. and Jornvall, H.: Class III human liver alcohol dehydrogenase: a novel structural type equidistantly related to the class I and class II enzymes. Biochemistry. 27 (1988) 1132-1140.

Koivusalo, M., Baumann, M. and Uotila, L.: Evidence for the identity of glutathione-dependent formaldehyde dehydrogenase and class III alcohol dehydrogenase. FEBS.Lett. 257 (1989) 105-109.

Kozak, M.: Structural features in eukaryotic mRNAs that modulate the initiation of translation. J.Biol.Chem. 266 (1991) 19867-19870.

Lin, Z., Edenberg, H.J. and Carr, L.G.: A novel negative element in the promoter of the mouse alcohol dehydrogenase gene *Adh-1*. J.Biol.Chem. 268 (1993) 10260-10267.

Pares, X. and Vallee, B.L.: New human liver alcohol dehydrogenase forms with unique kinetic characteristics. Biochem.Biophys.Res.Commun. 98 (1981) 122-130.

Seeley, T.L., Mather, P.B. and Holmes, R.S.: Electrophoretic analyses of alcohol dehydrogenase, aldehyde dehydrogenase, aldehyde reductase, aldehyde oxidase and xanthine oxidase from horse tissues. Comp.Biochem.Physiol.[B] 78 (1984) 131-139.

Sharma, C.P., Fox, E.A., Holmquist, B., Jornvall, H. and Vallee, B.L.: cDNA sequence of human class III alcohol dehydrogenase. Biochem.Biophys.Res.Commun. 164 (1989) 631-637.

Stewart, M.J., McBride, M.S., Winter, L.A. and Duester, G.: Promoters for the human alcohol dehydrogenase genes *ADH1, ADH2,* and *ADH3:* interaction of CCAAT/enhancer-binding protein with elements flanking the *ADH2* TATA box. Gene. 90 (1990a) 271-279.

Stewart, M.J., Shean, M.L. and Duester, G.: Trans activation of human alcohol dehydrogenase gene expression in hepatoma cells by C/EBP molecules bound in a novel arrangement just 5' and 3' to the TATA box. Mol.Cell.Biol. 10 (1990b) 5007-5010.

van Ooij, C., Snyder, R.C., Paeper, B.W. and Duester, G.: Temporal expression of the human alcohol dehydrogenase gene family during liver development correlates with differential promoter activation by hepatocyte nuclear factor 1, CCAAT/Enhancer binding protein α, liver activator protein, and D-element-binding protein. Mol.Cell.Biol. 12 (1992) 3023-3031.

Wagner, F.W., Pares, X., Holmquist, B. and Vallee, B.L.: Physical and enzymatic properties of a class III isozyme of human liver alcohol dehydrogenase: chi-ADH. Biochemistry. 23 (1984) 2193-2199.

CLASS I AND CLASS IV ALCOHOL DEHYDROGENASE (RETINOL DEHYDROGENASE) GENE EXPRESSION IN MOUSE EMBRYOS

Gregg Duester, Hwee Luan Ang, Louise Deltour, Mario H. Foglio, Terry F. Hayamizu, and Mirna Zgombic-Knight

La Jolla Cancer Research Foundation
10901 North Torrey Pines Road
La Jolla, California 92037

INTRODUCTION

Mammalian alcohol dehydrogenase (ADH) exists as several classes of related enzymes, all linked evolutionarily to the medium-chain ADHs present in bacteria, fungi, plants, and animals (Kaiser et al., 1993). One distinctive feature of the mammalian ADH family is the ability of some members to act as retinol dehydrogenases as well as ethanol dehydrogenases (Zachman and Olson, 1961; Mezey and Holt, 1971; Boleda et al., 1993). ADH classes I and IV have been demonstrated to function in this manner in both humans and rodents (Yang et al., 1993; Boleda et al., 1993). This feature may be important since it implies that these classes of ADH may participate in the conversion of retinol (vitamin A alcohol) to its active forms including retinal and retinoic acid, the latter functioning as a transcriptional regulatory ligand controlling cellular differentiation (De Luca, 1991). It further implies that excessive ethanol consumption may competitively inhibit retinol oxidation by ADH, thus leading to under-utilization of vitamin A. This point is particularly critical when one considers a potential connection between vitamin A utilization by developing embryos and fetal alcohol syndrome, a form of teratogenesis characterized by ethanol-induced brain and craniofacial abnormalities (Duester, 1991).

The mouse is the experimental animal of choice for studies of mammalian embryology. Embryos develop quickly and are born at about 19-20 days of gestation. Mouse embryos exhibit an asymmetry in their ability to convert retinol to retinoic acid by the gastrula stage (7.5-8.5 days) at which time the structure known as Hensen's node (anterior portion of the primitive streak) acquires a much greater ability to synthesize retinoic acid than surrounding anterior or posterior tissues (Hogan et al., 1992). Hensen's node controls development of the notochord and neural tube, and retinoic acid has been demonstrated to exist in the neural tube which later contributes to much of the brain tissue as well as the neural crest tissue known to produce most of the craniofacial structures (Wagner et al., 1990; Wagner et al.,

Enzymology and Molecular Biology of Carbonyl Metabolism 5
Edited by H. Weiner *et al.*, Plenum Press, New York, 1995

1992). Retinoic acid is known to regulate expression of several *Hox* homeobox genes which act as transcriptional regulatory factors for hindbrain and craniofacial development (Simeone et al., 1990; Marshall et al., 1992). Thus, there is good evidence that retinoic acid plays a regulatory role in these tissues during early mouse development. Since the brain and craniofacial region are the main teratogenic targets of ethanol it is of interest to determine if any forms of ADH are expressed in the embryonic neural tube and contributing to retinoic acid synthesis.

Mammalian ADH has been best studied in the human where it exists as a family of isozymes encoded by seven genes falling into five classes (Duester et al., 1986; Yasunami et al., 1991; Parés et al., 1992). All isozymes are dimeric, each monomer having a molecular weight of about 40,000 daltons, and the interclass amino acid sequence identity ranges from 58-69% between any two classes. The human class I ADH sub-family has been studied extensively at the gene level. Humans possess three closely related class I ADH genes (*ADH1*, *ADH2*, and *ADH3*) which have been cloned and characterized (Duester et al., 1986; Stewart et al., 1990). The embryonic expression pattern of the human *ADH3* gene has been estimated by analyzing the expression of an *ADH3-lacZ* transgene in mouse embryos as described herein and elsewhere (Zgombic-Knight et al., 1994). Interestingly, these results indicate that this human class I ADH gene is expressed in the neural tube as early as day 9.5 of gestation, particularly in the ventral floor plate which is enriched in retinoic acid compared to the dorsal neural tube. The other major alcohol/retinol dehydrogenase conserved acrosss many species is class IV ADH, and the human *ADH7* gene encoding this enzyme has recently been cloned in this laboratory (Satre et al., 1994). The mouse possesses only one class I ADH gene, *Adh-1*, which has been cloned (Ceci et al., 1987; Zhang et al., 1987), as well as a class IV ADH gene, *Adh-3*, which has also been cloned as described herein. Mouse *Adh-1* and *Adh-3* gene probes have been used to perform *in situ* hybridization studies to examine the normal expression patterns of these two alcohol/retinol dehydrogenases in mouse embryos. *Adh-1* was not expressed in early embryos, but was expressed in certain organs of older embryos starting on day 10.5 of gestation. On the other hand, *Adh-3* was expressed in early embryos undergoing gastrulation and neurulation (days 7.5-9.5 of gestation), being observed in the primitive streak, Hensen's node, neural tube, and hindbrain. These findings have implications for ADH function as a retinoic acid synthetic enzyme during embryogenesis and for the mechanism of fetal alcohol syndrome.

MATERIALS AND METHODS

ADH3-lacZz REPORTER GENE CONSTRUCTION AND ANALYSIS IN TRANSGENIC MICE

A 1.15 kilobase *Hind*III fragment containing the human *ADH3* promoter derived from the plasmid *ADH3-cat*(-1102) (Duester et al., 1991) was ligated upstream of the *E. coli lacZ* gene encoding β-galactosidase derived from the plasmid pCH110 (Pharmacia, Inc.). The AUG translation start codon for *lacZ* is the first AUG codon downstream of the *ADH3* transcription initation site, and downstream of *lacZ* there is a polyadenylation region derived from SV40. A 4.96 kilobase *Xho*I-*Bam*HI DNA fragment from this plasmid *ADH3-lacZ(-1102)* was purified by agarose gel electrophoresis, then injected (2 ng/μl) into the male pronucleus of fertilized mouse eggs (FVB female x C57BL6 male) which were then transferred to pseudopregnant FVB females (Hogan et al., 1986). Five offspring representing different integration sites were identified as carrying the transgene based upon Southern blot analysis of tail DNA. Embryos were staged by counting somites (Rugh, 1990), and stained for β-galactosidase activity as previously described using the chromophore X-gal (Mendelsohn et al., 1991). Embryos derived from two founders exhibited an identical staining pattern and are described herein.

Cloning of the Class IV ADH Gene

Our laboratory has recently reported the cloning of a cDNA encoding human class IV ADH (Satre et al., 1994). In order to clone this cDNA we first used the published partial amino acid sequence of purified rat stomach class IV ADH (Parés et al., 1990) to design and synthesize degenerate oligonucleotides able to hybridize to the predicted rat stomach mRNA for use as polymerase chain reaction (PCR) primers. A predicted PCR product of 740 base pairs was generated, radiolabeled and used to screen a human stomach cDNA library. A nearly full-length cDNA stretching from codon 8 to the 3'-untranslated region was isolated and sequenced. This cDNA was used to screen a human genomic DNA library, and a phage was characterized that contains exons 1 and 2 encoding the 5'-untranslated region to codon 39. The sequencing of these exons completed the full-length sequence of the human class IV ADH coding region. Using the human *ADH7* cDNA as a probe we were than able to clone and sequence a full-length cDNA for the mouse class IV ADH gene (*Adh-3*) from stomach RNA.

In Situ Hybridization

Mouse embryos from days 7.5-14.5 of gestation were fixed, embedded in paraffin, and sectioned at 6 µm. *In situ* hybridization was carried out on sections as described previously using ^{35}S-labeled riboprobes and autoradiography for detection (Wilkinson and Green, 1990). Unsectioned whole embryos from days 7.5-9.5 of gestation were also subjected to whole mount *in situ* hybridization as described using digoxygenin-labeled riboprobes and alkaline phosphatase-linked anti-digoxygenin antibodies for detection (Wilkinson, 1992). Antisense riboprobes were derived from T3 or T7 transcription of the *Adh-1* and *Adh-3* cDNAs cloned in Bluescript II KS. The sense riboprobe was also generated and used on some samples to serve as a control for non-specific binding of probe.

RESULTS

Introduction of an *ADH3-lacZ* Transgene Into Transgenic Mice

Initial interest in the embryonic expression patterns of ADH genes was sparked by the discovery of a retinoic acid response element in the promoter region of the human *ADH3* gene (Duester et al., 1991). *ADH3* encodes one of the three class I ADH isozymes of humans. The other two genes encoding human class I ADH isozymes, *ADH1* and *ADH2*, were shown to lack this retinoic acid response element. Since retinoic acid is involved in the regulation of early embryonic development we thought that it would be informative to examine the embryonic expression pattern of *ADH3* which may be serving a special role in retinoid metabolism due to its potential to participate in retinoic acid synthesis and be regulated by retinoic acid.

A 1.15 kilobase fragment of *ADH3* 5'-flanking DNA was fused to the *E. coli lacZ* gene encoding β-galactosidase as shown in Fig. 1. This region of the *ADH3* gene has previously been shown to contain promoter and enhancer sequences which direct transcription upon transfection into tissue culture cells (Duester et al., 1991; Van Ooij et al., 1992).

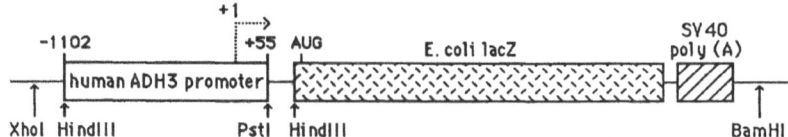

Figure 1. *ADH3-lacZ* fusion. The human *ADH3* promoter from position -1102 to +55 relative to the transcription initiation site is shown fused to the *E. coli lacZ* gene. For microinjection into fertilized eggs, an XhoI/BamHI DNA fragment was isolated after release from the plasmid.

In order to analyze expression of the human *ADH3* gene during early embryogenesis, we have produced lines of transgenic mice carrying *ADH3-lacZ* randomly integrated into their genomic DNA. Two independent transgenic founder mice produced embryos expressing the transgene in the same pattern indicating that expression in these lines was not effected by the site of transgene integration in mouse chromosomal DNA.

Transgene expression was detected from day 9.5 onwards in the central nervous system, head, limbs, and several organs (Table 1). Examination of embryos at day 9.5 revealed staining in the central nervous system, especially along the ventral midline of the neural tube including the floor plate. Neural tube staining extended from its most anterior position in the midbrain to more posterior positions in the hindbrain and future spinal cord. No staining was observed in the forebrain. This pattern of staining continued in older embryos (12.5 and 13.5 days), but no staining was observed in the central nervous system in embryos of 14.5 days or older.

Table 1. Expression of *ADH3-lacZ* in murine embryonic tissues

Tissues examined	day 9.5	day 12.5	day 13.5	day 14.5
Central Nervous System				
dorsal neural tube	-	-	-	-
ventral neural tube	+	+	+	-
spinal cord	+	+	+	-
hindbrain	+	+	+	-
midbrain	+	+	-	-
forebrain	-	-	-	-
Head				
craniofacial region	+	+	+	+
otic vesicles (ears)	+	+	+	+
optic vesicles (eyes)	+	+	+	+
Limbs				
forelimb buds	+	+	+	+
hindlimb buds	NA	+	+	+
Organs				
heart	+	+	+	-
kidney (metanephros)	NA	-	+	+
liver	-	-	-	-
lung	-	-	-	-
digestive tract (gut)	-	-	-	+

(+), β-galactosidase staining observed; (-), no staining observed;
NA, not applicable since structure not present at that stage

In the head, staining was observed in the branchial arches of day 9.5 embryos which contribute to craniofacial structures in later embryos. This data suggests that cephalic neural crest cells which migrate from the hindbrain and give rise to the branchial arches express the *ADH3* transgene. The otic vesicles (developing ears) and optic vesicles (eyes) stained from 9.5 to 14.5 days.

In embryos cut transversely at the level of the forelimb buds, a dorsoventral gradient of *ADH3* expression was observed in the neural tube with highest expression ventrally. In day 9.5 embryos, intense staining was observed in the ventral midline corresponding to the floor plate, and staining tapered off about halfway to the dorsal roof plate which was unstained. In a day 12.5 embryo the neural tube continued to show a dorsoventral gradient of *ADH3* transgene expression with the floor plate most intensely stained. Located laterally on both sides of the floor plate, the ventral horns of the neural tube which contain the motor neurons were stained, but regions dorsal to this were not.

Expression of the Mouse Class I ADH Gene *Adh-1* in Late Embryonic Tissues

Since the human class I ADH gene *ADH3* displayed an interesting pattern of expression in the neural tube of early transgenic mouse embryos, we decided to examine the expression pattern of the single endogenous mouse class I ADH gene *Adh-1*. *In situ* hybridization analyses failed to detect any *Adh-1* expression in early mouse embryos. However, starting at day 10.5 we did observe *Adh-1* expression in certain developing organs. Expression was noted in the mesonephros at day 10.5, in the liver and heart at day 12.5, and in the liver and adrenal gland at day 14.5 (Ang and Duester, unpublished data). These findings correlate with a recent *in situ* hybridization study performed by others (Vonesch et al., 1994).

Unlike the human *ADH3* transgene mentioned above, *Adh-1* expression was not observed in the neural tube. Analysis of the mouse *Adh-1* promoter has also provided further evidence that this gene differs from the human *ADH3* gene in that it does not possess a retinoic acid response element (Zgombic-Knight and Duester, unpublished data; Vonesch et al., 1994).

Characterization of the Class IV ADH Gene

Since the mouse class I ADH gene was not expressed in early embryos, we decided to examine the expression pattern of another ADH which acts as a retinol dehydrogenase. The class IV ADH isozyme, which is particularly abundant in the adult stomach and very scarce in the adult liver, has been demonstrated by others to be active for retinol oxidation (Yang et al., 1993; Boleda et al., 1993). Thus, in order to study the expression pattern of class IV ADH in mouse embryos we isolated a cDNA for this gene.

Since protein sequence information was first available for the rat class IV ADH enzyme (Parés et al., 1990), this information was used to construct oligonucleotides to produce a partial cDNA from rat stomach RNA using reverse transcription and polymerase chain reaction technologies. This partial rat cDNA was then used as a hybridization probe to isolate human cDNA and genomic clones, and the nucleotide sequence of the full coding region of human class IV ADH was determined (Satre et al., 1994). The sequence agreed for the most part with the partial amino acid sequence of human class IV ADH determined from various peptides (Parés et al., 1992; Stone et al., 1993). The predicted amino acid sequence of human class IV ADH is shown aligned with the recently determined full-length sequence

```
                10        20        30        40        50        60
ADH7 human   -GTAGKVIKCK AAVLWEQKQP FSIEEIEVAP PKTKEVRIKI LATGICRTDD HVIKGTMVSK
ADH IV rat   SNRV...... .....GTN.. ....D..... ..A....V.. ......G... ..........

                70        80        90       100       110       120
ADH7 human   FPVIVGHEAT GIVESIGEGV TTVKPGDKVI PLFLPQCREC NACRNPDGNL CIRSDITGRG
ADH IV rat   .........V .....V..E. ...R...... .......... .P....E... .....L....

               130       140       150       160       170       180
ADH7 human   VLADGTTRFT CKGKPVHHFM NTSTFTEYTV VDESSVAKID DAAPPEKVCL IGCGFSTGYG
ADH IV rat   .......... ......Q... .......... L......... AE.....A.. ..........

               190       200       210       220       230       240
ADH7 human   AAVKTGKVKP GSTCVVFGLG GVGLSVIMGC KSAGASRIIG IDLNKDKFEK AMAVGATECI
ADH IV rat   .....A..S. ....A..... .......... .A........ ..I.....Q. .LD.......

               250       260       270       280       290       300
ADH7 human   SPKDSTKPIS EVLSEMTGNN VGYTFEVIGH LETMIDALAS CHMNYGTSVV VGVPPSAKML
ADH IV rat   N.R.F..... ....D....T .Q.......R ....V...S. .......... ..A.......

               310       320       330       340       350       360
ADH7 human   TYDPMLLFTG RTWKGCVFGG LKSRDDVPKL VTEFLAKKFD LDQLITHVLP FKKISEGFEL
ADH IV rat   S......... .......... W......... .....E.... .G.....T.. .HN.......

               370
ADH7 human   LNSGQSIRTV LTF
ADH IV rat   .Y........ ...
```

Figure 2. Rat/human class IV ADH amino acid sequence alignment. The human class IV ADH amino acid sequence predicted from the *ADH7* gene is shown aligned with the sequence of the rat enzyme as determined by amino acid sequence analysis of peptides (Parés et al., 1994). Dots indicate sequence conservation.

of the rat class IV ADH enzyme (Parés et al., 1994) (Fig. 2). The rat/human interspecies sequence identity is 87.1%.

The full-length amino acid sequence of human class IV ADH predicted from the *ADH7* gene was aligned with the full-length sequences derived from the other six known human ADH genes, together comprising all five known classes of ADH. Within the 373 amino acid sequence of *ADH7* this alignment revealed sequence identities of about 69% with all three human class I ADH genes (*ADH1*, *ADH2*, and *ADH3*), 59% with the class II ADH

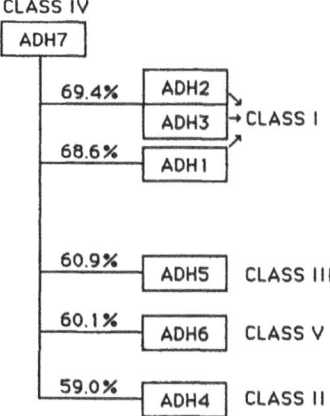

Figure 3. Human ADH sequence homology. The amino acid sequence identities between the human class IV enzyme and all other known human classes are shown. Indicated inside the boxes are the names of the genes. All full-length sequences are referenced in Satre et al.(1994).

gene *ADH4*, 61% with the class III ADH gene *ADH5*, and 60% with the class V ADH gene *ADH6* (Fig. 3). The sequence identity observed between the class IV ADH gene and the three class I ADH genes was 8-10% higher than that observed between class IV and the other classes, and from 6-13% higher than any other interclass sequence identities among all the human ADHs. Thus, the class IV ADH sequence derived from *ADH7* clearly fits into the human ADH family and is more closely related to the class I ADHs than to the other classes of ADH.

Isolation of the human class IV ADH cDNA has now enabled us to use it as a probe to clone the mouse class IV ADH cDNA from stomach RNA (Zgombic-Knight, Foglio, and Duester, unpublished data). The mouse class IV ADH cDNA was fully sequenced and the predicted amino acid sequence shares 89% sequence identity with human class IV ADH, and 93% with the rat enzyme. The mouse appears to have three ADH isozymes encoded by *Adh-1*, *Adh-2*, and *Adh-3* (Algar et al., 1983). Isolation of cDNA clones has been described previously for *Adh-1* encoding class I ADH (Edenberg et al., 1985; Ceci et al., 1986) and *Adh-2* encoding class III ADH (Hur et al., 1992). The mouse class IV ADH cDNA we have

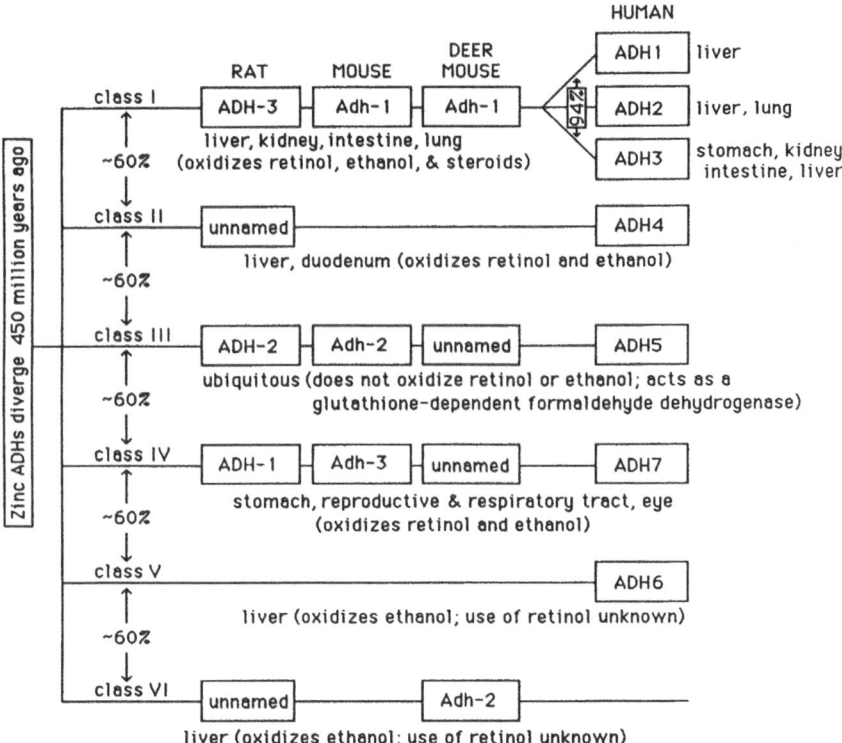

Figure 4. Human and rodent ADH genes. The mammalian zinc-containing medium-chain ADHs have all diverged from class III ADH, the most primitive form, about 450 million years ago during the time of early vertebrate evolution (Jörnvall et al., 1993). Five classes of ADH are known in humans with the interclass sequence identity in the 60% range (Satre et al., 1994). The human has three closely related class I ADH genes sharing about 94% sequence identity. Rodents have only one class I ADH gene, and less additional classes as compared with humans. The rat has been demonstrated to possess classes I, II, III, and IV (Parés et al., 1990), but not class V. The mouse has been shown to have classes I, III, and IV (Algar et al., 1983). The deermouse possesses classes I, III, and IV as well as a sixth class not yet seen in human or mouse (Zheng et al., 1993); this sixth class has recently been identified in the rat and called class VI ADH (J.-O. Höög and M. Brandt, see this book). Some of the enzymatic properties and sites of expression are noted for each class.

isolated is expressed highly in the stomach and very weakly in the liver (Zgombic-Knight, Ang, and Duester, unpublished data), thus corresponding to the expression pattern of the last known murine ADH gene, *Adh-3* (Holmes et al., 1983). A summary of the known mammalian ADH genes from humans and several rodents are listed in Fig. 4.

Expression of the Class IV ADH Gene *Adh-3* in Early Mouse Embryos

In order to compare the expression patterns for class I and class IV ADH in early mouse embryos, cDNAs for murine *Adh-1* and *Adh-3* were used as probes for whole mount *in situ* hybridization. *Adh-1* mRNA was not detected in early embryos undergoing gastrulation and neurulation; i.e. 7.5-9.5 days of gestation. However, *Adh-3* mRNA was detected at all of these early stages as summarized (Table 2). Embryos at 7.5 days exhibited *Adh-3* mRNA throughout all of the embryonic tissues, but not the extraembryonic tissues. Expression was low in the anterior headfold portion of the 7.5 day embryo, but was higher in the posterior portion of the embryo containing the primitive streak. At 8.5 days, expression of *Adh-3* was localized to only the central portion of the embryo including presumptive hindbrain, anterior trunk, and Hensen's node. No expression was observed on day 8.5 in either the anterior head (presumptive forebrain and midbrain) or the posterior trunk. The developing neural groove, extending from the midbrain to the posterior end of the embryo, was observed to express *Adh-3* only in its central portion at the level of the hindbrain and anterior trunk. Expression of *Adh-3* at 9.5 days was noticed to display a strong posterior dominance with a high level of expression in the entire trunk, forelimb buds, tail bud, and posterior neural tube. At 9.5 days, expression in the hindbrain disappeared, but a low level of expression was observed in the most anterior portion of the forebrain.

Table 2. Class I and Class IV ADH expression in early mouse embryos

Tissues examined	*Adh-1* class I	*Adh-3* class IV
7.5 day embryo		
headfold	-	+/-
primitive streak	-	+
8.5 day embryo		
forebrain	-	-
midbrain	-	-
hindbrain	-	+
Hensen's node	-	+
neural groove	-	+
anterior trunk	-	+
posterior trunk	-	-
9.5 day embryo		
forebrain	-	+/-
midbrain	-	-
hindbrain	-	-
neural tube (trunk)	-	+
trunk	-	+
forelimb bud	-	+
tailbud	-	+

Detection was by whole mount *in situ* hybridization.
(+), mRNA detected; (-), no mRNA detected;
(+/-), weak expression.

DISCUSSION

Class I and class IV ADH are known to function as cytosolic retinol dehydrogenases oxidizing retinol to retinal which can then be oxidized to retinoic acid. Since retinoic acid is the active form of vitamin A involved in regulating cellular differentiation, growth, and development, there is the implication that these classes of ADH may be performing an essential reaction in retinoid metabolism. Interest in the role of ADH in early embryonic retinoic acid synthesis has been sparked by the following findings: (a) retinoic acid is present in early embryos undergoing gastrulation and neurulation (Wagner et al., 1990; Hogan et al., 1992), (b) gastrulation is the most sensitive stage for teratogenesis by ethanol (Webster et al., 1983; Sulik et al., 1984), and (c) ethanol is a competitive inhibitor of retinol oxidation (Mezey and Holt, 1971; Julià et al., 1986). Since there is a lack of information on which forms of ADH are expressed during development, we have begun an analysis of class I and class IV ADH gene expression during embryogenesis.

The three human class I ADH genes (*ADH1*, *ADH2*, *ADH3*) were the first forms to be well characterized (Duester et al., 1986; Stewart et al., 1990). Studies on these three genes led to the discovery that *ADH3*, but not *ADH1* or *ADH2*, is regulated transcriptionally by retinoic acid, further implicating this form of ADH as a player in the control of retinoic acid synthesis (Duester et al., 1991). In human adults and fetuses, expression of the class I ADH genes has been observed to be highest in liver, lung, kidney, and intestine (Smith et al., 1971; Bilanchone et al., 1986), all containing epithelial cells that synthesize retinoic acid from retinol (Wolf, 1984). The human class IV ADH gene, which also encodes an enzyme that functions as a retinol dehydrogenase, has recently been characterized (Satre et al., 1994). Class IV ADH expression in the adult is found primarily in the stomach and esophagus (Moreno and Parés, 1991; Yin et al., 1993), also containing epithelial cells. There is a lack of information on expression patterns of human class I and class IV ADH in early embryos due to inaccessibility of tissue. Thus, we have decided to study ADH expression during mouse embryogenesis. One approach has been to introduce the human *ADH3* gene into transgenic mice to observe embryonic expression. Another approach has been to analyze the expression patterns of the endogenous mouse class I and class IV ADH genes, *Adh-1* and *Adh-3*, respectively. The availability of cDNAs for the human class I and class IV ADHs have led to the cloning of the mouse cDNAs for these two genes. We have used these cDNAs as *in situ* hybridization probes to study mouse embryonic ADH expression.

In our first approach we have analyzed the expression pattern of the human *ADH3* promoter in transgenic mouse embryos. Analysis of these transgenic mice has indicated several new sites of ADH gene expression which have not previously been reported. Expression of the *ADH3-lacZ* transgene was observed in the embryonic central nervous system, craniofacial region, heart, and limb buds suggesting that ADH may exist in these tissues and play a role in their development possibly as a retinoic acid synthetic enzyme. Expression in the embryonic central nervous system was particularly interesting. The *ADH3-lacZ* transgene was expressed along the entire neural tube from 9.5 d.p.c. onwards with highest expression in the ventral floor plate, exhibiting a ventral to dorsal gradient of decreasing expression. Previous studies have shown that the neural tube possesses the enzymatic machinery to convert retinol to retinoic acid via an NAD+ linked reaction, and that the ventral floor plate is enriched in this activity compared to dorsal neural tube tissue (Wagner et al., 1990). The dorsoventral gradient of *ADH3* transgene expression we noticed in the neural tube suggests that a gradient of ADH-catalyzed retinoic acid synthesis could be established with the high end located in the ventral floor plate. Our data suggests that human class I ADH is expressed in the correct location to serve as a retinoic acid synthetic enzyme for this system.

In our second approach for analyzing ADH gene expression, *in situ* hybridization, our initial studies focused upon the mouse class I ADH gene, *Adh-1*. These studies indicate that *Adh-1* is not expressed in early mouse embryos, but is expressed from day 10.5 and thereafter in various organs including the mesonephros, liver, lung, and adrenal gland. There was no expression in the neural tube as mentioned above for the human *ADH3-lacZ* transgene. These findings force us to conclude that the early embryonic expression pattern of the mouse class I ADH gene *Adh-1* does not correlate with that of the human class I ADH gene *ADH3*. This may not be too surprising if one considers that the human has three class I ADH genes, and that one of them, *ADH2*, has a fetal and adult pattern of expression that more closely matches that of mouse *Adh-1* (Smith et al., 1971; Balak et al., 1982). It is possible that human *ADH3* evolved as a human-specific class I ADH involved in early embryonic metabolism of retinoids, and that a different class of ADH functions as a retinol dehydrogenase for early mouse embryos.

Our *in situ* hybridization studies on the mouse class IV ADH gene, *Adh-3*, have clearly indicated that this gene is expressed in early mouse embryos. Expression of *Adh-3* mRNA was noted in embryos at days 7.5-9.5, whereas *Adh-1* mRNA was not detected at any of these stages. At day 7.5, during gastrulation, *Adh-3* was expressed along the primitive streak located in the posterior region of the embryo, with lower expression in the anterior headfold region. By day 8.5 *Adh-3* expression was absent in the head region (forebrain and midbrain) and posterior trunk region, and was localized to only the central portion of the embryo containing the hindbrain and anterior trunk. Within this central region lies Hensen's node (a transient structure) which gives rise to midline structures including the notochord and floor plate of the neural tube. Interestingly, Hensen's node tissue from day 8 mouse embryos has previously been found to synthesize retinoic acid at a greater rate than either anterior or posterior tissues (Hogan et al., 1992). Thus, expression of *Adh-3* (encoding a retinol dehydrogenase) correlates with a major site of retinoic acid synthesis in early embryos. *Adh-3* expression on day 9.5 was much more posteriorly restricted, being observed in the entire trunk region, but not the head except for a low level of expression in the forebrain. Neural tube expression was noted in the trunk, but not further anterior.

The results for mouse *Adh-3* embryonic expression suggest that class IV ADH is acting as a retinol dehydrogenase for early mouse development. The correlation of *Adh-3* expression with Hensen's node on day 8 is most indicative of an important role for class IV ADH, since the node is a major site of retinoic acid synthesis and plays an important role in central nervous system development. The embryonic midline structures that are derived from the node include the ventral floor plate of the neural tube which has been shown to synthesize retinoic acid at a greater rate than dorsal neural tube tissue (Wagner et al., 1990). Expression of *Adh-3* has not yet been analyzed sufficiently to determine dorsal/ventral differences, but ventral floor plate expression was noted for the human *ADH3-lacZ* transgene. In summary, observation of mouse class IV ADH and human class I ADH expression at sites of retinoic acid synthesis in early mouse embryos further supports the hypothesis that ethanol teratogenesis operates through inhibition of retinoic acid synthesis catalyzed by ADH.

ACKNOWLEDGMENTS

This work was supported by NIH grant AA07261 and a grant from the Alcoholic Beverage Medical Research Foundation.

REFERENCES

Algar, E.M., Seeley, T.-L., and Holmes, R.S., 1983, Purification and molecular properties of mouse alcohol dehydrogenase isozymes, *Eur. J. Biochem.* 137: 139-147.

Balak, K.J., Keith, R.H., and Felder, M.R., 1982, Genetic and developmental regulation of mouse liver alcohol dehydrogenase, *J. Biol. Chem.* 257: 15000-15007.

Bilanchone, V., Duester, G., Edwards, Y., and Smith, M., 1986, Multiple mRNAs for human alcohol dehydrogenase (ADH): developmental and tissue-specific differences, *Nucleic Acids Res.* 14: 3911-3926.

Boleda, M.D., Saubi, N., Farrés, J., and Parés, X., 1993, Physiological substrates for rat alcohol dehydrogenase classes: Aldehydes of lipid peroxidation, omega-hydroxyfatty acids, and retinoids, *Arch. Biochem. Biophys.* 307: 85-90.

Ceci, J.D., Lawther, R., Duester, G., Hatfield, G.W., Smith, M., O'Malley, M.P., and Felder, M.R., 1986, Androgen induction of alcohol dehydrogenase in mouse kidney: Studies with a cDNA probe confirmed by nucleotide sequence analysis, *Gene* 41: 217-224.

Ceci, J.D., Zheng, Y.-W., and Felder, M.R., 1987, Molecular analysis of mouse alcohol dehydrogenase: nucleotide sequence of the Adh-1 gene and genetic mapping of a related nucleotide sequence to chromosome 3, *Gene* 59: 171-182.

De Luca, L.M., 1991, Retinoids and their receptors in differentiation, embryogenesis, and neoplasia, *FASEB J.* 5: 2924-2933.

Duester, G., Smith, M., Bilanchone, V., and Hatfield, G.W., 1986, Molecular analysis of the human class I alcohol dehydrogenase gene family and nucleotide sequence of the gene encoding the β subunit, *J. Biol. Chem.* 261: 2027-2033.

Duester, G., 1991, A hypothetical mechanism for fetal alcohol syndrome involving ethanol inhibition of retinoic acid synthesis at the alcohol dehydrogenase step, *Alcohol. Clin. Exp. Res.* 15: 568-572.

Duester, G., Shean, M.L., McBride, M.S., and Stewart, M.J., 1991, Retinoic acid response element in the human alcohol dehydrogenase gene *ADH3*: Implications for regulation of retinoic acid synthesis, *Mol. Cell. Biol.* 11: 1638-1646.

Edenberg, H.J., Zhang, K., Fong, K., Bosron, W.F., and Li, T.-K., 1985, Cloning and sequencing of cDNA encoding the complete mouse liver alcohol dehydrogenase, *Proc. Natl. Acad. Sci. USA* 82: 2262-2266.

Hogan, B., Costantini, F., and Lacy, E., 1986, "*Manipulating the Mouse Embryo*," Cold Spring Harbor Laboratory, Cold Spring Harbor.

Hogan, B.L.M., Thaller, C., and Eichele, G., 1992, Evidence that Hensen's node is a site of retinoic acid synthesis, *Nature* 359: 237-241.

Holmes, R.S., Duley, J.A., and Burnell, J.N., 1983, The alcohol dehydrogenase gene complex on chromosome 3 of the mouse, in " *Isozymes, Vol. 8: Cellular Localization, Metabolism, and Physiology*," M.C. Rattazzi, J.G. Scandalios, and G.S. Whitt, eds., Alan R. Liss, Inc., New York, p. 155.

Hur, M.-W., Ho, W.-H., Brown, C.J., Goldman, D., and Edenberg, H.J., 1992, Molecular cloning of mouse alcohol dehydrogenase-B$_2$ cDNA: nucleotide sequences of the class III ADH genes evolve slowly even for silent substitutions, *DNA Sequence-J. DNA Seq. Map.* 3: 167-175.

Julià, P., Farrés, J., and Parés, X., 1986, Ocular alcohol dehydrogenase in the rat: Regional distribution and kinetics of the ADH-1 isoenzyme with retinol and retinal, *Exp. Eye Res.* 42: 305-314.

Jörnvall, H., Danielsson, O., Eklund, H., Hjelmqvist, L., Höög, J.-O., Parés, X., and Shafqat, J., 1993, Enzyme and isozyme developments within the medium-chain alcohol dehydrogenase family, *Adv. Exp. Med. Biol.* 328: 533-544.

Kaiser, R., Fernández, M.R., Parés, X., and Jörnvall, H., 1993, Origin of the human alcohol dehydrogenase system: Implications from the structure and properties of the octopus protein, *Proc. Natl. Acad. Sci. USA* 90: 11222-11226.

Marshall, H., Nonchev, S., Sham, M.H., Muchamore, I., Lumsden, A., and Krumlauf, R., 1992, Retinoic acid alters hindbrain *Hox* code and induces transformation of rhombomeres 2/3 into a 4/5 identity, *Nature* 360: 737-741.

Mendelsohn, C., Ruberte, E., LeMeur, M., Morriss-Kay, G., and Chambon, P., 1991, Developmental analysis of the retinoic acid-inducible RAR-β2 promoter in transgenic animals, *Development* 113: 723-734.

Mezey, E. and Holt, P.R., 1971, The inhibitory effect of ethanol on retinol oxidation by human liver and cattle retina, *Exp. Mol. Pathol.* 15: 148-156.

Moreno, A. and Parés, X., 1991, Purification and characterization of a new alcohol dehydrogenase from human stomach, *J. Biol. Chem.* 266: 1128-1133.

Parés, X., Moreno, A., Cederlund, E., Höög, J.-O., and Jörnvall, J., 1990, Class IV mammalian alcohol dehydrogenase: Structural data of the rat stomach enzyme reveal a new class well separated from those already characterized, *FEBS Lett.* 277: 115-118.

Parés, X., Cederlund, E., Moreno, A., Saubi, N., Höög, J.-O., and Jörnvall, H., 1992, Class IV alcohol dehydrogenase (the gastric enzyme): structural analysis of human σσ-ADH reveals class IV to be variable and confirms the presence of a fifth mammalian alcohol dehydrogenase class, *FEBS Lett.* 303: 69-72.

Parés, X., Cederlund, E., Moreno, A., Hjelmqvist, L., Farrés, J., and Jörnvall, H., 1994, Mammalian class IV alcohol dehydrogenase (stomach alcohol dehydrogenase): Structure, origin, and correlation with enzymology, *Proc. Natl. Acad. Sci. USA* 91: 1893-1897.

Rugh, R., 1990, "*The Mouse: Its Reproduction and Development,*" Oxford University Press, New York.

Satre, M.A., Zgombic-Knight, M., and Duester, G., 1994, The complete structure of human class IV alcohol dehydrogenase (retinol dehydrogenase) determined from the *ADH7* gene, *J. Biol. Chem.* 269: 15606-15612.

Simeone, A., Acampora, D., Arcioni, L., Andrews, P.W., Boncinelli, E., and Mavilio, F., 1990, Sequential activation of *HOX2* homeobox genes by retinoic acid in human embryonal carcinoma cells, *Nature* 346: 763-766.

Smith, M., Hopkinson, D.A., and Harris, H., 1971, Developmental changes and polymorphism in human alcohol dehydrogenase, *Ann. Hum. Genet.* 34: 251-271.

Stewart, M.J., McBride, M.S., Winter, L.A., and Duester, G., 1990, Promoters for the human alcohol dehydrogenase genes *ADH1*, *ADH2*, and *ADH3*: interaction of CCAAT/enhancer binding protein with elements flanking the *ADH2* TATA box, *Gene* 90: 271-279.

Stone, C.L., Thomasson, H.R., Bosron, W.F., and Li, T.-K., 1993, Purification and partial amino acid sequence of a high-activity human stomach alcohol dehydrogenase, *Alcohol. Clin. Exp. Res.* 17: 911-918.

Sulik, K.K., Lauder, J.M., and Dehart, D.B., 1984, Brain malformations in prenatal mice following acute maternal ethanol administration, *Int. J. Devl. Neuroscience* 2: 203-214.

Van Ooij, C., Snyder, R.C., Paeper, B.W., and Duester, G., 1992, Temporal expression of the human alcohol dehydrogenase gene family during liver development correlates with differential promoter activation by HNF1, C/EBPα, LAP, and DBP, *Mol. Cell. Biol.* 12: 3023-3031.

Vonesch, J.-L., Nakshatri, H., Philippe, M., Chambon, P., and Dollé, P., 1994, Stage and tissue-specific expression of the alcohol dehydrogenase 1 (*Adh-1*) gene during mouse development, *Dev. Dyn.* 199: 199-213.

Wagner, M., Thaller, C., Jessell, T., and Eichele, G., 1990, Polarizing activity and retinoid synthesis in the floor plate of the neural tube, *Nature* 345: 819-822.

Wagner, M., Han, B., and Jessell, T.M., 1992, Regional differences in retinoid release from embryonic neural tissue detected by an in vitro reporter assay, *Development* 116: 55-66.

Webster, W.S., Walsh, D.A., McEwen, S.E., and Lipson, A.H., 1983, Some teratogenic properties of ethanol and acetaldehyde in C57BL/6J mice: Implications for the study of the fetal alcohol syndrome, *Teratology* 27: 231-243.

Wilkinson, D.G. and Green, J., 1990, *In situ* hybridization and the three-dimensional reconstruction of serial sections, in " *Postimplantation Mammalian Embryos: A Practical Approach,*" A.J. Copp and D.L. Cockroft, eds., Oxford University Press, New York, p. 155.

Wilkinson, D.G., 1992, Whole mount *in situ* hybridization of vertebrate embryos, in " *In Situ Hybridization: A Practical Approach,*" D.G. Wilkinson, ed.,IRL Press, Oxford, p. 75.

Wolf, G., 1984, Multiple functions of vitamin A, *Physiol. Rev.* 64: 873-937.

Yang, Z.N., Davis, G.J., Hurley, T.D., Stone, C.L., Li, T.-K., and Bosron, W.F., 1993, Catalytic efficiency of human alcohol dehydrogenases for retinol oxidation and retinal reduction, *Alcohol. Clin. Exp. Res.* 17: 496.

Yasunami, M., Chen, C.-S., and Yoshida, A., 1991, A human alcohol dehydrogenase gene (*ADH6*) encoding an additional class of isozyme, *Proc. Natl. Acad. Sci. USA* 88: 7610-7614.

Yin, S.-J., Chou, F.-J., Chao, S.-F., Tsai, S.-F., Liao, C.-S., Wang, S.-L., Wu, C.-W., and Lee, S.-C., 1993, Alcohol and aldehyde dehydrogenases in human esophagus: comparison with the stomach enzyme activities, *Alcohol. Clin. Exp. Res.* 17: 376-381.

Zachman, R.D. and Olson, J.A., 1961, A comparison of retinene reductase and alcohol dehydrogenase of rat liver, *J. Biol. Chem.* 236: 2309-2313.

Zgombic-Knight, M., Satre, M.A., and Duester, G., 1994, Differential activity of the promoter for the human alcohol dehydrogenase (retinol dehydrogenase) gene *ADH3* in neural tube of transgenic mouse embryos, *J. Biol. Chem.* 269: 6790-6795.

Zhang, K., Bosron, W.F., and Edenberg, H.J., 1987, Structure of the mouse Adh-1 gene and identification of a deletion in a long alternating purine-pyrimidine sequence in the first intron of strains expressing low alcohol dehydrogenase activity, *Gene* 57: 27-36.

Zheng, Y.-W., Bey, M., Liu, H., and Felder, M.R., 1993, Molecular basis of the alcohol dehydrogenase-negative deer mouse. Evidence for deletion of the gene for class I enzyme and identification of a possible new enzyme class, *J. Biol. Chem.* 268: 24933-24939.

37

MOLECULAR EVOLUTION OF CLASS I ALCOHOL DEHYDROGENASES IN PRIMATES

MODELS FOR GENE EVOLUTION AND COMPARISON OF 3' UNTRANSLATED REGIONS OF cDNAS

Brenda Cheung,[1] Roger S. Holmes[1-3] and Ifor R. Beacham[1]

[1]School of Science
Griffith University
Brisbane, Qld. 4111, Australia
[2]Dept. of Genetics
Southwest Foundation for Biomedical Research
PO Box 28147
San Antonio, TX 78284
[3]Address for correspondence

INTRODUCTION

Human Class I alcohol dehydrogenases (ADHs) comprise homodimeric and heterodimeric isozymes of three highly homologous subunits (α, β and γ), which are encoded by distinct genes, namely *ADH1*, *ADH2* and *ADH3*, respectively (see Smith, 1986; Ikuta *et al*, 1986). These enzymes are all active in human liver, but are differentially distributed in other tissues of the body (Smith *et al*, 1971). The baboon (*Papio hamadryas*) genome also contains three Class I ADH genes (Trezise *et al*, 1991), however only one of these, *ADH2* encoding ADHβ subunits, is expressed in baboon liver (Holmes *et al*, 1986; Trezise *et al*, 1989). A second Class I ADH is known to be expressed in baboon kidney (Holmes *et al*, 1986; 1990).

In this paper, we report on the comparative cDNA and deduced amino acid sequence for baboon kidney Class I ADH, and on its common ancestral lineage with human ADHγ. In addition, a phylogenetic analysis of human (α, β, γ) baboon (β, γ) and rhesus monkey (α) Class I ADH subunits is described, and three models for primate ADH gene evolution discussed. Finally, we have compared the 3' untranslated regions of all ADH genes from baboon with each other and the corresponding human Class I ADH genes.

MATERIALS AND METHODS

Total RNA was isolated from frozen baboon kidney using the method of Chomczynski and Sacchi (1987). Poly(A$^+$) RNA was obtained using biotinylated oligo (dT) (Promega), and reverse transcribed using oligo (dT) as primer. The resulting cDNA was then subjected to PCR amplification using primers listed in Table 1. BAB081, BAB091 and BAB11 primers were derived from the baboon ADHβ sequence (Trezise *et al*, 1989); BAB10 and BAB12 were based on known baboon kidney ADH sequence; and RACE-2 from the RACE primer (Frohman *et al*, 1988).

Amplifications were performed using Taq polymerase and annealing temperatures of 55°C or 63°C. PCR products were cloned into pBluescript pSK+ (Stratagene, USA). Nucleotide sequencing was performed on both strands using a modified T7DNA polymerase (USB, USA), and the consensus sequence derived from three independent amplifications.

Coding sequences for human ADH α, β and γ subunits (Ikuta *et al*, 1986); baboon ADHβ (Trezise et al, 1989) and γ subunits, and rhesus monkey ADHα (Light *et al*, 1992) were aligned using Clustal V (Higgins, 1992). Phylogenetic relationships between the human and Old World monkey (baboon and rhesus) Class I ADHs were then determined using two methods, with the horse ADH E sequence (Park and Plapp, 1991) as the outgroup: a neighbour joining method, with bootstrapping probabilities for 1,000 replications (Saitou and Nei, 1987); and the maximum likelihood method (Felsenstein, 1981).

RESULTS AND DISCUSSION

Baboon Kidney Class I ADH is ADHγ

The baboon kidney Class I ADH cDNA sequence contained the entire coding region, 72 nucleotides (nts) of the 5' noncoding region and 271 nts of the 3' noncoding region (data not shown). Table 2 shows the amino acid and nucleotide differences between the three human (α, β, γ) and Old World monkey (α from rhesus; β and γ from baboon) coding regions for Class I ADH genes.

Table 1. Primers used in amplification of the baboon kidney
ADH gene (ADH3)

BAB 081	5' AGAGCTCTATGCACTCAAGCAGAGAAG 3' `+1` ... `+19` / SacI
BAB 091	5' AGAGCTCTCACCTGGTTTGACTGTAGT 3' `+335` ... `+316` / SacI
BAB 10	5' CCGAATTCACTTCGCATTAAGATGGTGG 3' `+179` ... `+198` / EcoR1
BAB 11	5' AGAGCTCTAGTCTTGAGGGTTGATGCA 3' `+812` ... `+793` / SacI
BAB 12	5' ATACCGCGGCTGTTATGGGCTGTAA 3' `+694` ... `+710` / SacII
RACE	5' GAGTCGACTCGAGATCGATT(16) 3' XhoI / SalI / ClaI
RACE-2	5' TAGGTACCGTCGACTCGAGATCGAT 3' KpnI / XhoI / SalI / ClaI

Table 2. Sequence Differences

		Sequence differences*				
		Human ADH subunits		Rhesus ADH	Baboon ADH subunits	
ADH	Subunit	β^1	γ^1	α	β	γ
Human	α	23 (55)	25 (69)	23 (53)	30 (73)	26 (75)
	β^1		20 (49)	21 (48)	11 (28)	24 (66)
	γ^1			30 (73)	25 (61)	20 (56)
Rhesus	α				25 (55)	31 (76)
Baboon	β					28 (68)

* Amino acid sequence differences (and nucleotide sequence differences in parentheses) between baboon ADHs, human αADH, β^1ADH, and γ^1ADH and rhesus αADH (and between nucleotide sequences of class I ADH-encoding genes). The amino acid and nucleotide sequences of human ADH subunits (α, β^1 and γ^1) are from previous reports.

A higher degree of sequence identity was observed for the baboon kidney ADH with human ADHγ (20 residues different), than with ADHα (26 differences) and ADHβ (24 differences), and is therefore designated as ADHγ.

Baboon and Human Class I ADH 3' Sequences

The 3' noncoding region for baboon kidney cDNA contains 271 nucleotides, and is highly homologous with the 3' noncoding regions of human ADHα, β and γ cDNA sequences, and with that of baboon ADHβ cDNA (Fig. 1) (Deuster *et al*, 1984; Ikuta *et al*, 1986; Heden *et al*, 1986; von Bahr-Lindstrom *et al*, 1986; Höög *et al*, 1986; Trezise *et al*, 1989). Alignment of these 3' sequences reveals a high degree of sequence similarity in the region encompassing nts 1141-1396, although there is an insertion of 18 nts in baboon ADHγ after nt 1213, which is not in the other mRNAs. Part of this insertion is a repeat of nt's 1216-1227. Otherwise, baboon ADHγ mRNA 3' sequence shows 89.1% sequence identity to baboon ADHβ; 92.2% to human ADHα; 88.3% to human ADHβ; and 95.3% to human ADHγ mRNA sequences. There are three polyadenylation signal consensus sequences (AATAAA), as in baboon ADHβ and human ADHα and ADHγ (see Trezise *et al*, 1989). The last signal sequence is 15 nt's from the polyadenylation site, and is therefore presumed to be involved in polyadenylation.

While the polyadenylation signal sequence AATAAA is necessary for cleavage and polyadenylation, other signals must also be involved in the selection of the polyadenylation site, as there are two additional such sequences upstream of the one used in baboon ADHβ and ADHγ, and in human ADHα and ADHγ mRNAs. A CAYTG consensus sequence has been proposed to be involved in the selection of the polyadenylation site (Berget, 1984). This consensus sequence may be located either upstream ('Class I' mRNA sequence), or downstream ('Class II' mRNA sequence) of the polyadenylation site. In baboon ADHβ mRNA, four of the five nt's immediately adjacent and 5' to the polyadenylation site match this sequence (Trezise *et al*, 1989), whereas for the baboon ADHγ mRNA, no such sequence is present. Thus, there are presumably other signal sequences involved in the selection of a polyadenylation site, perhaps downstream of the polyadenylation site.

Evolution of Primate Class I ADH Genes

Three models have been proposed for the evolution of primate Class I ADH genes, which involve successive duplications from a single 'proto-primate' ancestral Class I ADH gene, generating the three Class I ADH genes now present in humans (Yasunami *et al*, 1990)

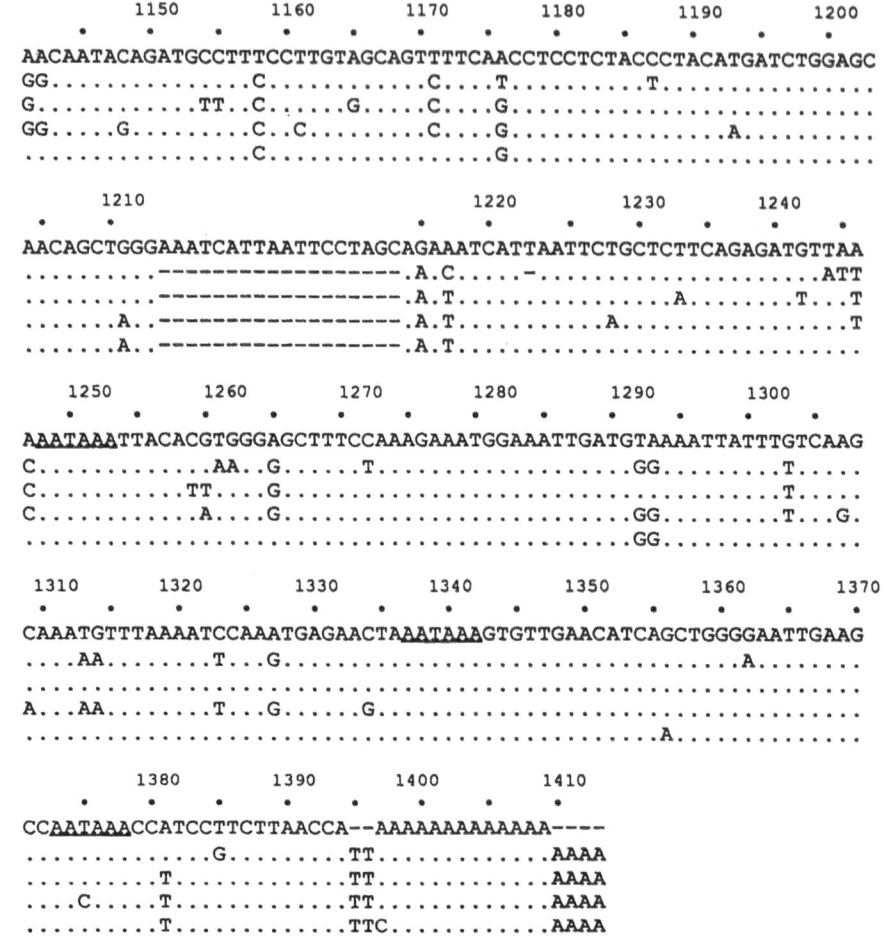

Figure 1.

and baboons (Trezise *et al*, 1991). These models differ according to whether the first duplication gave rise to ADHα (model A) (Ikuta *et al*, 1986; Trezise *et al*, 1991); ADHγ (model B) (Yokoyama and Yokoyama, 1987; Sun and Plapp, 1992; Yokoyama and Harry, 1993); or ADHβÊ(model C) (Yokoyama *et al*, 1990).

In this study, we have used the coding sequences of the following mammalian Class I ADH genes: human α, β and γ (Deuster *et al*, 1986; Ikuta *et al*, 1986; von Bahr-Lindstrom *et al*, 1986; Hg *et al* 1986); baboon β and γ (Trezise *et al*, 1989); rhesus α (Light *et al*, 1992); and horse E (Park and Plapp, 1991). Figure 2 shows phylogenetic trees generated by two methods, using the horse E sequence as the outgroup. Figure 2A was generated by the neighbour-joining method which is consistent with model A; whereas the maximum likelihood method (Figure 2B) generates a phylogenetic tree consistent with model B. Clearly, further data and analyses are required to distinguish these models, and to conclusively exclude model C.

Estimates of the timing of these gene duplications were obtained by using the number of synonomous nt substitutions to calibrate a molecular evolutionary 'clock', based upon the human α, β and γ , baboon β and γ , and rhesus α sequences, and using 25 mya as the

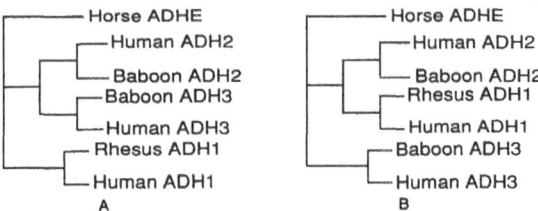

Figure 2.

time for the Catarrhini branch point (the common ancestor for human and Old World monkeys) (see Fleagle, 1988). Based upon an average rate of synonomous nt substitutions of 2.12 x 10^{-9} for the primate Class I ADH genes, the data suggests that the first duplication of human ADH genes occurred about 34 mya, and the second duplication at around 24 mya. This is consistent with the existence of three Class I ADH genes in baboon, as the second gene duplication occurred at, or near, the Catarrhini branchpoint.

In summary, we have isolated and sequenced baboon kidney cDNAs, generated from poly (A$^+$) RNA following reverse transcription and amplification using primers based on sequences of baboon liver and kidney Class I ADHs. The deduced amino acid sequence was most highly homologous with human ADHγ, and baboon kidney Class I ADH was therefore designated as ADHγ. Comparisons of human ADH α, β and γ mRNA 3' noncoding regions, with those for baboon ADH β and γ, revealed extensive homology and evidence for three polyadenylation signal consensus sequences, in each case. Phylogenetic analyses of human ADH α, β, and γ, baboon β and γ, and rhesus α coding sequences were consistent with successive gene duplications occurring at around 34 mya and 24 mya, although the sequence of such duplications remains to be clarified.

ACKNOWLEDGEMENT

The baboon tissues used in this study were kindly supplied by Dr J.L. VandeBerg, Southwest Foundation for Biomedical Research, San Antonio, Texas, USA.

REFERENCES

Berget, S.M., 1984, Are U4 small nuclear ribonucleoproteins involved in polyadenylation, *Nature* 309: 179-182.

Chomczynski, P., and Sacchi, N., 1987, Single-step method of RNA isolation by acid guanidinium thiocyanate-phenol-chloroform extraction, *Anal. Biochem.* 162: 156-159.

Deuster, G., Smith, M., Bilanchone, V., and Hatfield, W.G., 1986, Molecular analysis of the human Class I alcohol dehydrogenase gene family and nucleotide sequence of the gene encoding the β-subunit. *J. Biol. Chem.* 261: 2027-2033.

Deuster, G., Hatfield, G.W., Bhler, R., Hempel, J., Jornvall, H., and Smith, M., 1984, Molecular cloning and characterization of a cDNA for the β-subunit of human alcohol dehydrogenase. *Proc. Natl. Acad. Sci. USA* 81: 4055-4059.

Felsenstein, J., 1981, Evolutionary trees from DNA sequences: a maximum likelihood approach. *J. Mol. Evol.* 17: 368-376.

Fleagle, J.G., 1988, *Primate Adaptation and Evolution*, Academic Press, N.Y., p. 257.

Frohman, M.A., Dush, M.K., and Martin, G.R., 1988, Rapid production of full-length cDNAs from rare transcripts: amplification using a single gene-specific oligonucleotide primer. *Proc. Natl. Acad. Sci. USA* 85: 8998-9002.

Heden, L-O., Hg, J-O., Larsson, K., Lake, M., Lagerholm, E., Vallee, B.L., Jornvall, H., and von Bahr-Lind-strom, H., 1986, cDNA clones coding for the β-subunit of human liver alcohol dehydrogenase have differently sized 3'-noncoding regions. *FEBS Letts.,* 194: 327-332.

Higgins, D.G., Bleasby, A.J. and Fuchs, R., 1992, Clustal V: improved software for multiple sequence alignment. *CABIOS* 8: 189-191.

Holmes, R.S., Courtney, Y.R., and VandeBerg, J.L., 1986, Alcohol dehydrogenase isozymes in baboons: tissue distribution, catalytic properties and variant phenotypes in liver, kidney, stomach and testis. *Alcoholism: Clin. Exp. Res.,* 10: 623-630.

Holmes, R.S., Meyer, J., and VandeBerg, J.L., 1990, Baboon alcohol dehydrogenase isozymes: purification and properties of liver Class I ADH. Moderate alcohol consumption reduces liver Class I and Class II ADH activities, in Markert, C.L., ed. *Isozymes: Structure, Function and Use in Biology and Medicine.* New York: Wiley-Liss, pp 819-841.

Höög, J-O., Heden, L-O., Larsson, K., Jörnvall, H., and von Bahr-Lindstrom, H., 1986, The gamma 1 and gamma 2 subunits of human liver alcohol dehydrogenase. cDNA structures, two amino acid replacements and compatibility with changes in enzymatic properties. *Eur. J. Biochem.* 159: 215-218.

Ikuta, T., Szeto, S., and Yoshida, A., 1986, Three alcohol dehydrogenase subunits: cDNA structure and molecular and evolutionary divergence. *Proc. Natl. Acad. Sci. USA,* 81: 4055-4059.

Light, D.R., Dennis, M.S., Forsythe, I.J., Liu, C-C., Green, D.W., Kratzer, D.A., and Plapp, B.V., 1992. α-Isoenzyme of alcohol dehydrogenase from monkey liver. Cloning, expression, mechanism, coenzyme and substrate specificity. *J. Biol. Chem.* 267: 12592-12599.

Park, D-H., and Plapp, B.V., 1991, Isoenzymes of horse liver alcohol dehydrogenase active on ethanol and steroids. *J. Biol. Chem.* 266: 13296-13302.

Saitou, N., and Nei, M., 1987, The neighbour-joining method: a new method for reconstructing phylogenetic trees. *Mol. Biol. Evol.* 4: 406-425.

Smith, M., 1986, Genetics of human alcohol dehydrogenase and aldehyde dehydrogenase. *Adv. Human. Genet.,* 17: 261-281.

Smith, M., Hopkinson, D.A., and Harris, H., 1971, Developmental changes and polymorphism in human alcohol dehydrogenase. *Ann. Human Genet.,* 34: 251-271.

Sun, H.W., and Plapp, B.V., 1992, Progressive sequence alignment and molecular evolution of the Zn-containing alcohol dehydrogenase family. *J. Mol. Evol.* 34: 522-535.

Trezise, A.E.O., Cheung, B., Holmes, R.S., and Beacham, I.R., 1991, Evidence for three genes encoding Class I alcohol dehydrogenase subunits in baboon and sequence analysis of the 5' region of the gene encoding the ADHβ subunit. *Gene,* 103: 211-218.

Trezise, A.E.O., Godfrey, E.A., Holmes, R.S., and Beacham, I.R. 1989. Cloning and sequencing of cDNA encoding baboon liver alcohol dehydrogenase. Evidence for a common ancestral lineage with the human β-ADH and for Class I ADH gene duplications predating primate radiation. *Proc. Natl. Acad. Sci. USA* 86: 5454-5458.

von Bahr-Linstrom, H., Höög, J-O., Heden, L-O., Kaiser, R., Fleetwood, L., Larsson, K., Lake, M., Holmquist, B., Holmgren, A., Hempel, J., Vallee, B.L., and Jornvall, H., 1986, cDNA and protein structure for the alpha subunit of human liver alcohol dehydrogenase. *Biochem.,* 25: 2465-2470.

Yasunami, M., Kikuchi, I., Sarkata, D., and Yoshida, A., 1990, The human Class I alcohol dehydrogenase gene cluster: three genes are tandomly organised in an 80kb long segment of the genome. *Genomics* 7: 152-158.

Yokoyama, S., and Yokoyama, R., 1987, Molecular evolution of mammalian Class I alcohol dehydrogenase. *Mol. Biol. Evol.* 4: 504-513.

Yokoyama, S., and Harry, D.E., 1993, Molecular phylogeny and evolutionary rates of alcohol dehydrogenase in vertebrates and plants. *Mol. Biol. Evol.* 10: 1215-1226.

Yokoyama, S.R., Yokoyama, C.S., Kinlaw, D.E., and Harry, D.E., 1990, Molecular evolution of the zinc-containing long chain alcohol dehydrogenase genes. *Mol. Biol. Evol.* 7: 143-154.

38

THE ROLE OF LEUCINE 116 IN DETERMINING SUBSTRATE SPECIFICITY IN HUMAN B₁ ALCOHOL DEHYDROGENASE

Thomas D. Hurley[*] and David L. Vessell

Department of Biochemistry and Molecular Biology
Indiana University School of Medicine
Indianapolis, IN 46202

INTRODUCTION

There are multiple forms of human alcohol dehydrogenase (EC 1.1.1.1). Each of these isoenzymes is dimer of approximately 80,000 molecular weight and contains between 373-377 amino acids per subunit (Bosron *et al.*, 1993). There now appears to be at least seven separate genes in humans (*ADH₁-ADH₇*) with polymorphism observed at the *ADH₂* and *ADH₃* loci. The sequence identity between these isoenzymes ranges from greater than 93% between the *ADH₁-ADH₃* genes to about 60% between *ADH₄* and any of the other genes (Ehrig *et al.*, 1990). The individual isoenzymes exhibit large differences in both affinity and maximal oxidation rate for different alcohol substrates (Ehrig *et al.*, 1990).

The three-dimensional structures of both the Horse E and human β₁ (product of the *ADH₂* gene) isoenzymes are now known to greater than 2.5 Å resolution (Eklund *et al.*, 1976 and Hurley *et al.*, 1994). Site-directed mutagenesis has been used to probe the determinants of coenzyme binding as well as the catalytic efficiency of steroid, primary, and secondary alcohol oxidation in both these isoenzymes (Hurley *et al.*, 1990, Park and Plapp, 1992, Hurley and Bosron, 1992, and Höög *et al.*, 1992). These studies have identified specific residue substitutions among the isoenzymes that account for the observed functional differences toward these substrates.

The recently published structure of the human β₁ alcohol dehydrogenase complexed with NAD(H) and cyclohexanol suggested that Leu 116 may play an important role in the binding of bulky substrates such as secondary alcohols (Hurley *et al.*, 1994). Specifically, the side chain of Leu 116 was found to occupy a distinctly different conformation in the cyclohexanol complex than that observed in binary NAD⁺ complexes or ternary NAD⁺:4-iodopyrazole complexes. We used site-directed mutagenesis to mutate Leu 116 to Ala and expressed the mutated enzyme in *E. coli*. The purified enzyme was characterized by

---

[*] To whom correspondence should be addressed.

steady-state kinetic analyses to determine the effect of this substitution on substrate specificity.

METHODS

The mutagenesis and expression of the human β_1 cDNA was performed as described (Hurley et al., 1990). Purification of the recombinant enzyme was performed as described (Hurley and Bosron, 1992), with the exception that 10 μM ZnSO$_4$ was included in all buffers to prevent activity loss. To prevent the formation of the insoluble ZnPO$_4$ salt, 25 mM Tris-HCl, pH 7.5 was substituted for 20 mM sodium phosphate, pH 7.5 as the Affi-Gel Blue column buffer and the enzyme was stored in 15 mM Tris-HCl, pH 7.5 containing 10 μM ZnSO$_4$ at 4°C. Long-term storage (> 2 weeks) was best in 15 mM Tris-HCl, pH 7.5, 10 μM ZnSO$_4$, and 50% glycerol at -20°C. Enzyme activity was measured in 100 mM glycine, pH 10.0 containing 2.4 mM NAD$^+$ (Boehringer Mannheim, grade I) and 33 mM ethanol using an extinction coefficient of 6.22 cm^{-1}mM^{-1} for product NADH.

Apparent K_M values for different alcohol substrates were determined at 2.4 mM NAD$^+$ by varying the alcohol concentration in 100 mM sodium phosphate, pH 7.5. All alcohols were of the highest grade available (Aldrich or Schweizerhall) and were used without further purification. The kinetic constants were determined from the initial velocity data using Cleland's kinetic programs (Cleland, 1979). All other reagents were from Sigma and were the highest grade commercially available.

RESULTS AND DISCUSSION

The substitution of Leu 116 by Ala (β116A) creates an active enzyme with distinct substrate preference compared to β_1 (Table 1). The apparent K_M values for straight-chain alcohols exhibited by β116A decrease by almost 100-fold as the chain length increases from two to six carbons. The K_M values exhibited by β_1 for the same substrates varied only 4-fold. A linear relationship between V_{max}/K_M and the log of the octanol:water partition coefficient [log(P)] is observed for β116A (Figure 1). Linear regression of the log(P) versus V_{max}/K_M data yielded a line with a slope of 0.78. The V_{max}/K_M values of β116A range from 2- to 4-fold lower than β_1 for the short chain alcohols such as ethanol and 1-propanol to 3- to 6-fold

Table 1. K_M Values for Alcohols

Substrate	β116A	β_1	β93A94I
Ethanol	1,100 ± 300	50	270
1-Propanol	920 ± 140	19	130
1-Butanol	67 ± 17	12	92
1-Pentanol	46 ± 18	19	66
1-Hexanol	18 ± 10	22	65
Isopropanol	62,900 ± 9,500	1,500	49
(R) 2-Butanol	36,100 ± 4,000	2,200	9.2
(S) 2-Butanol	18,200 ± 800	360	29
Cyclohexanol	9,800 ± 2,800	3,900	10

The data for β_1 and β93A94I are from Stone et al., 1989 and Hurley and Bosron, 1992, respectively.

Table 2. V_{max}/K_M Values for Alcohols

Substrate	log(P)	β116A	β₁	β93A94I
Ethanol	-0.32	31	80	12
1-Propanol	0.34	35	160	29
1-Butanol	0.88	520	240	39
1-Pentanol	1.4	680	190	48
1-Hexanol	2.0	1200	200	49
Isopropanol	0.05	0.16	0.79	86
(R) 2-Butanol	0.61	0.19	2.9	490
(S) 2-Butanol	0.61	0.42	18	140
Cyclohexanol	1.2	0.16	0.76	420

The data for β₁ and β93A94I are from Stone *et al.*, 1989 and Hurley and Bosron, 1992, respectively.

higher than β₁ for the longer chain alcohols such as 1-pentanol and 1-hexanol (Table 2). This is intermediate between the β₁ data which exhibits a slope of 0.16 and the α isoenzyme data which exhibits a slope of 1.6 for substrates with less than 5 carbons (Figure 1). Clearly the substitution of Leu 116 by Ala in β₁ alters the manner in which the enzyme interacts with different straight-chain alcohols. One possible explanation is that the alanine side chain at position 116 increases the number of possible conformations accessible to smaller substrates, such as ethanol, within the binding pocket. This would translate into fewer productive

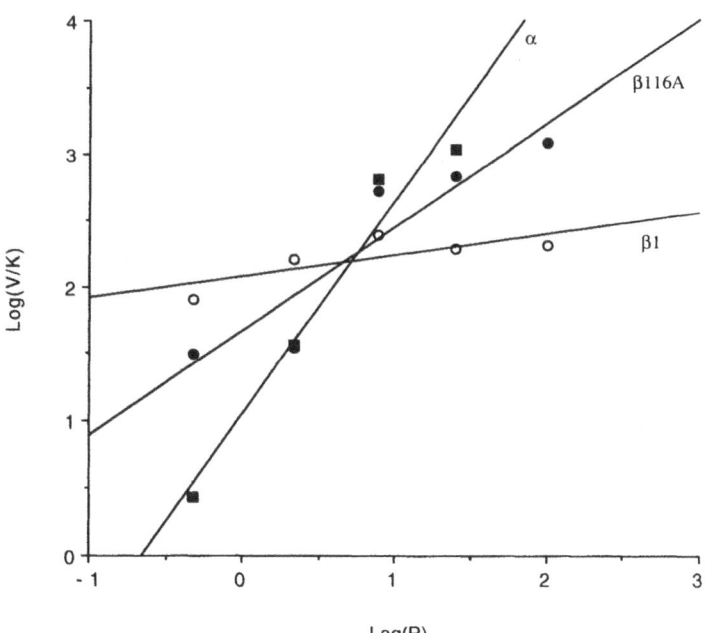

Figure 1. A plot of V_{max}/K_M versus Log(P) for primary alcohols for the different enzymes. The data for β116A, β₁, and α are displayed as (●), (○), and (■), respectively.

encounters between the enzyme and substrate, thus lowering the substrate's apparent V_{max}/K_M value (Table 2). More definitive answers must await determination of the alcohol association and dissociation rate constants and hydride transfer rates for these substrates by stopped-flow analyses.

The substitution of Leu 116 by Ala also produced a significant decrease in the efficiency of the mutant enzyme toward secondary alcohols (Tables 1 and 2). We and others have previously shown that substitutions at positions 48 and 93 in β_1, which widen the substrate pocket near the zinc atom, are correlated with an increased efficiency toward secondary alcohols (Hurley and Bosron, 1992, and Höög et al., 1992). Specifically, we reported that the substitution of Phe 93 and Thr 94 by Ala and Ile, respectively, created an enzyme with marked preference for secondary alcohols (β93A94I in Table 2). Our modelling experiments suggested that Leu 116 should not influence the oxidation of secondary alcohols in this manner, because it would not directly interact with these bulky substrates when they are bound to the catalytic zinc atom. However, the catalytic efficiency of β116A toward secondary alcohols is between 5- and 40-fold lower than β_1 (Table 2). This suggests that widening the entrance of the substrate pocket is detrimental to the oxidation of secondary alcohols in the β_1 isoenzyme.

A possible explanation for this phenomenon is that the side chain of Leu 116 acts as a bottleneck toward the binding and release of bulky substrates, such as secondary alcohols. We suggested that the position of cyclohexanol near Leu 116 in the three-dimensional structure of β_1 complexed with NAD(H) and cyclohexanol might represent the slow step in its binding pathway (Hurley et al., 1994). It is possible, therefore, that the side chain of Leu 116 may, in effect, trap cyclohexanol in the substrate binding pocket long enough for the oxidation of cyclohexanol. An important requirement for this mechanistic explanation is that no step for the oxidation of cyclohexanol is slower than the rate-limiting dissociation of NADH during ethanol oxidation, since the V_{max} for cyclohexanol oxidation is identical to that for ethanol oxidation for the β_1 isoenzyme (3 min^{-1} vs. 4 min^{-1}, Table 3). The subsitution of Ala for Leu 116 may remove this bottleneck since the apparent K_M values for secondary alcohols increase dramatically compared to β_1 (Table 1) and the V_{max} values exhibited by β116A for secondary alcohol oxidation are 3- to 10-fold lower than those for oxidation of primary alcohols (Table 3). The increased K_M values may indicate that secondary alcohols are not bound as effectively in β116A. Furthermore, the differences in the V_{max} values suggest that the rate-limiting step for β116A may be different for primary and secondary alcohols. It would be interesting to determine hydride transfer rates and individual associa-

Table 3. V_{max} Values for Alcohol Oxidation

Substrate	β116A	β_1	β93A94I
Ethanol	34.4 ± 3.6	4.0	3.2
1-Propanol	32.4 ± 4.0	3.0	3.8
1-Butanol	35.2 ± 3.6	2.8	3.6
1-Pentanol	31.2 ± 3.2	3.6	3.2
1-Hexanol	21.6 ± 3.2	4.4	3.2
Isopropanol	10.4 ± 0.4	1.2	4.2
(R) 2-Butanol	6.9 ± 0.3	6.3	4.5
(S) 2-Butanol	7.6 ± 0.2	6.5	4.1
Cyclohexanol	1.8 ± 0.1	3.0	4.2

The data for β_1 and β93A94I are from Stone et al., 1989 and Hurley and Bosron, 1992, respectively.

tion and dissociation rates for primary and secondary alcohols with this enzyme in order to better define the mechanistic possibilities.

ACKNOWLEDGEMENTS

This work was supported by K21-AA00150, R37AA07117, and the Grace M. Showalter Trust. We also express our thanks to Marvin Makinen for valuable discussions.

REFERENCES

Bosron, W.F., Ehrig, T., and Li, T.-K. (1993) Genetic factors in alcohol metabolism and alcoholism. *Semin. Liver Dis.* **13**, 126-135.

Cleland, W.W. (1979) Statistical analysis of enzyme kinetic data. *Methods Enzymol.* **63A**, 103-138.

Ehrig, T., Bosron, W.F., and Li, T.-K. (1990) Alcohol and Aldehyde Dehydrogenase. *Alcohol Alcoholism* **25**, 105-116.

Eklund, H., Nordström, B., Zeppezauer, E, Söderlund, G., Ohlsson, I.,

Boiwe, T., Söderberg, B.-O., Tapia, O., Brändén, C.-I., and Åkeson, Å. (1976) Three-dimensional structure of horse liver alcohol dehydrogenase at 2.4 Å resolution. *J. Mol. Biol.* **102**, 27-59.

Höög, J.-O., Eklund, H., and Jörnvall, H. (1992) A single residue exchange gives human recombinant ββ alcohol dehydrogenase γγ properties. *Eur. J. Biochem.* **205**, 519-526.

Hurley, T.D., Edenberg, H.J., and Bosron, W.F. (1990) Expression and kinetic characterization of variants of human β₁β₁ alcohol dehydrogenase containing substitutions at position 47. *J. Biol. Chem.* **265**, 16366-16372.

Hurley, T.D. and Bosron, W.F. (1992) Human alcohol dehydrogenase: Dependence of secondary alcohol oxidation on the amino acids at positions 93 and 94. *Biochem. Biophys. Res. Commun.* **183**, 93-99.

Hurley, T.D., Bosron, W.F., Stone, C.L., and Amzel, L.M. (1994) Structures of three human β alcohol dehydrogenase variants: Correlations with their functional differences. *J. Mol. Biol.* **239**, 415-429.

Park, D.-H. and Plapp, B.V. (1992) Interconversion of E and S isoenzymes of horse liver alcohol dehydrogenase. *J. Biol. Chem.* **267**, 5527-5533.

Stone, C.L., Li, T.-K., and Bosron, W.F. (1989) Stereospecific Oxidation of Secondary Alcohols by Human Alcohol Dehydrogenases. *J. Biol. Chem.* **264**, 11112-11116.

39

MUTATIONS OF HUMAN CLASS III ALCOHOL DEHYDROGENASE

Mats Estonius, Jan-Olov Höög, Olle Danielsson, and Hans Jörnvall

Department of Medical Biochemistry and Biophysics
Karolinska Institutet
S-171 77 Stockholm, Sweden

INTRODUCTION

The residues lining the active site pocket of class III alcohol dehydrogenase differ markedly from those of class I and other classes. Two charged residues in the vicinity of the catalytic site are characteristic of class III. One is Arg115, which is at the outer part of the substrate-binding cleft and is associated with major enzymatic characteristics of class III, *i.e.* fatty acid activation (Moulis et al., 1991; Holmquist et al., 1993) and glutathione-dependent formaldehyde dehydrogenase activity (Engeland et al., 1993). The other charged residue is Asp57, located in the middle part of the substrate pocket. In an attempt to study the relative importance of residues involved in differentiation of classes, we have by site-directed mutagenesis examined not only the function of Asp57, but also the roles of Tyr93 and Thr48 of class III (Estonius et al., 1994). The charge and polarity at positions 57 and 93 were altered by Asp57Leu and Tyr93Phe substitutions, respectively. As opposed to Asp in class III, Leu is found at position 57 in most class I alcohol dehydrogenases, and in contrast to Tyr in class III, Phe or Ala are found at position 93. Therefore, a combination of the Asp57Leu and Tyr93Phe exchanges in one mutant is of interest in order to see if a class I alcohol dehydrogenase substrate pocket, and class I enzymatic properties, can be mimicked by altering the active site structure in a direction toward those of class I enzymes. In addition, to check for differences in the oxidation of short-chain aliphatic alcohols *contra* S-hydroxymethylglutathione (HMGSH), we have replaced Thr48 with Ala, a mutation deleterious to class I alcohol dehydrogenase activity (Höög et al., 1992).

MATERIALS AND METHODS

A 1360 bp cDNA fragment harboring the entire coding part of a human class III alcohol dehydrogenase (Sharma et al., 1989), isolated after cleavage with *Eco*RI and *Ssp*I, was ligated into the plasmid pKK 223-3 (Pharmacia Biotech). This resulted in an expression vector for class III (Estonius et al., 1994), similar to those used for class I alcohol dehydrogenases (Höög et al., 1992). Mutagenesis was performed according to the phosphorothioate method (Taylor et al., 1985), and in all, five recombinant class III alcohol dehydrogenase enzymes were expressed in *E. coli*: rχwt (wild-type), rχT48A (Thr48 replaced with Ala), rχD57L (Asp57 replaced with Leu), rχY93F (Tyr93 replaced with Phe), and rχD57L:Y93F (Asp57 and Tyr93 replaced with Leu and Phe, respectively). The protein purifications utilized ion-exchange and affinity charomatographies, and protein amounts were determined colorimetrically (Bradford, 1976). Protein purity was confirmed by SDS/polyacrylamide gel electrophoresis. Oxidation rates of ethanol, octanol, and 12-hydroxydodecanoic acid were determined with 2.4 mM NAD$^+$ in 0.1 M glycine/NaOH, pH 10. Formaldehyde dehydrogenase activity was determined with 1 mM GSH/2.4 mM NAD$^+$ (rχwt and rχY93F) or 10 mM GSH/10 mM NAD$^+$ (rχT48A, rχD57L, and rχD57L:Y93F) in 0.1 M sodium pyrophosphate, pH 8.0. Inhibition of enzyme activity was measured with 50 mM 4-methylpyrazole and 500 mM ethanol at pH 10.0. All kinetic determinations were performed at 24°C. A weighted non-linear regression analysis program (Lutz et al., 1986) was used to calculate kinetic constants.

RESULTS AND DISCUSSION

For the recombinant proteins, except rχT48A which did not exhibit ethanol activity, plots of *v versus* [S] were linear up to 2 M ethanol, yielding k_{cat}/K_m values of 0.045 mM-1 × min^{-1}, which are in agreement with earlier reports for the native enzyme (Wagner et al., 1984). This suggests that the overall oxidation of ethanol remains unaltered with the Asp57Leu and Tyr93Phe substitutions. Kinetic constants obtained with other substrates are summarized in Table 1, and discussed further below.

Table 1. Kinetic constants for recombinant wild-type and mutated class III ADHs.

Enzyme	Octanol			12-HDA			HMGSH		
	K_m	k_{cat}	k_{cat}/K_m	K_m	k_{cat}	k_{cat}/K_m	K_m	k_{cat}	k_{cat}/K_m
	mM	min^{-1}	mM^{-1} min^{-1}	mM	min^{-1}	mM^{-1} min^{-1}	mM	min^{-1}	mM^{-1} min^{-1}
rχwt	0.5	160	320	0.06	160	2700	0.004	200	50000
rχT48A			<0.05			<0.015	4.4	30	7
rχD57L	0.3	160	530	0.06	160	2700	0.6	55	90
rχY93F	0.5	35	70	0.04	30	750	0.004	40	10000
rχD57L:Y93F	0.1	30	300	0.02	25	1250	0.5	20	40
rχR115A	1.2	150	125	0.68	122	180	0.28	97	350
rχR115D	2.2	240	110	1.30	97	75	0.91	36	40

Assay conditions were as given in the materials section. k_{cat} values were calculated per subunit with a molecular mass of 40 kDa. Values given for rχ115D are from Engeland et al. (1993). The k_{cat}/K_m value for octanol of rχT48A was determined at pH 8.0

Asp57Leu. The replacement of Asp57 with Leu does not considerably affect the oxidation of aliphatic alcohols (Table 1), but decreases the catalytic efficiency (k_{cat}/K_m) toward HMGSH by a factor of about 500, depending on an increased K_m and a reduced k_{cat} for this substrate. This decrease in k_{cat}/K_m of rχD57L in the oxidation of HMGSH is of the same order of magnitude as the decrease found for the Arg115Asp or Arg115Ala exchanges (Engeland et al., 1993; Table 1), which similarly affect both K_m and k_{cat}. Significantly, this suggests that Asp57 is equally crucial as Arg115 to the formaldehyde dehydrogenase activity.

Tyr93Phe. Essentially without influence on any of the K_m values, the main effect of the Tyr93Phe substitution is an overall decreased k_{cat} (Table 1). This may indicate that this residue replacement could affect a rate-limiting step common to both the aliphatic alcohol and HMGSH turnover of class III. Another finding is an increased inhibition of ethanol activity at high 4-methylpyrazole concentrations (Table 2). This enhanced inhibition presumably is related to an increase in hydrophobicity of the substrate-binding pocket, facilitating binding of the inhibitor.

Asp57Leu:Tyr93Phe. The net effect of the combined mutations is a catalytic efficiency which is lowered overall (Table 1). Although the K_m values for the aliphatic alcohols are largely unaltered, the k_{cat} with these substrates is reduced by a factor of 5, and therefore is roughly the same as with the Tyr93Phe exchange, indicating that it is the latter residue exchange that causes the reduction in k_{cat} (Table 1). The decrease in k_{cat} for HMGSH is of the same order of magnitude as for other alcohols tested, but a 100-fold increase in K_m for HMGSH is noticed. In this case, the Asp57Leu exchange of the double mutant appears to cause the increase in K_m for HMGSH, since the same effect is manifest with the sole Asp57Leu mutant (Table 1). This could indicate either different rate-limiting steps in the oxidation of aliphatic alcohols on the one hand, and of HMGSH on the other, or different modes of substrate binding. At high inhibitor concentrations, the ethanol activity of rχD57L:Y93F is abolished (Table 2). This enhanced inhibition could be explained partly by the Asp57Leu substitution, but evidently the specific inhibition of ethanol oxidation is mediated by the potentiating effect of Phe at position 93 (Table 2).

Thr48Ala. Ser or Thr are found at position 48 in all alcohol dehydrogenases, and are known to interact with both the substrate and the coenzyme via hydrogen bonding (Eklund et al., 1990). In class I, removal of the side-chain hydroxyl of this residue results in an inactive enzyme (Höög et al., 1992). This loss of classical alcohol dehydrogenase activity applies also to class III, where rχT48A exhibits no activity toward ethanol. In contrast, the

Table 2. 4-Methylpyrazole inhibition of recombinant class III ADHs

Protein	Residue at position		Ethanol activity at 50 mM 4MePz (% of uninhibited activity)
	57	93	
Wild-type class III, rχwt	Asp	Tyr	20
Mutated class III, rχD57L	**Leu**	Tyr	15
Mutated class III, rχY93F	Asp	**Phe**	10
Mutated class III, rχD57L:Y93F	**Leu**	**Phe**	0

As shown by the values for the mutant forms, they have gained an increased 4-methylpyrazole inhibition. Ethanol activity is expressed as % of uninhibited activity at 0.5 M ethanol. Bold-face indicates residue exchanges.

formaldehyde dehydrogenase activity is retained, although with roughly a 1000-fold decrease in k_{cat}/K_m (Table 1). The retention of formaldehyde dehydrogenase activity with the concomitant loss of class I type of alcohol dehydrogenase activity of rχT48A could be explained by enzyme-substrate interactions additional to those functional in the oxidation of primary aliphatic alcohols. This would suggest that the amino function of HMGSH interacts with the enzyme, presumably with Asp57, in a similar manner as has been suggested for the interactions of Arg115 of class III with a carboxyl group of HMGSH (Engeland et al., 1993). However, the rat class VI enzyme has Glu and Arg at positions 57 and 115, respectively, but is devoid of formaldehyde dehydrogenase activity (Höög, unpublished). Consequently, not only the types of residue at these two positions are important for the formaldehyde dehydrogenase activity, but also the protein fold and the exact side-chain orientation of the substrate pocket residues. In conclusion, we point out the possibility that the amino group of HMGSH interacts with Asp57 of class III, and show that the residues at positions 57 and 93 determine the degree of 4-methylpyrazole inhibition of enzymatic activity.

REFERENCES

Bradford, M. M. (1976) A rapid and sensitive method for the quantitation of microgram quantities of protein utilizing the principle of protein-dye binding, *Anal. Biochem. 72*, 248-254.

Engeland, K., Höög, J.-O., Holmquist, B., Estonius, M., Jörnvall, H. & Vallee, B. L. (1993) Mutation of Arg-115 of human class III alcohol dehydrogenase: A binding site required for formaldehyde dehydrogenase activity and fatty acid activation, *Proc. Natl. Acad. Sci. USA 90*, 2491-2494.

Estonius, M., Höög, J.-O., Danielsson, O. & Jörnvall, H. (1994) Residues specific for class III alcohol dehydrogenase. Class transitions and S-hydroxymethylglutathione binding. Site-directed mutagenesis of the human enzyme, *Biochemistry*, in press.

Höög, J.-O., Eklund, H. & Jörnvall, H. (1992) A single-residue exchange gives human recombinant ββ alcohol dehydrogenase γγ isozyme properties, *Eur. J. Biochem. 205*, 519-526.

Holmquist, B., Moulis, J.-M., Engeland, K. & Vallee, B. L. (1993) Role of arginine 115 in fatty acid activation and formaldehyde dehydrogenase activity of human class III alcohol dehydrogenase, *Biochemistry 32*, 5139-5144.

Lutz, R. A., Bull, C. & Rodbard, D. (1986) Computer analysis of enzyme-substrate-inhibitor kinetic data with automatic model selection using IBM-PC compatible microcomputers, *Enzyme 36*, 197-206.

Moulis, J.-M., Holmquist, B. & Vallee, B. L. (1991) Hydrophobic anion activation of human liver χχ alcohol dehydrogenase, *Biochemistry 30*, 5743-5749.

Sharma, C. P., Fox, E. A., Holmquist, B., Jörnvall, H. & Vallee, B. L. (1989) cDNA sequence of human class III alcohol dehydrogenase, *Biochem. Biophys. Res. Commun. 164*, 631-637.

Taylor, J. W., Ott, J. & Eckstein, F. (1985) The rapid generation of oligonucleotide-directed mutations at high frequency using phosphorothioate-modified DNA, *Nucleic Acids Res. 13*, 8765-8785.

Wagner, F. W., Parés, X., Holmquist, B. & Vallee, B. L. (1984) Physical and enzymatic properties of a class III isozyme of human liver alcohol dehydrogenase: χ-ADH, *Biochemistry 23*, 2193-2199.

40

HUMAN AND RAT CLASS IV ALCOHOL DEHYDROGENASES

Correlations of Primary Structures with Enzymatic Properties

Jaume Farrés[1], Alberto Moreno[1], Bernat Crosas[1], Ella Cederlund[2], Abdellah Allali-Hassani[1], Josep M. Peralba[1], Lars Hjelmqvist[2], Hans Jörnvall[2] and Xavier Parés[1]

[1] Department of Biochemistry and Molecular Biology
Faculty of Sciences
Universitat Autònoma de Barcelona, 08193 Bellaterra
Spain
[2] Department of Medical Biochemistry and Biophysics
Karolinska Institutet
S-171 77 Stockholm, Sweden

INTRODUCTION

Mammalian alcohol dehydrogenase (ADH, EC 1.1.1.1) is a complex enzymatic system composed of multiple molecular forms, which have been grouped into classes according to their enzymatic and structural characteristics. In addition to the three classes (I-III) early recognized (Vallee and Bazzone, 1983), more recent studies have led to the identification of three more classes: class IV (Parés et al., 1990, Parés et al., 1992), class V (Yasunami et al., 1991), and class VI (Zheng et al., 1993).

Human and rat stomach mucosa contain novel ADH forms which were purified and characterized at enzymatic level (Julià et al., 1987, Yin et al., 1990, Moreno and Parés, 1991, Boleda et al., 1993). Preliminary work at structural level revealed that the stomach enzymes belonged to a new class, named class IV (Parés et al., 1990, Parés et al., 1992). However, the relatively low amount of class IV ADH that is present in the gastrointestinal tract has precluded a more thorough structural study until recently, when we reported the complete amino acid sequence for the rat enzyme (Parés et al., 1994) and the cDNA sequence coding for the human enzyme (Farrés et al., 1994).

Here the structural and kinetic properties of both human and rat class IV enzymes are discussed and compared in relation to those of other ADH classes.

EXPERIMENTAL

Purification of Class IV Alcohol Dehydrogenase from Human and Rat Stomach

Human class IV ADH was purified as previously reported (Moreno and Parés, 1991), but two modifications were introduced. First, prior to the DEAE-Sepharose column, the dialyzed supernatant was filtered through a filter cake (200 ml total volume) of DEAE-Sepharose equilibrated with 10 mM Tris/HCl, 0.5 mM DTT, pH 9.0. Secondly, the AMP-Sepharose chromatography employed 100 mM Tris/HCl, 0.5 mM DTT, pH 8.0, as a loading buffer. Rat class IV ADH was isolated by improvement of a previous procedure (Julià et al., 1987), modified to use two consecutive chromatographic steps on Blue-Sepharose instead of on AMP-Sepharose, and by addition of a final Sephacryl S-300 chromatography in 100 mM Tris/HCl, pH 8.5 (Parés et al., 1994).

Enzyme Assays

Human and rat class IV enzymes were assayed as reported (Moreno and Parés, 1991, Julià et al., 1987, Boleda et al., 1993). ADH activity was determined spectrophotometrically in 0.1 M sodium phosphate, pH 7.5, or 0.1 M glycine/NaOH, pH 10.0, at 25°C, by measuring the formation or utilization of NADH. Glutathione-dependent formaldehyde dehydrogenase activity was determined according to the method of Koivusalo et al. (1989). The dissociation constant for NAD^+ (K_{ia}) was calculated from the plot of $1/v$ versus $1/[NAD^+]$ at different alcohol concentrations. NAD^+ concentrations ranged from 0.25 to 2 mM for the human enzyme and from 0.12 to 2.4 mM for the rat enzyme. Ethanol (25-200 mM) was used for the human enzyme and octanol (0.2-1 mM) for the rat enzyme.

Determination of the Amino Acid Sequence of Rat Class IV ADH

As reported by Parés et al. (1994), the purified protein was treated with dithiothreitol and carboxymethylated. Separate batches were digested with Lys-, Glu-, Gly- and Asp-specific proteases, as well as chymotrypsin, as described (Cederlund et al., 1991). The resulting peptide mixtures were fractionated by reverse-phase HPLC. The N-terminal residue was deblocked by treatment with nonaqueous methanolic trifluoroacetic acid, 1:1 (vol/vol)(for 16 h at room temperature) to remove the acetyl group. Sequence analyses were performed by degradation in a gas-phase (Applied Biosystems 477A with on-line 120A analyzer) or a solid-phase (Milligen 6600) sequencer.

Cloning and Sequencing of the cDNA Coding for Human Class IV ADH

A more detailed description of the methods employed can be found in Farrés et al. (1994). In summary, two oligonucleotide primers (primer 1, coding for amino acid residues 138 to 143, and primer 2, complementary to the sequence coding for amino acid residues 282 to 287) were designed based on the amino acid sequence of rat class IV ADH (Parés et al., 1994), and used to amplify a 449-bp cDNA fragment from rat stomach mucosa poly(A)-rich RNA. The PCR product was subsequently used as a probe to screen a human stomach cDNA library in λgt10 (Clontech Laboratories). To obtain the 5'-end of the human class IV ADH cDNA, direct amplification from the cDNA library was performed, using two sets of nested λgt10 and class IV-specific primers. DNA fragments were subcloned into pBluescript II SK(+). Sequence determination of double-stranded DNA was carried out by the dideoxynucleotide chain termination method, using fluorescently labeled T7 and T3 primers with the *Taq* dye primer cycle sequencing kit (Applied Biosystems), in an Applied Biosystems 373A DNA sequencer.

Results and discussion

Enzymatic Properties and Physiological Role of Class IV ADH

The most distinct kinetic feature of class IV enzymes is their high kcat values for ethanol (Table 1) and, in general, for medium- and long-chain aliphatic alcohols and aldehydes (Julià et al., 1987, Moreno and Parés, 1991, Boleda et al., 1993). Class IV kcat values are the highest for any ADH class, including the very active human class I $\beta_2\beta2$ and $\beta_3\beta3$ variants (800 min^{-1} and 600 min^{-1}, respectively, Burnell and Bosron, 1989). Double reciprocal plots of alternate substrate and product inhibition kinetics (not shown) were compatible with an ordered bi bi mechanism. Similar to other ADHs for which the rate-limiting step is coenzyme dissociation, a good correlation apprears to exist between higher kcat values and higher values of the kinetic constants for NAD$^+$ (Table 1).

The Km values for ethanol are also higher in class IV ADHs with respect to class I isoenzymes (Table 1). The Km for ethanol is even much higher in the rat class IV ADH (2.4 M) than in the human class IV enzyme (37 mM). In general, Km values for straight-chain aliphatic substrates are also higher, and therefore kcat/Km values are lower (Table 2), for rat class IV ADH than for the corresponding human enzyme. Thus, ethanol and acetaldehyde are poorly metabolized by rat class IV ADH. Human class IV still eliminates ethanol and acetaldehyde with only moderate efficiency. This fact, taken together with the relatively low abundance of the enzyme in the gastric mucosa (approximately 10 μg enzyme/g tissue), questions the involvement of class IV ADH in the first-pass metabolism of ethanol, although the overall contribution of human class IV ADH to the extrahepatic metabolism of ethanol cannot be underestimated. Furthermore, fundamental differences between the human and rat class IV enzymes should be taken into account when using the rat as a model to study the metabolism and effects of ethanol.

In terms of kcat/Km values, medium- to long-chain alcohols and aldehydes are among the best substrates of class IV ADH (Table 2, Julià et al., 1987, Moreno and Parés, 1991, Boleda et al., 1993). Aromatic compounds such as benzyl alcohol and *m*-nitrobenzal-

Table 1. Kinetic constants of class I and class IV ADHs at pH 7.5. Values for human class I ($\alpha\alpha$, $\beta_1\beta1$ and $\gamma_1\gamma1$) isoenzymes were taken from Burnell and Bosron (1989)

	Class I	Class IV	
		Human	Rat
Km ethanol (mM)	0.05-4.2	37	2400
kcat ethanol (min^{-1})	18-170	1510	2600
Km NAD' (μM)	7.9-13	180	300
K$_{ia}$ NAD' (μM)	32-90	1600	4000

Table 2. kcat/Km values of human and rat class IV ADH at pH 7.5. n.d., not determined

Substrate	Human kcat/Km (mM^{-1}.min^{-1})	Rat kcat/Km (mM^{-1}.min^{-1})
Ethanol	41	1.1
Butanol	n.d.	9.7
Pentanol	28330	n.d.
Octanol	34590	1560
Benzyl alcohol (pH 10.0)	4780	9990
Cyclohexanol (pH 10.0)	12	1.2
Acetaldehyde	1860	54
trans-2-Hexenal	117200	45300
Octanal	92410	202800
m-Nitrobenzaldehyde	723430	109200

dehyde are also very good substrates. Cyclohexanol and secondary alcohols are very poorly oxidized. Class IV enzymes do not show activity with formaldehyde in the presence of glutathione.

Among the physiologically relevant substrates, it was reported that the rat class IV enzyme is active with aldehydic products of lipid peroxidation, such as hexanal, *trans*-2-hexenal and 4-hydroxynonenal, with ω-hydroxyfatty acids, and with retinoids (Boleda et al., 1993). Human class IV is also very efficient with lipid peroxidation-derived aldehydes and ω-hydroxyfatty acids (Moreno and Parés, 1991, and unpublished results). Km values for retinol and retinal are in the 10-30 μM range for both the rat (Boleda et al, 1993) and the human (unpublished results) class IV enzymes. It has also been shown that human class IV has the highest kcat/Km values for retinoids among all human classes (Yang et al., 1994). A multiple role in the elimination of toxic aldehydes, the oxidation of ω-hydroxyfatty acids, and in the metabolic pathway of retinoic acid has been suggested for rat class IV ADH (Boleda et al., 1993). The presence of class IV ADH in the mucosa of the upper digestive tract, i.e., mouth, esophagus and stomach (Boleda et al., 1989, Moreno et al., 1994), cornea (Holmes, 1988) and other epithelia, which are known target tissues for retinoic acid and the sites of active lipid peroxidation, supports this hypothesis.

Interestingly, class 3 aldehyde dehydrogenase (ALDH) exhibits a similar specificity towards medium-chain and long-chain and aromatic compounds (but not retinal) that class IV ADH. In addition, class 3 ALDH is localized in the same organs as class IV ADH (i.e., cornea, mouth, esophagus, and stomach mucosa). This supports the idea that both enzymes constitute a metabolic system characteristic of external mucosa, with a role in the metabolism of xenobiotics and lipid peroxidation-derived products (Parés and Farrés, 1995).

Primary Structures of Rat and Human Class IV ADHs

Rat class IV ADH was digested by different proteases and 96 peptides were analyzed after purification by reverse-phase HPLC. The results gave overlapping fragments for all

Table 3. Pairwise sequence comparisons between human ADH classes (interclass identities), and between human and rat ADH classes (intraclass identities)

Human Class	% Interclass identities (human)				% Intraclass identities (human/rat)
	I	II	III	IV	
I		59	61	70	82
II			63	59	77
III				61	94
IV					88

regions and established that the complete amino acid sequence of rat class IV ADH contains 374 residues (Parés et al., 1994). The N-terminal residue is blocked by an acyl group as in all other mammalian ADHs analyzed so far.

The information provided by the rat structure allowed us to undertake the cloning of the human class IV ADH cDNA. Four independent clones containing overlapping fragments of human class IV cDNA were obtained from the human stomach cDNA library. A 520-bp fragment covering the 5' end of the cDNA was isolated by using PCR with nested primers. The complete nucleotide sequence, including 2055 bp has been published elsewhere (Farrés et al., 1994) and is available under the accession number X76342 from the EMBL/GenBank Data Bank. Three more research groups have reported independently the cloning of human class IV ADH DNA (Satre et al., 1994, Kedishvili et al., 1994, Yokoyama et al., unpublished).

Two in-frame ATG codons are present in the 5' region of the cDNA. The downstream ATG is found in a proper sequence for being the initiation codon. The existence of the additional upstream ATG codon, which is located in a non-optimal sequence context, has been confirmed by genomic DNA sequencing (Satre et al., 1994). With the second ATG as the initiation codon, the deduced amino acid sequence contains 374 residues. In analogy with rat class IV ADH and other mammalian ADHs, the N-terminal methionine is expected to be removed leaving the next residue, in this case a Gly, to constitute the acetylated N-terminus.

The deduced amino acid sequence includes the partial protein sequences previously reported for human class IV ADH (Parés et al., 1992, Stone et al., 1993), but differs at several positions. It also differs at three positions from the cDNA-deduced sequence of Yokoyama et al. (unpublished). None of the differences involves a functionally critical residue and thus, even if reflecting polymorphism, they are unlikely to affect enzymatic properties.

The identity between human and rat class IV ADHs at the amino acid level is 88 %, which is an intermediate value between that of the 'variable' class I (82 %) and that of the 'constant' class III (94 %) (Table 3, Yin et. al., 1991, Jörnvall et al., 1993). Table 3 also shows pairwise sequence comparisons between the various human ADH classes. Positional identities between any two human ADH classes usually are in the 60 % range. However, class IV and class I ADH show as much as 70 % sequence identity, indicating a closer evolutionary relationship between the two classes (Farrés et al., 1994).

Structure-Function Relationships in Class IV Enzymes

The overall conformation expected of class IV enzymes is typical of that of ADHs of other classes in general (Eklund et al., 1990, Parés et al., 1994), allowing the correlation of specific amino acid replacements with changes in enzymatic properties.

Both in the rat and in the human class IV enzymes, the residues involved in substrate binding are in general well conserved (with the exception of position 294 which is Ala in the rat enzyme) with respect to those found in class I isozymes, (Table 4). This fact and the overall higher sequence identity may explain similar substrate specificities and functional roles for class IV and class I enzymes. Residue 294 is located in the middle region of the substrate-binding cleft. The exchange Val/Ala in the rat class IV makes this region less hydrophobic and may provide an additional space at the active site which could explain the higher Km values of the rat enzyme for ethanol (Table 1) and other aliphatic alcohols and aldehydes (Julià et al., 1987, Boleda et al., 1993, Moreno and Parés, 1991).

In class IV ADH, most of the residues that may participate in coenzyme binding are identical to those found in class I forms (Eklund et al., 1984; 1990), with few exceptions that are summarized in Table 4. Arg47 interacts in class I enzymes with the pyrophosphate moiety of the coenzyme, and its presence has been correlated with relatively low kcat values (Hurley et al., 1990). However, human class IV, also with Arg47, is the human ADH form with the highest kcat values, demonstrating that Arg47 is not always associated with tight coenzyme binding and low kcat values. Rat class IV has Gly at position 47, but this exchange alone does not explain the kcat increase for the rat enzyme since even an opposite change is observed when Arg47 in $\beta_1\beta_1$ is changed to Gly47 by site-directed mutagenesis (Hurley et al., 1990). Amino acid exchanges at positions 230, 271 and 363 could contribute to a weakening of coenzyme binding and explain the higher kcat and K_{ia} values of human class IV ADH with respect to class I enzymes, and even the higher kcat and K_{ia} values of the rat class IV enzyme. In particular, the negative charge of Glu230 (Gln in the rat enzyme), is note worthy since this residue is non-polar or a positively-charged residue in other

Table 4. Amino acid exchanges in functionally important domains of class IV ADHs

region	residue	Class I	Class IV	
		human γ_1/rat	human	rat
substrate binding	294	Val	Val	Ala
coenzyme binding	47	Arg	Arg	Gly
	230	Ala	Glu	Gln
	271	Arg	His	Arg
	363	Arg	Asn	Tyr
coenzyme binding domain	259	Asp	Gly	Gly
	260	Gly	Asn	Asn
	261	Gly	Asn	Thr

human ADH classes. Furthermore, exchanges at residue 271 have been associated with differences in kcat values of the class I γγ isoenzymes (Bosron et al., 1983, Höög et al., 1986).

Position 363 (Arg in human γ_1 subunits and rat class I) is exchanged to Asn and Tyr in the human and rat class IV enzymes, respectively. It has been reported that this residue could affect the environment of Arg47 (Eklund et al., 1984; Hurley et al., 1991). A support for a distinct environment for Arg47 in class IV is the lack of activation by Cl⁻, in contrast to what is known for class I $\beta_1\beta_1$ isoenzymes (Burnell and Bosron, 1989, Hurley et al., 1990). Rat class IV is also insensitive to Cl⁻ ions, as expected from the presence of Gly at position 47.

Two highly conserved Gly residues (which constitute a reverse turn in class I isoenzymes) at positions 260 and 261 of the coenzyme binding domain are exchanged to larger residues in the human (two Asn) and rat (Asn and Thr) enzymes (Table 4). These important exchanges could further contribute to alter coenzyme binding in class IV ADHs.

In conclusion, some of the discussed amino acid exchanges may account for the differences observed in the kinetic properties and substrate specificities of class IV enzymes. Particularly, some structural features have resulted in a human class IV ADH more specific for ethanol than the rat class IV enzyme. Apart from this, the two class IV enzymes show similar substrate specificities and tissue distribution, suggesting common roles in both species, probably related to the metabolism of endogenous medium- and long-chain alcohols and aldehydes, such as the products of lipid peroxidation, ω-hydroxyfatty acids and retinoids. Thus, the overall physiological function of class IV ADHs would be the maintenance and protection of epithelial tissues.

ACKNOWLEDGEMENTS

This work was supported by grants from the Spanish Dirección General de Investigación Científica y Técnica (PB92-0624 and PM91-0173), Fondo de Investigaciones Sanitarias de la Seguridad Social (94/0796), the Swedish Medical Research Council (13X-3532), the Swedish Alcohol Research Fund and the Commission of the European Communities (BMH1-CT93-1601).

REFERENCES

Boleda, M.D., Julià, P., Moreno, A., and Parés, X. (1989) Role of extrahepatic alcohol dehydrogenase in rat ethanol metabolism, *Arch. Biochem. Biophys.* **274**, 74-81.

Boleda, M.D., Saubi, N., Farrés, J., and Parés, X. (1993) Physiological substrates for rat alcohol dehydrogenase classes: Aldehydes of lipid peroxidation, ω-hydroxyfatty acids, and retinoids, *Arch. Biochem. Biophys.* **307**, 85-90.

Bosron, W.F., Magnes, L.J., and Li, T.-K. (1983) Kinetic and electrophoretic properties of native and recombined isoenzymes of human liver alcohol dehydrogenase, *Biochemistry* **22**, 1852-1857.

Burnell, J.C., and Bosron, W.F. (1989) Genetic polymorphism of human liver alcohol dehydrogenase and kinetic properties of the isoenzymes, in *Human metabolism of alcohol*, vol. II (Crow, K.E. and Batt, R.D., eds.), CRC Press, Boca Raton, Florida, 1989, pp. 65-75.

Cederlund, E., Peralba, J.M., Parés, X., and Jörnvall, H. (1991) Amphibian alcohol dehydrogenase, the major frog liver enzyme. Relationships to other forms and assessment of an early gene duplication separating vertebrate class I and class III alcohol dehydrogenases, *Biochemistry* **30**, 2811-2816.

Eklund, H., Samama, J.P., and Jones, T.A. (1984) Crystallographic investigations of nicotinamide adenine dinucleotide binding to horse liver alcohol dehydrogenase, *Biochemistry* **23**, 5982-5996.

Eklund, H., Müller-Wille, P., Horjales, E., Futer, O., Holmquist, B., Vallee, B.L., Höög, J.-O., Kaiser, R., and Jörnvall, H. (1990) Comparison of three classes of human liver alcohol dehydrogenase. Emphasis on different substrate-binding pockets, *Eur. J. Biochem.* **193**, 303-310.

Farrés, J., Moreno, A., Crosas, B., Peralba, J.M., Allali-Hassani, A., Hjelmqvist, L., Jörnvall, H., and Parés, X. (1994) Alcohol dehydrogenase of class IV (σσ-ADH) from human stomach: cDNA sequence and structure/function relationships, *Eur. J. Biochem.* **224**, 549-557.

Holmes, R.S. (1988) Alcohol dehydrogenases and aldehyde dehydrogenases of anterior eye tissues from humans and other mammals, in *Biomedical and Social Aspects of Alcohol and Alcoholism* (Kuriyama, K., Takada, A., and Ishii, H., eds.), Elsevier Science Publishers, New York, pp. 51-57.

Höög, J.-O., Hedén, L.-O., Larsson, K., Jörnvall, H., and von Bahr Lindström, H. (1986) The γ_1 and γ_2 subunits of human liver alcohol dehydrogenase. cDNA structures, two amino acid replacements, and compatibility with changes in the enzymatic properties, *Eur. J. Biochem.* **159**, 215-218.

Hurley, T.D., Edenberg, H.J., and Bosron, W.F. (1990) Expression and kinetic characterization of variants of human $\beta_1\beta_1$ alcohol dehydrogenase containing substitutions at amino acid 47, *J. Biol. Chem.* **265**, 16366-16372.

Hurley, T.D., Bosron, W.F., Hamilton, J.A., and Amzel, L.M. (1991) Structure of human $\beta_1\beta_1$ alcohol dehydrogenase: Catalytic effects of non-active-site substitutions, *Proc. Natl. Acad. Sci. USA* **88**, 8149-8153.

Jörnvall, H., Persson, B., and Jörnvall, H. (1993) Variability patterns of dehydrogenases versus peptide hormones and proteases/antiproteases, *FEBS Lett.* **335**, 69-72.

Julià, P., Farrés, J., and Parés, X. (1987) Characterization of three isoenzymes of rat alcohol dehydrogenase. Tissue distribution and physical and enzymatic properties, *Eur. J. Biochem.* **162**, 179-189.

Kedishvili, N.Y., Stone, C.L., Thomasson, H.R., Carr, L.G., Edenberg, H.J., Bosron, W.F., and Li, T.-K. (1994) *Alcohol. Clin. Exp. Res.* **18**, 16A.

Koivusalo, M., Baumann, M., and Uotila, L. (1989) *FEBS Lett.* **257**, 105-109.

Moreno, A. and Parés, X. (1991) Purification and characterization of a new alcohol dehydrogenase from human stomach, *J. Biol. Chem.* **266**, 1128-1133.

Moreno, A., Parés, A., Ortiz, J., Enríquez, J., and Parés, X. (1994) Alcohol dehydrogenase from human stomach. Variability in normal mucosa and effect of age, gender, ADH$_3$ phenotype and gastric region, *Alcohol Alcohol.*, in press.

Parés, X., Moreno, A., Cederlund, E., Höög, J.-O., and Jörnvall, H. (1990) Class IV mammalian alcohol dehydrogenase. Structural data of the rat stomach enzyme reveal a new class well separated from those already characterized, *FEBS Lett.* **277**, 115-118.

Parés, X., Cederlund, E., Moreno, A., Saubi, N., Höög, J.-O., and Jörnvall, H. (1992) Class IV alcohol dehydrogenase (the gastric enzyme). Structural analysis of human σσ-ADH reveals class IV to be variable and confirms the presence of a fifth mammalian alcohol dehydrogenase class, *FEBS Lett.* **303**, 69-72.

Parés, X., Cederlund, E., Moreno, A., Hjelmqvist, L., Farrés, J., and Jörnvall, H. (1994) Mammalian class IV alcohol dehydrogenase (stomach alcohol dehydrogenase): Structure, origin, and correlation with enzymology, *Proc. Natl. Acad. Sci. USA* **91**, 1893-1897.

Parés, X. and Farrés, J. (1995) Alcohol and aldehyde dehydrogenases in the gastrointestinal tract, in *Ethanol and the Gastrointestinal Tract: Mechanisms in Disease* (Preedy, V.R. and Watson, R.R., eds.) CRC Press, in press.

Satre, M.A., _gombi_-Knight, M., and Duester, G. (1994) The complete structure of human class IV alcohol dehydrogenase (retinol dehydrogenase) determined from the *ADH7* gene, *J. Biol. Chem.* **269**, 15606-15612.

Stone, C.L., Thomasson, H.R., Bosron, W.F., and Li, T.-K. (1993) Purification and partial amino acid sequence of a high-activity human stomach alcohol dehydrogenase, *Alcohol. Clin. Exp. Res.* **17**, 911-918.

Vallee, B.L. and Bazzone, T.J. (1983) Isozymes of human liver alcohol dehydrogenase, *Isozymes: Curr. Top. Biol. Med. Res.* **8**, 219-244.

Yang, Z.N., Davis, G.J., Hurley, T.D., Stone, C.L., Li, T.-K., and Bosron, W.F. (1994) Catalytic efficiency of human alcohol dehydrogenases for retinol oxidation and retinal reduction, *Alcohol. Clin. Exp. Res.* **17**, 496.

Yasunami, M., Chen, C.-S., and Yoshida, A. (1991) A human alcohol dehydrogenase gene (*ADH 6*) encoding an additional class of isozyme, *Proc. Natl. Acad. Sci. USA* **88**, 7610-7614.

Yin, S.-J., Wang, M.-F., Liao, C.-S., Chen, C.-M., and Wu, C.-W. (1990) Identification of a human stomach alcohol dehydrogenase with distinctive kinetic properties, *Biochem. Int.* **22**, 829-835.

Yin, S.-J., Vagelopoulos, N., Wang, S.-L., and Jörnvall, H. (1991) Structural features of stomach aldehyde dehydrogenase distinguish dimeric aldehyde dehydrogenase as a 'variable' enzyme. 'Variable' and 'constant' enzymes within the alcohol and aldehyde dehydrogenase families, *FEBS Lett.* **283**, 85-88.

Yokoyama, H., Baraona, E., and Lieber, C.S. (1994) Molecular cloning of cDNA of human class IV ADH. Submission to the GenBank database (accession number L33179).

Zheng, Y.-W., Bey, M., Liu, H., and Felder, M.R. (1993) Molecular basis of the alcohol dehydrogenase-negative deer mouse. Evidence for deletion of the gene for class I enzyme and identification of a possible new enzyme class, *J. Biol. Chem.* **268**, 24933-24939.

41

CLONING AND EXPRESSION OF A HUMAN STOMACH ALCOHOL DEHYDROGENASE ISOZYME

Natalia Y. Kedishvili, William F. Bosron, Carol L. Stone, Cara F. Peggs, Holly R. Thomasson, Kirill M. Popov, Lucinda G. Carr, Thomas D. Hurley, Howard J. Edenberg, and Ting-Kai Li

Department of Biochemistry and Molecular Biology
Department of Medicine
635 Barnhill Drive
Indianapolis, IN 46202-5122

INTRODUCTION

In humans, there is a family of NAD^+- and zinc-dependent alcohol dehydrogenases (E.C. 1.1.1.1) that exhibit broad substrate specificity toward aliphatic alcohols (Vallee and Bazzone, 1983, Smith, 1986, and Bosron et al., 1993). The various isozyme subunits are encoded by at least 6 different gene loci (*ADH1* through *ADH6*). Most recently, a new isozyme called σ-ADH or μ-ADH has been isolated from human stomach tissue that has a high K_m (about 30 mM) and relatively high catalytic efficiency for ethanol ($k_{cat}/K_m \sim 52$ $min^{-1}mM^{-1}$)(Wang et al., 1990, Stone et al, 1993, Moreno and Parés, 1991). One or more isozymes with similar electrophoretic mobility are found in the esophagus (Yin et al., 1990).

Partial amino acid sequence of purified human σ-ADH indicates that it has about 60 to 75% sequence identity with the other six human ADH subunits and that it likely diverged after the avian/amphibian split (Stone et al., 1993). In lower animals, isozymes with electrophoretic mobilities and substrate specificities similar to human σ-ADH are found in the stomach of the rat (Julia et al., 1987) and mouse (Algar et al., 1983), and the retina of the rat (Parés et al., 1985).

In this report, we describe the isolation and complete nucleotide sequence of a cDNA encoding human stomach σ-ADH, and the expression of the recombinant enzyme in a heterologous system. A cDNA and partial genomic sequence for σ-ADH was recently published by Satre et al. (1994).

EXPERIMENTAL PROCEDURES

Degenerate oligonucleotides were synthesized based on peptide sequences of human σ-ADH (5): 15-21 (Primer 1: TGGCA^A/GCA^A/GAA^A/GCAACCITT, sense), 300-306

(Primer 2-1: CATIGG^A/GTC^A/GTAIGT*IA*GCAT, antisense), and 338-344 (Primer 3-1: TG^A/GTC*IA*G^A/GTC^A/GAA^T/CTT^T/CTT, antisense). The amino acid position number was based on the alignment of σ-ADH peptides with the β$_1$–ADH isozyme sequence. Additional Primers 2-2 and 3-2 differed from Primers 2-1 and 3-1 in the codon for leucine (AA^T/C versus GAI, underlined). Human stomach poly(A)+ RNA (Clontech) was reverse transcribed and used as a template for the initial PCR directed by Primers 3-1 and 3-2 paired with Primer 1 at the annealing temperature of 42°C. The products of the initial reaction were subjected to a subsequent round of PCR with the internally positioned Primers 2-1 and 2-2 paired with Primer 1. The annealing temperature was increased to 51°C. The PCR products were purified, subcloned into M13mp18RF (Bethesda Research Laboratories), and sequenced (United States Biochemicals).

A 5'-stretch human stomach cDNA library was screened using an [α-^{32}P]dATP-labeled, 873 bp, PCR product obtained with Primers 1 and 3-1. The probe was labeled by random oligonucleotide primer extension (United States Biochemicals). The hybridization was done in 5x Denhardt's solution, 5x SSPE, 0.1 mg/ml salmon sperm DNA, and 0.1% SDS at 65°C overnight. The final wash of the nitrocellulose filters was performed in 0.1x SSC, 0.1% SDS at 65°C for 30 min. Positive plaques were purified, and the cDNA inserts from phage DNA were subcloned into M13mp18RF digested with *Eco*R I. Sense and antisense single-stranded DNA was prepared and sequenced at least two times.

The coding region of σ–ADH was amplified by PCR with primers CTTTTTC*GGATCC*ATGGGCACTGCTGGAAAAG   (sense)   and CCACTT*GAATTC*TCAAAACGTCAGGACCGT (antisense). The primers contained recognition sequences for the restriction endonucleases *Bam*H I and *Eco*R I (underlined in the nucleotide sequence above), respectively. The amplified coding region of σ-ADH was subcloned into the expression vector pGEX-2T (Pharmacia LKB). The final construct of the σ-ADH fused with glutathione S-transferase encoded three extra amino acids (Gly, Ser, and starting Met) on the N-terminus of the σ-ADH cDNA. The expression of the fusion protein was performed as described by Guan and Dixon (1991). Cells were harvested by centrifugation and suspended in 10 ml PBST containing 0.1% β-mercaptoethanol, 10 μM ZnSO$_4$, and the protease inhibitors phenylmethylsulfonyl fluoride (50 μg/ml), leupeptin (10 μg/ml), and benzamidine (5 mM). Cells were sonicated and centrifuged. The fusion protein was purified using glutathione-agarose affinity chromatography. After cleavage with thrombin, σ-ADH was separated from glutathione S-transferase by elution from a Mono Q column (Pharmacia) with a linear salt gradient. Fractions containing active σ-ADH were combined and concentrated.

RESULTS

Construction of Oligonucleotide Probes and PCR Amplification of σ–ADH

The peptide sequences available for human σ–ADH did not allow design of oligonucleotide probes with low degeneracy and high melting temperature. Degeneracy was reduced, therefore, by incorporating inosine residues in the design of four oligonucleotides corresponding to amino acids 15-21, 148-154, 300-306, and 338-344 as reported by Stone *et al*. (1993) using the alignment of partial σ-ADH sequences to the β-ADH sequence. For primers containing leucine, two oligonucleotides were synthesized encoding for two separate Leu codons to minimize degeneracy. Agarose gel electrophoresis of the products of the second round of PCR amplification of reverse-transcribed poly(A)+ RNA from human

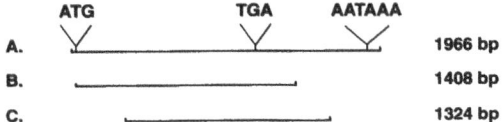

Figure 1. Scheme of three cDNA clones encoding human σ-ADH. A, 1966 bp clone; B, 1408 bp clone; and C, 1324 bp clone.

stomach revealed discrete bands migrating in positions corresponding closely to those predicted on the basis of published nucleotide sequences of cDNA clones encoding ADH isozymes. PCR amplification products between Primers 1 and 3, 2 and 3, and 2 and 4 were predicted to be 873 bp, 474 bp, and 588 bp, respectively, in size. PCR amplification products contained in the major bands, which migrated to these positions, were recovered from preparative-scale agarose gel electrophoresis and subcloned in M13mp18RF for nucleotide sequence analysis. All individual subclones derived from PCR-amplified human stomach cDNA were found to encode σ-ADH.

Analysis of Bacteriophage λ Clones Encoding Human Stomach σ–ADH

Three clones with the insert sizes of 1966 bp (A), 1324 bp (B), and 1408 bp (C), which contained sequences generated by PCR amplification of reverse-transcribed human stomach RNA, were isolated from a human stomach cDNA library constructed in bacteriophage λgt11. The ATG codon of the sequence obtained from the A clone was present at nucleotide 19 (Fig. 1).

The translated sequence from the ATG codon predicted a polypeptide of 373 amino acids with a M_r of 39,902, which is in good agreement with the subunit M_r of the purified human stomach σ-ADH (Moreno and Parés, 1991, and Stone *et al.*, 1993).

The sequence of the B clone, which was 1324 bp in length, was identical with the A clone except that it was a truncated version that started at nucleotide 327 and ended at nucleotide 1651 of the larger A clone. The C clone was identical with all of the coding region but it was missing the first 18 nucleotides of the 5' non-coding region found in the A clone and ended at nucleotide 1427. The 3'-untranslated region of clone A contained a single consensus polyadenylation signal at nucleotide 1877.

Expression of Recombinant σ–ADH

The coding region of human stomach σ–ADH cDNA was cloned into a pGEX-2T expression vector and expressed in TG-1 cells as a glutathione S-transferase fusion protein.

```
GTAGKVIKCKAAVLWEQKQPFSIEEIEVAPPKTKEVRIKILATGICRTDDHVIKGTMVSKFPVIVG - 66

HEATGIVESIGEGVTTVKPGDKVIPLFLPNCRECNACRNPEGNLCIRSDITGRGVLADGTTRFTCR - 132

GKPVHHFMNTSTFTEYTVVDESSVAKIDDAAPPEKVCLIGCGFSTGYGAAVKTGKVKPGSTCVVFG - 198

LGGVGLSVIMGCKSAGASRIIGIDLNKPKFEKAMAVGATECISPKDSTKPISEVLSEMTGNNVGYT - 264

FEVIGHLETMIDALASCHMNYGTSVVVGVPPSAKMLTYDPMLLFTGRTWKGCVFGGLKSRDDVPKL - 330

VTEFLAKKFDLDQLITHVLPFKKISEGFELLNSGQSIRTVLTF - 373
```

Figure 2. Deduced amino acid sequence of human stomach σ-ADH.

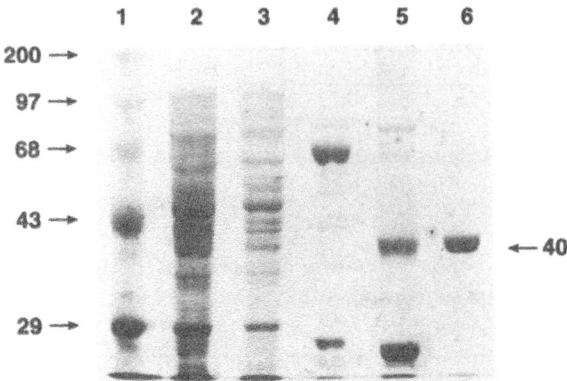

Figure 3. Purification of recombinant σ-ADH. Lane 1, molecular weight standards; lane 2, cell homogenate; lane 3, soluble protein fraction; lane 4, purified fusion protein, ~66 kD; lane 5, fusion protein cleaved with thrombin; lane 6, purified σ-ADH after separation on Mono Q column.

It was easily purified under non-denaturing conditions using a glutathione affinity column. Following the single purification step, the 66 kD fusion protein was nearly homogeneous as analyzed by SDS-PAGE (Fig. 3, lane 4).

Glutathione S-transferase (26 kD) was removed by specific cleavage of the fusion protein by thrombin (Fig. 3, lane 5), and ion-exchange chromatography on Mono Q (Fig. 3, lane 6). From 1 L of bacterial culture, approximately 3 mg of purified σ–ADH was obtained. The specific activity of the recombinant σ–ADH was similar to that of the stomach enzyme. The mobility of the recombinant protein was examined by starch gel electrophoresis. The expressed enzyme was more anodic than the ππ isozyme (isoelectric point 8.2), but more cathodic than σ-ADH in the stomach homogenate (Fig. 4).

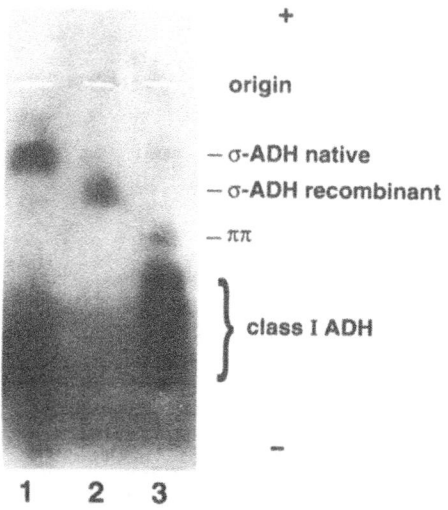

Figure 4. Starch gel electrophoresis of native and recombinant σ-ADH. Lane 1, stomach mucosal homogenate; lane 2, purified recombinant σ-ADH; lane 3, liver homogenate.

one S-transferase (26 kD) was removed by specific cleavage of the fusion protein by thrombin (Fig. 3, lane 5), and ion-exchange chromatography on Mono Q (Fig. 3, lane 6). From 1 L of bacterial culture, approximately 3 mg of purified σ–ADH was obtained. The specific activity of the recombinant σ–ADH was similar to that of the stomach enzyme. The mobility of the recombinant protein was examined by starch gel electrophoresis. The expressed enzyme was more anodic than the ππ isozyme (isoelectric point 8.2), but more cathodic than σ-ADH in the stomach homogenate (Fig. 4).

Alignment of the Complete Protein Sequence of σ-ADH with Other Human ADH Isozymes

Phylogenetic relationships of σ-ADH with other ADH isozymes were examined using the TREE program by Feng and Doolittle (1990). The amino acid sequence of human stomach σ-ADH was most similar to that of rat stomach ADH. It also shared a high percent of identity with human class I isozymes α (68.6%), β (69.4%), and γ (68.9%), and was more distantly related to the human class II π (58.6%) and class III χ (60.3%) isozymes.

DISCUSSION

In this paper, we describe the isolation and complete nucleotide sequence of a cDNA encoding human stomach σ-ADH. Our previous attempts to amplify a fragment of σ-ADH with degenerate oligonucleotide probes by PCR from whole stomach poly(A)$^+$ RNA resulted in amplification of π-ADH and not σ-ADH. Apparently, the abundance of π-ADH mRNA in stomach muscle tissue far exceeds that of σ-ADH in the mucosal layer. Our attempts to obtain a satisfactory mucosal mRNA sample from donors undergoing surgical gastrectomy or stomach biopsy were not successful. Therefore, an alternate strategy for amplifying a part of the σ-ADH message from the total stomach poly(A)$^+$ RNA was devised. The degeneracy was reduced by incorporating inosine residues in the design of the oligonucleotide primers. This allowed the use of higher annealing temperatures. To ensure the specific amplification of σ-ADH cDNA, several combinations of internally nested primers were used. As a result of successive rounds of PCR amplifications with increasingly higher annealing temperatures, several PCR products of expected sizes were obtained. When sequenced, they were positively identified as fragments of a cDNA encoding σ-ADH. The longest PCR product (~ 900 bp) was used to screen the stomach cDNA library. Although many signals of varying intensities were obtained after the first round of screening, only three plaques that gave the strongest signal were purified and sequenced. The longest of the three cDNA clones obtained in this study (1966 bp) encodes a putative protein of 373 amino acids. The 3'- noncoding region is 822 bp long and, unlike class I ADH cDNAs, it has only one polyadenylation signal. The deduced amino acid sequence of σ-ADH is identical with that reported by Stone et al. (1993).

Alignment of human σ-ADH with ADH isozymes shows that it has higher amino acid sequence identity with human class I ADH than with class II (π) or class III (χ) ADH. Alignment of the complete protein sequence of rat stomach class IV ADH (14) with that of human stomach ADH shows 87% identity between these two enzymes. Human and rat σ-ADH apparently diverged at about the same time, very close to the precursor of chicken class I ADH, as reported by Pares et al.(1994) and Satre et al.(1994).

To further characterize human σ-ADH, an expression vector was constructed. In this system, σ-ADH was produced in E. coli as a fusion protein with glutathione S-transferase, purified through glutathione affinity column, and cleaved with thrombin to release the recombinant σ-ADH. Because of the specificity of the thrombin cleavage site, two amino acids (Gly and Ser) became attached to the N-terminus of σ-ADH. Together with starting methionine they make the recombinant σ-ADH longer than the native stomach enzyme by 3 amino acids. Although this addition does not change the number of charged amino acids of the expressed protein, its mobility on the starch gel is slightly different from that of a native stomach σ-ADH. It is likely that the altered electrophoretic mobility of the expressed enzyme versus native σ-ADH is due to the unblocked N-terminal glycine in the recombinant protein. A similar observation was reported for χ-ADH, where the recombinant χ-ADH with an unacetylated N-terminus exhibited a different retention time on ion-exchange chromatography than native enzyme (Fairwell et al., 1984). The kinetic properties of recombinant ADH are identical to those of the native enzyme.

The protein sequence of human σ-ADH has ~ 70% identity with class I β ADH. All of the amino acids thought to be critical to the activity of alcohol dehydrogenases are conserved in σ-ADH. Therefore, we ventured to predict the tertiary structure of this isozyme, based on the known structure of β-ADH. An important difference in the structures between the two isozymes is the deletion at Gly-117 in σ-ADH. Gly-117 in β_1-ADH is located in a markedly variable region that forms a loop (amino acid residues 116, 117, 118). This loop appears to hinder the binding of large substrates such as retinol in the β-ADH active site as determined by the X-ray structure of several human β ADH complexes (Hurley *et al.*, 1994). The loop is located on the surface of the catalytic domain. Docking experiments of *all trans*-retinol with a class I β-ADH X-ray structure model show that the deletion of Gly-117 in σ-ADH causes a widening of the entrance to the substrate-binding barrel. This may account for the high catalytic efficiency of σ-ADH for *all trans*-retinol oxidation as compared with the α, β, γ , π or χ isozymes (Boleda *et al.*, 1993, Yang *et al.* 1994).

REFERENCES

Algar, E.M., Seeley, T.-L., and Holmes, R.S. (1983) Purification and molecular properties of mouse alcohol dehydrogenase isozymes. *Eur. J. Biochem.* **137**, 139-147

Boleda, M.D., Saubi, N., Farres, J., Pares, X. (1993) Physiological substrates for rat alcohol dehydrogenase classes: aldehydes of lipid peroxidation, ω-hydroxyfatty acids, and retinoids. *Arch. Biochem. Biophys.* **307**, 85-90

Bosron, W.F., Ehrig, T., and Li, T.-K. (1993) Genetic factors in alcohol metabolism and alcoholism. *Semin. Liver. Dis.* **13**, 126-13

Fairwell, T., Krutzsch, H., Hempel, J., Jeffery, J. and Jornvall, H. (1984) Acetyl-blocked N-terminal structures of sorbitol and aldehyde dehydrogenases. *FEBS Letters* **170**, 281-289

Feng D.-F. and Doolittle R.F. (1990) Progressive alignment and phylogenetic tree construction of protein sequences. *Methods. Enzymol.* **183**, 375-387

Guan, K.L., and Dixon, T.E. (1991) Eukaryotic proteins expressed in Escherichia coli: An improved thrombin cleavage and purification procedure of fusion proteins with glutathione S-transferase. *Anal. Biochem.* **192**, 262-267

Hurley, T.D., Bosron, W.F., Stone, C.L., and Amzel, L.M. (1994) Structures of three human β alcohol dehydrogenase variants. Correlations with their functional differences. *J. Mol. Biol.* **239**, 415-429

Julia, P., Farres, J., and Pares, X. (1987) Characterization of three isoenzymes of rat alcohol dehydrogenase. Tissue distribution and physical and enzymatic properties. *Eur. J. Biochem.* **162**, 179-189

Moreno, A. and Pares, X. (1991) Purification and characterization of a new alcohol dehydrogenase from human stomach. *J. Biol. Chem.* **266**, 1128-1133

Pares, X., Cederlund, E., Moreno, A., Hjelmqvist, L., Farres, J., and Jornvall, H. (1994) Mammalian class IV alcohol dehydrogenase (stomach alcohol dehydrogenase). Structure, origin, and correlation with enzymology. *Proc. Natl. Acad. Sci. USA* **91**, 1893-1897

Pares, X., Julia, P., and Farres, J. (1985) Properties of rat retina alcohol dehydrogenase. *Alcohol* **2**, 43-46

Satre, M.A., Zgombik-Knight, M., and Duester, G. (1994) The complete structure of human class IV alcohol dehydrogenase (retinol dehydrogenase) determined from the *ADH7* gene. *J. Biol. Chem.* **269**, 15606-15612

Smith, M. (1986) Genetics of human alcohol and aldehyde dehydrogenases. *Adv. Hum. Genet.* **15**, 249-290

Stone, C.L., Thomasson, H.R., Bosron, W.F., and Li, T.-K. (1993) Purification and partial amino acid sequence of a high-activity human stomach alcohol dehydrogenase. *Alcoholism: Clin. Expt. Res.* **17**, 911-918

Vallee, B.L. and Bazzone, T.J. (1983) Isozymes of human liver alcohol dehydrogenase. *Isozymes Curr. Top. Biol. Med. Res.* **8**, 219-244

Wang, S.-L., Wu, C.-W., Cheng, T.-C.,and Yin, S.-J. (1990) Isolation of high-K_M aldehyde dehydrogenase isoenzymes from human gastric mucosa. *Biochem. Int.* **22**, 199-204

Yang, Z.-N., Davis, G.J., Hurley, T.D., Stone, C.L., Li, T.-K., and Bosron, W.F. (1994) Catalytic efficiency of human alcohol dehydrogenases for retinol oxidation and retinal reduction. *Alcoholism: Clin. Expt. Res.* **18**, 587-591

Yin, S.-J., Wang, M.-F., Liao, C,-S., Chen, C.-M., and Wu, C.-W. (1990) Identification of a human stomach alcohol dehydrogenase with distinctive kinetic properties. *Biochem. Int.* **22**, 829-835

42

PURIFICATION AND PROPERTIES OF MURINE CORNEAL ALCOHOL DEHYDROGENASE

EVIDENCE FOR CLASS IV ADH PROPERTIES

John E. Downes and Roger S. Holmes[*]

School of Science
Griffith University
4111, Brisbane, Australia

INTRODUCTION

Alcohol dehydrogenases (ADH; E.C. 1.1.1.1), in the presence of nicotamide adenine dinucleotide (NAD), catalyse the reversible oxidation of alcohols to aldehydes and ketones. In the mouse there are three known isozymes of ADH designated as $ADH-A_2$, $ADH-B_2$ and $ADH-C_2$ which are encoded by three distinct structural genes *Adh-1*, *Adh-2* and *Adh-3*, respectively (Holmes et al, 1981; Holmes,1977; Holmes, 1978; Holmes, 1979). $ADH-A_2$ is the major liver isozyme and is a class I isozyme (Holmes, 1979). $ADH-B_2$ occurs mostly in the liver and kidney and is a class III isozyme (Algar et al, 1983). $ADH-C_2$ exhibits highest activity in the stomach and cornea and has been designated as a separate class of ADH (Algar et al, 1983), consistent with class IV properties (Parés et al, 1994).

The mammalian cornea provides protection for the interior tissues of the eye by absorbing ultraviolet (UV) light in the range of 290 to 320 nm (Boettner and Wolters, 1962; Pitts, 1978). High exposures to UV light may result in a number of pathological responses in the cornea (Kurzel et al, 1977), as well as the production of free-radical species which may be involved in initiating lipid peroxidation processes which may eventually result in the formation of a number of cytotoxic products including peroxidic aldehydes (Esterbauer, 1982; Kappus, 1985).

Recent studies from this laboratory (Downes et al, 1992; Downes et al, 1993) have indicated that corneal ADH activity (along with aldehyde dehydrogenase) is predominantly localized in the epithelial and endothelial layers of the mammalian cornea and may be involved in detoxifying peroxidic aldehydes.

This communication describes the isolation and characterization of mouse corneal alcohol dehydrogenase ($ADH-C_2$) with alcohol and aldehyde substrates. A dual role for

---

[*] Address for correspondence.

corneal ADH-C$_2$ has been proposed in assisting in the metabolism of peroxidic aldehydes generated through exposure to UV-B light and in regenerating NAD$^+$ for corneal aldehyde dehydrogenase to continually oxidize these aldehydes. .

MATERIALS AND METHODS

Extraction and Purification of ADH-C$_2$ from Murine Corneas

Corneal homogenates were prepared from SWR/J inbred mice which exhibit low levels of corneal aldehyde dehydrogenase and the 'a' allele for ADH-C$_2$ (Holmes et al, 1988). Corneas were removed from freshly killed mice and then kept at -70° until required. Pools of frozen corneas were then homogenized as a 10% w/v extraction in 20 mM NaOH-Mes, 0.1 mM EDTA, 10% (v/v) Glycerol, 1 mM DTT (pH 6.2) buffer on ice, using an Ultra-Turrax homogenizer, then centrifuged at 45,000 g for 20 minutes at 4°C. The supernatant was then applied directly to a CM sepharose CL-6B column (2.5 cm^2 x 6 cm) pre-equilibrated in 20 mM NaOH-Mes, 0.1 mM EDTA, 10% (v/v) Glycerol, 1 mM DTT (pH 6.2). The pH of the eluent was then adjusted to 7.0 with dilute NaOH and applied to a 5' AMP column (1.5 cm^2 x 10 cm) pre-equilibrated with 20 mM NaOH-Mops, 0.1 mM EDTA, 10% (v/v) Glycerol, 1 mM DTT (pH 7.0) buffer containing 50 mM NaCl. The column was then extensively washed in the same buffer until the absorbance at 280 nm of the eluent, was reduced to baseline level. ADH-C$_2$ activity was eluted from the column with 0.25 mM NAD$^+$ and concentrated using a TCA precipitation method prior to being loaded onto a SDS Gel or used directly for kinetic studies. Purity of the concentrate was assessed by SDS gel electrophoresis and isoelectric focusing.

Assays and Kinetic Studies

Corneal alcohol dehydrogenase (ADH) activity was routinely assayed in 100 mM glycine-NaOH buffer (pH 10.5) containing 1.5 mM NAD and 10 mM trans-2-hexen-1-ol at 30°C using an LKB UV/Visible spectrophotometer and monitoring the production of NADH at 340 nm. Kinetic studies on ADH-C$_2$ were performed in the same buffer for alcohol substrates or in 50 mM sodium phosphate buffer (pH 7.4) containing 0.2 mM NADH for aldehyde substrates. The K$_m$ values for both NAD$^+$ and NADH were obtained using 10 mM trans-2-hexenol and 10 mM benzaldehyde as substrates respectively. The effects of pyrazole and p-hydroxymercuribenzoate on ADH-C$_2$ were investigated using 100 mM glycine-NaOH buffer (pH 10.5) containing 0.5 mM NAD$^+$ and trans-2-hexenol as substrate. Various concentrations of inhibitor were incubated with enzyme for 2 minutes at 30°C prior to the addition of various concentrations of substrate to initiate the reactions.Enzyme activity is expressed in International Units (IU) or micromoles per minute. Protein concentrations in corneal extacts were determined using the method of Smith et al (1985). Polyacrylamide gels (10% resolving and 4% stacking) containing 1% SDS were prepared in a Mini Protean TM State Cell System according to the method of Laemmli (1970) as modified by Ames (1974) and stained for protein with 0.25% Coomassie Brilliant Blue G-250. Gel filtation of purified ADH-C$_2$ was performed according to the method of Andrews (1965) by using a Sephadex G-200 column (dimensions 1.78 cm^2 x 40 cm) in 20 mM NaOH-Mops, 0.1 mM EDTA, 10% (v/v) Glycerol, 1 mM DTT (pH 7.0). The column was calibrated with the following protein standards; catalase (240 kDa), lactate dehydrogenase (140 kDa), aldehyde dehydrogenase (110 kDa), isocitrate dehydrogenase (60 kDa) and myoglobin (17 kDa).

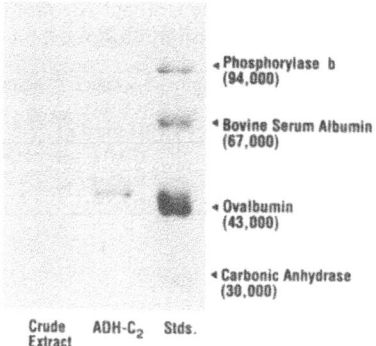

Figure 1. Sodium dodecyl sulfate polyacrylamide gel electrophoresis of fractions from the purification of ADH-C_2. Channel 1 is the crude extract, channel 2 is the purified corneal AHD-4 and channel 3 is the protein subunit standards.

RESULTS

Purification and Properties of ADH-C_2

Mouse corneal alcohol dehydrogenase (ADH-C_2) has been isolated from the corneas of SWR/J inbred mice as a homogenous protein using a two step purification proceedure involving ion-exchange chromatography followed by affinity chromatography. The purified enzyme has been characterized in terms of its subunit molecular weight, native molecular weight, substrate and cofactor specificity. Table 1 summarizes the purification procedure adopted with details of the activity at each step. Figure 1 illustrates SDS gel electrophoretic analysis of the homogeneity of the purified enzyme and indicates an apparent subunit molecular weight of 47 kDa. Gel exclusion chromatography of ADH-C_2 indicated that the native molecular mass of the protein was 97 kDa and it would appear that in solution ADH-C_2 behaved as a dimer. Corneal ADH-C_2 constitutes approximately 0.016% of the total soluble protein in the cornea and 0.007% per milligram of wet weight corneal tissue.

Kinetic Analysis

Table 2 summarizes some of the kinetic properties of mouse corneal ADH-C_2 with a range of alcohol and aldehyde substrates and provides a comparison with other class IV ADHs so far examined. The aldehyde substrates were chosen to include some naturally occurring 'peroxidic' aldehydes likely to be produced in the cornea through light illumination, as well as acetaldehyde, the primary product of ethanol metabolism and benzaldehyde, an aromatic aldehyde found in the diet. Kinetic studies using the aldehyde substrates were performed at pH 7.4 to preserve 'in vivo' physiological pH conditions. For the alcohol substrates chosen, in terms of comparative ratios of kcat to Michaelis constant, the highest value for catalytic efficiency was obtained for trans-2-hexenol, whilst for the aldehyde substrates the high values were exhibited for the peroxidic aldehydes hexanal and 4-hydroxynonenal, however no activity was detected using trans-2-hexenal or malondialdehyde as substrate. Inhibition studies revealed that ADH-C_2 was sensitive to both pyrazole and p-hydroxymercuribenzoate, with K_i values calculated as 0.3 mM and 0.59 mM respectively.

Table 1. Purification of Mouse Corneal Alcohol Dehydrogenase
using a CM sepharose CL-6B column followed by a 5'AMP affinity

Purification Step	Activity (I.U.)	Protein (mg)	Specific Activity (I.U.mg^{-1})	%Yield	Purification Factor
Extract	3.048	29.92	0.102	100	1
CM CL-6B	2.357	7.28	0.324	77	3
5'AMP	1.808	0.029	62.34	59	611

DISCUSSION

This study appears to be the first report on the purification of a mammalian corneal ADH. Murine corneal ADH-C$_2$ exhibited class IV ADH properties, having 'high' Km properties with ethanol (mM range), is active with aromatic and peroxidic aldehydes, and uses both NAD$^+$ and NADH but not NADP$^+$ and NADPH as cofactors. These properties are comparable to the mammalian stomach class IV ADHs previously characterized (see Table 2). Table 2 indicates that the human stomach σ-ADH exhibits a much lower Km value for ethanol compared to baboon, rat, and mouse stomach ADHs and mouse corneal ADH. Mouse stomach and corneal ADHs appear to have similar kcat values for ethanol, however kcat values for the other substrates examined particularly the aldehyde substrates, appear to be greater in the stomach isozyme. The high kcat values for baboon stomach ADH3 indicate the greatest overall catalytic efficiency compared to the other class IV ADHs, in regard to metabolizing both alcohol and aldehyde substrates.

ADH-C$_2$ is not active with trans-2-hexenal or malondialdehyde. It exhibits very high specific activity as do the stomach ADHs (Algar et al, 1983; Moreno and Pares,1991; Algar et al, 1992; Julia et al, 1987 and Boleda et al, 1993) and was inhibited by both pyrazole and p-hydroxymercuribenzoate. In common with all other murine ADH isozymes, corneal ADH-C$_2$ appeared to be a dimer and also exhibited the highest specific activity (using 10 mM trans-2-hexenol as substrate) of all murine ADH isozymes so far examined (Algar et al, 1983). This study has also demonstrated that murine corneal ADH is particularly active with peroxidic aldehydes. Peroxidic aldehydes may result from UV-induced lipid peroxidation processes occurring in the unsaturated lipid bilayers of cells. Corneal cells are particularly vulnerable to these processes as they function to transmit incoming light. Previous studies from this laboratory have indicated that corneal ADH may be implicated in protecting the eye from the pathological effects of ultraviolet light, particularly UVB (Downes et al, 1992; Downes et al, 1993; Downes et al, 1994). We have shown that in a number of mammalian species, ADH is predominantly located in the epithelial and endothelial layers of the cornea where it is ideally located to metabolize peroxidic aldehydes (generated through UV-induced lipid peroxidation processes) in the presence of increasing cellular levels of NADH generated through the activity of corneal aldehyde dehydrogenase (Downes et al, 1992). As well, previous studies of UV exposure on mouse eyes from C57BL/6J mice have shown that levels of ADH decrease and recover with increases and recovery from corneal opacification resulting from exposures to UVB radiation (Downes et al, 1993). UVB exposure is known to produce free radical species. Lipid peroxidation results from the attack by a free radical on an unsaturated lipid whereby a hydrogen atom is abstracted from the α methylene group of a lipid chain creating a lipid radical, and in the presence of oxygen the lipid undergoes a series of degradative steps which eventually results in the formation of peroxidic aldehydes amongst other products. A role for corneal ADH-C$_2$ has been proposed in assisting in the metabolism of peroxidic aldehydes generated through exposure to UV-B light and in

Table 2. Kinetic constants for class IV alcohol dehydrogenase from different mammals. Km values are in mM, kcat values are in min^{-1}, and kcat/Km values are in mM^{-1}min^{-1}. Alcohol substates and NAD$^+$ were assayed at pH 10.0 whilst aldehyde substrates and NADH were assayed at either pH 7.4 or 7.5. Data for mouse corneal ADH was from this study, for mouse stomach ADH from Algar et al, 1983; for human σ-ADH from Moreno and Pares,1991; for baboon stomach ADH from Algar et al, 1992; and for Rat stomach ADH from Julia et al, 1987 and Boleda et al, 1993. (*= x10^4; ns = unable to saturate)

Substrate	Mouse Corneal ADH-C$_2$			Mouse Stomach ADH-C$_2$			Human Stomach σ-ADH			Baboon Stomach ADH3			Rat Stomach ADH1		
	Km	kcat	kcat/Km	Km	kcat	kcat/Km	Km	kcat	kcat/Km	Km	kcat	kcat/Km	Km	kcat	kcat/Km
Alcohols															
Ethanol	177	3262	18.4	232	3160	13.6	11	590	52	539	2.2*	40	340	3760	11
Hexanol	4.2	1796	427	0.8	2505	3131				1.3	2.1*	1.6*			
Trans2Hexenol	0.1	1669	1.8*	.05	2635	5.3*				0.1	2.1*	21*			
4OH-nonenal	0.4	583	1495												
Aldehydes															
Benzaldehyde	1.8	2201	1251	1.6	6180	3863	0.4	3000	6900	.33	4.4*	13*			
Acetaldehyde	37.6	1830	48.7	48.2	8060	167	12.7	4300	340				ns	-	50
Hexanal	1.5	4205	2766										2.4	7.8*	3.3*
Trans2Hexenal	0.0						0.2	550	3200	0.3	3.3*	11*	1.1	4.6*	4.2*
4OH-nonenal	0.3	643	2296										1.3	2800	2150
Malondialdhyde	0.0														
Cofactors															
NAD$^+$	0.15						0.23			0.12			.03		
NADH	0.02						0.18						.05		

regenerating NAD$^+$ for corneal aldehyde dehydrogenase to continually oxidize these aldehydes.

ACKNOWLEDGEMENTS

This research was supported in part by a grant from the National Health and Medical Research Council of Australia.

REFERENCES

Algar, E.M., Seeley, T.L., and Holmes R.S., (1983) Purification and molecular properties of mouse alcohol dehydrogenase isozymes. *Eur J Biochem.* 137:139-147.

Algar, E.M., VandeBerg J.L., and Holmes R.S., (1992) A gastric alcohol dehydrogenase in the baboon: Purification and properties of a 'high-Km' enzyme, consistent with a role in 'first pass' alcohol metabolism. *Alcohol Clin Exp Res.* 16:922-927.

Ames, G.F-L., (1974) Resolution of bacterial proteins by polyacrylamide gel electrophoresis on slabs. Membrane, soluble, and periplasmic fractions. *J Biol Chem.* 249:634-644.

Andrews P., (1965) The gel filtration behaviour of proteins related to their molecular weight over a wide range. *Biochem J.* 96:595-606.

Boettner E.A., and Wolters J.R., (1962) Transmittance of the ocular media. *Invest Ophthalmol.* 1:776-783.

Boleda, M.D.,Saudi, N., Farrés, J., and Parés, X., (1993) Physiological substrates for rat alcohol dehydrogenase classes: aldehydes of lipid peroxidation, ω-hydroxyfatty acids, and retinoids. *Arch Biochem Biophys.* 307:85-90.

Downes, J.E., VandeBerg, J.L., Hubbard, G.B., and Holmes, R.S., (1992) Regional distribution of mammalian corneal aldehyde and alcohol dehydrogenase. *Cornea.* 11:560-566.

Downes, J.E., Swann P.G., and Holmes, R.S. (1993) Ultraviolet light-induced pathology in the eye: associated changes in ocular aldehyde dehydrogenase and alcohol dehydrogenase activities. *Cornea.* 12:241-248.

Downes, J.E., Swann P.G., ,and Holmes, R.S., (1994) Differential corneal sensitivity to ultraviolet light among inbred strains of mice. *Cornea.* 13:67-72.

Esterbauer, H., (1982). Aldehydic products of lipid peroxidation. In *'Free Radicals, Lipid Peroxidation and Cancer'* (Ed. McBrian D.C.H. and Slater, T.F.) Academic Press, New York, pp 101-128.

Holmes, R.S., (1977) The genetics of α-hydroxyacid oxidase and alcohol dehydrogenase in the mouse: evidence for multiple loci and linkage between *Hao-2* and *Adh-3*. *Genetics* 87:709-716.

Holmes, R.S., (1978) Electrophoretic analyses of alcohol dehydrogenase, aldehyde dehydrogenase, aldehyde oxidase, sorbitol dehydrogenase and xanthine oxidase from mouse tissues. *Comp Biochem Physiol.* 61B:339-349.

Holmes, R.S., (1979) Genetics and ontogeny of alcohol dehydrogenase isozymes in the mouse: evidence for a cis-acting regulator gene (*Adt-1*) controlling C_2 isozyme expression in reproductive tissues and close linkage of *Adh-3t* and *Adt-1* on chromosome 3. *Biochem Genet.* 17:461-472.

Holmes, R.S., Albanese, R., Whitehead, F.D., and Duley, J.A., (1981) Mouse alcohol dehydrogenase isozymes in the mouse: products of closely localized duplicated genes exhibiting divergent kinetic properties. *J Exp Zool.* 215:151-157.

Holmes, R.S., Popp, R.A., and VandeBerg, J.L., (1988) Genetics of ocular NAD-dependent alcohol dehydrogenase and aldehyde dehydrogenase in the mouse: evidence for genetic identity with stomach isozymes and localization of Ahd-4 on chromosome 11 near trembler. *Biochem Genet.* 26:191-205.

Julia, P., Farrés, J., and Pares, X., (1987) Characterization of three isozymes of rat alcohol dehydrogenase. *Eur J Biochem.* 162:179-189.

Kappus, H., (1985) Lipid peroxidation: mechanisms, analysis enzymology and biological relevance. In *Oxidative Stress.* (Sies, H. Ed.) Academic Press, New York. pp 273-310.

Kurzel, RB.,Wolbarsht, ML. and Yomanashi, BS., (1977) Ultraviolet radiation effects on the human eye. In *Photochemical and Photobiological Reviews.* (Smith KC. Ed.) Plenum Press, New York. pp133-166.

Laemmli, U.K., (1970) Cleavage of Structural protein during the assembly of the head of Bacteriophage T4. *Nature.* 227:680-685.

Moreno, A., and Pares, X. (1983) Purification and characterization of a new alcohol dehydrogenase from human stomach *J Biol Chem.* 266:1128-1133.

Parés, X., Cederlund, E., Moreno A., Hjelmqvist L., Farrés J. (1994) Mammalian class IV alcohol dehydrogenase (stomach alcohol dehydrogenase): Structure, origin, and correlation with enzymology. *Proc Natl Acad Sci. USA.* 91, 5:1893-1897.

Pitts, D.G., (1978) The ocular effects of ultraviolet radiation. *Am J Optom Physiol Optics.* 55:19-35.

Smith, P.K., Krohn, R.I., Hermanson, G.T., Mallia, A.K., Gartner, F.H. Provenzana, M.D., Fujimoto, E.K., Goeke, N.M., Olson, B.J., Klenk, D.C., (1985) Measurement of Protein Using Bicinchoninic Acid. *Anal. Biochem.* 150:76-85.

43

MAMMALIAN CLASS VI ALCOHOL DEHYDROGENASE

Novel Types of the Rodent Enzymes

Jan-Olov Höög and Margareta Brandt

Department of Medical Biochemistry and Biophysics
Karolinska Institutet
S-171 77 Stockholm, Sweden

INTRODUCTION

The number of mammalian alcohol dehydrogenases (ADHs) has increased in recent years, with six classes currently being defined according to structural differences (Jörnvall & Höög, 1984; cf. Jörnvall et al., this volume). In the rat, a widely used model in studies on the metabolism of alcohol, three distinct forms have been characterized (Julià et al., 1987), while the human ADHs have been grouped earlier into three classes according to kinetic and electrophoretic properties (Vallee & Bazzone, 1983).

The three rat ADHs correspond to classes I, III, and IV, with the latter form being referred to as the stomach enzyme (Parés et al., 1990), which was earlier thought to correspond to the human class II ADH because of similar kinetic properties. Sequence analysis of the stomach enzyme and the isolation of a cDNA clone coding for a rat class II ADH showed that at least four classes of ADH exist in the rat (Parés et al., 1990; Höög, 1990). In human, a fifth class was verified after the isolation of a genomic clone that coded for a protein that was not identical to any of the other defined classes of mammalian ADH (Yasunami et al., 1991; Parés et al., 1992). That started the investigation for further types of rat ADHs. All the different types of mammalian ADHs, differ in enzymatic properties and tissue distribution, beside the structural differences that have been used to define the classes. The class V ADH is the only class that has not been proven directly at protein level in man, but the isolated DNA was used to express recombinant enzyme showing that the protein had ethanol dehydrogenase activity (Chen & Yoshida, 1992). The K_m-values for ethanol for the different mammalian ADHs spans from 50 mM, for the human bb isozyme (Bosron et al., 1983), to the class III enzyme that is not possible to saturate. Class III ADH is further distinguished from the other types in the respect that this is the only type that shows glutathione-dependent formaldehyde dehydrogenase activity (Koivusalo et al., 1989). The least studied form of the well-established classes of mammalian ADHs is class II, which up until recently, had been characterized only in man (Li et al., 1977; Höög et al., 1987). Now a class II type structure has been determined from rat that shows that class II is the least conserved ADH in mammals (Höög, 1990; unpublished). So far, description of five classes of ADH is limited to man. Recently, a further new ADH structure was isolated from deer-mouse, Adh-2, that showed large differences to any of the other mammalian forms (Zheng et al., 1993).

We have now shown that five classes of ADH also exist in rodents and that the fifth type in rat forms ADH class VI, which is highly homologous to the deer-mouse Adh-2 enzyme.

MATERIALS AND METHODS

Cloning of cDNA

A 5'extended cDNA-library with mRNA isolated from liver of male rats was from Clontech Inc., CA; Hybond-N$^+$ nylon filters, [a-^{32}P]dCTP, and [a-^{35}S]dATP were from Amersham, UK; Glassmilk from BIO 101, CA; Qiagen ion exchange columns from Diagen, FRG; DNA polymerase and DNase were from International Biotechnology Inc, CT; *Taq* DNA polymerase was from Perkin-Elmer Cetus, CA; restriction endonucleases and dideoxy sequencing chemicals, including T7 DNA polymerase, were from Pharmacia Biotech, Sweden.

A 500 bp fragment of a cDNA coding for human class V ADH was PCR amplified after the published sequence (Yasunami et al., 1991) using two synthesized primers that included *Eco*RI sites. A human liver cDNA-library (Clontech Inc., CA) was used as cDNA source and the PCR reaction was run in a total volume of 50 ml using 2.5 U of *Taq* polymerase for 35 cycles. Amplified PCR fragment was gel purified with glassmilk before cloning into a pEMBL9 vector, and was analyzed with the dideoxy chain termination method (Sanger et al., 1977) on double-stranded DNA. The obtained fragment was nick-translated with [a-^{32}P]dCTP prior to screening the rat liver cDNA library according to standard procedures (Sambrook et al., 1989), and cDNA-inserts from isolated l-phages were size estimated by agarose gel electrophoresis after cleavage with *Eco*RI. The cDNA was purified with glassmilk and subcloned into a pEMBL9 vector before mapping with restriction endonucleases. Plasmids were purified with the Qiagen ion exchange procedure and selected

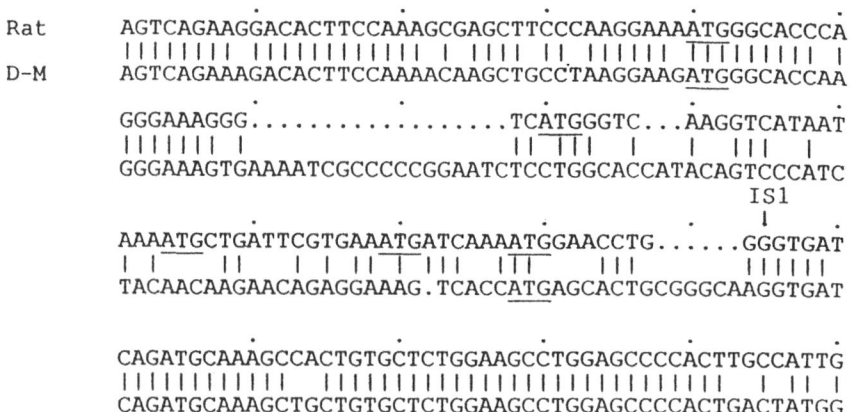

Figure 1. Alignment of cDNAs coding for 5'ends of rat and deer-mouse class VI ADH. Possible ATG start codons are underlined. IS1 indicate the splicing point for intron 1 in all investigated mammalian ADHs.

Table 1. Positional identities between the rat class VI ADH and the different human ADHs

	Human				
	class I	class II	class III	class IV	class V
Rat class VI	54.6%	50.8%	59.2%	50.7%	65.1%

fragments were further purified with glassmilk and subcloned into M13mp18 and mp19 vectors.

Sequence Analysis

DNA sequence analysis was carried out by the dideoxynucleotide chain termination method (Sanger et al., 1977) both on single-stranded DNA subcloned in M13 vectors and on double-stranded DNA using T7 DNA polymerase in combination with [a-^{35}S]dATP. Universal and reverse M13 primers were used in combination with sequence specifc primers that were synthesized. The entire rat cDNA was sequenced on both strands and the PCR amplified cDNA coding for human class V ADH was sequenced on two independent cloned fragments.

Expression of Recombinant Protein

A 5' 400 bp cDNA fragment was subcloned into M13mp8 prior to introducing an *Eco*RI site just ahead the ATG codon corresponding to the initiator ATG codons in other mammalian ADHs. For this purpose a 38-mer oligonucleotide was synthesized and the phosphorothioate method (Taylor et al., 1985) was used for site-directed mutagenesis with an Amersham mutagenesis kit. The 3' cDNA fragment was liberated with *Eco*RI and *Bgl*II, and the *Bgl*II site was converted into a *Hin*dIII site. Both fragments were gel purified with glassmilk before ligation into *Eco*RI and *Hin*dIII digested vector pKK 223-3 (Pharmacia Biotech, Sweden). The same type of construction has earlier been used for expression of other mammalian ADHs (Höög et al., 1993).

Recombinant protein was expressed in 2 l cultures using an *E. coli lac*I^q strain, TG1, with a burst of 0.2 mM isopropyl-b-thiogalactopyranoside. After harvesting, the cells were disrupted in 1 mM dithiothreitol, 10 mM Tris-HCl, pH 8, and sonicated intermittently for 10 min before centrifugation for 60 min at 48,000x*g*. The supernatant was applied on a DEAE column (DE-52, Whatman; 150 ml) for purification. The main first peak contained ADH activity and the recombinant protein was analyzed for purity by SDS polyacrylamide gel electrophoresis.

Enzymatic activity was determined spectrophotometrically at 340 nm using a Beckman DU-8 spectrophotometer to follow the NADH formation in 0.1 M glycine-NaOH buffer, pH 10, 25°C or in 0.1 M sodium phosphate buffer, pH 8.0, 25°C. The NAD$^+$ concentration used was 2.4 mM, and for analysis of glutathione-dependent formaldehyde dehydrogenase activity 1 mM glutathione and 1 mM formaldehyde (Ladd Research Industries, VT) were used. Ethanol and octanol of analysis grade were used without further purification. NAD$^+$ grade III and glutathione were from Sigma. A weighted non-linear regression analysis computer program was used to fit all lines to data points and to calculate K$_m$-values.

RESULTS

A total of 100,000 plaques derived from a rat liver cDNA-library were screened with a cDNA-fragment coding for human class V ADH. The probe fragment was obtained with PCR amplification of a human liver cDNA library using two oligonucleotide primers separated with 500 bp. Three positive signals were obtained after the screening and the corresponding clones were isolated and analyzed. The cDNA-inserts were subcloned into pEMBL9 vectors. All cDNA-clones harboured inserts of about 1400 bp and showed the same restriction pattern. One of the clones was subjected to DNA sequence analysis giving a sequence of 1350 bp which harbours the entire coding region and also 5'non-coding and 3'non-coding regions. The 3'non-coding region was not isolated in its complete form though no poly-A-signal and poly-A-tail were found. In the 5'end, a short open reading frame was found resulting in a polypeptide of eleven amino acid residues. This open reading frame is terminated by a double stop codon, but it overlaps the main open reading frame. Three additional upstream ATG codons are in the same reading frame as the codon corresponding to the ATG start codon found in other mammalian ADHs (Fig. 1). The cDNA structure determined is highly homologous to other ADH cDNA structures from rodents, but upstream of the corresponding splicing point for intron one no homology can be detected. Compared to the closely related cDNA coding for deer-mouse Adh-2 a deletion of 26 bp was found in the rat cDNA, but 62 bp further upstream the two structures are highly homologous again (Fig. 1).

The deduced amino acid sequence from the cDNA forms a polypeptide of 388 amino acid residues, if the first ATG start codon in the main open reading frame is used, and is homologous to other mammalian ADHs. Alignment of the rat amino acid sequence derived from the isolated cDNA to different human ADHs shows that the rat sequence is between 50% to 65% identical (Table 1), with the highest score for human class V. The deer-mouse Adh-2 (class VI) and the now investigated rat class VI ADH are 79% identical at the protein level and 85% identical at the DNA level (Table 2).

The rat amino acid sequence shows an elongated N-terminus compared to all mammalian ADHs and the amino acid residues corresponding to exon one show no homology to any mammalian ADH. An insertion at position 60 and a deletion at position 120, as compared to class I ADH, are also found in the rat enzyme. The residues lining the active site pocket are not identical to human class V, but constitute a mixture of the other classes, and at four positions the rat enzyme shows unique residues, Glu57, Cys93, Ile110 and Ser294 (Table 3). The coenzyme interacting residue at position 47 is Gly, in contrast to Arg and His in most mammalian forms. Thr at position 48 is identical with ADHs without steroid dehydrogenase activity; Lys51 corresponds to the residue in class I ADH (His), which is proposed to interact in a charge relay system; and Arg115 is identical with the residue found among class III ADHs, exhibiting glutathione-dependent formaldehyde dehydrogenase activity.

Table 2. Comparison of class I and VI ADH from rat and deer-mouse at protein and DNA levels

Class	I	VI
Protein	88.8%	78.6%
DNA	85.4%	83.3%

Expression of recombinant protein in *Escherichia coli*, using the ATG codon corresponding to the translation start codon used in other mammalian ADHs, yielded active protein. The protein was partly pure (>50%) after a one-step purification with ion-exchange chromatography. K_m-values were determined for ethanol (20 mM) and octanol (10 mM) at pH 10. The recombinant rat protein showed no activity with hydroxymethyl-glutathione as substrate.

DISCUSSION

A fifth type of ADH in rat has been isolated at cDNA level using a PCR amplified human cDNA coding for class V ADH as screening probe. The deduced amino acid sequence consists of 388 amino acid residues with a high homology to other mammalian ADHs. The sequence is unique with an elongated N-terminus that has not been found in any other ADH. An insertion at position 60, compared to class I, like in other mammalian ADHs (Jörnvall et al., 1987) is found in this type and a deletion at position 120. The latter has been observed in human class V ADH and in the deer-mouse Adh-2. Amino acid residues lining the active site pocket (Table 3) are not identical to any other ADH, but at several positions, identities are observed. The ADH now characterized is classified as class VI and also includes the deer-mouse Adh-2 enzyme (Zheng et al., 1993). Unique residues in the active site pocket are found at four positions. Glu57 occupies the middle part of the pocket, which will give this ADH an extra negative charge that has been shown to be of importance in class III ADH (Asp; Estonius et al., 1994; Estonius et al., this volume). A completely unique residue is found at position 93 (Cys) at the inner part of the pocket. Only Ala, Phe and Tyr have been reported previously at this position; Phe is the typical class I residue, which forms the hydrophobic pocket; whereas Tyr is found among class III ADHs, enabling a more hydrophilic pocket (Eklund et al., 1990). Cys is a much smaller residue, like Ala (in the fetal form), resulting in more space in the pocket. The Cys will further make the enzyme more sensitive to oxidation. Ile110 is also unique, but other hydrophobic residues occupy this position in other ADHs. Finally, Ser294 will further change the characteristics of the active site pocket for the class VI ADH. Val is commonly reported at this position for most other ADH classes, and has been ascribed a structural function (Jörnvall et al., 1989; Eklund et al., 1990). In a fish species (cod) position 294 is occupied by Trp preceding a deletion (Danielsson et al., 1992). However, in the recently characterized class II type ADH from rat, both residues at positions 294 and 295 are deleted (Höög, unpublished). From all these results, it is obvious that position 294 can be highly variable. These four unique residues in the active site pocket of class VI, except Ile110, change the properties of the pocket from a very hydrophobic into a very hydrophilic one. Predictions of the substrate specificity for this enzyme will probably be close to the specificity for class III ADH/glutathione-dependent formaldehyde dehydrogenase, as a result of the hydrophilic pocket and the presence of Arg at position 115, which have been shown to be essential for formaldehyde dehydrogenase activity (Engeland et al., 1993). An acidic residue at position 57 is from molecular modelling studies characteristic for the class III ADH pocket (Eklund et al., 1990), and shown to heavily influence the formaldehyde dehydrogenase activity (Estonius et al., 1994). The two unique residues Cys93 and Ser294 are more difficult to predict, but both change the active site pocket into a more polar one compared to the classical ADHs (Jörnvall et al., 1989; Eklund et al., 1990). All unique residues are identical to the deer-mouse Adh-2 (class VI) except the residue at position 57, which is occupied by Lys instead of Glu, resulting in an opposite charge. Expression of the recombinant class VI enzyme showed that this enzyme had no glutathione-dependent formaldehyde dehydrogenase activity, as could have been expected, and thereby demonstrates that there are more determinants in the class III ADH for this unique activity.

Table 3. Amino acid residues lining the substrate-binding cleft of human and rodent ADHs indicates a deletion

| Position | coenzyme binding and inner part of binding cleft | | | | | | | middle and outer part of binding cleft | | | | | |
	47	51	48	93	140	141		57	110	115	116	294	318
Human a	Gly	His	Thr	Ala	Phe	Leu		Met	Tyr	Asp	Val	Val	Ile
Human b	Arg	His	Thr	Phe	Phe	Leu		Leu	Tyr	Asp	Leu	Val	Val
Human g	Arg	His	Ser	Phe	Phe	Val		Leu	Tyr	Asp	Leu	Val	Ile
Rat	Arg	His	Ser	Phe	Phe	Leu		Leu	Tyr	Asn	Leu	Val	Ile
Deer Mouse	Arg	His	Ser	Phe	Phe	Ile		Leu	Leu	Asp	Leu	Val	Ile
Human II	His	Thr	Thr	Tyr	Phe	Phe		Phe	Phe	Ser	Asn	Val	Phe
Rat II	Pro	Asn	Thr	Phe	Phe	Met		Lys	Leu	Arg	Asn	D	Phe
Human III	His	Tyr	Thr	Tyr	Tyr	Met		Asp	Leu	Arg	Val	Val	Ala
Rat III	His	Tyr	Thr	Tyr	Phe	Met		Asp	Leu	Arg	Val	Val	Ala
Rat IV	Gly	His	Thr	Phe	Phe	Met		Met	Leu	Asp	Leu	Ala	Val
Human V	Gly	Lys	Thr	Phe	Phe	Gly		His	Phe	Lys	Gln	Val	Val
Rat VI	Gly	Lys	Thr	Cys	Tyr	Ile		Glu	Ile	Arg	Leu	Ser	Ile
Deer Mouse VI	Gly	Lys	Thr	Cys	Tyr	Ile		Lys	Ile	Arg	Leu	Ser	Val

However, the high K_m-value for octanol (10 mM) is compatible with the hydrophilic active site pocket. This has been observed earlier with the hydrophilic active site pocket in class III ADH. The change from a hydrophobic to a hydrophilic active site pocket between class I and III ADH results in an increase in the K_m-value for longer alcohols like octanol (Eklund et al., 1990).

A variant of alternative splicing is found in the rat class VI ADH. Upstream of the corresponding splicing junction of intron one, the rat class VI cDNA has a sequence that is not homologous to the other ADH structures. However, 62 bp upstream of this splicing point a high degree of homology is found again. Compared to the deer-mouse cDNA coding for Adh-2, a deletion of 26 bp has occurred. This deletion/alternative splicing introduces three extra ATG start codons that elongate the deduced rat amino acid sequence by 14 residues as compared to the deer-mouse structure. Furthermore, a short open reading frame is found in the rat cDNA, and a corresponding open reading frame is also found in the cDNA coding for deer-mouse Adh-2, which partly overlaps the three first ATG start codons. This will probably weaken the translation initiation and it is difficult to state which ATG start codon is used. Interestingly, the short 5' open reading frame translates into amino acid residues common in exon one for other mammalian ADHs. In other ADHs, alternative splicing and point mutations at stop codons have been found. In a human cDNA coding for the human a-subunit, a region of 149 bp was deleted corresponding exactly to exon eight (Jörnvall et al., 1989), and this deletion was explained by slipped base pairing. In a cDNA coding for human class II ADH, a mutated stop codon was found, resulting in a C-terminal elongated with twelve amino acid residues (Höög et al., 1987).

The region after the unique Ser294 in class VI ADH and the corresponding deletion in the rat class II type ADH shows very low positional identity. In the region of residues 294-315, a positional identity of 23% is found between the human and rat class II ADHs, which is close to the background value. At position 60, the human class II and VI ADHs have an insertion, although this residue is missing in the rat class II type enzyme. The region

Table 4. Positional identities between human and rat ADHs

CLASS I	82%
CLASS II	73%
CLASS III/GSH-dependent formaldehyde dehydrogenase	94%
CLASS IV	87%
CLASS V-VI	65%

around this deletion shows no homology between the different forms. Both these deletions with almost non-homologous regions are consistent with identified variable segments in ADHs around amino acid residues 50-62 (V_1) and 294-312 (V_3) (Persson et al., 1993), where the latter region is forming the subunit interaction area. The region of amino acid residues 112-130 forms another variable segment (V_2). This is the region where the four residue insertion in class II ADHs is found, as well as the deletion in class VI. Thus, all three variable segments are affected among these mammalian ADHs. The shape of the active site pocket in the rat class II type ADH will also differ, firstly because of the deletion at position 294-295, and secondly, and more importantly because of the introduction of a proline residue at position 47 (Höög, unpublished). Residue 47 is interacting with the pyrophosphate in the coenzyme, NAD^+ (Eklund et al., 1976; 1990). In most investigated ADHs this position is occupied by Arg or His (Table 3; Jörnvall et al., 1987), both positively charged residues that form ionic bonds to the negatively charged pyrophosphate. Amino acid residues 46 to 53 in the class I horse enzyme form an α-helix (Eklund et al., 1976) and the same is true for the human class I b-subunit (Hurley et al., 1991). Gly47, which is present within human class I (a-subunit) and classes IV and VI ADHs, is incapable of forming ionic bonds with the pyrophosphate, and results in highly altered kinetic parameters (Hurley & Bosron, 1992, Parés et al., 1994). Pro found in the rat class II type enzyme cannot like Gly interact with the coenzyme in a similar manner as in most class I ADHs. Furthermore, the Pro can change the structure of the active site pocket probably by breaking the a-helix. This will, as for class VI ADH, change the substrate specificity caused by structural changes.

The results obtained from comparison of human and rat ADHs at protein level clearly shows that the class II enzymes are the least conserved class of the mammalian class I - VI enzymes (Table 4; Höög unpublished). Class III is the far most conserved structure, with 94% positional identity, which supports the proposal that the class III alcohol dehydrogenase/glutathione-dependent formaldehyde dehydrogenase has a more strict function in protecting the cells from formaldehyde generated in different metabolic reactions (Uotila & Koivusalo, 1989). Classes I and IV ADHs show intermediate positional identity. Class VI shows only 65% positional identity to the human class V, and the low identity demonstrates that the two enzymes are of different classes. The identity between the two class VI enzymes, rat and deer-mouse, is 79%. However, the two rodents are fairly closely related, indicating that this figure should be lower for a corresponding human structure, if such a human ADH exists (cf. Tables 2 and 4). Thus, the two extremes are class II and III with positional identities of 73% and 94%, respectively. Thus far, class V ADH has been described only in man (Yasunami et al., 1991). The meaning of the large structural difference of the class II enzymes is not known, but in contrast to class III ADH the function seems not to be strictly specific. The human class II enzyme can catalyze a broad range of hydroxyl-containing compounds (Li et al., 1977), and the best substrates found are 4-hydroxyalkenals (Sellin et al., 1991). Alignment of the coding parts of the class VI cDNAs from rat and deer-mouse yields 83.3% identity, and the corresponding value for class I ADHs is 85.4%. The higher values at protein level compared to values at DNA level for the class I enzymes (Table 2) shows that the wobbling position in the codons are affected in a higher frequency compared to the positions that yields another amino acid residue. For the class VI enzymes the opposite is observed indicating a more strict function for the class I enzymes than for the class VI enzymes. This observation can further indicate that the substrate specificity for the class VI ADHs, like for the class II ADHs, is very broad with a function as a general alcohol oxidizing/aldehyde reducing enzyme.

The tissue distribution of the rat class VI ADH is limited to kidney and liver (Höög, unpublished), which differs from the rat class I, II, and III ADHs (Estonius et al., 1993). Class I ADH is found in almost every tissue, but with a 100-fold expression difference between liver and brain. Class III is found in every tissue examined at a fairly even level, whereas class II ADH exhibits a high expression limited to liver, duodenum and testis. mRNA for human class V ADH has been detected in a number of tissues including liver and stomach (Yasunami et al., 1991; Estonius, unpublished) and rat class IV has been shown to be expressed mainly in stomach, esophagus and eye (Parés et al., 1994). All these investigations show that no other ADH than class VI is limited to one or two tissues. However, the deer-mouse class VI ADH is found mainly in liver and very small amounts seems to be present in lung, but no mRNA or enzyme is found in kidney (Zheng et al., 1993). This finding may reflect that the class VI ADH is strictly regulated, resulting in different tissue distribution patterns for different species.

In summary, this new rat ADH shows a low positional identity as compared with known human ADHs, the active site pocket is very hydrophilic and harbours several unique amino acid residues, and the tissue distribution is very limited. Therefore, this ADH is defined as a new class, class VI.

ACKNOWLEDGEMENTS

This work was supported by grants from the Swedish Medical Research Council and the Karolinska Institute. A personal fellowship to JOH from Procordia AB is acknowledged.

REFERENCES

Bosron, W.F., Magnes, L.J., and Li T.-K., 1983, Kinetic and electrophoretic properties of native and recombined isoenzymes of human liver alcohol dehydrogenase, *Biochemistry* 22:1852-1857.

Chen, C.-S. and Yoshida, A., 1991, Enzymatic properties of the protein encoded by newly cloned human alcohol dehydrogenase ADH6 gene, *Biochem. Biophys. Res. Commun.* 181:743-747.

Danielsson, O., Eklund, H., and Jörnvall, H., 1993, The major piscine liver alcohol dehydrogenase has class-mixed properties in relation to mammalian alcohol dehydrogenases of classes I and III, *Biochemistry* 31:3751-3759.

Eklund, H., Nordström, B., Zeppezauer, E., Söderlund, G., Ohlsson, I., Boiwe, T., Söderberg, B.-O., horse liver alcohol dehydrogenase at 2.4 Å resolution, *J. Mol. Biol.* 102:27-59.

Eklund, H., Müller-Wille, P., Horjales, E., Futer, O., Holmquist, B., Vallee, B.L., Höög, J.-O., Kaiser, R., and Jörnvall, H., 1990, Comparison of three classes of human liver alcohol dehydrogenase. Emphasis on different substrate binding pockets, *Eur. J. Biochem.* 193:303-310.

Engeland, K., Höög, J.-O., Holmquist, B., Estonius, M., Jörnvall, H., and Vallee, B.L., 1993, Mutation of Arg-115 of human class III alcohol dehydrogenase. A binding site required for formaldehyde dehydrogenase activity and fatty acid activation, *Proc. Natl. Acad. Sci. USA* 90:2491-2494.

Estonius, M., Danielsson, O., Karlsson, C., Persson, H., Jörnvall, H., and Höög, J.-O., 1993, Distribution of alcohol and sorbitol dehydrogenases: Assessment of mRNAs in rat tissues, *Eur. J. Biochem.* 215:497-503.

Estonius, M., Höög, J.-O., Danielsson, O., and Jörnvall, H., 1994, Residues specific for class III alcohol dehydrogenase. Class transition and S-hydroxymethylglutathione binding. Site-directed mutagenesis of the human enzyme, *Biochemistry*, in press.

Hurley, T.D., Bosron, W.F., Hamilton, J.A., and Amzel, L.M., 1991, Structure of human b_1b_1 alcohol dehydrogenase: catalytic effects of non-active site substitutions, *Proc. Natl. Acad. Sci. USA* 88:8149-8154.

Hurley, T. and Bosron, W.F., 1992, Human alcohol dehydrogenase: dependence of secondary alcohol oxidation on the amino acids at positions 93 and 94, *Biochem. Biophys. Res. Commun.* 183:93-99.

Höög, J.-O., 1990, Mammalian class II alcohol dehydrogenase: Species and class comparisons at genomic and protein levels, *in:* Enzymology and Molecular Biology of Carbonyl Metabolism 3. Edited by H. Weiner, B. Wermuth, and D.W. Crabb, Plenum Press, New York, pp. 285-292.

Höög, J.-O., von Bahr-Lindström, H., Hedén, L.-O., Holmquist, B., Larsson, K., Hempel, J., Vallee, B.L., and Jörnvall, H., 1987, Structure of the class II enzyme of human liver alcohol dehydrogenase. Combined cDNA and protein sequence determination of the p-subunit, *Biochemistry* 26:1926-1932.

Höög, J.-O., Karlsson, C., Eklund, H., Shapiro, R., and Jörnvall, H., 1993, Site-directed mutagenesis of mammalian alcohol and sorbitol dehydrogenases map functional differences within the enzyme family, *in:* Enzymology and Molecular Biology of Carbonyl Metabolism 4. Edited by H. Weiner, D.W. Crabb, and T.G. Flynn, Plenum Press, New York, pp. 439-450.

Julià, P., Farrés, J., and Parés, X., 1987, Characterization of three isozymes of rat alcohol dehydrogenase. Tissue distribution and enzymatic properties, *Eur. J. Biochem.* 162:179-189.

Jörnvall, H. and Höög, J.-O, 1994, Nomenclature of alcohol dehydrogenases, *Alcohol and Alcoholism*, in press.

Jörnvall, H., Persson, B., and Jeffery, J., 1987, Characteristics of alcohol/polyol dehydrogenases. The zinc-containing long-chain alcohol dehydrogenases, *Eur. J. Biochem.* 167:195-201.

Jörnvall, H., von Bahr-Lindström, H., and Höög, J.-O., 1989, Alcohol dehydrogenases - structure, *in:* Human Metabolism of Alcohol vol. 2. Edited by R.D. Batt and K.E. Crow, CRC press, Boca Raton pp.43-64.

Koivusalo, M., Baumann, M., and Uotila, L., 1989, Evidence for the identity of glutathione-dependent formaldehyde and class III alcohol dehydrogenase, *FEBS Lett.* 257:105-109.

Li, T.-K., Bosron, W.F., Dafeldecker, W.P., Lange, L.G., and Vallee, B.L., 1977, Isolation of P-alcohol dehydrogenase of human liver: Is it a determinant of alcoholism? *Proc. Natl. Acad. Sci. USA* 74:4378-4381.

Parés, X., Moreno, A., Cederlund, E., Höög, J.-O., and Jörnvall, H., 1990, Class IV mammalian alcohol dehydrogenase. Structural data of the rat stomach enzyme reveal a new class well separated from those already characterized, *FEBS Lett.* 277:115-118.

Parés, X., Cederlund, E., Moreno, A., Saubi, N., Höög, J.-O., and Jörnvall, H. ,1992, Class IV alcohol dehydrogenase (the gastric enzyme). Structural analysis of human ss-ADH reveals class IV to be variable and confirms the prescence of a fifth mammalian alcohol dehydrogenase class, *FEBS Lett.* 303:69-72.

Parés, X., Cederlund, E., Moreno, A., Hjelmqvist, L., Farrés, J., and Jörnvall, H., 1994, Mammalian class IV alcohol dehydrogenase (stomach alcohol dehydrogenase): Structure, origin, and correlation with enzymology, *Proc. Natl. Acad. Sci. USA* 91:1893-1897.

Persson, B., Bergman, T., Keung, W.M., Waldenström, U., Holmquist, B., Vallee, B.L., and Jörnvall, H., 1993, Basic features of class-I alcohol dehydrogenase: variable and constant segments coordinated by inter-class and intra-class variability. Conclusions from characterization of the alligator enzyme, *Eur. J. Biochem.* 216:49-56.

Sambrook, J., Fritsch, E.F., and Maniatis, T., 1989, *in:* Molecular Cloning: a laboratory manual, Cold Spring Harbor Laboratory, Cold Spring Harbor, New York.

Sanger, F., Nicklen, S., and Coulson, A.R., 1977, DNA sequencing with chain termination inhibitors, *Proc. Natl. Acad. Sci. USA* 74:5463-5467.

Sellin, S., Holmquist, B., Mannervik, B., and Vallee, B.L., 1991, Oxidation and reduction of 4-hydroxyalkenals catalyzed by isozymes of human alcohol dehydrogenase, *Biochemistry* 30:2514-2518.

Taylor, J.W., Ott, J., and Eckstein, F., 1985, The rapid generation of oligonucleotide-directed mutations at high frequency using phosphorothioate-modified DNA, *Nucleic Acids Res.* 13:8765-8785.

Uotila, L. and Koivusalo, M., 1989, Glutathione-dependent oxidoreductases: Formaldehyde dehydrogenase, *in*: Coenzymes and Cofactors, Glutathione. Chemical, Biochemical and Medical Aspects, vol III, part A. Edited by D.Dolphin *et al.*, John Wiley & Sons, New York pp. 517-551.

Vallee, B.L. and Bazzone, T.J., 1983, Isozymes of human liver alcohol dehydrogenase, *Isozymes* 8:219-244.

Yasunami, C.C., Chen, C.-S., and Yoshida, A., 1991, A human alcohol dehydrogenase gene (ADH6) encoding an additional class of isozyme, *Proc. Natl. Acad. Sci. USA* 88:7610-7614.

Zheng, Y.-W., Bey, M., Liu, H., and Felder, M.R., 1993, Molecular basis of the alcohol dehydrogenase-negative deer mouse. Evidence for deletion of the gene for class I enzyme and identification of a possible new enzyme class. *J. Biol. Chem.* 268:24933-24939.

CRYSTALLIZATIONS OF NOVEL FORMS OF ALCOHOL DEHYDROGENASE

Mustafa El-Ahmad, Ramaswamy S., Olle Danielsson, Christina Karlsson, Mats Estonius, Jan-Olov Höög, Hans Eklund and Hans Jörnvall

Department of Medical Biochemistry and Biophysics
Karolinska Institutet
S-171 77 Stockholm, Sweden, and
Department of Molecular Biology
Swedish University of Agricultural Sciences
S-751 24 Uppsala, Sweden

INTRODUCTION

Alcohol dehydrogenases in nature are derived from different protein families (cf Jörnvall et al., 1993). The mammalian alcohol dehydrogenases constitute what presently appears to be six classes within a large family of medium-chain dehydrogenases and reductases, MDR, with at least seven characterized activity types (Persson et al., 1994). The classes of the mammalian alcohol dehydrogenases differ in structure (55-68% sequence identity), substrate pockets, subunit interactions, and other properties. Much of these aspects has been interpreted by comparisons and modelling studies based on crystallographic analyses of just two of the enzymes of one class, the class I horse (Eklund et al., 1976) and human (Hurley et al., 1991) enzymes, constituting the classical liver enzyme with ethanol activity. It is desirable to get direct crystallographic data on the conformations of more of the enzymes, especially since differences exist between the classes, class-hybrid properties have been found in early enzyme forms (of fish), and the two well-characterized classes, I and III, differ in internal variability patterns (Danielsson et al., 1994a). We have therefore crystallized five novel forms of these dehydrogenases, which should enable further structural characterizations and hence evaluation of the conclusions from modelling and from natural variants.

NEW FORMS CRYSTALLIZED

The Five New Forms Crystallized Are

* The fish class I alcohol dehydrogenase obtained from cod liver. This enzyme has partly class I/III hybrid properties (Danielsson et al., 1992), interpreted to reflect its evolution

of class I activity from a class III ancestral form (Danielsson & Jörnvall, 1992). The enzyme exhibits ethanol activity and pyrazole inhibition much like the mammalian class I alcohol dehydrogenases, but an amino acid sequence more related to class III sequences than to class I sequences. It also has differences at functionally important positions. Three such differences noted are a His (instead of Leu and Asp in the human class I and III forms, respectively) at the substrate-interacting position 57, a Trp residue at position 294, interpreted to influence relationships at the inter-domain cleft, and a Tyr residue (like in mammalian class III enzymes) replacing the coenzyme-interacting His-51 of the mammalian class I enzymes (Danielsson et al., 1992). These and other relationships are essential to check by direct analysis.

* The fish class III alcohol dehydrogenase obtained from cod liver. This is a typical class III protein, with essentially unaltered properties versus the mammalian class III enzymes (Danielsson & Jörnvall, 1992). Since no class III structure has been crystallographically verified, such a structure is important to solve, in order to finally establish all conclusions based on the glutathione-dependent formaldehyde dehydrogenase activity of this enzyme. The relationship to class I has been studied for the mammalian enzymes by modelling (Eklund et al., 1990) and site directed mutagenesis (Engeland et al., 1993; Estonius et al., 1994), but are important to verify.

* The cyclostome class III alcohol dehydrogenase obtained from hagfish. This is also a typical class III protein. In addition, it can participate *in vitro* in extensive hybridizations with distantly related enzyme forms (Danielsson et al., 1994b), illustrating the conservation of subunit interacting segments in the class III protein at large (Danielsson et al., 1994a).

* The class III alcohol dehydrogenase of a human recombinant wild-type form (Engeland et al., 1993). Apart from a missing N-terminal acetyl group, this enzyme is identical to the native human class III alcohol dehydrogenase and its structure would nicely complement those of the class I and III types in early vertebrates (above) and of the class I mammalian structures (Eklund et al., 1976; Hurley et al., 1991).

* The sorbitol dehydrogenase of a rat recombinant form (Karlsson et al., 1991). This would constitute the first sorbitol dehydrogenase structure. Although this structure, too, has been interpreted from modelling (Eklund et al., 1985), differences are extensive toward the mammalian alcohol dehydrogenases, including in sorbitol dehydrogenase a missing 21-residue loop, a missing structural zinc atom, and only 25% overall residue identity with mammalian class I alcohol dehydrogenase. Furthermore, the modelling has suggested a third zinc liganding position to the active site zinc atom that is now known not to be the true ligand (Karlsson et al., 1994). Hence, novel structural details are to be expected in this region, constituting a further reason why determination of the sorbitol dehydrogenase fold is important.

MATERIALS AND METHODS: CRYSTALLIZATIONS

The cod class I enzyme. This protein was prepared from cod liver as described (Danielsson et al., 1992). The enzyme was crystallized as an NAD$^+$/pyrazole complex using the hanging drop method. Crystals were first obtained in drops over a reservoir of 50 mM TES buffer, pH 6.9, 10 mM NAD$^+$, 1 mM pyrazole and 15.2% PEG 2000 at a protein concentration in the drop of 0.5 mg/ml. The crystals grew as thin plates and the very small crystals initially obtained were improved in size by macroseeding (Thaller et al., 1981). The average size of the crystals used for data collection was 0.5 x 0.1 x 0.05 mm. These crystals are monoclinic, with two subunits per asymmetric unit, and have been described recently (Ramaswamy et al., 1994a).

The cod class III enzyme. The cod enzyme is thus far the only class III alcohol dehydrogenase where isozymes have been detected and characterized (Danielsson and Jörnvall, 1992). The homodimeric *ll* isozyme (*l* for low-activity type, in distinction from the other subunit type, with *h* subunits of higher activity) was prepared as described and used for crystallization. The protein was crystallized as an NAD^+ complex using the hanging drop method. Crystals were first obtained in drops over a reservoir of 50 mM TES buffer, pH 6.9, 10 mM NAD^+, 100 mM NaCl, and 23.6% PEG 4000, at a protein concentration in the drops of 1 mg/ml. The average size of the crystals used for data collection was 0.4 x 0.3 x 0.075 mm. The crystals were monoclinic, also with two subunits per asymmetric unit.

The hagfish class III enzyme. This enzyme was recently characterized and used for distant cross-species hybridizations, showing an extensive conservation of the subunit interacting areas in all class III enzymes, from *Drosophila*, through hagfish, to mammals and humans (Danielsson et al., 1994b). The enzyme was now prepared from hagfish liver and crystallized as an NAD^+ complex using the hanging drop method. Crystals were obtained in drops using two different conditions. In the initial attempts, the reservoir consisted of 100 mM Hepes, pH 6.9, 10 mM NAD^+, 1% isopropanol, 100 mM NaCl, and 25% PEG 4000, and the protein concentration in the drop was 1 mg/ml. The average size of these crystals was 0.5 x 0.15 x 0.1 mm. The other condition utilized 100 mM Hepes, pH 6.0, 10 mM NAD^+, 100 mM NaCl, 900 mM sodium citrate, at a protein concentration in the drop of 3 mg/ml and the average size of the crystals obtained was 0.2 x 0.1 x 0.05 mm.

Human recombinant class III enzyme. The protein was prepared as described (Engeland et al., 1993) and was crystallized as an NAD^+ complex using the hanging drop method over a reservoir of 150 mM Tris-HCl, pH 8.5, 10 mM NAD^+, 10 mM $CdCl_2$, 50% PEG 4000. The protein concentration in the drop was 5 mg/ml and the average size of the crystals 0.25 x 0.15 x 0.05 mm.

Rat recombinant sorbitol dehydrogenase. The protein was prepared using the expression vector pSDHex in *E. coli* strain TG1, as described (Karlsson & Höög, 1993) and was crystallized using the hanging drop method. Crystals were first obtained over a reservoir of 0.4 M Hepes, pH 6.75, 10 mM NAD^+, 50 mM NaCl, 1.6 M sodium citrate. The protein concentration in the drop was 2 mg/ml, and the crystals obtained were very thin but with the other two dimensions 0.4 x 0.3 mm.

RESULTS

Crystallizations. Many different conditions were tested using various buffers, pH, salt, and PEG concentrations. Crystallizations were slow and crystals obtained frequently of small size. In most cases, crystallizations as NAD^+ complexes were most successful, and addition of NaCl or other salts was found frequently to be useful (cf. above). In the case of the cod class III enzyme, more than one crystal form was obtained, but overall the methods listed were found to give reproducible results of one crystal type.

All crystals obtained are shown in Fig 1. The two forms giving the largest crystals were the cod class I and III enzymes, which both produced monoclinic crystals.

Diffraction analysis. All of the crystals listed above have been used in initial diffraction tests. For the two cod enzymes, these diffractions have been extended further. They both contain a dimer per asymmetric unit, and the cell dimensions are a=103.3 Å, b=47.4 Å, c=80.7Å, =104.6° for class I, and a=127.5 Å, b=76.6 Å, c=93.4 Å, =99.4° for class III. For the cod class I enzyme, the structure was solved by molecular replacement, first producing a model based on the horse enzyme according to the molecular replacement solution (Ramaswamy et al., 1994a). This full chain

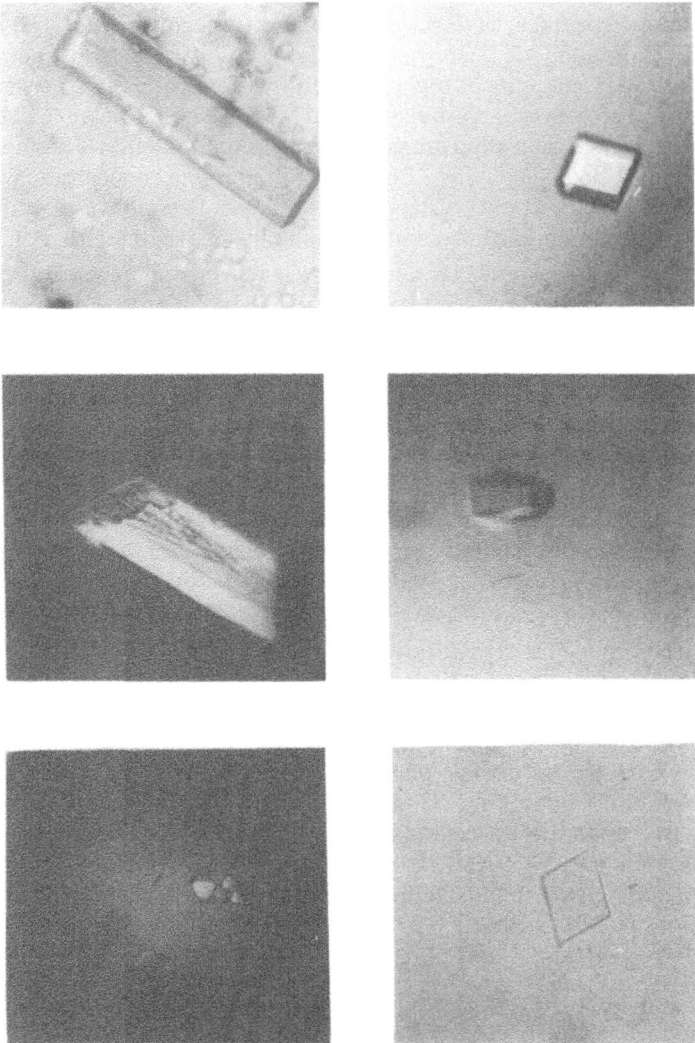

Figure 1. Crystals of cod liver alcohol dehydrogenase class I (top, left) and class III (top, right), of hagfish liver alcohol dehydrogenase (a class III enzyme, Danielsson et al., 1994b) under two different conditions as stated in the text (middle, left from PEG, and right from citrate), and of recombinant class III human alcohol dehydrogenase (bottom, left) and rat sorbitol dehydrogenase (bottom, right). All crystals were obtained by the hanging drop method as described in the text.

model was then used for further refinement and has now been refined to an R-factor of 20%, allowing detailed analysis.

Conformational interpretation. A summary on the chain-fold is given in Fig. 2, and demonstrates major deviations from the mammalian enzyme in two respects. First, three loop structures are different, corresponding to the three loops known to differ between the classes of the enzyme (Persson et al., 1993) and already interpreted in respect to location with the conformation (Danielsson et al., 1994a). In these regions the cod class I enzyme with its hybrid class III structure has loops clearly different from those of the mammalian class I enzyme (Ramaswamy et al., 1994b), reinforcing

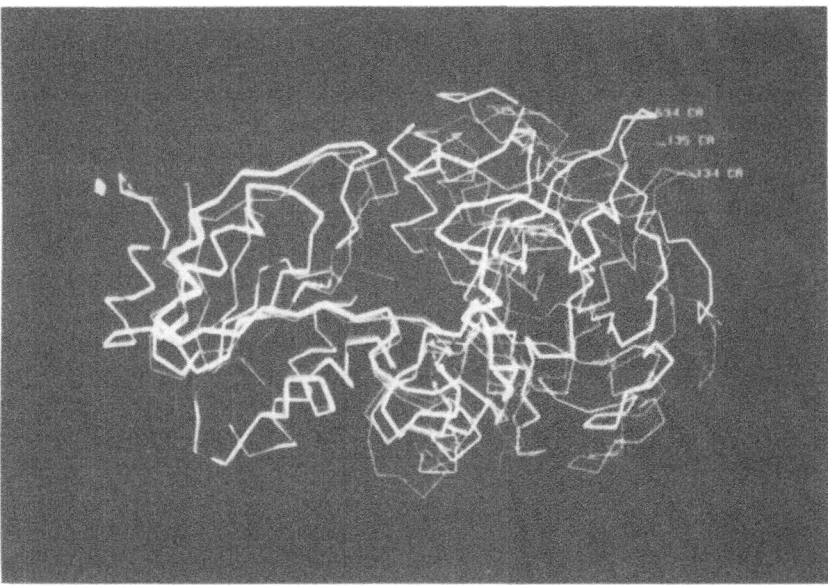

Figure 2. Tracing of the chain-fold of the conformation of the holoenzyme of cod liver class I alcohol dehydrogenase in relation to the chain-folds of the apo- and holoenzymes of mammalian alcohol dehydrogenase in the form of the class I equine enzyme (Eklund et al., 1984). The intermediate position of the cod enzyme (tracing marked 135 CA), in between the apo- (tracing marked 134 CA) and holo- (tracing marked 634 CA) enzyme forms of the mammalian protein, is clearly visible, suggesting a less pronounced domain movement in the cod enzyme upon coenzyme binding.

the conclusions based on model building (Danielsson et al., 1992) and on comparison of the class differences (Persson et al., 1993; Danielsson et al, 1994a). The other principal difference between the new structure and the mammalian class I structure known before is that the domains have a different relative position in respect to each other in the holoenzyme. Thus, the mammalian holoenzyme domain rotation versus that of the apoenzyme structure (Eklund et al., 1984) appears to be twice as large as that for the cod holoenzyme (Fig. 2).

The loop differences involve the coenzyme-interacting His-51 of the mammalian class I enzyme replaced by Tyr in the cod enzyme. The tyrosine side-chain does not form a hydrogen bond to the ribose of NAD^+ and points outward at the surface. Water molecules occupy the position of His-51. This result was unexpected and not anticipated from previous modelling experiments.

Another local difference affects the large Trp residue at position 294 of the cod enzyme. This addition affects the inter-domain cleft region, and constitutes another clear difference of functional consequence in the domain rotations associated with the catalytic activity. In this case, though, the modelling suggested the same conclusion (Danielsson et al., 1992), which is therefore now verified.

In conclusion, these results with a class I structure of some hybrid class III nature demonstrate that conformational differences are quite extensive in localized regions between the classes, but that the basic conformational fold is clearly related. Hence, the results support overall similarities previously modelled, but highlight the value of direct analyses in order to obtain the deviating structures in localized regions.

DISCUSSION

Of the five enzymes crystallized, analysis have been most extensive thus far with the cod class I enzyme that has class I/III hybrid properties. The tentative interpretations presently available show local conformational deviations from the mammalian class I structures in the three loop regions known to differ between classes and in the holoenzyme domain location. Both deviations have functional consequences, affecting coenzyme binding and inter-domain spaces available for domain movements. In principle the crystallographic data verify the conclusions from sequence analysis and modelling, since exactly these segments were those highlighted already from pre-crystallographic comparisons (Danielsson et al., 1994a). Nevertheless, the exact answers on the differences were slightly different in one case, regarding the His/Tyr exchange at position 51. In this case, modelling suggested a ligand change from His to Tyr, keeping the position, while the crystallographic analysis now suggests instead that Tyr are not binding to NAD^+. Hence, as expected, sequence comparisons, enzyme correlations, and modelling are useful in highlighting interesting segments, but the exact nature of the differences is difficult to establish until direct crystallographic investigations of the deviating species.

Since local differences at the active site are already known also for sorbitol dehydrogenase, it will be of great interest to see the exact nature of those deviations also. In the case of the class III enzyme, deviations at the active site appear necessary to bind the glutathione of the hydroxymethylglutathione adduct with formaldehyde, and two residues have already been singled out, Arg-115 and Asp-57 (Estonius et al., 1994). In the case of sorbitol dehydrogenase, all except one of the extra hydroxyls of the substrate have been suggested to hydrogen bond to the altered residues at the substrate-binding pocket (Eklund et al., 1985). With crystals now available, it will be of interest to see how these predicions will be fulfilled. Finally, the SDH structure will be important to correlate with the active site zinc binding as deduced from site-directed mutagenesis now further treated in a recent summary (Karlsson et al., 1994). As regards class III alcohol dehydrogenase, direct model transfers suggest that the class I orientation cannot be kept if Arg-115 and Asp-57 are to bind the glutathione adduct (Estonius et al., 1994), hence supporting an altered loop conformation in the class III enzymes as also found in the cod enzyme now analyzed.

ACKNOWLEDGMENTS

This work was supported by grants from the Swedish Medical Research Council, the Swedish Natural Research Council, the Swedish Alcohol Research Fund, the Swedish Employment Board, Peptech Ltd, and Pfizer Inc.

REFERENCES

Danielsson, O. and Jörnvall, H., 1992, "Enzymogenesis": Classical liver alcohol dehydrogenase origin from the glutathione-dependent formaldehyde dehydrogenase line. *Proc. Natl. Acad. Sci. USA* 89:9247-9251.

Danielsson, O., Eklund, H. and Jörnvall, H., 1992, The major piscine liver alcohol dehydrogenase has class-mixed properties in relation to mammalian alcohol dehydrogenases of classes I and III. *Biochemistry* 31:3751-3759.

Danielsson, O., Atrian, S., Luque, T., Hjelmqvist, L., Gonzàlez-Duarte, R. and Jörnvall, H., 1994a, Fundamental molecular differences between alcohol dehydrogenase classes. *Proc. Natl. Acad. Sci. USA* 91:4980-4984.

Danielsson, O., Shafqat, J., Estonius, M. and Jörnvall, H., 1994b, Alcohol dehydrogenase, class III contrasted to class I: Characterization of the Cyclostome enzyme, multiplicity like for the human form, and distant cross-species hybridization. *Eur. J. Biochem.*, in press.

Eklund, H., Nordström, B., Zeppezauer, E., Söderlund, G., Ohlsson, I., Boiwe, T., Söderberg, B.-O., Tapia, O., Brändén, C.-I. and Åkeson, Å., 1976, Three-dimensional structure of horse liver alcohol dehydrogenase at 2.4 Å resolution. *J. Mol. Biol.* 102:27-59.

Eklund, H., Samama, J.-P. and Jones, T.A., 1984, Crystallographic investigations of nicotinamide adenine dinucleotide binding to horse liver alcohol dehydrogenase. *Biochemistry* 23:5982-5996.

Eklund, H., Horjales, E., Jörnvall, H., Brändén, C.-I. and Jeffery, J., 1985, Molecular aspects of functional differences between alcohol and sorbitol dehydrogenases. *Biochemistry* 24:8005-8012.

Eklund, H., Müller-Wille, P., Horjales, E., Futer, O., Holmquist, B., Vallee, B.L., Höög, J.-O., Kaiser, R. and Jörnvall, H., 1990, Comparison of three classes of human liver alcohol dehydrogenase. Emphasis on different substrate-binding pockets. *Eur. J. Biochem.* 193:303-310.

Engeland, K., Höög, J.-O., Holmquist, B., Estonius, M., Jörnvall, H. and Vallee, B.L., 1993, Mutation of Arg-115 of human class III alcohol dehydrogenase: A binding site required for formaldehyde dehydrogenase activity and fatty acid activation. *Proc. Natl. Acad. Sci. USA* 90:2491-2494.

Estonius, M., Höög, J.-O., Danielsson, O. and Jörnvall, H., 1994, Residues specific for class III alcohol dehydrogenase. S-hydroxymethylglutathione-binding and site-directed mutagenesis of the human enzyme. *Biochemistry*, submitted.

Hurley, T.D., Bosron, W.F., Hamilton, J.A. and Amzel, L.M., 1991, Structure of human ₁ ₁ alcohol dehydrogenase: catalytic effects of non-active site substitutions. *Proc. Natl. Acad. Sci. USA* 88:8149-8153.

Jörnvall, H., Danielsson, O., Eklund, H., Hjelmqvist, L., Höög, J.-O., Parés, X. and Shafqat, J., 1993, Enzyme and isozyme developments within the medium-chain alcohol dehydrogenase family, *in*: "Enzymology and Molecular Biology of Carbonyl Metabolism 4", H. Weiner, ed. , Plenum, New York, pp. 533-544.

Karlsson, C. and Höög, J.-O., 1993, Zinc coordination in mammalian sorbitol dehydrogenase: Replacement of putative zinc ligands by site-directed mutagenesis. *Eur. J. Biochem.* 216:103-107.

Karlsson, C., Jörnvall, H. and Höög, J.-O., 1991, Sorbitol dehydrogenase: cDNA coding for the rat enzyme. Variations within the alcohol dehydrogenase family independent of quaternary structure and metal content. *Eur. J. Biochem.* 198: 761-765.

Karlsson, C., Jörnvall, H. and Höög, J.-O., 1994, Zinc binding of alcohol and sorbitol dehydrogenases. *in*: This vol., in press.

Persson, B., Bergman, T., Keung, W.M., Waldenström, U., Holmquist, B., Vallee, B.L. and Jörnvall, H., 1993, Basic features of class-I alcohol dehydrogenase: variable and constant segments coordinated by inter-class and intra-class variability. Conclusions from characterization of the alligator enzyme. *Eur. J. Biochem.* 216: 49-56.

Persson, B., Zigler, J.S., Jr. and Jörnvall, H., 1994, A super-family of medium-chain dehydrogenase/reductase (MDR): sub-lines including -crystallin, alcohol and polyol dehydrogenases, quinone oxidoreductases, enoyl reductases, VAT-1 and further proteins. *Eur. J. Biochem.*, in press.

Ramaswamy, S., El-Ahmad, M., Danielsson, O., Jörnvall, H. and Eklund, H., 1994a, Crystallisation and crystallographic investigations of cod alcohol dehydrogenase class I and class III enzymes. *FEBS Lett.* 350: 122-124.

Ramaswamy, S., El-Ahmad, M., Danielsson, O., Jörnvall, H. and Eklund, H., 1994b, High-resolution structure of a piscine class I/III hybrid liver alcohol dehydrogenase. *Biochemistry*, manuscript.

Thaller, C., Weaver, L.H., Eichele, G., Wilson, E., Karlsson, R. and Jansonius, J.N., 1981, Repeated seedings technique for growing large single crystals of proteins. *J. Mol. Biol.* 147:465-469.

45

HUMAN SORBITOL DEHYDROGENASE GENE
cDNA Sequence and Expression

Takeshi Iwata and Deborah Carper

Laboratory of Mechanisms of Ocular Diseases
National Eye Institute
National Institutes of Health
9000 Rockville Pike
Bethesda MD 20892

INTRODUCTION

Sorbitol dehydrogenase (EC 1.1.1.14) catalyzes the conversion of sorbitol to fructose with NAD^+ as coenzyme. It is a member of the multigene family of alcohol dehydrogenases (Jörnvall et al., 1984) which includes ζ -crystallins (Borras et al., 1989). Sorbitol dehydrogenase (SDH, the second enzyme) and aldose reductase (AR, the first enzyme) comprise the polyol pathway. This pathway is believed to be involved in osmotic regulation (Burg, 1988) and diabetic complications such as neuropathy (Gabbay et al., 1973), retinopathy (Robison et al., 1983), and cataracts (Kinoshita et al., 1974). Recently SDH involvement in the cytosolic ratio of $NADH/NAD^+$ effecting several metabolic pathways (Williamson et al., 1993) has been described. The SDH activity in normal human lens has been reported to be higher than in rat, rabbit and calf lens (Jedziniak et al., 1981). The SDH activity also has been linked to cataract formation in nondiabetics (Vaca et al., 1982; Shin et al., 1984).

The cDNA sequence for SDH has been determined for rat (Karlsson et al., 1991: Wen et al., 1993), bacteria *Bacillus subtilis* (Ng et al., 1992), yeast *Saccharomyces cerevisiae* (Sarthy et al., 1993), silkworm *Bombx mori* (Niimi et al., 1993), and human (Iwata et al., 1993, 1994a, 1994b: Lee et al., 1994). The cloning of the gene from *Bacillus subtilis* gut B (Ng et al., 1992) and human (Iwata et al., 1993, 1994a, 1994b) has also been reported recently. The human SDH gene has been localized to chromosome 15q21.1 by fluorescence in situ hybridization (Iwata et al., 1993, 1994a, 1994b). The tissue distribution of SDH expression has been performed on rat (Estonius et al., 1993). Here we report the human tissue distribution of SDH mRNA in comparison with AR mRNA and the level of SDH mRNA expression in retina and a retinoblastoma cell line (Y79).

Enzymology and Molecular Biology of Carbonyl Metabolism 5
Edited by H. Weiner *et al.*, Plenum Press, New York, 1995

MATERIALS AND METHODS

Isolation of Human SDH cDNA Clones

A partial rat cDNA clone (λ SDH2, Karlsson et al., 1991) was used as a probe to screen a human liver cDNA library (Stratagene). The probe was labeled by random priming (GIBCO BRL) with [α -^{32}P]dCTP (~ 3,000 Ci/mmol, Du Pont NEN). 3 x 10^5 phage clones were screened by the method of Young et al., 1983. Isolated clones were directly sequenced in both directions by cycle sequencing (_mol DNA sequencing system, Promega; dsDNA Cycle Sequencing System, GIBCO BRL) with [γ -^{33}P]ATP (~3,000 Ci/mmol, Du Pont NEN) end labeled primers or by the fluorescence autosequencing system (Taq DyeDeoxy Terminator Cycle Sequencing Kit, 370A DNA Sequencer, Applied Biosystems).

5'-RACE Method

The 5' RACE (Rapid amplification of cDNA ends) and primer extension method was used to obtain the 5' end sequence of the cDNA and to determine the transcriptional initiation site. Two 5'-RACE kits (5' RACE System, GIBCO BRL; 5'-AmpliFINDER_ RACE Kit, CLONTECH) were combined and modified for usage. One microgram of poly(A) RNA from human liver and 10 pmol of specific antisense primer were used for reverse trancription. The cDNA-RNA complex was digested with RNase H followed by spin column purification (GlassMAX_ DNA Isolation Spin Cartridge System, GIBCO BRL). The single strand cDNA was tailed with dCTP or ligated with the anchor oligonucleotide (AmpliFINDER anchor, CLONTECH). After purification by spin column, cDNAs were amplified by two rounds of PCR (Ampli Taq, Perkin Elmer Inc.) for 30 cycles each using the anchor primer and two other specific primers. The PCR products were purified with a spin column (Magic PCR Prep, Promega) and subcloned into the pCR_II vector (Invitrogen) for sequencing. The 5' end sequence was determined using the M13 forward and reverse primers (Invitrogen).

Northern Blot Analysis

Total RNA from human lens and retinoblastoma cell line (Y79) were isolated by the acid guanidinium isothiocyanate-phenol-chloroform extraction method (Chomcznski et al., 1987). Total RNAs from human brain, heart, kidney, liver, lung, placenta, retina, skeletal muscle, and testis were obtained from Clontech. Approximately 10 µg of RNA was separated on a 1% agarose gel run at 30 V for 12 h. The separated RNAs were blotted onto a nylon membrane (Boehringer Mannheim Co.) in 20 x SSC (Boehringer Mannheim Co.) and pre-hybridized with 5 x SSCP, 2.75 % Dextran sulfate (Oncor), 40 % deionized formamide, 4 x Denhardt's solution, 2 % SDS, sheared salmon sperm DNA (100 µg/ml, Research Genetics) in 10 mM Tris-HCl buffer (pH 8.0) for 12 h at 42 °C. One hundred nanograms of partial human cDNA insert or 18S ribosomal DNA probe (gift of Dr.B.Holmes, Cornell Univ., Ithaca, NY) was labeled by random priming with [α -^{32}P]dCTP (~ 3,000 Ci/mmol, Du Pont NEN) and purified by Sephadex G-50 column (NICK™ Column, Pharmacia Biotech). The probe was denatured at 97 °C for 5 min and added to the membrane for hybridization for 12 h at 42 °C. The probed membrane was washed in 2 x SSC, 0.1 % SDS solution for 30 min at room temperature and twice in 0.1 x SSC, 0.1 % SDS at 42 °C. The washed membrane was exposed to X-OMAT AR film (Kodak) for 3 to 12 h at -70 °C. The bands were digitally recorded into a Macintosh® Quadra950 computer (Apple Computer Inc.) by scanning (UC1200S, UMAX Data System Inc.) using PhotoShop software (Adobe

Systems Inc.) and the intensity of each band was quantified using the NIH Image software (developed by Reisband, W., NIH, Bethesda, MD).

RESULTS

The Full Length Human SDH mRNA and Sequence Comparison

Two cDNA clones were isolated from the 3×10^5 plaques screened and were sequenced on both strands. The deduced amino acid sequence was compared with protein sequence previously reported (Karlsson et al., 1989). The assembled nucleotide sequence of the two clones covered 80 % of the coding region, and a complete 3'-untranslated region. The 5'-RACE method was performed to obtain additional 5' end sequence of the cDNA. From three 5'-RACE products the nucleotide sequence of the human SDH cDNA was completed (2,471 bp). The open reading frame encodes 356 amino acid residues (Fig.1). Four differences were observed in the deduced amino acid sequence as compared to that previously reported (Karlsson et al., 1989). One additional alanine (GCG) at codon 1, glutamine to methionine difference at codon 185, serine to threonine difference at codon 280, and threonine to isoleucine difference at codon 288 were found (Fig. 1).

The SDH nucleotide sequence of human was compared with other species previously published using the Gap and PileUp (GCG) programs. The nucleotide sequence homology between human and rat, silkworm, *Bacillus subtilis*, and *Saccharomyces cerevisiae* were 79 %, 49 %, 47 %, and 43 % respectively (Fig. 2).

The deduced amino acid sequence of human SDH was also compared for homology between sheep, rat, silkworm, *Bacillus subtilis*, and *Saccharomyces cerevisiae* using the PileUp (GCG) program (Fig. 3).

SDH and AR Expression in Human Tissues

The level of SDH transcripts in 10 different human tissues were examined by northern blot analysis (Fig. 4). A 2.5 Kb transcript was observed for SDH, which is consistent with the full length sequence of the cDNA (2,471 bp). After normalization of the blot by hybridization to the 18S ribosomal DNA probe, the highest level of steady state expression for SDH was observed in the lens and kidney.

These results differ from those obtained in a study of the rat tissues, where the highest levels of SDH transcript were detected in liver and kidney (Estonius et al., 1993). The AR transcripts were also determined on the same blot. A 1.5 Kb transcript was observed which agrees with previous data (Bohren et al., 1989; Nishimura et al., 1990). The highest expression was observed in heart followed by kidney, brain, and muscle.

SDH Expression in Retinoblastoma Cell Line (Y79)

SDH mRNA expression was also compared between retina and Y79. Two additional bands were observed on northern blot in Y79 at 1.4 and 2.9 Kb which were not observed in the retina mRNA extract.

DISCUSSION

The full length cDNA for human SDH has been cloned and sequenced. The 89 bp of 5'-untranslated sequence and 1,263 bp of 3'-untranslated sequence encompases 60 % of the

Figure 1. Nucleotide sequence and deduced amino acid sequence of human sorbitol dehydrogenase and 5', 3' untranslated sequences. The nucleotide sequence extends from the transcription initiation site to poly(A). The numbering on the right refers to nucleotides. The first nucleotide of each exon are indicated by boldface.

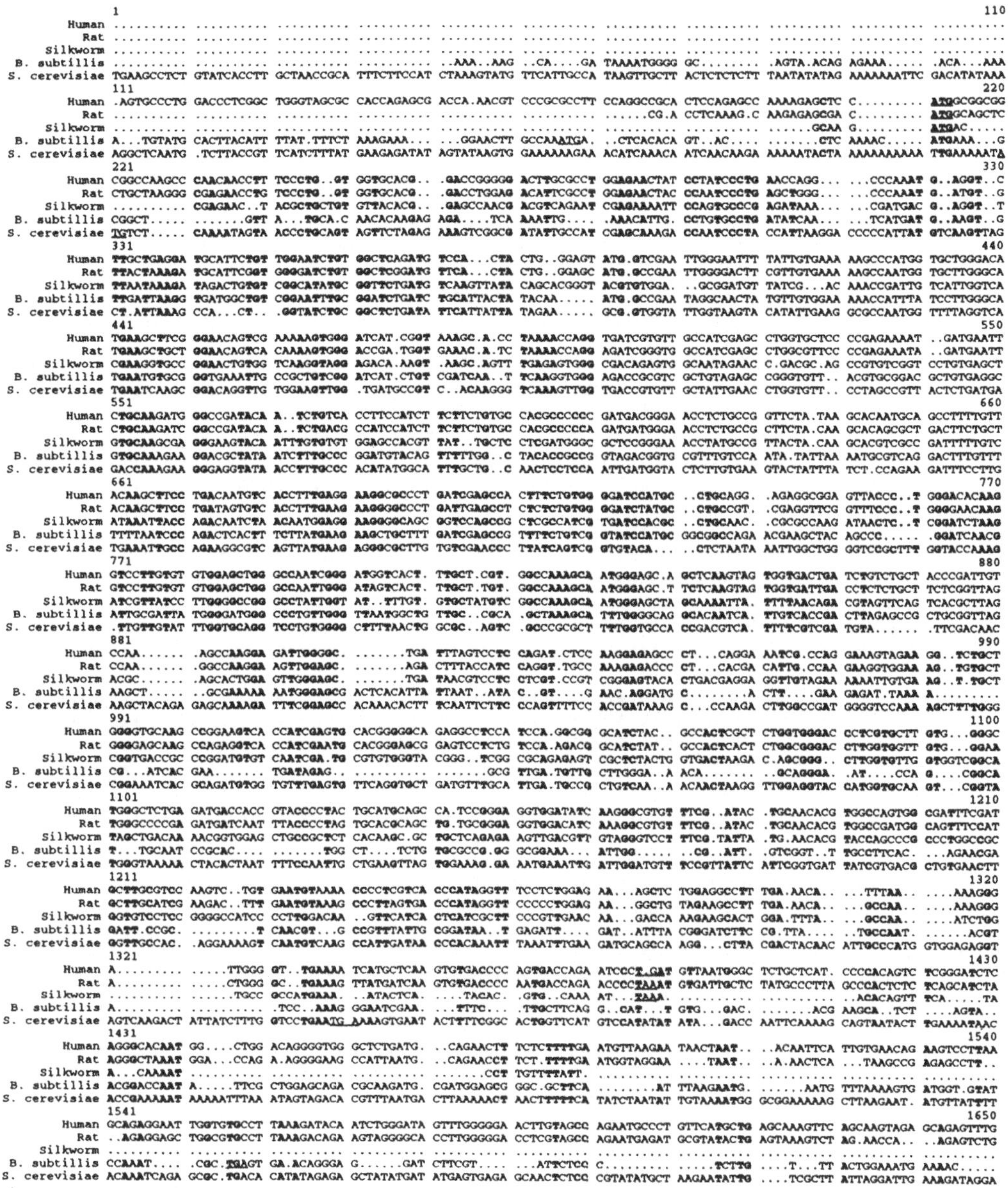

```
              1                                                                                                110
    Human     ..........  ..........  ..........  ..........  ..........  ..........  ..........  ..........  ..........  ..........  ..........
      Rat     ..........  ..........  ..........  ..........  ..........  ..........  ..........  ..........  ..........  ..........  ..........
 Silkworm     ..........  ..........  ..........  ..........  ..........  ..........  ..........  ..........  ..........  ..........  ..........
B. subtillis  ..........  ..........  ..........  ..........  .....AAA...AAG  ..CA....GA  TAAAATGGGG  GC........  .AGTA.ACAG  AGAAA.....  .ACA...AAA
S. cerevisiae TGAAGCCTCT  GTATCACCTT  GCTAACCGCA  TTTCTTCCAT  CTAAAGTATG  TTCATTGCCA  TAAGTTGCTT  ACTCTCTCTT  TAATATATAG  AAAAAAATTC  GACATATAAA
              111                                                                                               220
    Human     .AGTGCCCTG  GACCCTCGGC  TGGGTAGCGC  CACCAGAGCG  ACCA..AACGT  CCCGCGCCTT  CCAGGCCGCA  CTCCAGAGCC  AAAAGAGCTC  C.......... ATGGCGGCGG
      Rat     ..........  ..........  ..........  ..........  ..........  .....CG.A  CCTCAAAG.C  AAGAGAGCGA  C.......... ATGGCAGCTC
 Silkworm     ..........  ..........  ..........  ..........  ..........  ..........  ..........  ..........  .....GCAA  G.......... ATGAC.....
B. subtillis  A....TGTATG  CACTTACATT  TTAT.TTTCT  AAAGAAA....  ...GGAACTT  GCCAAATGA.  ..CTCACACA  GT..AC....  ....CTC  AAAAC..... ATGAAA....G
S. cerevisiae AAGGCTCAATG  .TCTTACCGT  TCATCTTTAT  GAAGAGATAT  AGTATAAGTG  GAAAAAGAA  ACATCAAACA  ATCAAACAAGA  AAAAATAGTA  AAAAAAATA
              221                                                                                               330
    Human     CGGCCAAGCC  CAACAACCTT  TCCCTG..GT  GGTGCACG..  .GACCGGGGG  ACTTGCGCCT  GGAGAACTAT  CCTATCCCTG  AACCAGG....  ....CCCAAAT  G..AGGT..C
      Rat     CTGCTAAGGG  CGAGAACCTG  TCCCTG..GT  GGTGCACG..  .GACCTGGAG  ACATTCGCCT  GGAGAACTAC  CCAATCCCTG  AGCTGGG....  ....CCCAAAT  G..ATGT..G
 Silkworm     ..........  CGAGAAC..T  ACGCTGCTGT  GTTACACG..  .GAGCCAACG  ACGTCAGAAT  CGAGAAAATT  CCAGTGCCCG  AGATAAA....  ...CGATGAC  G..AGGT..T
B. subtillis  CGGCT.....  CAAAATAGTA  ..T..TGCA..C  AACACAAGAG  AGA....TCA  AAATTG....  AAACATTG.  CCTGTGCCTG  ATATCAA....  ...TCATGAT  G..AAGT..G
S. cerevisiae TGTCT.....  CAAAATAGTA  ACCCTGCAGT  AGTTCTAGAG  AAGTCGGCG  ATATTGCCAT  CGAGCAAAGA  CCAATCCCTA  CCATTAAGGA  CCCCCATTAT  GTCAAGTTAG
              331                                                                                               440
    Human     TTGCTGAGGA  TGCATTCTGT  TGGAATCTGT  GGCTCAGATG  TCCA...CTA  CTG..GGAGT  ATG.GTCGAA  TTGGGAATTT  TATTGTGAAA  AAGCCCATGG  TGCTGGGACA
      Rat     TTACTAAAGA  TGCATTCCGT  GGGGATCTGT  GGCTCGGATG  TTCA...CTA  CTG..GGAGC  ATG.GCCGAA  TTGGGGACTT  CGTTGTGAAA  AAGCCCATGG  TGCTTGGGCA
 Silkworm     TTAATAAAGA  TAGACTGTGT  CGGCATATGC  GGTTCTGATG  TCAAGTTATA  CAGCACGGGT  ACGTGTGGA.  ..GCCGATGT  TATCG...AC  AAACCGATTG  TCATTGGTCA
B. subtillis  TTGATTAAGG  TGATGGCTGT  CGGAATTTGC  GGATCTGATC  TGCATTACTA  TACAA.....  ATG.GCCGAA  TAGGCAACTA  TGTTGTGGAA  AAACCATTTA  TCCTTGGGCA
S. cerevisiae CT.ATTAAAG  CCA...CT..  .GGTATCTGC  GGCTCTGATA  TTCATTATTA  TAGAA.....  GCG.GTGGTA  TTGGTAAGTA  CATATTGAAG  GCGCCAATGG  TTTTAGGTCA
              441                                                                                               550
    Human     TGAAGCTTCG  GGAACAGTCG  AAAAAGTGGG  ATCAT.CGGT  AAAGC.A.CC  TAAAACCAGG  TGATCGTGTT  GCCATCGAGC  CTGGTGCTCC  CCGAGAAAAT  ..GATGAAATT
      Rat     TGAAGCTGCT  GGAACAGTCA  CAAAAGTGGG  ACCGA.TGGT  GAAAC.A.TC  TAAAACCAGG  AGATCGGGTG  GCCATCGAGC  CTGGCGTTCC  CCGAGAAATA  ..GATGAAATT
 Silkworm     CGAAAGTGGT  GGAACTGTGG  TCAAGGTAGG  AGACA.AAGT  .AAGC.AGTT  TGAGAGTGGA  CGACAGAGTG  GCAATAGAAC  C.GACGC.AG  CCGTGTCCGGT  CCTGTGAGGT
B. subtillis  TGAATGTGCG  GGTGAAATTG  CCGCTGTCGG  ATCAT.CTGT  CGATCAA..T  TCAAGGTGGG  AGACCGCGTC  GCTGTAGAGC  CGGGTGTT..  ACGTGCGGAC  GCTGTGAGGC
S. cerevisiae TGAAATCAAGC  GGACAGGTTG  TGGGAGTGG  .TGATGCCGT  C..ACAAGGG  TCAAAGTTGG  TGACCGTGTT  GCTATTGAAC  CTGGTGTT..  CCTAGCCGTT  ACTCTGATGA
              551                                                                                               660
    Human     CTGCAAGATG  GGCCGATACA  A..TCTGTCA  CCTTCCATCT  TCTTCTGTGC  CACGCCCCC  GATGACGGGA  ACCTCTGCCG  GTTCTA..TAA  GCACAATGCA  GCCTTTTGTT
      Rat     CTGCAAGATC  GGCCGATACA  A..TCTGACG  CCATCCATCT  TCTTCTGTGC  CACGCCCCA  GATGATGGGA  ACCTCTGCCG  CTTCTA..CAA  GCACAGCGCT  GACTTCTGCT
 Silkworm     GTGCAAAGAA  GGAGTGTGTG  GGAGCCACGT  TAT..TGCTC  CTCGATGGGC  GCTCGGGGAA  ACCTATGCCG  TTACTA..CAA  GCACGTCGCC  GATTTTTGTC
B. subtillis  GTGCAAAGAA  GGACGCTATA  ATCTTTGCCC  GGATGTACAG  TTTTTGG..C  TACACCGCCG  GTAGACGGTG  CGTTTGTCCA  ATA.TATTAA  AATGCGTCAG  GACTTTGTTT
S. cerevisiae TGAAATGGCC  GGAGGTATA  TCTTTGCCC  ACATATGGCA  TTGCTG..C  AACTCCTCCA  ATTGATGGTA  CTCTTGTGAA  GTACTATTTA  TCT..CCAGAA  GATTTCCTTG
              661                                                                                               770
    Human     ACAAGCTTCC  TGACAATGTC  ACCTTTGAGG  AAGGCGCCCT  GATCGAGCCA  CTTTCTGTGG  GGATCCATGC  ..CTGCAGG.  .AGAGGCGGA  GTTACCC..T  GGGACACAAG
      Rat     ACAAGCTTCC  TGATAGTGTC  ACCTTTGAAG  AAGGGGGCCT  GATTGAGCCT  CTCTCTGTGG  GGATCTATGC  ..CTGCCGT.  .CGAGGTTCG  GTTTCCC..T  GGGGAACAAG
 Silkworm     ATAAATTACC  GGACAATCTA  ACAATGGAGA  AAGGGGCAAC  GGTCCAGCCG  CTCGGCCATCG  TGATCCACGC  ..CTGCAAG  ATAAGCTC..T  CGGATCTAAG
B. subtillis  TTTTAATCCC  AGACTCACTT  TCTTATGAAG  AAGCTGCTTT  GATCGAGCCG  TTTTCTGTCG  GTATCCATGC  GGCGGCCAGA  ACGAAGCTAC  AGCCC.....  .GGATCAACG
S. cerevisiae TGAAATTGCC  GAAAGGCGTC  ACTTATGAAG  AGGGCGCTTG  TGTCGAACCC  TTATCAGTCG  GTATCAATGC  ..CTCAATA  AATTGGCTGG  GGTCGGCTTT  GGTACCAAAG
              771                                                                                               880
    Human     CTCCTTGTGT  GTGGAGCTGG  GCCAATCGGG  ATGGTCACT.  TTGCT.CGT.  GGCCAAAGCA  ATGGGAGC.A  GGTCAAGTAG  TGGTGACTGA  TCTGTCTGCT  ACCCGATTGT
      Rat     GTCCTTGTGT  GTGGAGCTGG  GCCAATTGGG  ATAGTCACT.  TTGCT.TGT.  GGCCAAAGCA  ATGGGAGC.T  TCTTCAAGTAG  TGGTGATTGA  CCTCTCTGCT  TCTCGGTTAG
 Silkworm     ATCGTTATCC  TTGGGGCCGG  GCCTATTGGT  AT..TTTGT.  GTGGTATGTC  GGCCAAAGCA  ATGGGAGCTA  GCAAAATTA.  TTGTCACCGA  CTTAGAGCCG  CTGCGGTTAG
B. subtillis  ATTGCGAATTA  TCGGGGATGGG  CCCTGTTGGG  TTAATGGCTG  TTGGC.CGCA  .GCCGCGCT  TTTGGGGCAG  GCACAATCA.  TTGTCACCGA  CTTAGAAGCC  CTGCGGTTAG
S. cerevisiae .TTGTTGTAT  TTGGTGCAGG  TCCTGTGGGG  CTTTTAACTG  GCGC..AGTC  .GCCCGCGCT  TTTGGTGCAG  CCGAGCTCA.  TTTTCGTCGGA  TGTA......  .TTCGACGAC
              881                                                                                               990
    Human     CCAA......  .AGCCAAGGA  GATTGGGGC.  ........TGA  TTTAGTCCTC  CAGAT.CTCC  AAGGAGAGCC  CT...CAGGA  AATCG.CCAG  GAAAGTAGAA  GG..TCTGCT
      Rat     CCAA......  .GGCCAAAGA  GGTTGGAGC.  ........AGA  CTTTACCATC  CAGGT.TGCC  AAAGAGACCC  CT...CACGA  CATTG.CCAA  GAAGGTGGAA  AG..TGTGCT
 Silkworm     ACGC......  .AGCACTGAA  GTTGGGAGC.  ........TGA  TAACGTCCTC  CTCGT..CCGT  CGGGAGTACA  CTGACGAGGA  GGTTGTAGAA  AAAATTGTGA  AG..T.TGCT
B. subtillis  AAGCT.....  .GCGAAAAA  AGTGGGAGCG  ACTCACATTA  TTAAT..ATA  G..GT...G  AAC.AGGATG  C.......A  CTT...GAA  GAGAT.TAAA  A.......
S. cerevisiae AAGCTACAGA  GAGCAAAAGA  TTTCGGAGCC  ACAAACACTT  TCAATTCTTC  CCAGTTTTCC  ACCGATAAAG  C...CCAAGA  CTTGGCCGAT  GGGGTCCAAA  AGCTTTTGGG
              991                                                                                               1100
    Human     GGGGTGCAAG  CCGGAAGTCA  CCATCGAGTG  CACGGGGGCA  GAGGCCTCCA  TCCA.GGCGG  GCATCTAC..  GCCACTCGCT  CTGGTGGGAC  CCTCGTGCTT  GTG....GAGC
      Rat     GGGGAGCAAG  CCAGAGGTCA  CCATCGAATG  CACGGGAGCA  GAGTCCTCTG  TCCA.AGAGA  GCATCTAT..  GCCACTCACT  CTGGCGGGAC  CTTGGTGGTT  GTG....GAGC
 Silkworm     CGGTGACCGC  CCGGATGTGT  CAATCGA.TG  CGTGTGGGTA  CGGG...TCGG  CGCAGAGAGT  CGCTCTACTG  GTGACTAAGA  C.AGCGGG..  CTTGGTGTTG  GTGGTCGGCA
B. subtillis  CG...ATCAC  GAA......  TGAATAGAG.  ........G  CGTTGA.TGTTG  CTTGGGA..A  AT...CCA  G...CGGCA
S. cerevisiae CGAAAATCAC  GCAGATGTGG  TGTTTGAGTG  TTCAGGTGCT  GATGTTTGCA  TTGA.TGCCG  CTGTCAA..A  ACAACTAAGG  TTGGAGGTAC  CATGGTGCAA  GT...CGGTA
              1101                                                                                              1210
    Human     TGGGCCTCTGA  GATGACCACC  GTGGACCTAC  TGCACCTAG  CA.TCCGGGA  GGTGGATATG  AAGGGCGTGT  TTCG..ATAC  .TGCAACACG  TGGGCCGATGG  CGATTTCGAT
      Rat     TGGGCCCCGA  GATGACCAAT  TTACCCCTAG  TGCACGGCAGC  TG.TGCGGGA  GGTGGACATC  AAAGGCGTGT  TTCG..ATAC  .TGCAACACG  TGGGCGATGG  CAGTTTCCAT
 Silkworm     TAGTCGACAA  AACGGTGGAG  CTGCGGCTCT  CACAAGC.GC  TGCTCAGAGA  AGTTTGACGT  GTAGGGTGCCT  TCG..TGTA  .TG..AACAG  TACCAGCGCG  CCCTGGCCGC
B. subtillis  T...TGCAAT  CCGCAC.....  ...TGG.CT.  ..TCTG  TGCGCCG.GG  GCGGAAA..  ATTGG.....  ...CG..ATT  .GTCGGT..T  TGCCTTCAC.  ..AGAAGGA
S. cerevisiae TGGGTAAAAA  CTACACTAAT  TTTCCAATTG  CTGAAGTTAA  TGAAAG.GA  AATGAAATTG  ATTGGATGTT  TCCGTTATTC  ATTCGGTGAT  TATCGTACG  CTGTGAACTT
              1211                                                                                              1320
    Human     GCTTGCGGTCC  AAGTC..TGT  GAATGTAAAA  CCCCTCGTCA  CCATAGTTAAA  TCCTCTGGAG  AA...GGCTC  TGGAGGCCTT  TGA.AACA..  ...TTTAA..  ...AAAGGG
      Rat     GCTTGCATCG  AAGAC..TTT  GAATGTAAAG  CCCTTAGTGA  CCATAGGTT  CCCCCTGGAG  AA...GGCTG  TAGAAGCCTT  TGA.AACA..  ...GCCAA..  ...AAAGGG
 Silkworm     GGTGTCCTCC  GGGGCCATCC  CCTTGGACAA  .GTTCATCA  CTCATCGCTT  CCCTTGAAC  AA...GACCA  AAGAAGACCT  GGA.TTTA..  ...GCCAA..  ...ATCTGG
B. subtillis  GATT.CCGC.  ...T.  CAACGT...G  CCGTTTATTG  CGGATAA..T  GAGATT..  ...GAT..ATTTA  CGGGATCTTC  CG.TTA....  ...TGCCAAT.  ......ACGT
S. cerevisiae .AGGAGAAGA  CTACACTAAT  TTTCCAATTG  CTGAAGTTAA  TGAAAG.GA  AATGAAATTG  ATTGGATGTT  TCCGTTATTC  ATTCGGTGAT  TATCGTACG  CTGTGAACTT
              1321                                                                                              1430
    Human     A.........  .....TTGGG  GT..TGAAAA  TCATGTCTCAA  GTGTGACCCC  AGTGACCAGA  ATCCCT.GAT  GTTAATGGCTC  TCTGCTCAT.  CCCCACAGTC  TCGGGATCTC
      Rat     A.........  ....CTTGGG  GC..TGAAAG  TTATGATCAA  GTGTGACCCC  AATGACCAGA  ACCCCTAAAT  GTGATTGGCTC  TATGCCCTTA  CCCCACTCTC  TCAGCATCTA
 Silkworm     .........  .....TGCC  CATGAAA  GGAATCGAA.  ..TTC...  .GTG..CAAA  AT...TAAA.  ..........  ..........  ..ACACAGTT  TCA....TA
B. subtillis  A.........  .....TCC..  AAAG  GGAATCGAA.  .....TTTC.  ..TTGCTTCAG  G..CAT...T  GTG...GAC.  ........ACG  AAGCA..TCT  ...AGTA..
S. cerevisiae AGTCAAGACT  ATTATCTTTG  ACTCCTGAATG  GACTTCGGCT  ACTTTTGGGC  TGTCCATATT  ATA...AGCC  AATTCAAAAG  CAGTAATACT  TGAAAAATAAC
              1431                                                                                              1540
    Human     AGGGACAAT  GG....CTGG  ACAGGGGTGG  GCTCTGATG.  ....CAGAACTT  TCTCTTTTGA  ATGTTAAGAA  TAACTAAT..  ...ACAATTCA  TTGTGAACAG  AAGTCCTTAA
      Rat     AGTGACTAAAT  GGA...CCAG  A.AGGGGAAG  CCATTAAATG.  ....CAGAACCT  TCT.CTTTGA  ATGGTAGGAA  ....TAAT..  .A..AACTCA  ..TAAGCCG  AGAGCCTT..
 Silkworm     A..CAAAAT  ..........  ..........  ..........  ....CCT  TGTTTTATT.
B. subtillis  AACGACCAAT  A....TTCG  CCGAGTAGCAGA  CGCAAGATG.  CGATGGAGGG  GGC.GCTTCA  ........AT  ATTAAGAAG.  ....AATG  TTTAAAAGTG  ATGGT.GTAT
S. cerevisiae ACCGAAAAAT  AAAAATTTAA  ATAGTAGACA  CGTTTAATGA  CTTAAAAACT  AACTTTTTCA  TATCTAAATA  TGTAAAATGG  GCGGAAAAAG  CTTAAGAAT  ATGTTATTTT
              1541                                                                                              1650
    Human     GCAGAGGAAT  TGGTGTGCCT  TAAAGATACA  ATCTGGGATA  GTTTGGGGGA  ACTTGTAGCC  AGAATGCCCT  GTTCATGCTG  AGCAAAGTTC  AGCAAGTAGA  GCAAGAGTTTG
      Rat     ..AGAGGAGC  TGGCGTGCCT  TAAAGACAGA  AGTAGGGGCA  CCTTGGGGGA  CCTCGTAGCC  AGAATGAGAT  GCGTATACTG  AGTAAAGTCT  AG.AACCA..  ...AGAGTCTG
 Silkworm     ..........  ..........  ..........  ..........  ..........  ..........  ..........  ..........  ..........  ..........  ..........
B. subtillis  CCAAAT.....  .CGC.TGACT  GA.ACAGGGA  G.......GAT  CTTCGT....  ..ATTCTCC  C.......  ........  ...T...TT  ACTGGAAATG  AAAAC.....
S. cerevisiae ACAAATCAGA  GCGC.TGACA  CATATAGAGA  GCTATATGAT  ATGAGTGAGA  GCAACTCTCC  CGTATATGCT  AAGAAATATTG  ...TCGCTT  ATTAGGATTG  AAAGATAGGA
```

Figure 2. Nucleotide sequence comparison of human SDH with rat, silkworm (*Bombyx mori*), *B.subtilis*, and *S.cerevisiae*. Nucleotide conserved in more than 4 species are highlighted in boldface. The translation start site (ATG) and stop site (TGA or TAA) are underlined.

total cDNA sequence. The long stretch of 3'-untranslated sequence might be involved with mRNA stability or unknown translational regulation. Unlike the rat, a pre-SDH mRNA generated before translational modification (Wen et al, 1993) was not observed in human SDH. One additional and three changed amino acids do not strongly effect the previous

```
                              1                                                    50
                   Human     .......... .......... .......... .......... ...AAAAKPN
                   Sheep     .......... .......... .......... .......... ....AAAKPE
                     Rat     MVFSSRVFFF SRVPLLQTLG GLTSRNTSSP PDPADTSKQE SDMAAPAKGE
                Silkworm     .......... .......... .......... .......... .......MTE
   Saccharomyces cerevisiae  .......... .......... .......... .......... ......MSQNS
        Bacillus subtilis    .......... .......... .......... .......... ..MAAAAKPN
                              51                                                   100
                   Human     NLSLVVHGPG DLRLENYPIP EP.GPNEVLL RMHSVGICGS DVHYWEYGRI
                   Sheep     NLSLVVHGPG DLRLENYPIP EP.GPNEVLL KMHSVGICGS DVHYWQ.GRI
                     Rat     NLSLVVHGPG DIRLENYPIP EL.GPNDVLL KMHSVGICGS DVHYWEHGRI
                Silkworm     NYAAVLHGAN DVRIEKIPVP EI.NDDEVLI KIDCVGICGS DVKLYSTGTC
   Saccharomyces cerevisiae  NPAVVLEKVG DIAIEQRPIP TIKDPHYVKL AIKATGICGS DIHYYRSGGI
        Bacillus subtilis    NLSLVVHGPG DLRLENYPIP EP.GPNEVLL RMHSVGICGS DVHYWEYGRI
                              101                                                  150
                   Human     GNFIVKKPMV LGHEASGTVE KVGSSVKHLK PGDRVAIEPG APRENDEFCK
                   Sheep     GBFVVKKPMV LGHEASGTVV KVGSLVRHLQ PGDRVAIQPG APRQTDEFCK
                     Rat     GDFVVKKPMV LGHEAAGTVT KVGPMVKHLK PGDRVAIEPG VPREIDEFCK
                Silkworm     GADVIDKPIV IGHEGAGTVV KVGDKVSSLR VGDRVAIEPT QPCRSCELCK
   Saccharomyces cerevisiae  GKYILKAPMV LGHESSGQVV EVGDAVTRVK VGDRVAIEPG VPSRYSDETK
        Bacillus subtilis    GNFIVKKPMV LGHEASGTVE KVGSSVKHLK PGDRVAIEPG APRENDEFCK
                              151                                                  200
                   Human     MGRYNLSPSI FFCATPPDD. ......GNLC RFYKHNAAFC YKLPDNVTFE
                   Sheep     IGRYNLSPTI FFCATPPDD. ......GNLC RFYKHNANFC YKLPDNVTFE
                     Rat     IGRYNLTPSI FFCATPPDD. ......GNLC RFYKHSADFC YKLPDSVTFE
                Silkworm     RGKYNLCVEP RYCSSMGAP. ......GNLC RYYKHVADFC HKLPDNLTME
   Saccharomyces cerevisiae  EGRYNLCPHM AFAATPPID. ......GTLV KYYLSPEDFL VKLPEGVSYE
        Bacillus subtilis    MGRYNLSPSI FFCATPPDD. ......GNLC RFYKHNAAFC YKLPDNVTFE
                              201                                                  250
                   Human     EGALIEPLSV GIHACRRGGV TLGHKVLVCG AGPIGMVTLL VAKAMGAAQV
                   Sheep     EGALIEPLSV GIHACRRAGV TLGNKVLVCG AGPIGLVNLL AAKAMGAAQV
                     Rat     EGALIEPLSV GIYACRRGSV SLGNKVLVCG AGPIGIVTLL VAKAMGASQV
                Silkworm     EGAAVQPLAI VIHACNRAKI TLGSKIVILG AGPIGILCAM SAKAMGASKI
   Saccharomyces cerevisiae  EGACVEPLSV GVHSNKLAGV RFGTKVVVFG AGPVGLLTGA VARAFGATDV
        Bacillus subtilis    EGALIEPLSV GIHACRRGGV TLGHKVLVCG AGPIGMVTLL VAKAMGAAQV
                              251                                                  300
                   Human     VVTDLSATRL SKAKEIGAD. ..LVLQISKE .SPQEIARKV EGLLGCK.PE
                   Sheep     VVTDLSASRL SKAKEVGAD. ..FILEISNE .SPEEIAKKV EGLLGSK.PE
                     Rat     VVIDLSASRL AKAKEVGAD. ..FTIQVAKE .TPHDIAKKV ESVLGSK.PE
                Silkworm     ILTDVVQSRL DAALELGAD. ..NVLLVRRE YTDEEVVEKI VKLLGDR.PD
   Saccharomyces cerevisiae  IFVDVFDNKL QRAKDFG.AT NTFNSSQFST DKAQDLADGV QKLLGGNHAD
        Bacillus subtilis    VVTDLSATRL SKAKEIGAD. ..LVLQISKE .SPQEIARKV EGLLGCK.PE
                              301                                                  350
                   Human     VTIECTGAEA SIQAGIYATR SGGTLVLVGL GSEMTTVPLL HAAIREVDIK
                   Sheep     VTIECTGVET SIQAGIYATH SGGTLVLVGL GSEMTSVPLV HAATREVDIK
                     Rat     VTIECTGAES SVQTGIYATH SGGTLVVVGM GPEMINLPLV HAAVREVDIK
                Silkworm     VSIDACGYGS AQRVALLVTK TAGLVLVVGI ADKTVELPLS QALLREVDVV
   Saccharomyces cerevisiae  VVFECSGADV CIDAAVKTTK VGGTMVQVGM GKNYTNFPIA EVSGKEMKLI
        Bacillus subtilis    VTIECTGAEA SIQAGIYATR SGGTLVLVGL GSEMTTVPLL HAAIREVDIK
                              351                                                  400
                   Human     GVFRYC.NTW PVAISMLASK ........SV NVKPLVTHRF PLEKALEAFE
                   Sheep     GVFRYC.NTW PMAISMLASK ........SV NVKPLVTHRF PLEKALEAFE
                     Rat     GVFRYC.NTW PMAVSMLASK ........TL NVKPLVTHRF PLEKAVEAFE
                Silkworm     GSFRIM.NTY QPALAAVSSG ........AI PLDKFITHRF PLNKTKEALD
   Saccharomyces cerevisiae  GCFRYSFGDY RDAVNLVAT. .......GKV NVKPLITHKF KFEDAAKAYD
        Bacillus subtilis    GVFRYC.NTW PVAISMLASK ........SV NVKPLVTHRF PLEKALEAFE
                              401            423
                   Human     ..TFKKGLGL KIMLKCDPSD QNP
                   Sheep     ..TSKKGLGL KVMIKCDPSD QNP
                     Rat     ..TAKKGLGL KVMIKCDPND QNP
                Silkworm     ..LAKSGAAM KILIHVQN.. ...
   Saccharomyces cerevisiae  YNIAHGGEVV KTIIFGPE.. ...
        Bacillus subtilis    ..TFKKGLGL KIMLKCDPSD QNP
```

Figure 3. Amino acid sequence comparison of human SDH with sheep, rat, silkworm (*Bombyx mori*), *B.subtilis*, and *S.cerevisiae*. Amino acids conserved in more than 5 species are highlighted in boldface.

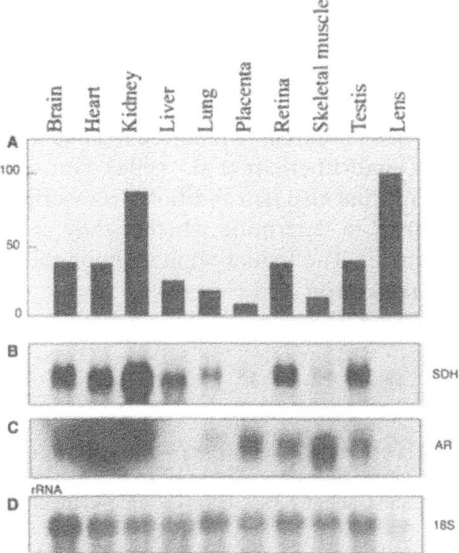

Figure 4. Expression of the SDH gene in various human tissues. (A) The relative amount of SDH expression in various tissues is indicated by the ratio of SDH mRNA/18S ribosomal RNA setting lens at 100 %. (B) Northern blot using human SDH cDNA probe. (C) Northern blot using probe human AR cDNA probe. (D) Northern blot standarization using the human 18S ribosomal DNA probe.

homology analysis performed with other species (Karlsson et al., 1991) nor have any likely impact on the proposed function of the enzyme.

The northern blot analysis clearly demonstrates high expression of SDH mRNA in lens and kidney. By comparing the expression of SDH in different human tissues to rat (Estonius et al., 1993) a significant difference was observed in liver and lens. A previous study demonstrated higher SDH enzyme activity in human lens compared to calf, rabbit, rat, and mouse (Jedziniak et al., 1981). The high expression of SDH in the human lens is of great interest in view of previous reports showing abnormal SDH activity in red blood cells in cataract families (Vaca et al., 1982; Shin et al., 1984). These data suggest that SDH may play an important role in human lens and dysfunction of this enzyme may lead to alterations in the polyol pathway.

Although northern analysis of AR on human tissues has been previously shown (Bohren et al., 1989; Nishimura et al., 1990) analysis on 10 different tissues were performed

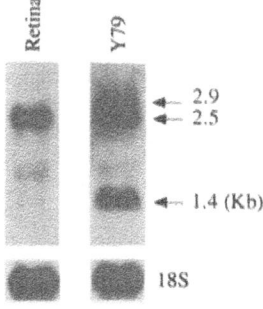

Figure 5. Expression of the SDH gene in retina and retinoblastoma cell line (Y79). Northern blot standarization is performed by human 18S ribosomal DNA probe.

for the first time. Surprisingly, the highest expression was observed in heart which has never been reported before and expression in testis was low among the tissues we have tested. Both SDH and AR gave coordinate high expression in brain, heart, and kidney but differential in placenta and skeletal muscle.

Aldose reductase has been previously reported to have high expression in tumor cell lines (Limjoco et al., 1991; Zeindl-Eberhart et al., 1994). Our result demonstrated not only high SDH expression in the Y79 but also two additional transcripts. The sequence of the two additional bands are of interest to determine whether they are from the SDH gene or a homologous gene family member. The induction in a tumor cell line suggests SDH may be involved in cell growth and regulation.

REFFERENCES

Borras T, Persson B, and Jörnvall H (1989). Eye lens ζ -crystallin relationships to the family of "long-chain" alcohol/polyol dehydrogenases. *Biochem.* **28**: 6133-6139

Bohren K, Bullock B, Wermuth B, and Gabbay KH (1989). The aldo-keto reductase superfamily. *J.Biol.Chem.* **264**: 9547-9551

Burg MB (1988). Role of aldose reductase and sorbitol in maintaining the medullary intracellular milieu. *Kidney Int.* **33**: 635-641

Chomczynski P, and Sacchi N (1987). Single-step method of RNA isolation by acid guanidinium thiocyanate-phenol-chloroform extraction. *Anal.Biochem.* **162**: 156-159

Estonius M, Danielsson O, Karlsson C, Persson H, Jörnvall H, and Höög J-O (1993). Distribution of alcohol and sorbitol dehydrogenase. *Eur.J.Biol.* **215**: 497-503

Gabbay KH (1973). The sorbitol pathway and the complications of diabetes. *N.Eng.J.Med.* **288**: 831-836

Iwata T, Höög J-O, Reddy VN, and Carper D (1993). Cloning of the human sorbitol dehydrogenase gene. *Invest.Ophthal.Vis.Sci.Supple.* **34**: 712

Iwata T, Vaca G, Rodriguez IR, and Carper D (1994a). Human sorbitol dehydrogenase gene: Candidate gene for congenital cataract. *Invest.Ophthal.Vis.Sci.Supple.* **35**: 1493

Iwata T, Popescu NC, Zimonjic DB, Vaca G, Karlsson C, Höög J-O, Rodriguez IR and Carper D (1994b). Submitted for publication.

Jedziniak JA, Chylack LT Jr, Cheng H MM, Grillis K, Kalustian AA, and Tung WH (1981). The sorbitol pathway in the human lens. *Invest.Ophthalmol.Vis.Sci.* **20**: 314-326

Jörnvall, H., von Bahr-Lindström, H., and Jeffery, J. (1984). Extensive variations and basic features in the alcohol dehydrogenase-sorbitol dehydrogenase family. *Eur.J.Biochem.* **140**: 17-23

Karlsson C, Maret W, Auld DS, Höög J-O, and Jörnvall H (1989). Variability within mammalian sorbitol dehydrogenases. *Eur.J.Biochem.* **186**: 543-550

Karlsson C, Jörnvall H, and Höög J-O (1991). *Eur.J.Biochem.* Sorbitol dehydrogenase: cDNA coding for the rat enzyme. **198**: 761-765

Kinoshita, JH (1974). Mechanisms initiating cataract formation. Proctor Lecture. *Invest.Ophthalmol.* **13**: 713-724

Lee FK, Cheung MC, and Chung S (1994). The human sorbitol dehydrogenase gene: cDNA cloning, sequence determination, and mapping by fluorescence in situ hybridization. *Genomics* **21**: 354-358

Limjoco TI, Carper D, Bondy C, and Chepelinsky AB (1991). Accumulation and spatial location of aldose reductase mRNA in a lens tumor of an alpha A-crystallin/SV40 T antigen transgenic mouse line. *Exp.Eye Res.* **52**: 759-762

Niimi T, Yamashita O, and Yaginuma T (1993). A cold-inducible *Bombyx* gene encoding a protein similar to mammalian sorbitol dehydrogenase. *Eur.J.Biochem.* **213**: 1125-1131

Nishimura C, Matsuura Y, Kokai Y, Akera T, Carper D, Morjana N, Lyons C, and Flynn TG (1990). Cloning and expression of human aldose reductase. *J.Biol.Chem.* **265**: 9788-9792

Ng K, Ye R, Wu X, and Wong S (1992). Sorbitol dehydrogenase from *Bacillus subtillis*. *J.Biol.Chem.* **267**: 24989-24994

Robison WG Jr, Kador PF, and Kinoshita JH (1983). Retinal capillaries: basement membrane thickening by galactosemia prevented with aldose reductase inhibitor. *Science* **221**: 1177-1179

Sarthy AV, Schopp C, and Idler KB (1993). Cloning and sequence determination of the encoding sorbitol dehydrogenase from *saccharomyces-cerevisiae*. *Gene* **140**: 121-126

Shin YS, Rieth M, and Endres W (1984). Sorbitol dehydrogenase deficiency in a family with congenital cataracts. *J.Inher.Metab.Dis.* **7**: 151-152

Vaca G, Ibarra B, Bracamontes M, Garcia-Cruz D, Sanchez-Corona J, Medina C, Wunsch C, Gonzales-Quiroga G, and Cantu JM (1982). Red blood cell sorbitol dehydrogenase deficiency in a family with cataract. *Hum.Genet.* **61**: 338-341

Wen Y, and Bekhor I (1993). Full-length cDNA sequencing reveals a mRNA coding for a protein containing an additional 42 amino acids at the N-terminal end. *Eur.J.Biochem.* **217**: 83-87

Williamson JR, Chang K, Frangos M, Hasan KS, Ido Y, Kawamura T, Nyengaard JR, van den Enden M, Kilo C, and Tilton RG (1993). Hyperglycemia pseudohypoxia and diabetic complications. *Diabetes* **42**: 801-813

Young RA, and Davis RW. (1983). Efficient isolation of genes by using antibody probes. *Proc.Natl.Acad.Sci.U.S.A.* **80**: 1194-1198

Zeindl-Eberhart E, Jungblut PR, Otto A, and Rabes HM (1994). Identification of tumor-associated proteins during rat hepatocarcinogenesis. Aldose reductase. *J.Biol.Chem.* **20: 14589-14594**

46

SHORT-CHAIN
DEHYDROGENASES/REDUCTASES

Bengt Persson[1,2], Maria Krook[1] and Hans Jörnvall[1]

[1]Department of Medical Biochemistry and Biophysics
Karolinska Institutet
S-171 77 Stockholm, Sweden, and
[2]European Molecular Biology Laboratory
Meyerhofstrasse 1,Postfach 10.2209
D-69012 Heidelberg, Germany

INTRODUCTION

The family of short-chain dehydrogenases/reductases (SDR) with subunits of typically 250-odd amino acid residues now encompasses 57 different characterised proteins, representing a wide variety of enzyme activities. The first characterised member of this family was the fruit-fly alcohol dehydrogenase (Schwartz and Jörnvall, 1976; Thatcher, 1980). This alcohol dehydrogenase is different from the classical liver alcohol dehydrogenase, which has larger subunits of about 370 residues, is zinc-dependent, and belongs to a family of medium-chain dehydrogenases/reductases (MDR) (Jörnvall et al., 1981; Persson et al., 1994). In the early 80's, two additional structures were shown to be related to the *Drosophila* alcohol dehydrogenase (Jörnvall et al., 1981, 1984), thus establishing a new enzyme family. These structures were glucose dehydrogenase (Jany et al., 1984) and ribitol dehydrogenase (Morris et al., 1974; Dothie et al., 1985). Already at this stage, it was seen that the molecular architecture was different between this short-chain dehydrogenase family and the medium-chain dehydrogenases. The coenzyme-binding region is located at the beginning of the C-terminal domain in the medium-chain dehydrogenases, with a classical Rossmann fold (Rossmann et al., 1975) and a conserved Gly-X-Gly-X-X-Gly pattern. In contrast, the short-chain dehydrogenases have a similar pattern of Gly-X-Gly-X-X-X-Gly at the N-terminal region. In addition, secondary structure predictions suggested this region to have a β-turn-α-turn-β motif, compatible with a Rossmann fold (Thatcher & Sawyer, 1980). Consequently, it was early clear that the two different families of dehydrogenases have different molecular architectures. The medium-chain dehydrogenases have an N-terminal, catalytic domain and a C-terminal, coenzyme-binding domain, while in the short-chain dehydrogenases the coenzyme-binding is located N-terminally and the catalytic site toward the C-terminal half of the molecule (Jörnvall et al., 1981).

The first characterised mammalian enzymes of the SDR family were the NAD+-linked 15-hydroxyprostaglandin dehydrogenase (Krook et al., 1990) and the 17β-hydroxys-

Enzymology and Molecular Biology of Carbonyl Metabolism 5
Edited by H. Weiner *et al.*, Plenum Press, New York, 1995

teroid dehydrogenase (Peltoketo et al., 1988), the latter, though, first not recognised as a short-chain dehydrogenase member. Those mammalian structures showed that the enzymes of the SDR family also are of interest in the metabolism of higher organisms, and not just present in insects and bacteria. Since then, the family has been shown to have members of nearly all kingdoms. At the beginning of the 90's, the number of known primary structures of the family had increased substantially (Fig. 1), giving possibilities for multiple alignment studies revealing conserved regions of the enzymes and for secondary structure predictions (Persson et al., 1991). Thus, the importance of Tyr-152 and Lys-156 (residue numbering according to 3α/20β-hydroxysteroid dehydrogenase), although recognised already at the first comparisons (Jörnvall et al., 1981), was even further emphasised in being two of only six strictly conserved residues within the 20 structures aligned at that time of the short-chain dehydrogenases then characterised.

During the last few years, the number of available amino acid sequences of SDR family members has tripled. The enzyme family presently encompasses 57 members with less than 70% residue identity in pairwise comparisons. These structures have now been compared and a multiple alignment is constructed, making it possible to draw general conclusions regarding the structural and functional characteristics of the family.

In addition, the three-dimensional structures of two SDR members have become available, those of 3α/20β-hydroxysteroid dehydrogenase (Ghosh et al., 1991) and of dihydropteridine reductase (Varughese et al., 1992). This opens possibilities to investigate the location of the conserved amino acid residues in the light of established conformations.

THE SHORT-CHAIN DEHYDROGENASE/REDUCTASE (SDR) FAMILY

After elimination of sequences with more than 70 % residue identity in pairwise comparisons (Table 1), the family of short-chain dehydrogenases/reductases (SDR) at present encompasses 57 different proteins. This elimination was undertaken in order not to bias the conclusions in favour of structures for which several isoforms or species variants are known. The members have been detected utilising protein comparisons, screenings in protein sequence databases, and literature searches. Successive alignments of the SDR

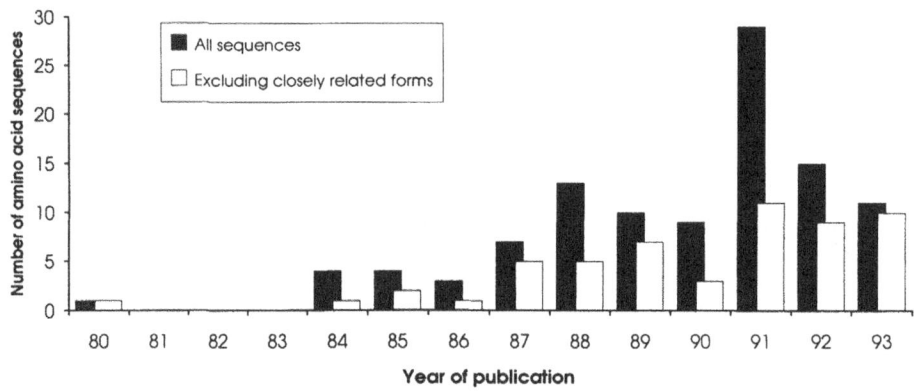

Figure 1. Completely determined amino acid sequences of short-chain dehydrogenases/reductases (SDR) published annually during the period 1980-1993. Dark bars give the total number of structures, while light bars indicate structures having less than 70 % residue identity with other SDR members.

Table 1. Proteins of the SDR family (short-chain dehydrogenases/reductases). Columns show identification code in the Swissprot database (except the codes 3adh_Psepu, Ardh_Cantr, Ba74_Eubsp, Dhes2_Hum, Dhes3_Hum, Hmtx_Leima, Ke6_Mouse, P29x_Dastr, P6_Drome, Scdh_Picab, Spre_Human,Ts2_Maize, Tr1_Dastr, Tr2_Dastr and Ver1_Asppa, not yet characterised. The multi-function proteins Fox2_Yeast and Hde_Cantr have each two domains that belong to the SDR family. Abbreviations: ACP, available in the Swissprot database), enzyme names and the source of enzyme/DNA 1992), release 28, March 1994, except the structures given in (Aziz et al., 1993; Baron et al., 1991; Bauer et al., 1993; Callahan and Beverley, 1992; De Long et al., 1993; Geissler et al., 1994; Ichinose et al., 1991; Nakajima et al., 1993; Rat et al., 1991; Skory et al., 1992; Suzuki et al., 1993; Wong, 1993; Wu et al., 1993).

Code	Enzyme	Species
25kd_Sarpe	Development-specific 25 kd protein	*Sarcophaga peregrina*
2bhd_Strex	3α/20β-Hydroxysteroid dehydrogenase	*Streptomyces exfoliatus*
3adh_Psesp	3-α-Hydroxysteroid dehydrogenase	*Pseudomonas* sp. B-0831
3bhd_Comte	3-β-Hydroxysteroid dehydrogenase	*Comamonas testosteroni*
Act3_Strco	Putative ketoacyl reductase	*Streptomyces coelicolor*
Adh_Drome	Alcohol dehydrogenase	*Drosophila melanogaster*
Ap27_Mouse	Adipocyte p27 protein	Mouse
Ardh_Cantr	Arabinitol dehydrogenase	*Candida tropicalis*
Ba72_Eubsp	7-α-Hydroxysteroid dehydrogenase	*Eubacterium* sp.
Ba74_Eubsp	7-α-Hydroxysteroid dehydrogenase	*Eubacterium* sp.
Bdh_Rat	D-β-Hydroxybutyrate dehydrogenase	Rat
Bend_Acica	*cis*-1,2-Dihydroxy-3,4-cyclohexadiene-1-carboxylate DH	*Acinetobacter calcoaceticus*
Bnze_Psepu	*cis*-1,2-Dihydrobenzene-1,2-diol DH	*Pseudomonas putida*
Bphb_Pseps	Biphenyl-*cis*-diol dehydrogenase	*Pseudomonas pseudoalcaligenes*
Budc_Klete	Acetoin dehydrogenase	*Klebsiella terrigena*
Dhca_Human	Carbonyl reductase (NADPH)	Human
Dhes_Human	Estradiol 17 β-dehydrogenase	Human
Dhes2_Hum	Estradiol 17 β-dehydrogenase type 2	Human
Dhes3_Hum	Estradiol 17 β-dehydrogenase type 3	Human
Dhgb_Bacme	Glucose 1-dehydrogenase	*Bacillus megaterium*
Dhii_Rat	Corticosteroid 11-β-DH	Rat
Dhk2_Strvn	Granaticin polyketide synthase putative ketoacyl red 2	*Streptomyces violaceoruber*
Dhma_Flas1	*N*-Acylmannosamine 1-dehydrogenase	*Flavobacterium* sp. 991
Dhpr_Rat	Dihydropteridine reductase	Rat
Enta_Ecoli	2,3-Dihydro-2,3-dihydroxybenzoate DH	*Escherichia coli*
Fabg_Cupla	3-Oxoacyl-ACP reductase	*Cuphea lanceolata*
Fabg_Ecoli	3-Oxoacyl-ACP reductase	*Escherichia coli*
Fixr_Braja	FixR protein	*Bradyrhizobium japonicum*
Fox2_Yeast	Hydratase-dehydrogenase-epimerase	*Saccharomyces cerevisiae*
Gutd_Ecoli	Sorbitol-6-phosphate 2-dehydrogenase	*Escherichia coli*
Hde_Cantr	Hydratase-dehydrogenase-epimerase	*Candida tropicalis*
Hdha_Ecoli	7-α-Hydroxysteroid dehydrogenase	*Escherichia coli*
Hmtx_Leima	Aldoketo reductase	*Leeishmania major*
Ke6_Mouse	Ke 6 protein	Mouse
Ligd_Psepa	Cα-dehydrogenase, gene ligD	*Pseudomonas paucimobilis*
Mas1_Agrra	Agropine synthesis reductase	*Agrobacterium rhizogenes*
Nodg_Azobr	Nodulation protein G	*Azospirillum brasilense*
Nodg_Rhime	Nodulation protein G	*Rhizobium meliloti*
P29x_Dastr	Troponine reductase homologue	*Datura stramonium*
P6_Drome	Fat body protein P6	*Drosophila melanogaster*
Pcr_Horvu	Protochlorophyllide reductase	*Hordeum vulgare*
Pgdh_Human	15-Hydroxyprostaglandin DH (NAD⁺)	Human
Phbb_Alceu	Acetoacetyl-CoA reductase	*Alcaligenes eutrophus*
Phbb_Zoora	Acetoacetyl-CoA reductase	*Zoogloea ramigera*
Ridh_Kleae	Ribitol 2-dehydrogenase	*Klebsiella aerogenes*
Scdh_Picab	Short-chain dehydrogenase	*Picea abies*
Spre_Human	Sepiapterin reductase	Human
Tr1_Dastr	Troponine reductase I	*Datura stramonium*
Tr2_Dastr	Troponine reductase II	*Datura stramonium*
Ts2_Maize	Sex determination gene product TS2	Maize
Ver1_Asppa	ver-1 gene product	*Aspergillus parasiticus*
Xyll_Psepu	*cis*-1,2-Dihydroxy-3,4-cyclohexadiene-1-carboxylate DH	*Pseudomonas putida*
Yinl_Lismo	Hypoth 26.8 kD prot in INLA 5'reg (orfA)	*Listeria monocytogenes*
Yrtp_Bacsu	Hypoth 25.3 kD prot in RTP 5'reg (orf238)	*Bacillus subtilis*
Yura_Myxxa	Hypoth prot in uraa 5'region	*Myxococcus xanthus*

Table 2. Residue identities in pairwise comparisons of the SDR members. Numbers give values in per cent, excluding gaps

	1 2bhd_Strex	2 Dhpr_Rat	3 Pgdh_Human	4 Dhca_Human	5 Pcr_Horvu	6 Hmtx_Lm	7 3bhd_Comte	8 Ts2_Maize	9 Mas1_Agrra	10 Dhii_Rat	11 Bdh_Rat	12 Dhes2_Hum	13 25kd_Sarpe	14 P6_Dme	15 Adh_Drome	16 Bnze_Psepu	17 Bphb_Pseps	18 Budc_Klter	19 Gutd_Ecoli	20 Bend_Acica	21 Xyll_Psepu	22 Fox2a_Yea	23 Hdea_Cantr	24 Fox2b_Yea	25 Hdeb_Cantr	26 Yinl_Lismo	27 Yrtp_Bacsu	28 Dhes3_Hum
2 Dhpr_Rat	15																											
3 Pgdh_Human	25	15																										
4 Dhca_Human	22	13	20																									
5 Pcr_Horvu	17	10	13	19																								
6 Hmtx_Lm	25	13	24	16	13																							
7 3bhd_Comte	36	15	30	21	16	25																						
8 Ts2_Maize	33	16	22	21	17	22	35																					
9 Mas1_Agrra	25	15	24	19	14	21	21	19																				
10 Dhii_Rat	24	11	18	16	15	14	19	19	18																			
11 Bdh_Rat	24	10	18	16	14	14	15	13	17	20																		
12 Dhes2_Hum	20	13	19	23	13	14	22	14	17	17	34																	
13 25kd_Sarpe	21	14	28	14	14	14	17	16	17	16	19	17																
14 P6_Dme	18	13	26	15	12	13	16	13	17	15	16	16	57															
15 Adh_Drome	18	15	23	14	15	13	20	17	15	17	16	17	33	29														
16 Bnze_Psepu	30	18	23	23	15	26	29	26	23	18	18	20	17	14	15													
17 Bphb_Pseps	30	20	23	19	15	25	27	26	18	15	19	18	18	18	18	58												
18 Budc_Klter	31	17	26	22	20	24	30	32	22	22	20	23	19	18	19	27	23											
19 Gutd_Ecoli	27	17	23	19	17	25	25	21	19	15	16	16	18	14	17	25	22	28										
20 Bend_Acica	31	16	24	18	13	26	26	25	16	19	18	15	19	17	15	25	22	25	23									
21 Xyll_Psepu	28	15	24	18	14	26	27	25	20	17	17	16	18	18	15	24	22	20	25	58								
22 Fox2a_Yea	27	16	20	15	16	16	20	18	15	19	14	15	18	15	17	24	22	24	24	26	25							
23 Hdea_Cantr	25	16	19	16	17	15	19	19	18	20	15	21	21	19	20	21	22	27	19	23	21	59						
24 Fox2b_Yea	24	14	23	17	16	18	22	19	19	18	17	16	17	14	17	20	20	23	20	23	23	43	43					
25 Hdeb_Cantr	28	14	23	18	17	16	24	21	19	17	18	16	17	15	16	21	23	25	20	23	20	43	40	56				
26 Yinl_Lismo	27	13	23	18	19	15	21	21	22	21	18	19	19	15	23	23	22	26	15	24	22	28	29	24	25			
27 Yrtp_Bacsu	30	15	26	20	19	19	26	27	25	24	21	20	23	21	23	26	25	27	25	29	27	30	28	25	30	31		
28 Dhes3_Hum	20	12	19	17	16	14	20	19	16	20	17	18	15	13	20	17	17	22	15	18	19	18	17	17	19	23	27	
29 Ligd_Psepa	28	12	20	16	15	16	22	24	18	16	17	17	19	17	16	18	22	24	23	20	20	20	21	17	18	24	28	14
30 Ridh_Kleae	23	18	26	18	18	18	26	25	18	20	20	17	13	17	19	24	25	24	19	25	28	20	21	23	21	30	28	21
31 Enta_Ecoli	28	18	26	21	19	24	26	27	16	16	18	21	19	19	18	25	24	25	24	27	25	24	22	24	24	22	24	19
32 Fixr_Braja	28	12	25	19	21	25	26	22	22	15	20	17	15	17	17	28	23	25	21	28	28	18	21	21	22	23	24	18
33 Ap27_Mouse	27	18	23	20	17	24	29	27	19	21	18	19	16	16	18	26	24	27	25	27	25	21	18	22	22	17	26	21
34 Dhma_Flas1	27	13	24	19	16	23	28	28	14	19	15	18	19	18	20	28	27	25	24	28	29	20	18	23	21	20	26	15
35 3adh_Psesp	25	16	21	16	16	24	22	26	18	15	14	16	13	15	15	25	22	21	20	26	25	19	18	19	19	20	22	18
36 Ba74_Eubsp	34	14	27	20	16	25	30	28	24	17	14	16	20	18	20	26	25	27	22	29	27	23	22	22	27	25	29	22
37 Hdha_Ecoli	33	13	23	22	18	23	30	29	19	22	15	18	17	14	18	26	26	28	27	27	27	28	27	29	29	23	24	22
38 Dhgb_Bacme	35	15	24	17	17	23	29	26	17	19	17	18	20	21	16	30	28	26	25	30	29	29	31	29	29	30	24	21
39 Ba72_Eubsp	34	14	26	18	17	24	30	28	22	17	18	18	20	20	19	28	28	28	25	25	28	26	23	23	27	27	30	20
40 Act3_Strco	34	15	27	24	17	26	28	26	26	21	25	22	20	19	23	30	28	31	30	28	28	22	23	24	26	24	31	22
41 Tr1_Dastr	26	16	25	23	15	26	29	22	17	20	18	18	16	17	17	31	29	24	24	27	26	22	22	19	22	22	25	20
42 Tr2_Dastr	28	17	24	20	15	26	28	24	21	20	22	18	18	16	17	27	28	26	21	26	28	22	22	20	21	23	23	18
43 P29x_Dastr	28	17	21	22	15	28	28	26	17	21	19	18	15	16	16	27	26	23	21	25	26	18	19	20	19	20	25	19
44 Ver1_Asppa	28	17	26	21	17	25	30	28	21	20	17	20	15	17	19	28	25	28	27	22	25	24	23	23	26	24	26	20
45 Nodg_Azobr	33	15	29	21	16	25	27	26	23	19	23	20	23	18	23	30	27	29	29	27	30	26	28	31	32	25	30	24
46 Phbb_Alceu	30	16	27	22	16	22	24	26	20	18	22	20	24	21	25	24	24	26	24	25	24	27	27	26	27	21	28	21
47 Phbb_Zoora	31	15	29	22	18	25	30	26	25	17	21	21	22	22	23	27	26	28	30	30	28	30	29	29	31	25	33	25
48 Nodg_Rhime	30	14	28	17	16	26	23	27	19	21	27	17	19	19	21	26	25	26	30	25	28	29	24	30	30	23	28	22
49 Ardh_Cantr	28	18	23	17	15	20	25	22	15	19	16	15	18	16	23	23	23	18	22	24	26	21	19	18	20	20	24	16
50 Scdh_Picab	28	15	25	20	19	25	29	28	19	20	18	19	17	18	20	26	24	29	27	26	25	22	24	22	23	24	29	20
51 Fabg_Cupla	35	14	28	18	17	25	28	26	24	21	23	18	22	21	25	30	28	31	29	28	27	33	34	34	33	29	30	24
52 Fabg_Ecoli	35	15	29	20	18	28	29	26	21	24	24	19	19	19	22	32	28	26	29	33	32	29	28	33	32	25	27	21
53 Ke6_Mouse	32	15	27	21	18	26	29	30	19	20	23	21	19	16	20	28	27	33	32	29	27	25	28	25	29	21	30	20
54 Dhk2_Strvn	33	14	24	24	19	29	29	29	26	21	22	17	19	21	18	31	26	25	25	30	30	22	24	22	24	21	25	19
55 Dhes_Human	21	15	20	21	17	20	19	16	21	19	21	14	20	16	17	21	20	25	19	15	19	20	18	18	17	24	24	16
56 Spre_Human	19	10	19	16	15	19	22	18	15	18	17	19	16	15	18	19	20	20	17	18	18	17	17	18	17	17	19	18
57 Yura_Myxxa	18	10	15	16	16	15	16	16	14	13	13	19	16	15	16	19	16	17	15	17	15	15	15	15	16	18	20	20

	29 Ligd_Psepa	30 Ridh_Kleae	31 Enta_Ecoli	32 Fixr_Braja	33 Ap27_Mouse	34 Dhma_Flas1	35 3adh_Psesp	36 Ba74_Eubsp	37 Hdha_Ecoli	38 Dhgb_Bacme	39 Ba72_Eubsp	40 Act3_Strco	41 Tr1_Dastr	42 Tr2_Dastr	43 P29x_Dastr	44 Ver1_Asppa	45 Nodg_Azobr	46 Phbb_Alceu	47 Phbb_Zoora	48 Nodg_Rhime	49 Ardh_Cantr	50 Scdh_Picab	51 Fabg_Cupla	52 Fabg_Ecoli	53 Ke6_Mouse	54 Dhk2_Strvn	55 Dhes_Human	56 Spre_Human
30 Ridh_Kleae	21																											
31 Enta_Ecoli	21	25																										
32 Fixr_Braja	21	23	34																									
33 Ap27_Mouse	22	24	29	27																								
34 Dhma_Flas1	20	23	28	28	32																							
35 3adh_Psesp	18	26	29	29	23	27																						
36 Ba74_Eubsp	21	23	28	28	26	23	22																					
37 Hdha_Ecoli	21	28	28	29	30	28	27	35																				
38 Dhgb_Bacme	20	26	26	30	23	27	25	30	31																			
39 Ba72_Eubsp	26	24	32	25	26	26	23	30	26	31																		
40 Act3_Strco	21	28	27	27	25	28	27	30	28	30	29																	
41 Tr1_Dastr	19	24	27	28	27	27	27	30	31	30	27	27																
42 Tr2_Dastr	19	23	24	28	29	27	26	30	33	30	27	28	64															
43 P29x_Dastr	20	21	26	26	24	24	27	31	27	28	28	29	61	57														
44 Ver1_Asppa	23	28	27	28	34	31	25	30	34	27	28	30	28	30	30													
45 Nodg_Azobr	22	23	26	30	27	30	24	34	30	30	28	39	28	28	32	33												
46 Phbb_Alceu	21	19	25	27	28	26	21	29	26	31	27	35	26	27	28	28	60											
47 Phbb_Zoora	25	22	29	32	28	30	24	31	25	32	31	38	30	30	31	35	54	51										
48 Nodg_Rhime	21	27	30	28	27	29	22	29	29	29	31	36	33	31	31	31	41	39	40									
49 Ardh_Cantr	19	20	24	22	27	22	24	25	29	26	27	23	28	27	28	24	26	23	25	27								
50 Scdh_Picab	22	25	31	27	28	31	24	29	30	32	28	29	27	29	26	38	30	29	31	34	21							
51 Fabg_Cupla	21	26	30	29	27	29	25	33	31	37	34	39	28	29	28	32	44	38	42	43	28	34						
52 Fabg_Ecoli	22	24	32	29	29	29	28	33	32	32	33	38	31	30	29	33	47	42	44	48	27	34	49					
53 Ke6_Mouse	21	24	29	31	31	30	26	33	30	31	29	34	27	29	27	33	37	36	39	38	26	34	38	38				
54 Dhk2_Strvn	21	25	27	32	27	24	29	34	29	33	25	33	30	30	31	31	29	28	30	29	24	30	33	31	35			
55 Dhes_Human	18	22	18	19	18	19	18	18	17	20	23	25	18	17	15	21	23	21	22	21	16	20	23	22	22	24		
56 Spre_Human	15	18	21	19	22	15	18	19	19	21	16	19	19	20	18	17	19	19	20	20	14	20	20	21	20	21	17	
57 Yura_Myxxa	18	18	18	15	19	17	17	14	16	20	18	20	17	16	14	19	17	17	19	16	15	16	17	18	16	20	19	18

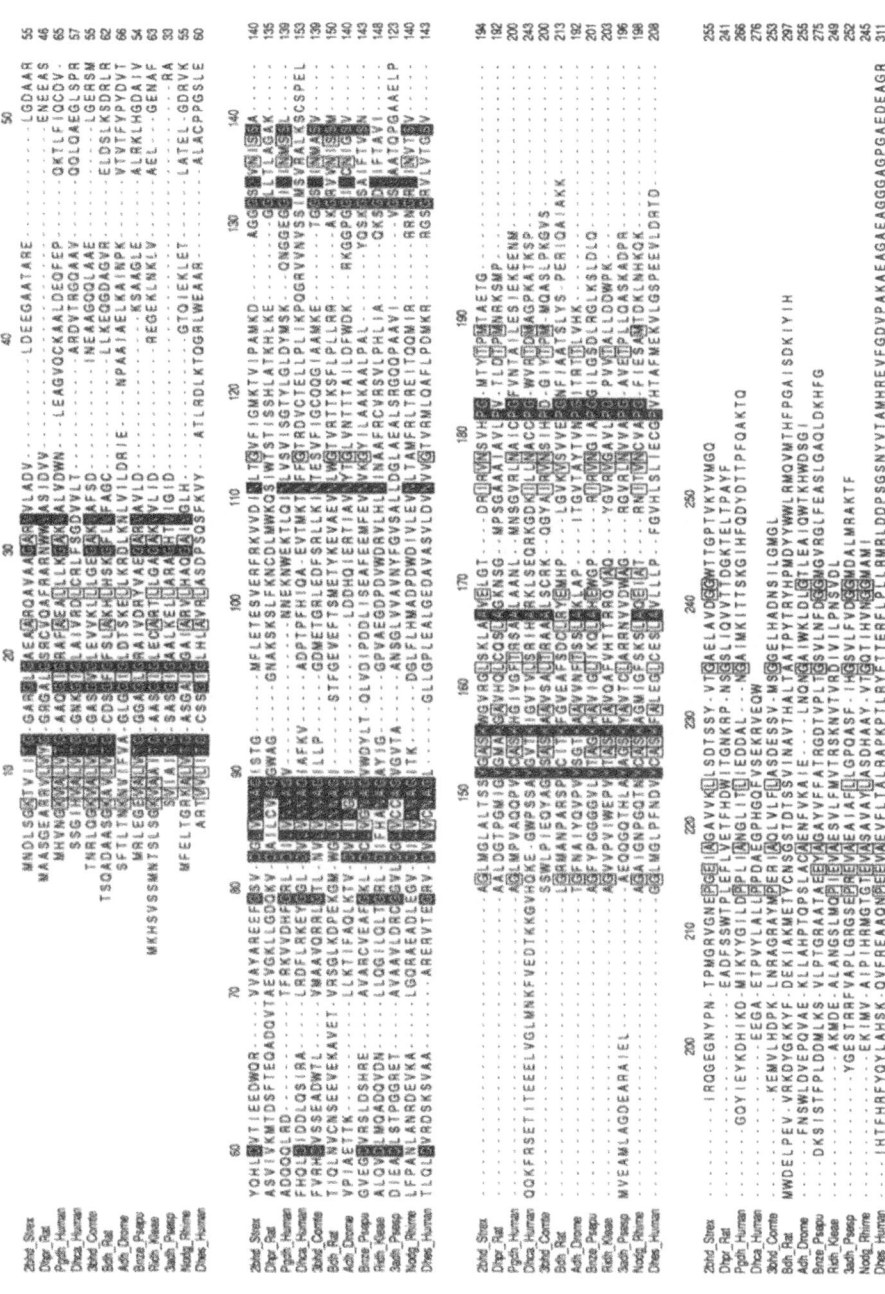

Figure 2. Subset showing 12 amino acid sequences of the multiple sequence alignment of 57 short-chain dehydrogenases/reductases. Residues given against a black background are those conserved in more than 70 % of the sequences aligned, while boxed residues are conserved in more than half of the structures. Sequence abbreviations as in Table 1. A few alignment positions have gaps in all sequences because this is a subset of the complete alignment with 57 structures, and those positions are not gaps in other structures.

Table 3. Amino acid residues conserved in the structures aligned in Fig. 2. Ordinary printing indicate those conserved in more than 70 % of the structures aligned, while bold residues indicate those conserved in more than 90 % of all family members. Residue numbers refer to the 3α/20β-hydroxysteroid dehydrogenase (Marekov et al., 1990)

Thr-12	Asp-82	Ile-134
Gly-13	Leu-84	**Ser-139**
Gly-17	Asn-86	**Tyr-152**
Gly-19	Asn-87	**Lys-156**
Gly-30	Ala-88	Ala-167
Val-33	Gly-89	**Pro-182**
Asp-60	Asn-111	Gly-183
Gly-79	**Gly-132**	

members revealed sequence patterns, which in turn were used for further scrutiny of the databases.

From Table 1, the SDR family can be seen to encompass a wide variety of enzyme activities, including alcohol and polyol dehydrogenases, steroid dehydrogenases, pro-staglandin dehydrogenases, carbonyl reductases, dihydropteridine reductases, and ketoacyl reductases. Table 1 also reveals that this enzyme family has representatives from different kingdoms, i.e. animals, plants, yeasts and bacteria. Thus, these enzymes constitute an evolutionary extremely old entity of importance in several metabolic pathways.

Most of the members of the SDR family are quite distantly related. Residue identities in pairwise comparisons typically amount to only 15-30 % (Table 2). This level of identity, together with the big number of enzymes involved, makes utilisation of multiple alignments an ideal method to detect conserved residues and sequence motifs in common for the whole

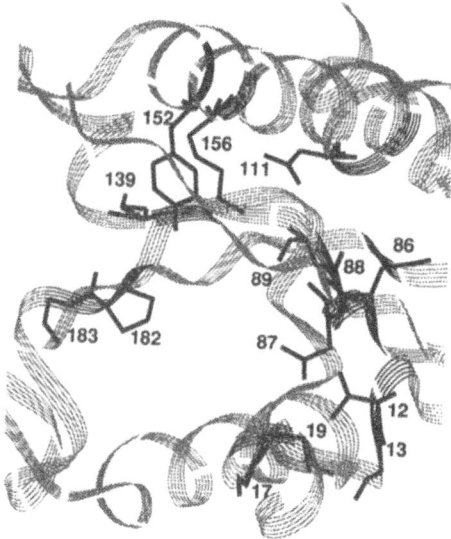

Figure 3. Active site of 3α/20β-hydroxysteroid dehydrogenase. Residues that are conserved in more than 70 % of all SDR sequences are indicated with numbers and side-chain. Apart from the highly conserved Tyr-152 and Lys-156, also other side-chains of importance for the enzymatic activity are seen in the figure.

family. A multiple alignment has been calculated for the 57 SDR members, of which a subset of 12 different structures is shown in Fig. 2.

CONSERVED RESIDUES

From the multiple alignment, it can be seen that only 23 residues are conserved in more than 70 % of the structures (Table 3). These residues are given in white against a black background in Fig. 2. Among these residues, we find the strictly conserved Tyr-152, which has been shown by chemical modification experiments (Krook et al., 1992) and by site-directed mutagenesis studies (Ensor and Tai, 1991; Albalat et al., 1992; Chen et al., 1993)

to be of critical importance for several members of the enzyme family. Ser-139 and Lys-156, which are closely positioned, are deduced also to be critical for the enzyme activity.

Among the highly conserved residues (boldface in Table 3), three glycine residues located close to the N terminus (Gly-13, Gly-17, and Gly-19) are involved in coenzyme-binding. They are positioned between βA and αB, participating in the Rossmann fold, as early anticipated from amino acid sequence comparisons (Thatcher and Sawyer, 1980).

Apart from these two patterns, the consecutive residues Asn-86-Asn-87-Ala-88-Gly-89 seem to be important for the formation of the active site (Fig. 3). In the tertiary structure, these residues are also found in the vicinity of Tyr-152. In addition, Pro-182 and Gly-183 are located close to Tyr-152.

In conclusion, there are four sequence motifs that are characteristic for the SDR family (Table 4). They can be utilised to screen subsequent issues of the databases, in order to find additional members within the family.

Of the well-conserved residues, about one third are glycine (Table 3). A high proportion of glycine is typical of distantly related proteins with well-conserved tertiary structures and therefore supports the view that all enzymes of the SDR family have similar folds.

Utilising the known three-dimensional structure of 3α/20β-hydroxysteroid dehydrogenase (Ghosh et al., 1991), the spatial position of the residues conserved can be investigated. In fact, all but one of the residues highly conserved (>>90% of all structures) are located closely together. Thus, Gly-13, Gly-17 and Gly-19 (numbers according to the 3α/20β-hydroxysteroid dehydrogenase) are found on one side of a cavity (Figure 3), while Tyr-152 and Lys-156 are positioned on the other side, in an α-helix, with the well-conserved Ser-139 in the close vicinity, and Pro-182 also lining the cleft. The single exception is Gly-132 on the other side of the molecule.

Table 4. List of well-conserved sequence motifs typical of the short-chain dehydrogenases/reductases. Residue numbers according to the 3α/20β-hydroxysteroid dehydrogenase. 'x' indicates any amino acid residue

Sequence	Position
GxxxGxG	13–19
NNAG	86–89
YxxxK	152–156
PG	182–183

HSD   DHR

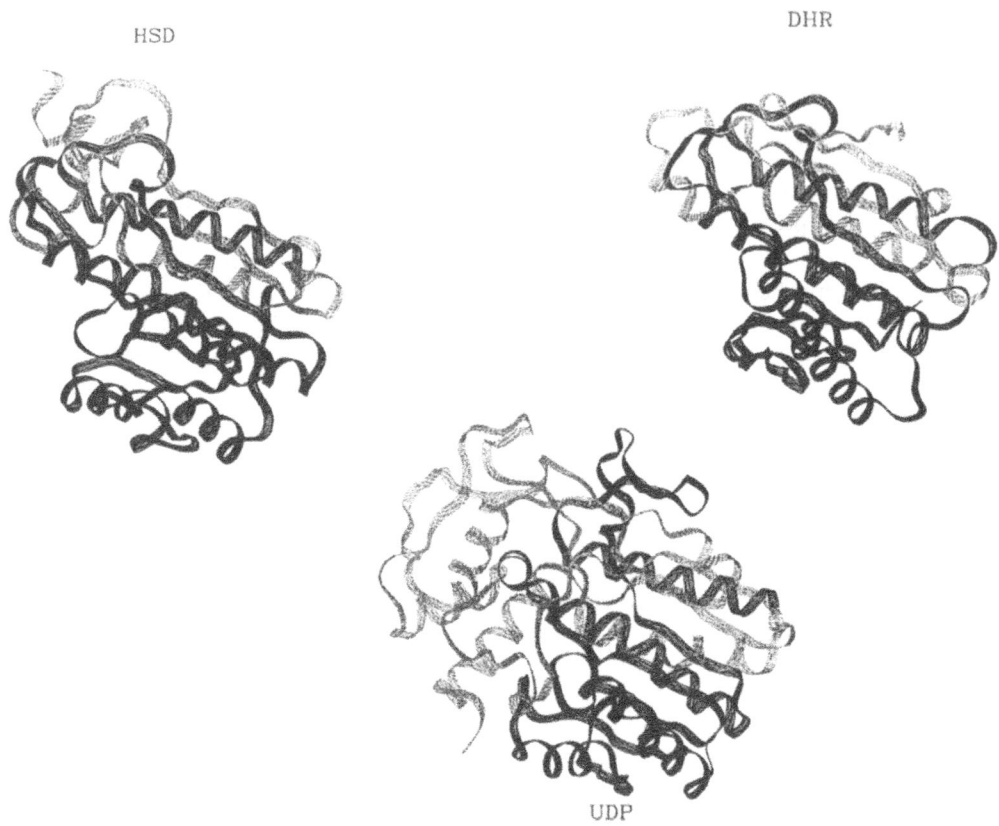

UDP

Figure 4. The three known tertiary structures of SDR members. They are 3α/20β-hydroxysteroid dehydrogenase (HSD; Ghosh et al., 1991), dihydropteridin reductase (DHR; Varughese et al., 1992) and UDP-glucose
4-epimerase (UDP; Bauer et al., 1992). The parts with most extensive homologies are shown in black, while
remaining parts are in gray. The three folds are homologous, which is seen especially clearly for the HSD and
DHR structures. The UDP molecule has an extra domain not present in the other two folds. However, it has
the basic pattern with alternating α/β structures and a six-stranded β-sheet in common with the other two
structures.

A reaction mechanism involving Tyr-152 and Lys-156 has been postulated (McKinley-McKee et al., 1991), with Tyr in the ionised form, stablised via the side-chain of Lys-156,
and giving a nucleophilic attack on the substrate.

EXTENDED SDR FAMILY

The family relationships of the SDR enzymes can be extended to include also
3β-hydroxy-5-ene-steroid dehydrogenase, UDP-glucose 4-epimerase and 20 more proteins
(Table 5). These are distantly related to the SDR family, as was indicated already in a previous
multiple alignment study (Persson et al., 1991), but is even more clearly evident after
three-dimensional structural alignment of the epimerase and two members of the SDR family
(Holm et al., 1994).

The tertiary structures of the three enzymes are given in Fig. 4, showing the similar
folds. The UDP-glucose 4-epimerase has a longer protein chain than the other two enzymes,

creating an extra domain (top left part of the UDP molecule in Fig. 4). This molecule consists of two domains with the domain border at about position 180 (Bauer et al., 1992). Furthermore, beyond this position in UDP-glucose 4-epimerase, there are no similarities

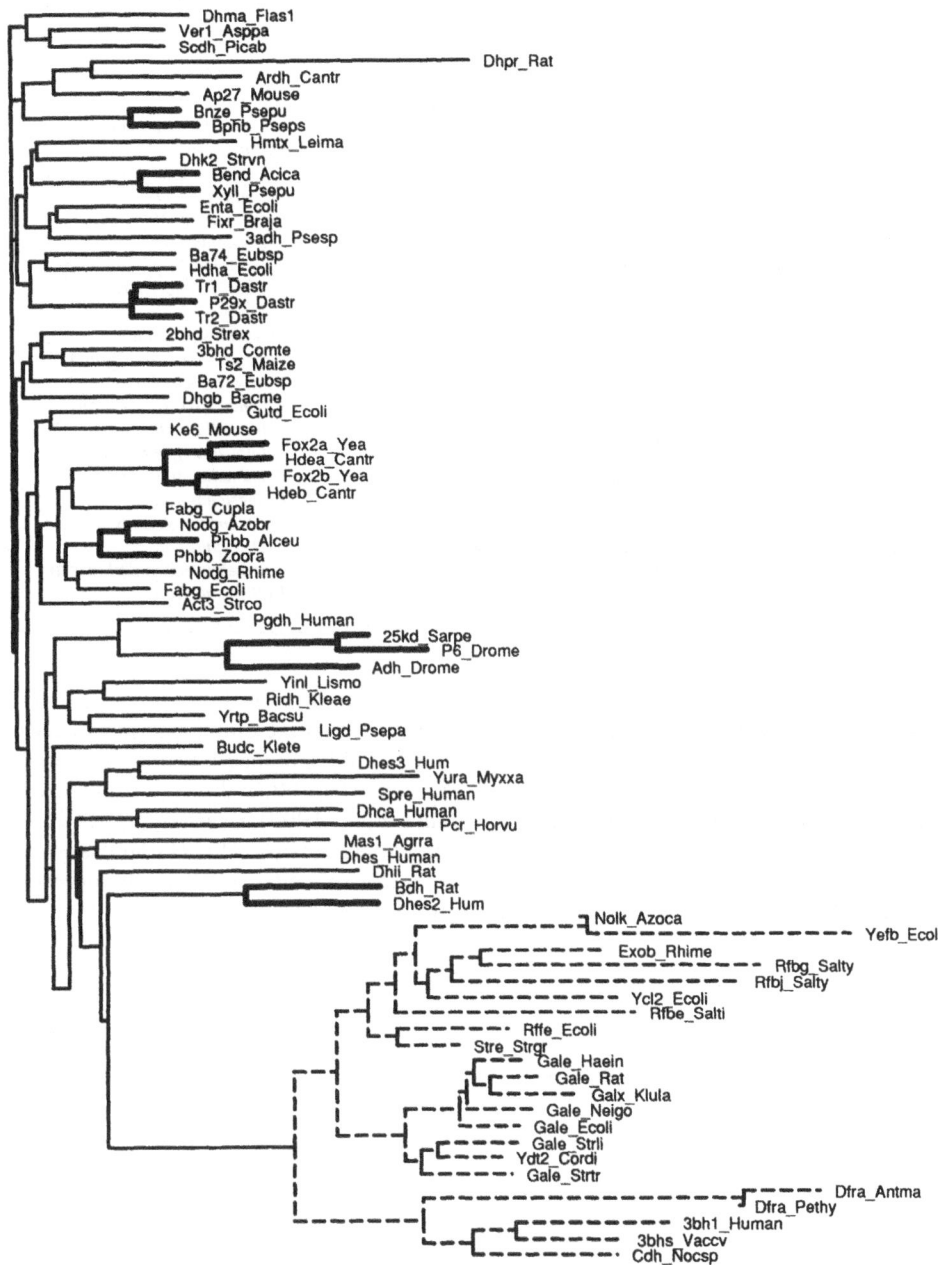

Figure 5. Unrooted evolutionary tree of the SDR family. Bold branches indicate those statistically significant as judged by bootstrap trials, i.e. they cluster on the same branch in more than 99% of the test trees (Thompson et al., 1994). The branches of the 22 epimerase-related proteins are dashed. The scale bar gives the distance, corresponding to an apparent sequence divergence of 10 %. Abbreviations as in Table 1 and 5.

Table 5. Proteins related to UDP-glucose 4-epimerase,
extending the SDR family. Columns show identification code in
the Swissprot database (except Cdh_Nocsp; Horinouchi et al.,
1991), enzyme names and the source of enzyme/DNA
characterised

Code	Enzyme	Species
3bh1_Human	3-β Hydroxy-5-ene steroid dehydrogenase type I	Human
3bhs_Vaccv	3-β Hydroxy-5-ene steroid dehydrogenase	*Vaccinia* virus (strain wr).
Cdh_Nocsp	Cholesterol dehydrogenase	*Nocardia* sp.
Dfra_Antma	Dihydroflavonol-4-reductase	*Antirrhinum majus*
Dfra_Pethy	Dihydroflavonol-4-reductase	*Petunia hybrida*
Exob_Rhime	UDP-Glucose 4-epimerase	*Rhizobium meliloti*
Gale_Ecoli	UDP-Glucose 4-epimerase	*Escherichia coli*
Gale_Haein	UDP-Glucose 4-epimerase	*Haemophilus influenzae*
Gale_Neigo	UDP-Glucose 4-epimerase	*Neisseria gonorrhoeae*
Gale_Rat	UDP-Glucose 4-epimerase	Rat
Gale_Strli	UDP-Glucose 4-epimerase	*Streptomyces lividans*
Gale_Strtr	UDP-Glucose 4-epimerase	*Streptococcus thermophilus*
Galx_Klula	UDP-Glucose 4-epimerase	*Kluyveromyces lactis*
Nolk_Azoca	Nodulation protein NolK	*Azorhizobium caulinodans*
Rfbe_Salti	CDP-Tyvelose-2-epimerase	*Salmonella typhi*
Rfbg_Salty	CDP-Glucose 4,6-dehydratase	*Salmonella typhimurium*
Rfbj_Salty	CDP-Abequose synthase	*Salmonella* sp. (Group b)
Rffe_Ecoli	UDP-N-Acetylglucosamine epimerase	*Escherichia coli*
Stre_Strgr	dTDP-Glucose 4,6-dehydratase	*Streptomyces griseus*
Ycl2_Ecoli	Hypothetical 37.6 kD protein in cld 5'region (orf2)	*Escherichia coli*
Ydt2_Cordi	Hypothetical 35.4 kD protein in dtxr 3'region	*Corynebacterium diphtheriae*
Yefb_Ecoli	Hypothetical 36.1 kD protein in cpsb 5'region (orf 0.9)	*Escherichia coli*

between this enzyme and the steroid dehydrogenase or dihydropteridin reductase. It could
therefore be concluded that enzymes of the SDR family and of the epimerase family have a
building block of 180 amino acid residues in common. In UDP-glucose 4-epimerase, Tyr-149
and Lys-153, corresponding to Tyr-152 and Lys-156 in the 3α/20β-hydroxysteroid dehydrogenase, are also located in the binding pocket (Bauer et al., 1992). Consequently, it seems
likely that both enzyme families have similar catalytic mechanisms. This is impressive with
respect to the number of different enzyme activities that the extended SDR family represents,
encompassing EC numbers from three different groups (Table 6).

Table 6. Enzyme activity types
represented by the members of the
extended SDR family

Enzyme activity type	EC class
Dehydrogenases	EC 1
Reductases	EC 1
Dehydratases	EC 4.2
Epimerases	EC 5.1
Isomerases	EC 5.3

EVOLUTION

From the information in the multiple sequence alignment (Fig. 2), evolutionary trees were calculated with the program Clustal W (Thompson et al., 1994) using the neighbour-joining method (Saitou and Nei, 1987) with corrections for multiple substitutions and distances calculated according to Kimura's empirical method (1983). The tree arrived at is depicted in Fig. 5. Most of the enzymes of the SDR family are equidistantly related, reflected by an early branching at similar levels. It is also clearly seen that the UDP-glucose 4-epimerase and related enzymes form a separate sub-family (Fig. 5, dashed branches). Furthermore, a limited number of sub-groups within the SDR family can be shown to be evolutionary more closely related, as indicated by bold branches in Fig. 5. This applies for instance to the *Drosophila* alcohol dehydrogenase, *Drosophila* P6 protein, and *Sarcophaga* development-specific 25 kD protein, which clearly form a subgroup.

CONCLUSIONS

Amino acid sequence comparisions and multiple sequence alignments of the short-chain dehydrogenases/reductases have shown that these enzymes belong to the SDR super-family, and that this family also is related to another family of dehydrogenases, epimerases, isomerases and dehydratases. A limited number of conserved residues are distinguished, of which several line up closely in space in the three known three-dimensional structures, thus defining residues likely to create the active site of these enzymes. It can be concluded that the different members, in spite of their different enzyme activities, most probably have one reaction mechanism in common.

ACKNOWLEDGEMENTS

This work was supported by grants from the Swedish Medical Research Council (project no 03X-3532 and 13F-10248), the Swedish Society of Medicine, the Swedish Alcohol Research Fund, the Foundation Blanceflor Boncompagni-Ludovisi, née Bildt, and the Swedish Society for Medical Research.

REFERENCES

Albalat, R., Gonzàlez-Duarte, R., and Atrian, S., 1992, Protein engineering of *Drosophila* alcohol dehydro-genase. Thy hydroxyl group of Tyr[152] is involved in the active site of the enzyme, *FEBS Lett.* 308:235-239.

Aziz, N., Maxwell, M. M., St.-Jacques, B., and Brenner, B. M., 1993, Downregulation of Ke 6, a novel gene encoded within the major histocompatibility complex, in murine polycystic kidney disease, *Mol. Cell. Biol.* 13:1847-1853.

Bairoch, A., and Boeckmann, B., 1992, The SWISS-PROT protein sequence data bank, *Nucleic Acids Res.* 20 Supplement:2019-2022.

Baron, S. F., Franklund, C. V., and Hylemon, P. B., 1991, submitted to the EMBL databank.

Bauer, A. J., Rayment, I., Frey, P. A., and Holden, H. M., 1992, The molecular structure of UDP-galactose 4-epimerase from *Escherichia coli* determined at 2.5 Å resolution, *Proteins* 12:372-381.

Bauer, S., Galliano, H., Pfeiffer, F., Meßner, B., Sandermann, Jr, H., and Ernst, D., 1993, Isolation and characterization of a cDNA clone encoding a novel short-chain alcohol dehydrogenase from norway spruce (*Picea abies* L. Karst), *Plant Physiol.* 103:1479-1480.

Callahan, H. L., and Beverley, S. M., 1992, A member of the aldoketo reductase family confers methotrexate resistance in Leishmania, *J. Biol. Chem.* 267:24165-24168.

Chen, Z., Jiang, J. C., Lin, Z.-G., Lee, W. R., Baker, M. E., and Chang, S. H., 1993, Site specific mutagenesis of *Drosophila* alcohol dehydrogenase: evidence for involvement of tyrosine-152 and lysine-156 in catalysis, *Biochemistry* 32:3342-3346.

DeLong, A., Calderon-Urrea, A., and Dellaporta, S. L. (1993) Sex determination gene *TASSELSEED2* of maize encodes a short-chain alcohol dehydrogenase required for stage-specific floral organ abortion, *Cell* 74:757-768.

Dothie, J. M., Giglio, J. R., Moore, C. B., Taylor, S. S., and Hartley, B. S., 1985, Ribitol dehydrogenase of *Klebsiella aerogenes*. Sequence and properties of wild-type and mutant strains, *Biochem. J.* 230:569-578.

Ensor, C. M., and Tai, H.-H., 1991, Site-directed mutagenesis of the conserved tyrosine 151 of human placental NAD⁺-dependent 15-hydroxyprostaglandin dehydrogenase yields a catalytically inactive enzyme, *Biochem. Biophys. Res. Commun.* 176:840-845.

Geissler, W. M., Davis, D. L., Wu, L., Bradshaw, K. D., Patel, S., Mendonca, B. B., Elliston, K. O., Wilson, J. D., Russell, D. W., and Andersson, S. (1994) Male pseudohermaphroditism caused by mutations of testicular 17β-hydroxysteroid dehydrogenase 3, *Nature Genetics* 7:34-39.

Ghosh, D., Weeks, C. M., Grochulski, P., Duax, W. L., Erman, M., Rimsay, R. L., and Orr, J. C., 1991, Three-dimensional structure of holo 3α,20β-hydroxysteroid dehydrogenase: A member of a short-chain dehydrogenase family, *Proc. Natl Acad. Sci.* 88:10064-10068.

Holm, L., Sander, C., and Murzin, A., 1994, Three sisters, different names, *Nature Structural Biology* 1:146-147.

Horinouchi, S., Ishizuka, H., and Beppu, T., 1991, Cloning, nucleotide sequence and transcriptional analysis of the NAD(P)-dependent cholesterol dehydrogenase gene from a *Nocardia* sp. and its hyperexpression in *Streptomyces* sp., *Appl. Environ. Microbiol.* 57:1386-1393.

Ichinose, H., Katoh, S., Sueoka, T., Titani, K., Fujita, K., and Nagatsu, T., 1991, Cloning and sequencing of cDNA encoding human sepiapterin reductase, *Biochem. Biophys. Res. Commun.* 179:183-189.

Krook, M., Marekov, L., and Jörnvall, H., 1990, Purification and structural characterization of placental NAD⁺-linked 15-hydroxyprostaglandin dehydrogenase. The primary structure reveals the enzyme to belong to the short-chain alcohol dehydrogenase family, *Biochemistry* 29:738-743.

Jany, K.-D., Ulmer, W., Fröschle, M., and Pfleiderer, G., 1984, Complete amino acid sequence of glucose dehydrogenase from *Bacillus megaterium, FEBS Lett.* 165:6-10.

Jörnvall, H., Persson, M., and Jeffery, J., 1981, Alcohol and polyol dehydrogenases are both divided into two protein types, and structural properties cross-relate the different enzyme activities within each type, *Proc. Natl Acad. Sci. USA* 78:4226-4230.

Jörnvall, H., von Bahr-Lindström, H., Jany, K.-D., Ulmer, W., and Fröschle, M., 1984, Extended superfamily of short alcohol-polyol-sugar dehydrogenases: Structural similarities between glucose and ribitol dehydrogenases, *FEBS Lett.* 165:190-196.

Kimura, M., 1983, The neutral theory of molecular evolution. Cambridge University Press, Cambridge, England.

Krook, M., Prozorovski, V., Atrian, S., Gonzàlez-Duarte, R., and Jörnvall, H., 1992, Short-chain dehydrogenases. Proteolysis and chemical modification of prokaryotic 3α/20β-hydroxysteroid, insect alcohol and human 15-hydroxyprostaglandin dehydrogenases, *Eur. J. Biochem.* 209:233-239.

Marekov, L., Krook, M., and Jörnvall, H., 1990, Prokaryotic 20β-hydroxysteroid dehydrogenase is an enzyme of the 'short-chain, non-metalloenzyme' alcohol dehydrogenase type, *FEBS Lett.* 266:51-54.

McKinley-McKee, J. S., Winberg, J.-O., and Pettersson, G., 1991, Mechanism of action of *Drosophila melanogaster* alcohol dehydrogenase, *Biochem. Internat.* 25:879-885.

Morris, H. R., Williams, D. H., Midwinter, G. G., and Hartley, B. S., 1974, A mass-spectrometric sequence study of the enzyme ribitol dehydrogenase from *Klebsiella aerogenes, Biochem. J.* 141:701-713.

Nakajima, K., Hashimoto, T., and Yamada Y., 1993, Two tropinone reductases with different stereospecificities are short-chain dehydrogenases evolved from a common ancestor, *Proc. Natl Acad. Sci. U.S.A.* 90:9591-9595.

Peltoketo, H., Isomaa, V., Mäentausta, O., and Vihko, R., 1988, Complete amino acid sequence of human placental 17β-hydroxysteroid dehydrogenase deduced from cDNA, *FEBS Lett.* 239:73-77.

Persson, B., Krook, M., and Jörnvall, H., 1991, Characteristics of short-chain alcohol dehydrogenases and related enzymes, *Eur. J. Biochem.* 200:537-543.

Persson, B., Zigler, Jr, J. S., and Jörnvall, H., 1994, A super-family of medium-chain dehydrogenases/reductases (MDR): Sub-lines including ζ-crystallin, alcohol and polyol dehydrogenases, quinone oxidoreductases, enoyl reductases, VAT-1 and further proteins, *Eur. J. Biochem.*, submitted.

Rat, L., Veuille, M., and Lepesant, J. A., 1991, *Drosophila* fat body protein P6 and alcohol dehydrogenase are derived from a common ancestral protein, *J. Mol. Evol.* 33:194-203.

Rossmann, M. G., Liljas, A., Brändén, C.-I., and Banaszak, L. J., 1975, Evolutionary and structural relationships among dehydrogenases, *The Enzymes, 3rd edn* (Boyer, P. D., ed.), vol. 11, pp. 61-102, Academic Press, New York.

Saitou, N., and Nei, M., 1987, The neighbor-joining method: A new method for reconstructing phylogenetic trees, *Mol. Biol. Evol.* 4:406-425.

Schwartz, M. F., and Jörnvall, H., 1976, Structural analyses of mutant and wild-type alcohol dehydrogenses from *Drosophila melanogaster, Eur. J. Biochem.* 68:159-168.

Skory, C. D., Chang, P. K., Cary, J., and Linz, J. E., 1992, Isolation and characterization of a gene from *Aspergillus parasiticus* associated with the conversion of versicolorin A to sterigmatocystin in aflatoxin biosynthesis, *Appl. Environ. Microbiol.* 58:3527-3537.

Suzuki, K., Ueda, S., Sugiyama, M., and Imamura, S., 1993, Cloning and expression of a *Pseudomonas* 3α-hydroxysteroid dehydrogenase-encoding gene in *Escherichia coli, Gene* 130:137-140.

Thatcher, D. R., 1980, The complete amino acid sequence of three alcohol dehydrogenase alleloenzymes (Adh^{N-11}, Adh^S and Adh^{UF}) from the fruitfly *Drosophila melanogaster, Biochem. J.* 187:875-883.

Thatcher, D. R., and Sawyer, L., 1980, Secondary structure prediction from the sequence of *Drosophila melanogaster* (fruitfly) alcohol dehydrogenase, *Biochem. J.* 187:884-886.

Thompson, J. D., Higgins, D. G., and Gibson, T. J., 1994, Improved sensitivity of profile searches through the use of sequence weights and gap excision, *Comput. Appl. Biosci.* 10:19-29.

Varughese, K. I., Skinner, M. M., Whiteley, J. M., Matthews, D. A., and Xuong, N. H., 1992, Crystal structure of rat liver dihydropteridine reductase, *Proc. Natl Acad. Sci. USA* 89:6080-6084.

Wong, B., 1993, submitted to the EMBL/GenBank/DDBJ databases.

Wu, L., Einstein, M., Geissler, W. M., Chan, H. K., Elliston, K. O., and Andersson, S. (1993) Expression cloning and characterization of a human 17β-hydroxysteroid dehydrogenase type 2, a microsomal enzyme possessing 20α-hydroxysteroid dehydrogenase activity, *J. Biol. Chem.* 268:12964-12969.

ZINC BINDING OF ALCOHOL AND SORBITOL DEHYDROGENASES

Christina Karlsson, Hans Jörnvall, and Jan-Olov Höög

Department of Medical Biochemistry and Biophysics
Karolinska Institutet
S-171 77 Stockholm, Sweden

INTRODUCTION

Zinc is an essential component of many enzymes, serving a role for catalytic activity or structural stability. The removal of catalytic zinc results in an inactive apoenzyme which, however, often retains the native tertiary structure. Structural zinc frequently contributes to the maintenance of the structure of oligomeric enzymes. The removal of zinc from such proteins therefore prevents subunit association. As discerned from zinc analysis of structurally investigated zinc metalloenzymes, the characteristics of a catalytic zinc-binding motif, in many cases, is a combination of three His/Glu/Asp/Cys residues and an activated H_2O-molecule (Vallee & Auld, 1990). The spacers between the first and second ligands are short, typically 1-3 amino acids long (alcohol dehydrogenase and sorbitol dehydrogenase are exceptions with 21-25 residues). The second spacer, longer in nature, separates the second and third ligands by about 20-120 amino acid residues (Vallee & Auld, 1989). The structural zinc is necessary for activity only to the extent that the overall conformation of the enzyme affects its action. It serves as a cross-linking agent to stabilize structures (Berg, 1987). The observed pattern of ligands frequently encompasses four cysteine residues, closely spaced in the linear amino acid sequence (Vallee & Auld, 1990). Sorbitol dehydrogenase (SDH) harbours one catalytic zinc atom per subunit (Jeffery et al., 1984a). Because of the structural relationship between SDH and alcohol dehydrogenase (ADH), two of the three ligands to the zinc could be established early in SDH (Cys44 and His69, Jeffery et al., 1984b). The homology in the area around the third zinc ligand in SDH, though, was low toward ADH. The modelled structure of sheep SDH using the crystal structure of horse ADH as reference, gave the best fit with a glutamic acid residue (Glu155) as the third zinc ligand (Eklund et al., 1985). In order to establish the exact nature of the third zinc ligand, a set of five different potential ligands were mutated to Ala or Gln, residues not able to ligand zinc, and the proteins were expressed in E. coli with subsequent purification and determination of enzyme activities and zinc contents (Karlsson, 1994). ADH contains two zinc atoms per subunit, one catalytic and one structural (Åkeson, 1964; Drum et al., 1969; Eklund et al., 1976). The binding site of the structural zinc atom in ADH involves four cysteine residues at positions 97, 100, 103, and 111. Though it has been designated as structural since long, the exact functions, as well as the question which structural property it maintains, are unclear. To investigate the contribution of each of the second zinc ligands to the overall conformation, we have performed *in vitro* mutagenesis of class I and III ADH (Jeloková *et al.*, 1994). The ligands were mutated, in separate constructs, to non-zinc liganding counterparts, Ala or Ser. Proteins expressed were found to be labile and therefore, were detectable only from crude extracts upon Western blot analysis. Confirmation of correctly working transcription processes was ascertained by positive Northern blot analyses. The recently published crystal structure of glucose dehydrogenase from the archaeon *Thermoplasma acidophilum*, reveals structural homology (in spite of low sequence identity) to ADH from horse liver and SDH from sheep liver (John *et al.*, 1994). Glucose dehydrogenase is a tetramer and posesses a structural zinc atom, contained within a loop similar to that of ADH. However, the orientation of this structural loop with respect to the subunit is markedly different from that of ADH. This further illustrates the role of the zinc loop in the quaternary structures of several of the enzymes within the medium- chain dehydrogenase/reductase super-family, MDR (Persson *et al.* 1994)

MATERIALS AND METHODS

Plasmid Constructions and Mutagenesis

A plasmid for expression of rat SDH, pSDHex, was constructed from the vector pKK223-3 (Pharmacia Biotech), with a *tac* promoter, and a full-length cDNA coding for rat SDH (Karlsson *et al.*, 1991; Karlsson & Höög, 1993). All mutagenic replacements were carried out with single-stranded DNA (with an M13mp8 vector harbouring the target DNA) according to the method of Taylor *et al.* (1985), using an Amersham mutagenesis kit. The mutated cDNA-fragments were liberated with *Hind*III and ligated into the expression vector pSDHex, that had been rendered free of the particular fragment. DNA fragments were liberated from expression plasmids (originating from pKK223-3) from human class I and III ADH (Höög *et al.*, 1987; Estonius *et al.*, 1994), *Bst*EII was used for class I ADH and *Kpn*I and *Eco*RI for class III ADH. The resulting fragments were ligated into M13mp8, and mutagenesis was performed as described above, for SDH. Mutagenic primers were synthesized on Applied Biosystem 380A or 381 instruments. The mutated DNA fragments were checked with dideoxy sequence analysis (Sanger *et al.*, 1977) using T7 DNA polymerase. Plasmids were purified using Qiagen ion exchange columns (Diagen) or Magic plasmid preps (Promega). Liberated fragments were size fractionated on 1.5% agarose gels prior to Geneclean treatment (Bio 101).

Protein Expression and Purification

Sorbitol dehydrogenase. The recombinant plasmids were transformed into *E. coli* TG1, *lac*I^q, cells and expression of protein was induced by addition of isopropyl-β-D-thiogalactopyranoside to a final concentration of 0.3 mM. Cells from 2 or 3 l cultures were harvested by centrifugation, and the pellet was resuspended in 10 mM Tris-HCl, pH 8. Cells were disrupted by sonication, followed by centrifugation. The supernatant was applied to a DEAE-cellulose column and the enzymes were eluted with the flow-through. The eluate was dialyzed against 20 mM potassium phosphate buffer, pH 6.8, and applied to a CM-cellulose column. The enzyme bound was eluted by raising the pH to 7.6 with a potassium phosphate buffer of 20 mM. An additional gel filtration step on a Superose 12 column (Pharmacia Biotech; fast protein liquid chromatography) was performed when needed. The purities of the pooled fractions were assessed by SDS polyacrylamide gel electrophoresis. Protein concentrations were measured according to Bradford (1976) with serum albumin or sheep SDH as standards. SDH activity was estimated by monitoring the production of NADH at 340 nm in a Beckman DU-8 spectrophotometer at room temperature. The standard assay contained 1 mM NAD$^+$ and 50 mM sorbitol in 0.1 M glycine-NaOH, pH 10. A weighted non-linear regression analysis program (Lutz et al., 1986) was used to calculate kinetic parameters. *Alcohol dehydrogenase*. The starting buffer, after sonication and centrifugation, has 1 mM dithiothreitol, 10 mM Tris-HCl, pH 8. In analogy with SDH, recombinant ADH also eluted with the flow-through after DEAE-cellulose chromatography. The pooled fractions were applied to a 5'AMP-Sepharose column (Pharmacia Biotech), washed with 0.5 mM dithiothreitol, 100 mM Tris-HCl, pH 8. The recombinant enzymes were eluted with 1 mM NAD$^+$ in 0.5 mM dithiothreitol, 100 mM Tris-HCl, pH 8. All cultures, containing the different expression plasmids, were at least once grown in the presence of protease inhibitors, 0.1 mM 3,4-dichloroisocoumarin and 0.5 mM EDTA, in attempts to increase protein yields. Fractions were analyzed for the presence of ADH activity as described above, but with the following assays, 2.4 mM NAD$^+$, 1 mM ethanol in 0.1 M glycine-NaOH, pH 10, for class I ADH and 2.4 mM NAD$^+$, 5 mM octanol in 0.1 M glycine-NaOH, pH 10, or 1 mM glutathione/1 mM formaldehyde in 0.1 M sodium phosphate, pH 8.5, for class III ADH. For both SDH and the two classes of ADH, crude extracts and all concentrated fractions were submitted to Western blot analysis. The proteins were size-fractionated by SDS polyacrylamide gel electrophoresis and transferred to polyvinylidene triflouride or nylon membranes and subsequently treated with affinity purified antibodies raised against sheep SDH and class I and III ADH, respectively.

Zinc Analysis

Zinc contents of the pure, mutant SDHs were determined by electrothermal atomic absorption spectrometry with a Perkin-Elmer 5000 instrument equipped with a HGA furnance and an AS 40 autosampler. All zinc determinations were performed at two or three dilutions, each in duplicate. Buffer control samples, representing the flow-through collected during the latter part of the final centrifugation, were also analyzed. Protein concentrations were measured by amino acid analysis.

mRNA Preparation and Northern Blot Analysis

Frozen cells, harbouring the different ADH expression plasmids, were resuspended in 4 M guanidine isothiocyanate, 0.1 M β-mercaptoethanol, 0.025 M sodium citrate, pH 7.0, and homogenized twice with a Polytron instrument for 15-20 s. Each homogenate was layered over a cushion of 5.7 M CsCl in 0.025 M sodium acetate, pH 5.5, and centrifuged at 80 000 x g overnight in a Beckman SW 50 rotor at 15°C (Chirgwin et al., 1979). Material recovered was quantified spectrophotometrically before blot analysis. About 20μg total RNA from each *E. coli* extract was subjected to electrophoresis in 1% agarose containing ethidium bromide and 0.7% formaldehyde. The total RNA was blotted to nitrocellulose filters and hybridized with [α-^{32}P]dCTP labelled probes (Mega-prime kit, Amersham). The 390 bp *Eco*RI/*Kpn*I probe used for detecting the class III mRNA was derived from the 5'-coding region of the corresponding human cDNA (Sharma et al., 1989). The 450 *Bst*EII probe used for detecting the class I mRNA was derived from the middle part of the human class I β-subunit cDNA (Hedén et al., 1986). Hybridizations were performed at 42°C overnight in 40% formamide, 4 X SSC (1 X SSC is 0.15 M NaCl, 0.015 M sodium citrate, pH 7.0), 1 X Denhardts solution and 10% dextran sulphate. Filters were washed at high stringency (0.1 X NaCl/citrate, 0.1% SDS, 54°C) before exposure to Kodak X-AR5 films for 16 h, using

Table 1. Kinetic constants for recombinant, unmutated SDH, SDH155Cys, and native SDH. Data for native, rat liver SDH are from Leissing & McGuinness (1978). *No detectable activity with 0.7 M fructose

| Substrate | Recombinant SDH | | | SDH155C | | | Native SDH |
	K_m (mM)	k_{cat} (min⁻¹)	k_{cat}/K_m (min⁻¹mM⁻¹)	K_m (mM)	k_{cat} (min⁻¹)	k_{cat}/K_m (min⁻¹mM⁻¹)	K_m (mM)
Sorbitol	0.34	42.6	125	170	182	1.07	0.35
Xylitol	0.13	36.6	282	89	147	1.6	-
Fructose	32	312	9.8	ND*	ND*	-	110
NAD⁺	0.041			0.56			-

intensifying screens. rRNA markers, 16S and 23S, corresponding to 1.5 kb and 2.9 kb, respectively, were used for size determination.

RESULTS

Catalytic Zinc Atom in Sorbitol Dehydrogenase

Three protein ligands coordinate the catalytic zinc atom in SDH. Two, Cys44 and His69, have been known since the first structure for a SDH became available (Jeffery *et al.*, 1984b). The third, though, has been an enigma. Five different positions were subjected to

Table 2. Zinc stoichiometries for recombinant, unmutated SDH, native SDH, and mutants. The value for native, sheep SDH is from Jeffery *et al.* (1984b)

Sample	Zn (μM)	Protein (μM)	Zn (atom/subunit)
Recombinant SDH	26.7	31.9	0.84
Native SDH			0.65
SDH44A	0.48	22.7	0.02
SDH69Q	4.6	22.8	0.20
SDH149Q	0.49	0.50	1.0
SDH150Q	1.02	4.1	0.25
SDH155A	6.7	7.9	0.85
SDH155C	7.0	7.2	0.97
SDH155Q	3.0	0.79	0.75

mutational exchanges in SDH, Cys44 to Ala, His69 to Gln, Glu149 to Gln, Glu150 to Gln, and Glu 155 to Ala, Cys or Gln. Ala and Gln are residues not able to ligand zinc. The codons for Cys44 and His69 were mutated in order to study the effect upon loss of an established zinc ligand. Three glutamic acids, at positions 149, 150 and 155 are in the proposed area of the third zinc ligand. The homology toward ADH is low in this region and all the alternatives must therefore be targeted for mutation. Six out of the seven mutants were inactive upon purification. Therefore, in the preparation of the mutant proteins, the isolation scheme for native, recombinant SDH had to be followed blindly without activity measurements. The mutant Glu155Cys showed activity, but the catalytic efficiency was decreased 120-fold (k_{cat}/K_m), the ability to reduce fructose was totally abolished, and the K_m for NAD$^+$ was increased about 15-fold (Table 1).

Six of the mutants were obtained in good yield but the mutant Glu149Gln was labile and only small amounts of that protein could be isolated (about 10 µg from five pooled 3 1 preparations). Zinc analysis was performed by atomic absorption spectrometry (Table 2). From these results, we conclude that the mutants Cys44Ala, His69Gln, and Glu150Gln contain no appreciable amounts of zinc, while the variants Glu149Gln, Glu155Ala, Glu155Cys, and Glu155Gln still harbour about one zinc atom per subunit. We therefore now propose the third zinc ligand in SDH to be Glu150, resulting in the zinc coordinating triad Cys44, His69, Glu150. Replacement of any one of the ligands by non-zinc liganding counterparts abolish the capacity of SDH to coordinate the catalytic zinc atom (Table 2). The previously postulated zinc ligand Glu155 is shown to have a strong impact on enzyme catalysis (Eklund *et al.*, 1985; Karlsson & Höög, 1993) and the residue Glu149 is considered to be of structural importance.

Structural Zinc Atom in Alcohol Dehydrogenase

The structural zinc binding site in ADH is composed of four, closely located cysteine residues (Fig. 1). To investigate the importance of this structural zinc atom which is absent in SDH, all putative ligands in ADH were subjected to exchanges to non-zinc liganding

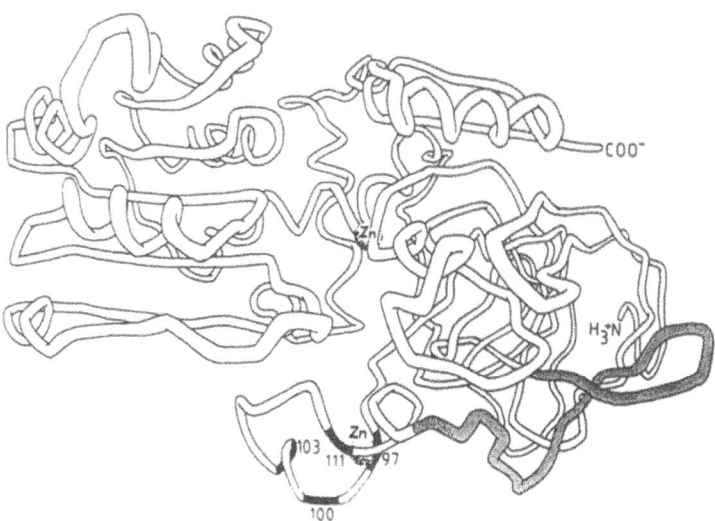

Figure 1. Mammalian alcohol dehydrogenase subunit, showing the main-chain positions of the four Cys ligands to the structural zinc atom labelled in black, and the position of the catalytic zinc atom. The loop structure, which is stippled, is missing in the sorbitol dehydrogenase structure.

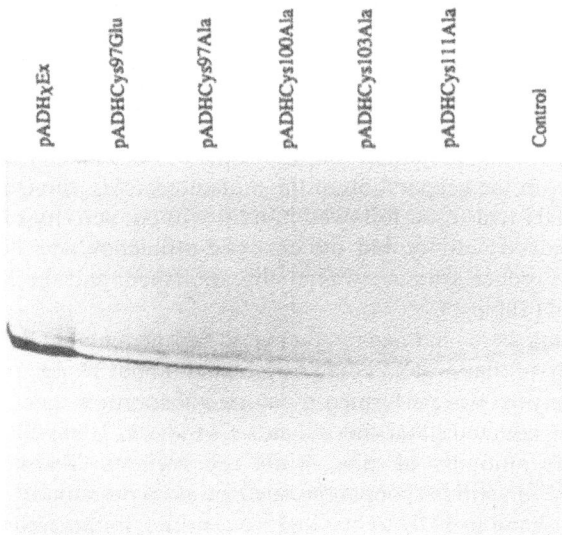

Figure 2. Western blot analysis of crude extracts from different class III ADHs. 50 ml cultures were pelleted, sonicated, and centrifuged. Aliquots of the supernatants were applied to an SDS polyacrylamide gel and electrophoresed. Proteins were blotted to a polyvinylidene triflouride membrane, which was submerged into an antibody solution (affinity purified antibodies). pADHexχ served as a positive control, and *E. coli* TG1 cells served as a negative control.

counterparts by site-directed mutagenesis. Recombinant, mutated ADHs were expressed from seven different plasmids, five derived from human class III ADH (Estonius *et al.*, 1994) and two from human class I ADH (Höög *et al.*, 1987). The four Cys-ligands (numbered 97, 100, 103, 111; Fig. 1) were mutated, in separate constructs, to Ala in class III, and all four in one construct to Ser in class I. Additionally, Cys97 was changed to Glu in class III, and Cys111 to Ser in class I. Ala and Ser are residues unable to ligand a zinc atom. Expression of all plasmids with the different mutants resulted in inactive ADHs. The amounts obtained were too small for isolation, but the native, recombinant class I and class III ADHs served as positive controls. In spite of the low recovery, all class III ADH mutants were possible to visualize upon Western blot analysis of crude extracts, using affinity purified antibodies (Fig. 2). To further verify the expression constructs, total RNA was purified from harvested cells containing all different plasmids. The total RNA was subjected to electrophoresis and subsequent Northern blot analysis. mRNA transcripts of correct size and abundance were obtained from each plasmid and provided evidence of successful constructs (Fig. 3). Degradation seems to occur during the production of recombinant proteins in the bacterial cell since the addition of different protease inhibitors to harvested cell extracts did not increase the amount of protein. This work proves, with site-directed mutagenesis, the direct importance of this structural zinc atom for the stability of dimeric ADHs. The results confirm that the area around the structural zinc atom is involved in the regulation of the quaternary structure. Such a role is compatible with the absence of this zinc in SDH, which has additional subunit interactions since it is a tetrameric enzyme.

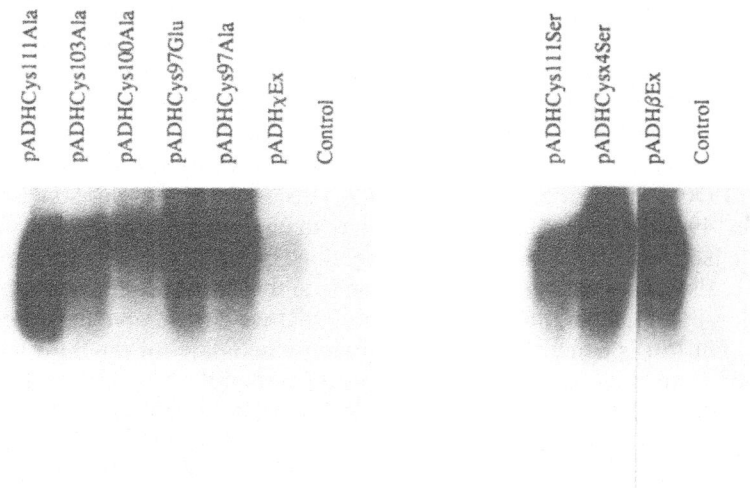

Figure 3. Autoradiograms from Northern blot analysis of recombinant ADH mRNA. Total RNA (20 µg/lane) was electrophoresed, transferred to filters and hybridized to human class I and class III specific probes. pADHexχ and pADHexβ are positive controls, giving recombinant class III and class I ADH, respectively. *E. coli* TG1 cells is a negative control.

DISCUSSION

The Zinc Binding Motif of SDH

The first primary structure of an SDH was determined by analysis of the enzyme from sheep liver (Jeffery *et al.*, 1984b). It revealed clear relationships to ADH from which functional implications could be made (Jörnvall *et al.*, 1984). In 1985, the sheep liver structure was modelled into the horse liver ADH crystal structure by computer graphics (Eklund *et al.* 1985). One important feature that resulted from the model was that the third zinc ligand agreed better with a glutamic acid residue (Glu155) than with a previously suggested cysteine residue. The use of EXAFS (extended X-ray absorption fine structure; Feiters & Jeffery, 1989) and Zn/Co replacement (Maret, 1989) also indicated that a Glu was compatible with being a zinc ligand in SDH. However, the position or definite nature of this third zinc ligand in SDH was not established, and initiated site-directed mutagenesis, made possible by the cloning of a rat cDNA (Karlsson *et al.*, 1991). The homology of SDH toward ADH is low in the area around the third zinc ligand, and we therefore studied a set of different putative ligands. Cys164 was an early candidate but has been ruled out before (Karlsson & Höög, 1993). Mutation of Cys44 to Ala, and of His69 to Gln was made in order to create a loss of zinc-binding capacity. Glu155 was mutated to Cys, to mimic the third ligand in ADH, to Ala in order to exclude zinc ligation, and to Gln since this exchange has been observed in *Bombyx mori* (Niimi *et al.*, 1993). With Ala and Gln at position 155, the catalytic activity was abolished, while the Cys-variant showed a significant decrease in activity. The exchange of Glu149 and Glu150 for Gln, in different constructs, resulted in complete loss of activity. The zinc analyses, performed by atomic absorption spectrometry, showed that three of the mutants had lost their zinc atoms, Cys44, His69, and Glu150. All other variants, including Glu149Gln, Glu155Ala, Glu155Cys, and Glu155Gln (Table 2), possessed one zinc atom per subunit. Thus, the zinc binding triad in SDH is concluded to be Cys44, His69, and Glu150.

This shift of the third zinc ligand, from Cys174 in ADH to Glu150 in SDH, results in a shorter spacer between the second and third ligands in SDH (79 residues) than in ADH (105-109 residues), which is consistent with the spacers in other proteins (Vallee and Auld, 1989). Further observations, apart from the zinc analysis *per se*, to strengthen Glu150 as the third zinc ligand, is the fact that it is followed by a Gly, which is also the situation in ADH. On the other hand, xylitol dehydrogenase (closely related to SDH) has a Leu at position 150 (Kötter *et al.*, 1990) but still a Glu at position 149. This weakens the Glu150-case and could imply that Glu149 is the proper ligand. Glu149 has an obvious structural role and very little protein can be isolated after mutation and expression. Notably, this pattern of recovery of little protein is also observed when the zinc ligands to the structural zinc atom of ADH are removed (below). Altogether, however, presuming that we should give most impact to the zinc analysis, Glu150 is concluded to be the obvious candidate for the third ligand to the active site zinc atom in SDH. It can also be concluded that the activity of SDH might be dependent on a carboxyl function at position 155, possibly by interaction with one of the many hydroxyl groups of the substrate. Cys, but not Ala or Gln, can to some extent replace the carboxylate at the active site. A negative charge may be needed for correct active-site relationships. With Gln at position 155 no activity can be detected in spite of correct size and abilities for hydrogen bonding. The *Bombyx mori* SDH indeed has Gln at this position, though with a very high K_m-value for sorbitol (150 mM; Niimi *et al.*, 1993). The structure must thus have been compensated elsewhere for enabling this activity. The Cys, with one methylene group less than Glu, causes a decrease in activity due to a non-optimal distance toward the substrate.

Role of the Structural Zinc Atom in Alcohol Dehydrogenase

The binding site of the structural zinc atom in ADH is composed of four Cys residues at positions 97, 100, 103, and 111, contained within a loop region that projects out from the catalytic domain (Eklund *et al.*, 1976; Fig. 1). Each of the ligands to the structural zinc atom of class III ADH were mutated to Ala in separate constructs and Cys97 additionally to Glu. Similarly in class I ADH, Cys111 was mutated to Ser and all four zinc ligands in a single construct to Ser. The results obtained are clear. All mutated proteins are expressed but labile and not possible to isolate in appreciable yield. Confirmation of correct transcription and translation processes was proven by positive Western and Northern blots (Fig. 1; 2). Thus, the conclusion is that each of the ligands is crucial in binding the structural zinc. The stability of the enzyme is maintained by keeping a surface that is needed for correct subunit interactions. This conclusion confirms early suggestions obtained by modelling and comparison with the SDH structure (Jörnvall *et al.* 1978; Eklund *et al.* 1985) and by sensitivities to proteases (Roumi *et al.* 1993). In the absence of even one of the ligands, the protein appears to be unstable and subject to rapid degradation. The 4-Cys zinc-binding motif is absent from mammalian SDHs but can be found in for instance *Bacillus subtilis* SDH (Ng *et al.*, 1992). Determination of the zinc content of this bacterial SDH reveals, however, only one atom per subunit (the catalytic atom, not bound at this site). This indicates that other functional changes have made the 4-Cys site unable to coordinate a structural zinc atom. Yeast ADH is, in analogy with SDH, a tetramer but contains two zinc atoms per subunit (Magonet *et al.*, 1992) and has a deletion (relative to ADH) adjacent to the Zn-loop (Jörnvall *et al.* 1978). One of the zinc atoms can be selectively removed by treatment with high concentrations of dithiothreitol, without loss of catalytic activity. Glucose dehydrogenase from *Thermoplasma acidophilum*, a tetramer - like SDH and yeast ADH, has two zinc atoms per subunit. Also in this enzyme, the loop containing the structural zinc atom has been suggested to be involved in maintenance of the quaternary structure (John *et al.*, 1994). The orientation of the structural zinc loop with respect to the subunit is markedly different from that of horse ADH.

This fact might be explained by differences in the fold of the two enzymes beyond this loop. Altogether, the presence of a structural zinc atom, as seen in dehydrogenases, seems to protect a surface loop that contributes to govern the quaternary structure of the enzymes, giving different subunit interactions, depending on its orientation, and presence or absence.

ACKNOWLEDGEMENTS

This work was supported by grants from the Swedish Medical Research Council, Pfizer Central Research, and Bengt Lundquist's Minne.

REFERENCES

Berg, J. M. (1987) Metal ions in proteins: Structural and functional roles. *Quant. Biol.* 52, 579-585.

Bradford, M. M. (1976) A rapid and sensitive method for the quantitation of microgram quantities of protein utilizing the principle of protein-dye binding. *Anal. Biochem.* 72, 248-254.

Chirgwin, J., Aeyable, A., McDonald, R. & Rutter, W. (1979) Isolation of biologically active ribonucleic acid from sources enriched in ribonuclease. *Biochemistry* 18, 5294-5299.

Drum, D. E., Li, T.-K. & Vallee, B. L. (1969) Considerations in evaluating the zinc content of horse liver alcohol dehydrogenase preparations. *Biochemistry* 8, 3783-3791.

Eklund, H., Nordström, B., Zeppezauer, M., Söderlund, G., Ohlsson, I., Boiwe, T., Söderberg, B.-O., Tapia, O. & Brändén, C.-I. (1976) Three-dimensional structure of horse liver alcohol dehydrogenase at 2.4 Å resolution. *J. Mol. Biol.* 102, 27.59.

Eklund, H., Horjales, E., Jörnvall, H., Brändén, C.-I. & Jeffery, J. (1985) Molecular aspects of functional differences between alcohol and sorbitol dehydrogenases. *Biochemistry* 24, 8005-8012.

Estonius, M., Höög, J.-O., Danielsson, O. & Jörnvall. H. (1994) Residues specific for class III alcohol dehydrogenase. Class transitions and S-hydroxymethylglutathione binding. Site-directed mutagenesis of the human enzyme. *Biochemistry*, in press.

Feiters, M. C. & Jeffery, J. (1989) Zinc environment in sheep liver sorbitol dehydrogenase. *Biochemistry* 28, 7257-7262.

Hedén, L.-O., Höög, J.-O., Larsson, K., Lake, M., Lagerholm, E., Holmgren, A., Vallee, B.L., Jörnvall, H. & von Bahr-Lindström, H. (1986) cDNA clones coding for the β-subunit of human liver alcohol dehydrogenase have differently sized 3'-non-coding regions. *FEBS Lett.* 194, 327-332.

Höög, J.-O., Weis, M., Zeppezauer, M., Jörnvall, H. & von Bahr-Lindström, H. (1987) Expression in *Escherichia coli* of active human alcohol dehydrogenase lacking N-terminal acetylation. *Biosci. Rep.* 7, 969-974.

Jeffery, J., Cummins, L., Carlquist, M. & Jörnvall, H. (1981) Properties of sorbitol dehydrogenase and characterization of a reactive cysteine residue reveal unexpected similarities to alcohol dehydrogenases. *Eur. J. Biochem.* 120, 229-234.

Jeffery, J., Chesters, S. J., Mills, C., Sadler, P. J. & Jörnvall, H. (1984a) Sorbitol dehydrogenase is a zinc enzyme. *EMBO J.* 3, 357-360.

Jeffery, J., Cederlund, E. & Jörnvall, H. (1984b) Sorbitol dehydrogenase. The primary structure of the sheep-liver enzyme. *Eur. J Biochem.* 140, 7-16.

Jeloková, J., Karlsson, C., Estonius, M., Jörnvall, H. & Höög, J.-O. (1994) Features of structural zinc in mammalian alcohol dehydrogenase. *Eur. J. Biochem.*, in press.

John, J., Crennell, S. J., Hough, D. W., Danson, M. J. & Taylor, G. L. (1994) The crystal structure of glucose dehydrogenase from *Thermoplasma acidophilum. Structure* 2, 385-393.

Jörnvall, H., Eklund, H. & Brändén, C.-I. (1978) Subunit conformation of yeast alcohol dehydrogenase. *J. Biol. Chem.* 253, 8414-8419.

Jörnvall, H., von Bahr-Lindström, H. & Jeffery, J. (1984) Extensive variations and basic features in the alcohol dehydrogenase - sorbitol dehydrogenase family. *Eur. J. Biochem.* 140, 17-23.

Karlsson, C., Jörnvall, H. & Höög, J.-O. (1991) Sorbitol dehydrogenase: cDNA coding for the rat enzyme. Variations within the alcohol dehydrogenase family independent of quaternary structure and metal content. *Eur. J. Biochem.* 198, 761-765.

Karlsson, C. & Höög, J.-O. (1993) Zinc coordination in mammalian sorbitol dehydrogenase. Replacement of putative zinc ligands by site-directed mutagenesis. *Eur. J. Biochem.* 216, 103-107.

Karlsson, C. (1994) Dissertation, Karolinska Institutet, Stockholm, Sweden.

Kötter, P., Amore, R., Hollenberg, C. P. & Ciriacy, M. (1990) Isolation and characterization of the *Pichia stipitis* xylitol dehydrogenase gene, XYL2, and construction of a xylose-utilizing *Saccharomyces cerevisiae* transformant. *Curr. Genet.* 18, 493-500.

Leissing, N. & McGuinness E. T. (1978) Rapid affinity purification and properties of rat liver sorbitol dehydrogenase. *Biochim. Biophys. Acta* 524, 254-261.

Lutz, R. A., Bull, C. & Rodbard, D. (1986) Computer analysis of enzyme-substrate-inhibitor kinetic data with automatic model selection using IBM-PC compatible microcomputers. *Enzyme* 36, 197-206.

Magonet, E., Hayen, P., Delforge, D., Delaive, E. & Remacle, J. (1992) Importance of the structural zinc atom for the stability of yeast alcohol dehydrogenase. *Biochem. J.* 287, 361-365.

Maret, W. (1989) Cobalt(II)-substituted class III alcohol dehydrogenase and sorbitol dehydrogenase from human liver. *Biochemistry* 28, 9944-9949.

Ng, K., Ye, R., Wu, X.-C. & Wong, S.-L. (1992) Sorbitol dehydrogenase from *Bacillus subtilis*. Purification, characterization, and gene cloning. *J. Biol. Chem.* 267, 24989-24994.

Niimi, T., Yamashita, O. & Yaginuma, T. (1993) A cold-inducible *Bombyx* gene encoding a protein similar to mammalian sorbitol dehydrogenase. Yolk nuclei-dependent gene expression in diapause eggs. *Eur. J. Biochem.* 213, 1125-1131.

Persson, B., Zigler J. S. Jr. & Jörnvall H. (1994) A super-family of medium-chain dehydrogenases/reductases (MDR): sub-lines including ζ-crystallin, alcohol and polyol dehydrogenases, quinone oxidoreductases, enoyl reductases, VAT-1 and further proteins. *Eur. J Biochem.*, submitted.

Roumi, P. Loomes, K. & Jörnvall, H. (1993) Comparative proteolysis of sorbitol and alcohol dehydrogenases. *Eur. J. Biochem.* 213, 487-492.

Sanger, F., Nicklen, S. & Coulson, A. R. (1977) DNA sequencing with chain-terminating inhibitors. *Proc. Natl. Acad. Sci.* 74, 5463-5467.

Sharma, C. P., Fox, E. A., Holmquist, B., Jörnvall, H. & Vallee, B. (1989) cDNA sequence of human class III alcohol dehydrogenase. *Biochem. Biophys. Res. Commun.* 164, 631-637.

Taylor, J. W., Ott, J. & Eckstein, F. (1985) The rapid generation of oligonucleotide-directed mutations at high frequency using phosphorothioate-modified DNA. *Nucleic Acids Res.* 13, 8765-8785.

Vallee, B. L. & Auld, D.S. (1989) Short and long spacer sequences and other structural features of zinc binding sites in zinc enzymes. *FEBS Lett.* 257, 138-140.

Vallee, B. L. & Auld, D.S. (1990) Zinc coordination, function, and structure of zinc enzymes and other proteins. *Biochemistry* 29, 5647-5659.

Åkeson, Å. (1964) On the zinc content of horse liver alcohol dehydrogenase. *Biocem. Biophys. Res. Commun.* 17, 211-214.

48

HORSE LIVER ALCOHOL DEHYDROGENASE-CATALYZED ALDEHYDE OXIDATION

The Sequential Oxidation Of Alcohols To Carboxylic Acids Under NADH Recycling Conditions

Norman J. Oppenheimer and Gary T. M. Henehan

Department of Pharmaceutical Chemistry S-926
University of California, San Francisco
San Francisco, CA 94143-0446

Horse liver alcohol dehydrogenase (HL-ADH) is a broad specificity enzyme that catalyzes the reversible oxidation of a wide range of both primary and secondary alcohols to their corresponding aldehydes and ketones (Figure 1). The reaction mechanism and kinetics of the enzyme have been extensively characterized and its 3-dimensional structure is known (for a review see Pettersson, 1987).

There have been sporadic reports in the literature concerning a second catalytic activity of HL-ADH, namely the oxidation of aldehydes to their corresponding carboxylic acids (Abeles & Lee, 1960; Dalziel & Dickinson, 1965; Hinson & Neal, 1975; Anderson & Dahlquist, 1982; Henehan & Oppenheimer, 1993; Shearer, et al., 1993). One manifestation of this activity is the aldehyde dismutation reaction (Figure 2).

Dismutation involves the binding of aldehyde to two different enzyme forms in the catalytic cycle. But unlike other dehydrogenase-catalyzed redox reactions, dismutation cannot be monitored by changes in A_{340} because there is no net change in NADH during a complete transit of the catalytic cycle. Kinetic constants for aldehyde dismutation have been determined using a titrimetric assay to measure acid production (Dalziel & Dickinson, 1965), or using an HPLC assay to measure the acid and alcohol produced (Shearer, et al., 1993). The ability of HL-ADH to oxidize aldehydes was originally ascribed to contamination of the enzyme preparations with an aldehyde dehydrogenase, however, Abeles & Lee (1960) demonstrated this activity to be an intrinsic property of HL-ADH.

The ability of an alcohol dehydrogenase to oxidize aldehydes appears remarkable given the high specificity and selectivity normally associated with enzyme catalysis. This ability can be rationalized because most aliphatic aldehydes exist in aqueous solution in equilibrium with their hydrated form (Bell & Evans, 1966) where the latter is a structural analogue of a secondary alcohol (Figure 3) and the presumed substrate for the reaction.

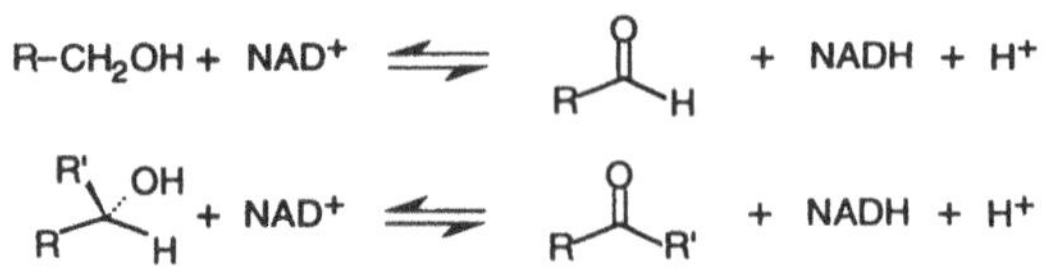

$$R-CH_2OH + NAD^+ \rightleftharpoons \underset{R}{\overset{O}{\|}}\!\!H + NADH + H^+$$

$$\underset{R}{\overset{R'\ OH}{\diagup}}\!\!H + NAD^+ \rightleftharpoons \underset{R}{\overset{O}{\|}}\!\!R' + NADH + H^+$$

Figure 1. The "normal" alcohol dehydrogenase reactions; the reversible oxidation of primary and secondary alcohols to their respective aldehydes and ketones.

Aldehyde oxidation or dismutation by alcohol dehydrogenases has received little attention and kinetic values for dismutation have been reported for only a narrow range of substrates. In retrospect, this can be understood given the properties of the reaction. Incubating an alcohol dehydrogenase with saturating concentrations aldehyde and NAD^+ (the obvious assay) does not give rise to changes in A_{340}, but instead results in the redox silent dismutation of the aldehyde. This makes assays for aldehyde dismutation much more difficult, than the simple spectrophotometric assay for alcohol oxidation. Moreover, kinetic values for aldehyde oxidation or dismutation, where they have been measured, appear much poorer than those for the oxidation of the corresponding alcohol; e.g., the K_m for acetaldehyde was reported to be 100 mM whereas that for ethanol is 0.18 mM (Dalziel & Dickinson, 1965). As a consequence, this chemistry has been regarded as an insignificant side reaction of only a single alcohol dehydrogenase, HL-ADH.

If HL-ADH can oxidize aldehydes, then why has the sequential oxidation of alcohols to their carboxylic acids (Figure 4) not been widely reported? The reason is that normal assays for ADH activity, where changes in A_{340} are followed, mask such sequential oxidations. Under initial rate conditions (<<10% substrate depletion), aldehyde concentrations will not accumulate to support significant aldehyde oxidation. Even if aldehydes are oxidized, the NADH produced cannot readily be distinguished from that produced by oxidation of the alcohol. Finally, most alcohol dehydrogenase assays employ trapping of the aldehyde product; at neutral pH by forming semicarbazones, or at higher pH by forming carbinolamines with amine-containing buffers, such as Tris or Glycine.

We have employed an NADH recycling system in the assay mixture using rabbit muscle lactate dehydrogenase and pyruvate to keep NADH levels well below 1.0 µM. This allows the irreversible conversion of alcohol to aldehyde without the use of trapping reagents. We note that these conditions are similar to those found *in vivo* where NADH is maintained at a relatively constant level, reportedly as low as 0.5 µM (see Hardman, *et al.*, 1991). With an NADH recycling system, there will be no change in A_{340}, however, the

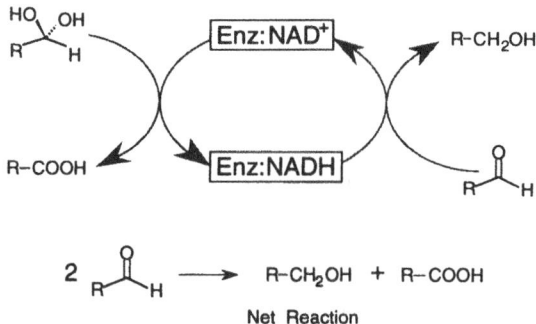

Figure 2. Proposed mechanism of the aldehyde dismutation reaction.

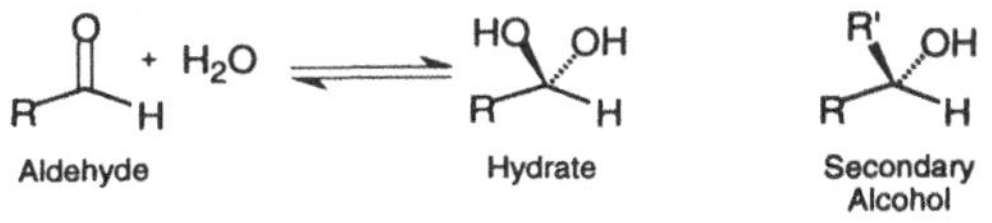

Figure 3. Structural similarity between hydrated form of aldehydes and secondary alcohols.

reaction progress can be monitored using high field ^{1}H NMR spectra to follow the time-dependent changes in concentrations of all the reaction components (Henehan & Oppenheimer, 1993). Moreover, this assay system allows a direct competition between alcohol and aldehyde for Enz•NAD$^+$, thus valid comparisons of their relative catalytic efficiencies can be made.

In this paper we present a quantitative analysis of the ability of HL-ADH to conduct sequential oxidation of the aliphatic alcohols, ethanol and octanol, to their corresponding acids under conditions where NADH is maintained at a low level using an NADH recycling system. The reaction progress curves establish that HL-ADH can oxidize alcohols and aldehydes with at least comparable efficiency.

MATERIALS AND METHODS

HL-ADH, rabbit muscle lactate dehydrogenase, sodium pyruvate and NAD$^+$ were obtained from the Sigma Chemical Company. Octanol and ethanol were obtained from Aldrich Chemical Company. HL-ADH was dialyzed for 14 hours against three changes of 0.1 M sodium phosphate buffer pH 7.5 before use. NAD$^+$ was standardized by absorbance measurement at 260 nm using an extinction coefficient of 1.8×10^4 l mol^{-1} cm^{-1} in distilled water. HL-ADH concentration was standardized using a spectrophotometric assay conducted at 37 °C in 0.1 M sodium phosphate buffer, pH 7.5 and contained 2 mM NAD$^+$ and 5 mM ethanol.

Sequential oxidation was monitored by a proton nuclear magnetic resonance assay method. Spectra were acquired at 500 MHz on a General Electric GN-500 instrument using a probe temperature of 37°C and 5 mm NMR tubes. Each time point consisted of 64 scans acquired with a spectral width of ±3000 Hz, using 16K data points. The pulse width was set to correspond to a 45° tip angle. The reactions were carried out in aqueous buffers with 10% D$_2$O to provide a lock signal. Suppression of the water resonance was achieved by presaturation with a decoupler pulse.

Assays were performed in a final volume of 0.5 mL and contained 0.1M sodium phosphate buffer pH 7.5, 2.0 mM NAD$^+$, 10% D$_2$O, alcohol at the stated concentration and an NADH recycling system consisting of 100 mM sodium pyruvate and 300 μg rabbit muscle lactate dehydrogenase (255 units). The assay mix (480 μL), without HL-ADH, was incubated for at least 10 min to achieve temperature equilibration and a control spectrum was taken. The reaction was initiated by addition of HL-ADH to give a final volume of 500 μL. Spectra were taken every 8.05 min. The resonances monitored to determine changes in concentration

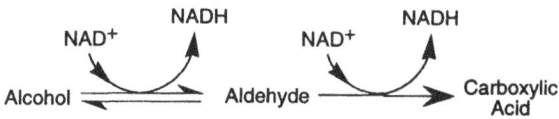

Figure 4. Scheme for the alcohol dehydrogenase-catalyzed sequential oxidation of alcohol to acid.

of assay components were as follows: ethanol was monitored by following the disappearance of its methyl group, a 3H triplet at 1.17 ppm.; acetaldehyde was monitored by following the appearance of the aldehydic proton, a 1H quartet at 9.67 ppm; acetate was monitored by following the appearance of its methyl group at 1.90 ppm.; octanol was monitored by following the disappearance of its methylene group, a 2H triplet at 3.65 ppm; octanal was monitored by following the appearance of the aldehydic proton, a 1H quartet at 9.71 ppm; octanoate was monitored by following the appearance of its α-CH_2 group at 2.15 ppm. Lactate formation was monitored by following the appearance of the CH quartet at 4.05 ppm and the CH_3 doublet at 1.27 ppm. The integrated areas of these resonances were converted to concentrations by comparison with the integrated areas of the nicotinamide ring protons of NAD^+ whose concentration was determined spectrophotometrically. The reported aldehyde concentrations represent the sum of free and hydrated species. In parallel reactions conducted in a uv spectrophotometer, no changes in A_{340} were observed during the course of the reaction.

RESULTS

A representative series of ^{1}H NMR spectra (Figure 5) are presented for an ethanol oxidation assay mixture with an NADH recycling system, showing the time-dependent changes in the resonances of the reaction components. The resulting changes in concentration, determined by integration of these resonances, are shown in Figure 6. In the initial stages of the reaction, acetaldehyde accumulates to a steady-state ratio of aldehyde to alcohol of 1:1.9. Production of acetate shows an initial lag before reaching a steady-state rate that is 17% of the maximal initial rate of alcohol oxidation. The NADH recycling system converts NADH to NAD^+ and in the process generates one mole of lactate per mole of NADH reoxidized. No acetate is produced in the absence of the NADH recycling system, instead there is a rapid rise in NADH to its anticipated equilibrium value of 138 µM which shuts down the reaction and leaves ethanol in vast excess over the aldehyde. Likewise, no aldehyde or acetate production is observed in the absence of HL-ADH; i.e., LDH does not catalyze observable oxidation of aldehydes or primary alcohols.

A plot of the time dependent changes of all the assay components for the sequential oxidation of octanol to octanoate (Figure 7) demonstrates that this reaction is fundamentally different than that with ethanol as a substrate. During the course of the reaction, the

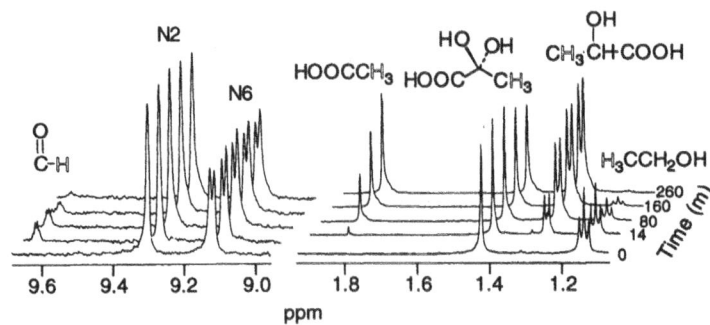

Figure 5. A representative set of ^{1}H NMR spectra shows the clearly resolved resonances of the assay components acquired during the sequential oxidation of ethanol to acetate. Note, the vertical scale of the downfield spectral region is six times that of the upfield region. The integrated area of the methyl group of lactate is always twice that of acetate confirming the stoichiometry of the reaction. The resonances of the hydrated form of acetaldehyde, which represents ca. 50% of the total, is not shown.

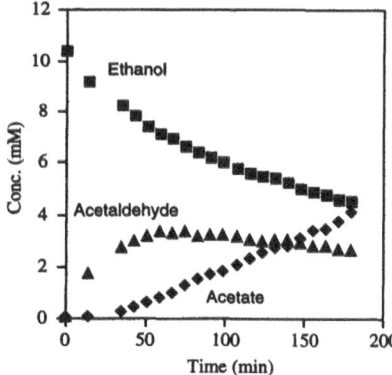

Figure 6. Plot of the time-dependent changes in concentration of reaction components for the sequential oxidation of ethanol to acetate in the presence of an NADH recycling system. The changing concentrations of ethanol (■), acetaldehyde (▲) and acetate (◆) were followed as described under materials and methods. For clarity the curve for lactate is not shown. Initially the assay contained: 10.4 mM ethanol, 2.0 mM NAD⁺, 100 mM sodium pyruvate and 255 units of rabbit muscle lactate dehydrogenase. The reaction was initiated by addition of 0.14 units of HL-ADH.

concentration of octanol declines with a concomitant increase in octanoate. At no time can free aldehyde be detected in solution. The stoichiometry of the reaction was the same as for the sequential oxidation of ethanol to acetate with two moles of lactate produced per mole of octanol oxidized to octanoate.

DISCUSSION

The progress curves shown in Figures 6 and 7 provide unequivocal evidence that HL-ADH is fully capable of catalyzing the oxidation of both aliphatic alcohols and aldehydes with at least comparable catalytic efficiencies. These experimental findings are in sharp contrast with the perception from the literature that aldehyde oxidation/dismutation is an

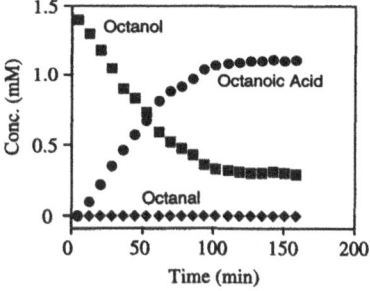

Figure 7. Plot of the time-dependent changes in concentration of reaction components for the sequential oxidation of octanol to octanoate in the presence of an NADH recycling system. The changing concentrations of octanol (■), octanal (▲) and octanoate (●) were followed as described under materials and methods. Initially the assay contained: 1.4 mM octanol, 2.0 mM NAD⁺, 100 mM sodium pyruvate and 255 units of rabbit muscle lactate dehydrogenase. The reaction was initiated by addition of 0.14 units of HL-ADH.

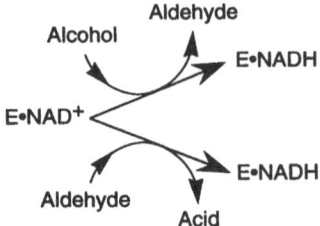

Figure 8. Scheme illustrating the direct competition of alcohol and aldehyde for E•NAD$^+$ in a sequential oxidation.

insignificant and irrelevant side reaction. The origins of this discrepancy will be discussed below.

Ethanol oxidation: For HL-ADH-catalyzed oxidation of ethanol, the NADH recycling system allows acetaldehyde to accumulate to a level where it directly competes with ethanol for the E•NAD$^+$ binary complex (see Figure 8). At steady-state (see Figure 6), ethanol is still present in a two-fold molar excess over acetaldehyde and both are being oxidized at essentially the same rate.

The maximum velocity of alcohol oxidation is known to be limited by the slow rate of NADH dissociation (Pettersson, 1987) whereas aldehyde dismutation bypasses coenzyme association and dissociation. Nevertheless, it has been argued (Shearer, et al., 1993) that aldehyde dismutation is inefficient because despite higher V_{max} values, the poor K_m makes the k_{cat}/K_m much lower than for the oxidation of the corresponding alcohol; e.g., k_{cat}/K_m for acetaldehyde dismutation is 119-fold lower than k_{cat}/K_m for ethanol oxidation (Dalziel & Dickinson, 1965). This direct comparison, however, is inappropriate.

$$\frac{V_{alc} \times [ald]}{V_{ald} \times [alc]} = \frac{\left(k_{cat}/K_m\right)^{alc}}{\left(k_{cat}/K_m\right)^{ald}}$$

The value of k_{cat}/K_m is normally interpreted as determining the relative velocity for competing substrates according to Equation 1 (Fersht, 1984), where V_{alc} and V_{ald} are the rates of alcohol and aldehyde oxidation in a mixture of both substrates. Based on the data shown in Figure 2, when $V_{alc}/V_{ald} = 1$ (at steady-state, $V_{alc} \cong V_{ald}$), and [ald]/[alc] = 0.53, then the ratio of their respective values of k_{cat}/K_m will equal 0.53; i.e., acetaldehyde oxidation is actually more efficient than ethanol oxidation! This analysis is used to estimate the efficiency of acetaldehyde oxidation, with the knowledge that Eqn. 1 is based on initial rate arguments. Nonetheless, the fact that acetaldehyde concentrations remain well below those of the alcohol provides conclusive evidence that the acetaldehyde is being oxidized efficiently.

Why is there such a discrepancy between our findings and the literature interpretation of k_{cat}/K_m for alcohol and aldehyde oxidation? A comparison of k_{cat}/K_m values requires kinetically comparable reactions where each substrate acts as a competitive inhibitor of the other (Fersht, 1984). This condition does not hold for a comparison of aldehyde dismutation with alcohol oxidation. Aldehydes are not strictly competitive with alcohols because aldehydes bind productively at two points in the catalytic cycle (see Figure 2) whereas ethanol binds only to E•NAD$^+$. Moreover, V_{max} and K_m for the two reactions are composed of entirely different sets of fundamental rate constants; alcohol oxidation includes terms for coenzyme binding and dissociation whereas dismutation does not. Therefore k_{cat}/K_m cannot be used to compare the efficiencies of alcohol oxidation with aldehyde dismutation. The assay conditions with NADH recycling provides a system whereby the relative efficacy of these two substrates can be determined based on their direct competition for the same E•NAD$^+$ complex. From the sequential oxidation progress curve shown in Figure 6, we clearly see that the relative efficiencies of ethanol and acetaldehyde oxidation must be at least comparable.

The fact that the kinetic values for acetaldehyde oxidation are more favorable than previously recognized does not necessarily mean that the liver ADH plays a major role in the oxidation of ethanol-derived acetaldehyde, just that it can no longer be dismissed out of hand. For example, it could participate in situations where high concentrations of acetaldehyde are achieved; e.g., in individuals with impaired acetaldehyde metabolism.

Oxidation of octanol: The progress curves for octanol oxidation demonstrate a fascinating aspect of the sequential oxidation reaction. For the direct conversion of octanol to octanoic acid we observe no detectable accumulation in solution of octanal as an intermediate. We propose the following model to account for these results (see Figure 9). To initiate the reaction octanol binds to E•NAD$^+$ and following hydride transfer produces E•NADH•Octanal. NADH dissociates from this complex rather than octanal, giving rise to a tight E•Octanal complex. At some point, the aldehyde becomes hydrated on the enzyme active site and NAD$^+$ binds to give E•NAD$^+$•Octanal that yields E•NADH•Octanoate.

We therefore find that the oxidation of alcohols to carboxylic acids by HL-ADH can follow two kinetically distinct pathways. In the first pathway (see Figure 8), the two-step oxidation of an alcohol to a carboxylic acid occurs with release of the aldehyde intermediate into solution where it must compete with alcohol for E•NAD$^+$; e.g., ethanol oxidation. We will use the term sequential oxidation to denote this process. The second pathway (see Figure 9) involves the same overall reaction, but where the aldehyde intermediate remains bound to the enzyme; e.g., octanol oxidation. In this case, the coenzyme exchanges and there is no direct competition between the aldehyde and the alcohol. We will use the term 'didehydrogenation' to denote this process.

The ability of HL-ADH to catalyze didehydrogenation of hydrophobic aldehydes was previously noted by Hinson and Neal (1975) who demonstrated that even in the absence of NADH recycling, octanoic acid was produced as an initial product of octanol oxidation and that semicarbazide was not efficient in trapping the aldehyde intermediate. These results point out the pitfalls involved with assessing the alcohol oxidation activity of an ADH when

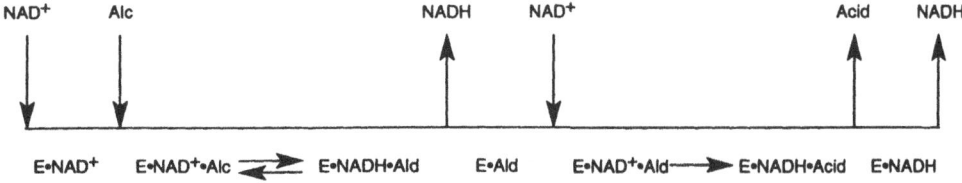

Figure 9. Kinetic scheme for the didehydrogenation of alcohol to acid where coenzyme affinity is less than that for aldehyde and aldehyde does not dissociate from the E•ald complex before NAD$^+$ binds. The step for aldehyde hydration is omitted from this scheme and could occur either before or after NADH release.

only changes in A_{340} are measured and no product analyses are conducted. There is no simple way to distinguish NADH produced from the oxidation of alcohol to aldehyde, from that produced by the two-step didehydrogenation of an alcohol to a carboxylic acid, yet the kinetic analysis will differ in detail for the two cases.

Mechanistic implications of didehydrogenation: The ability to oxidize an alcohol to a carboxylic acid at the same active site without release of the intermediate aldehyde is well established for several enzymes including; UDP glucose dehydrogenase, histidinol dehydrogenase, and HMG CoA reductase. For these enzymes, the aldehyde intermediate is thought to remain sequestered by forming a thiohemiacetal with an active site cysteine in analogy to the mechanism proposed for aldehyde dehydrogenases. This idea has been challenged by recent mutagenesis studies with histidinol dehydrogenase whereby changing "essential" cysteine residues to alanine does not abolish its ability to conduct didehydrogenation (Teng, *et al.*, 1993). The formation of covalently bound intermediates need not be invoked to explain didehydrogenation; all that is needed is for the aldehyde intermediate to bind more tightly than NADH, become hydrated, then it can be oxidized to a carboxylate.

Physiological advantages of didehydrogenation: We have found for HL-ADH that along with octanol, histidinol is oxidized to histidine without release of detectable concentrations of the intermediate histidinal (data not shown). Therefore, the primary advantage of didehydrogenation is that it allows conversion of an alcohol to its acid while sequestering the aldehyde intermediate. This tight binding obviates the need to establish finite concentrations of the aldehyde in the cytoplasm where it could covalently modify cellular constituents while awaiting further oxidation by aldehyde dehydrogenases. Moreover, if the aldehyde is biologically active, then didehydrogenation prevents its accumulation and thus unwanted responses. Note that Pocker & Li (1993) have shown that HL-ADH converts retinal to retinoic acid, and in separate experiments, retinol to retinal. It is therefore highly likely that HL-ADH is capable of the sequential oxidation, if not the didehydrogenation, of retinol to retinoate. We are currently investigating this intriguing possibility.

How general is the ability of alcohol dehydrogenases to oxidize aldehydes?: Initially, the oxidation of aldehydes was thought to be a unique property of HL-ADH. Yeast-ADH, which is highly homologous and mechanistically identical to HL-ADH, reportedly does not exhibit this activity (Dickinson & Monger, 1973) We find that yeast alcohol dehydrogenase does, in fact, oxidize aldehydes but at a rate ca. 0.1% that of HL-ADH (unpublished data). This observation demonstrates that even quite closely related enzymes can differ substantially in their ability to oxidize aldehydes. On the other hand, the alcohol dehydrogenase from *Drosophila* (D-ADH) is reported to catalyze aldehyde oxidation (Moxon, *et al.*, 1985), and an analogous reaction has been demonstrated for lactate dehydrogenase with the oxidation of glyoxylate to oxalate (Duncan, 1980). The finding that these structurally and mechanistically distinct enzymes can catalyze the facile oxidation of alcohols and aldehydes suggests that aldehyde oxidation is an intrinsic property of the dehydrogenation reaction, irrespective of the chemical mechanism by which oxidation is accomplished. This property will have far ranging consequences regarding the mechanism, kinetics, biochemistry and physiological function of the enzymes we call "alcohol" dehydrogenases.

REFERENCES

Abeles, R. H. & Lee, H. A. (1960). The Dismutation of Formaldehyde by Liver Alcohol Dehydrogenase. *J. Biol. Chem. 235,* 1499-1503.

Anderson, D. C. & Dahlquist, F. W. (1982). [19]F Nuclear Magnetic Resonance Observations of Aldehyde Dismutation Catalysed by Horse Liver Alcohol Dehydrogenase. *Arch. Biochem. Biophys. 217,* 226-235.

Bell, R. P. & Evans, P. G. (1966). Kinetics of the Dehydration of Methylene Glycol in Aqueous Solution. *291*, 297-323.

Dalziel, K. & Dickinson, F. M. (1965). Aldehyde Mutase. *Nature 206*, 255-257.

Dickinson, F. M. & Monger, G. P. (1973). A Study of the Kinetics and Mechanism of Yeast Alcohol Dehydrogenasewith a Variety of Substrates. *Biochem. J. 131*, 261-270.

Duncan, R. J. S. (1980). The Disproportionation of Glyoxylate by Lactate Dehydrogenase. *Arch. Biochem. Biophys. 201*, 128-136.

Fersht, A. (1984). *Enzyme Structure and Mechanism*. Freeman, New York. 400-404.

Hardman, M. J., Page, R. A., .Wiseman, M. S. & Crow, K. E. (1991). Regulation of Rates of Ethanol Metabolism and Liver [NAD⁺]/[NADH] Ratio. *Alcoholism: A Molecular Perspective*. T. N. Palmer, Ed. Plenum Press, New York. 27-33.

Henehan, G. T. M. & Oppenheimer, N. J. (1993). Horse Liver Alcohol Dehydrogenase-Catalyzed... *Biochem 32*, 735-738.

Hinson, J. A. & Neal, R. A. (1975). An Examination of Octanol and Octanal Methabolism to Octanoic Acid by Horse Liver Alcohol Dehydrogenase. *Biochimica et Biophysica Acta 384*, 1-11.

Moxon, L. N., Holmes, R. S., Parsons, P. A., Irving, M. G. & Doddrell., D. M. (1985). Purification and Molecular Properties of Alcohol Dehydrogenase from *Drosophila Melanogaster* : Evidence from NMR and Kinetic Studies for Function as an Aldehyde Dehydrogenase. *Comp. Biochem. Physiol. 80B, No. 3*, 525-535.

Pettersson, G. (1987). Liver Alcohol Dehydrogenase. *CRC Crit. Rev. Biochem. 21*, 349-389.

Pocker, Y. & Li, H. (1993). The Catalytic Specificity of Liver Alcohol Dehydrogenase: Vitamin A Alcohol and Vitamin A Aldehyde Activities. *Enzymology and Molecular Biology of Carbonyl Metabolism 4*. H. Weiner, D. W. Crabb and T. G. Flynn, Ed. Plenum Press, New York. 411-418.

Shearer, G. L., Kim, K., Le, K. M., Wang, C. K. & Plapp, B. V. (1993). Alternative Pathways and Reactions of Benzyl Alcohol and Benzaldehyde with Horse Liver Alcohol dehydrogenase. *Biochem 32*, 1186-11194.

Teng, H., Segura, E. & Grubmeyer, C. (1993). Conserved Cysteine Residues of Histidinol Dehydrogenase Are Not Involved in Catalysis. Novel Chemistry Required for Enzymatic Aldehyde Oxidation. *J. Biol. Chem. 268*, 14182-8.

49

ANALYTICAL APPROACHES TO ALCOHOL DEHYDROGENASE STRUCTURES

Madalina T. Gheorghe, Ingemar Lindh, William J. Griffiths, Jan Sjövall and Tomas Bergman

Department of Medical Biochemistry and Biophysics
Karolinska Institutet
S-171 77 Stockholm, Sweden

INTRODUCTION

Sequence analysis of polypeptides by Edman degradation is dependent on a free N-terminal α-amino group for the reaction with phenylisothiocyanate. However, alcohol dehydrogenases, like many other proteins, contain an acetylated N-terminal residue which blocks degradation (cf. Tsunasawa and Hirano, 1993). The conventional approach for blocked proteins involves enzymatic or chemical cleavage and reverse-phase HPLC-separation of fragments, followed by internal sequence analysis. The drawbacks associated with this technique are high protein consumption, long handling times and the fact that the N-terminal fragment remains inaccessible to Edman degradation. In this paper, we have tested direct chemical deblocking and applied it to both a synthetic peptide corresponding to the N-terminal segment of horse liver alcohol dehydrogenase and to the intact protein.

An alternative technique for structural analysis of peptides is mass spectrometry which is both sensitive and independent of whether the N-terminus is blocked or not. With the introduction of fast-atom bombardment (FAB) ionization (Barber et al., 1981) and the recent developments in electrospray (ES) ionization (Fenn et al., 1989), the possibilities of sequence analysis by mass spectrometry have been greatly increased (cf. Carr et al., 1991). Sequence information is usually obtained from collision-induced dissociation (CID) spectra. However, both positive- and negative-ion CID spectra are complex due to extensive fragmentation. By localising the charge at a specific site on the peptide via derivatization, the fragmentation becomes more predictable. Since fragmentation takes place remote to the charge, this process is known as charge-remote fragmentation. The resultant mass spectra are much easier to interpret, which facilitates determination of the amino acid sequence. In the present work, derivatization with 4-aminonaphthalenesulphonic acid (Lindh et al., 1994) has been applied to a small synthetic model peptide and to a synthetic segment of alcohol dehydrogenase for mass spectrometric structure analysis.

In addition to N-terminal acetylation, mammalian alcohol dehydrogenases are characterized by their zinc-content. These medium-sized enzymes are dimeric, and each subunit (molecular mass about 40 kDa) contains two zinc atoms. One is part of the catalytic site, whereas the other appears to have a structural role (Åkeson, 1964; Drum et al., 1967; Drum and Vallee, 1970). The second, or structural zinc, is bound to a separate loop of the polypeptide chain that contributes to subunit interactions (Eklund et al., 1976), but the precise function of this non-catalytic zinc is not known. It is tetrahedrally coordinated by four cysteine residues that are located fairly close to each other in the polypeptide backbone (positions 97, 100, 103 and 111 in horse liver alcohol dehydrogenase, cf. Eklund et al., 1976). This zinc site constitutes a segment of the protein which is known as a highly variable region of medium-chain (Persson et al., 1991) alcohol dehydrogenases (Jörnvall, 1985; Jörnvall et al., 1987a), where only the zinc-liganding cysteines and their spacing is conserved according to the pattern Cys-(Xaa)$_2$-Cys-(Xaa)$_2$-Cys-(Xaa)$_7$-Cys (Jörnvall et al., 1987b; Vallee and Auld, 1990; Vallee and Auld, 1991). We have shown that a 23-residue synthetic peptide corresponding to the exact equivalent of the segment in horse liver alcohol dehydrogenase encompassing the loop with the liganding cysteine residues, binds zinc in a manner that closely resembles the conditions in the native enzyme (Bergman et al., 1992; Bergman et al., 1993). This model system is now expanded to analogues with amino acid exchanges in liganding positions to determine the importance of individual cysteine residues for metal interaction and polypeptide conformation.

To summarize, in this report we present data on three analytical approaches to structural characterization of alcohol dehydrogenases: chemical deblocking of the N-terminal acetyl group, mass spectrometric structure determination and synthetic peptides as models for the binding and coordination of the structural zinc atom.

EXPERIMENTAL PROCEDURES

Synthetic peptides employed in the respective studies were prepared with an Applied Biosystems 430A instrument using side-chain-protected tertiary butyloxycarbonyl amino acid derivatives (cf. Kent, 1988). Cleavage from the resin and deprotection was accomplished by treatment with hydrogen fluoride for 1 h at 0°C in the presence of scavengers (HF/dimethylsulphide/anisole, 10/1/1 ml, complemented with p-thiocresol, 0.2 g, for peptides containing cysteine). The cleavage-products were washed with diethylether, followed by extraction of peptides with 30% acetic acid and lyophilization. The crude peptide preparations were purified by reverse-phase HPLC (Vydac C$_{18}$, 250 x 22 mm, 10 ml/min) using a linear gradient of acetonitrile (0 - 60%, 45 min) in aqueous 0.1% trifluoroacetic acid. N-terminal acetylation was performed before cleavage and deprotection, by treatment with a mixture of acetic anhydride/triethylamine/dichloromethane (9:4:87, by vol.) for 10 min at room temperature. Horse liver alcohol dehydrogenase was purchased from Sigma.

Deacetylation of both peptide and intact protein was carried out with a mixture of trifluoroacetic acid and methanol (1:1) for three days at 43°C. The samples were carefully dried in small (500 µl) plastic tubes with caps. The acid/alcohol solution (100 µl) was added, and after a short vortex the tubes were closed and incubated as described above. After this treatment, the reagents were removed under vacuum and the products were analyzed using capillary electrophoresis and sequence analysis. The capillary electrophoresis employed a Beckman P/ACE 2000 system operated as described (Bergman et al., 1991), and the sequence analysis was performed using an Applied Biosystems 470A instrument with reverse-phase HPLC of phenylthiohydantoin amino acids essentially as described (Kaiser et al., 1988).

Derivatization of peptides for mass spectrometry was performed as described (Lindh et al., 1994). Briefly, to a solution of the peptide in pyridine/HCl (pH 5), 1-ethyl-3-(3-dimethylaminopropyl) carbodiimide and 4-aminonaphthalenesulphonic acid were added. After incubation at 25°C for 2 h, the reaction was quenched with acetic acid and the naphthalenesulphonated peptide was purified using reverse-phase HPLC. Mass spectrometry was performed with a VG AutoSpec-QFPD double-focusing instrument equipped with an array detector. Mass spectra were recorded in the negative-ion mode with FAB or ES ionization. CID spectra were generated using helium as the collision gas in the first field free region gas cell and daughter ion B/E linked scans were recorded on the derivatized peptides (cf. Lindh et al., 1994).

Metal-binding experiments were performed at pH 7.5 in 50 mM phosphate buffer. Reduction of peptides was accomplished with a 20-fold molar excess of dithiothreitol (DTT) and incubation for 6 h at 37°C, after which the preparations were stored at -20°C until used. Spectral grade octa-hydrate of cobalt chloride was obtained from Johnson Matthey Chemicals. Exclusion chromatography on Bio-Gel P4 (BioRad) was used to remove DTT before

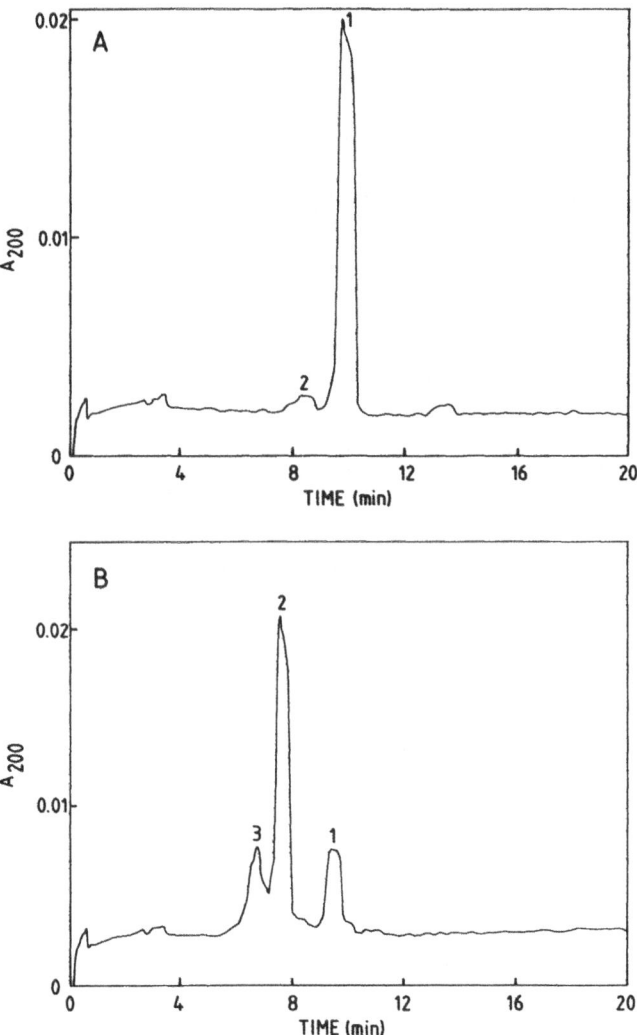

Figure 1. Capillary electrophoresis of the N-terminal alcohol dehydrogenase segment before deblocking (A), and after deacetylation for three days at 43°C (B). Peak 1 corresponds to the acetyl blocked peptide, peak 2 to the deacetylated peptide and peak 3 represents a cleavage after Gly (cf. Fig. 2).

incubation with metal. Absorption spectra were recorded at 25°C with a Beckman DU-64 spectrophotometer.

RESULTS AND DISCUSSION

A combination of trifluoroacetic acid and methanol was found to be efficient for N-terminal deacetylation of both a 14-residue synthetic peptide corresponding to the N-terminal segment of horse liver alcohol dehydrogenase and for the intact protein (374 residues), without appreciable cleavage of internal peptide bonds. The effect of temperature and reaction time on deacetylation was evaluated using capillary electrophoresis and sequence analysis. The results indicate that optimal deacetylation is achieved using a 1:1 mixture of trifluoroacetic acid and methanol for three days at 43°C. For the peptide, capillary electrophoresis before (Fig. 1A) and after (Fig. 1B) deblocking reveals a situation well suited for successful Edman degradation.

The deacetylated peptide fraction corresponds to the major peak, and represents a 65% removal of the acetyl group (peak 2, Fig. 1B), while the extent of undesirable internal peptide bond cleavage is low and corresponds to 18% of the total sample (peak 3, Fig. 1B). The sequencer initial yield was 37%, and the ratio of deblocking over unspecific cleavage was 7:1 calculated on the major side-cleavage after residue 4, Gly (Fig. 2). The intact alcohol dehydrogenase was treated according to the parameters found for the peptide, and sequence analysis revealed an initial yield of 60%.

Interestingly, despite the larger size (374 instead of 14 residues), the ratio of deblocking over unspecific cleavage of peptide bonds was similar to that of the peptide, 8:1 (Fig. 2). The results thus indicate that N-terminally acetylated proteins can be directly analyzed using Edman degradation after deblocking of the intact protein.

A synthetic segment of horse liver alcohol dehydrogenase, Ser-Phe-Glu-Val-Ile-Gly-Arg-Leu-Asp-Thr-Met-Val (peptide 1) corresponding to residues 265-276 of the parent molecule (cf. Eklund et al., 1976) and a synthetic tripeptide, Leu-Ala-Leu (peptide 2), were studied by negative-ion fast-atom bombardment and electrospray mass spectrometry. We have recently shown that by derivatizing peptides with aminosulphonic acids in a peptide linkage at the C-terminus, negative-ion formation can be enhanced and fragmentation in collision-induced dissociation reactions controlled (Lindh et al., 1994). When analyzing peptide 1 derivatized with 4-aminonaphthalenesulphonic acid using FAB ionization (Fig. 3A), the sensitivity for detection, compared to the underivatized peptide, is improved by a

	1 (Ac) Ser	2 - Thr	3 - Ala	4 - Gly	5 - Lys	6 - Val	7 - Ile	8 - Lys	
Peptide	196	342	1286	682	1317	834	619	734	(pmol)
Protein	221	313	1089	824	1238	1780	1177		(pmol)

	Peptide	Protein
Initial yield in the sequence analysis calculated for Ala	37%	60%
Estimated ratio of deblocking over unspecific cleavage	7:1	8:1

Figure 2. Sequence analysis of deacetylated polypeptides.

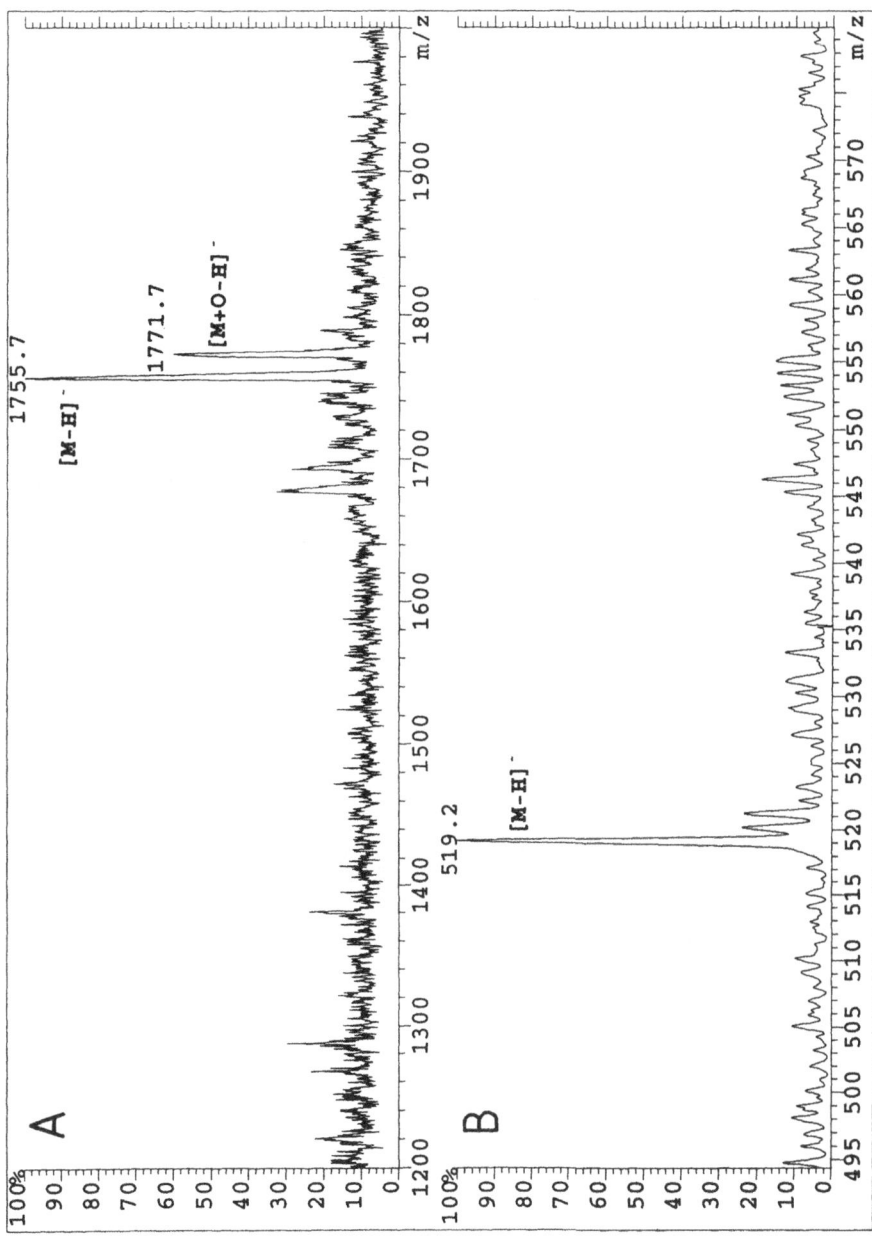

Figure 3. Negative-ion fast-atom bombardment mass spectrum of 50 picomole dinaphthalenesulphonated and dehydrated peptide 1 (A), and negative-ion electrospray mass spectrum (1 femtomole consumed) of naphthalenesulphonated peptide 2 recorded at the array detector (B).

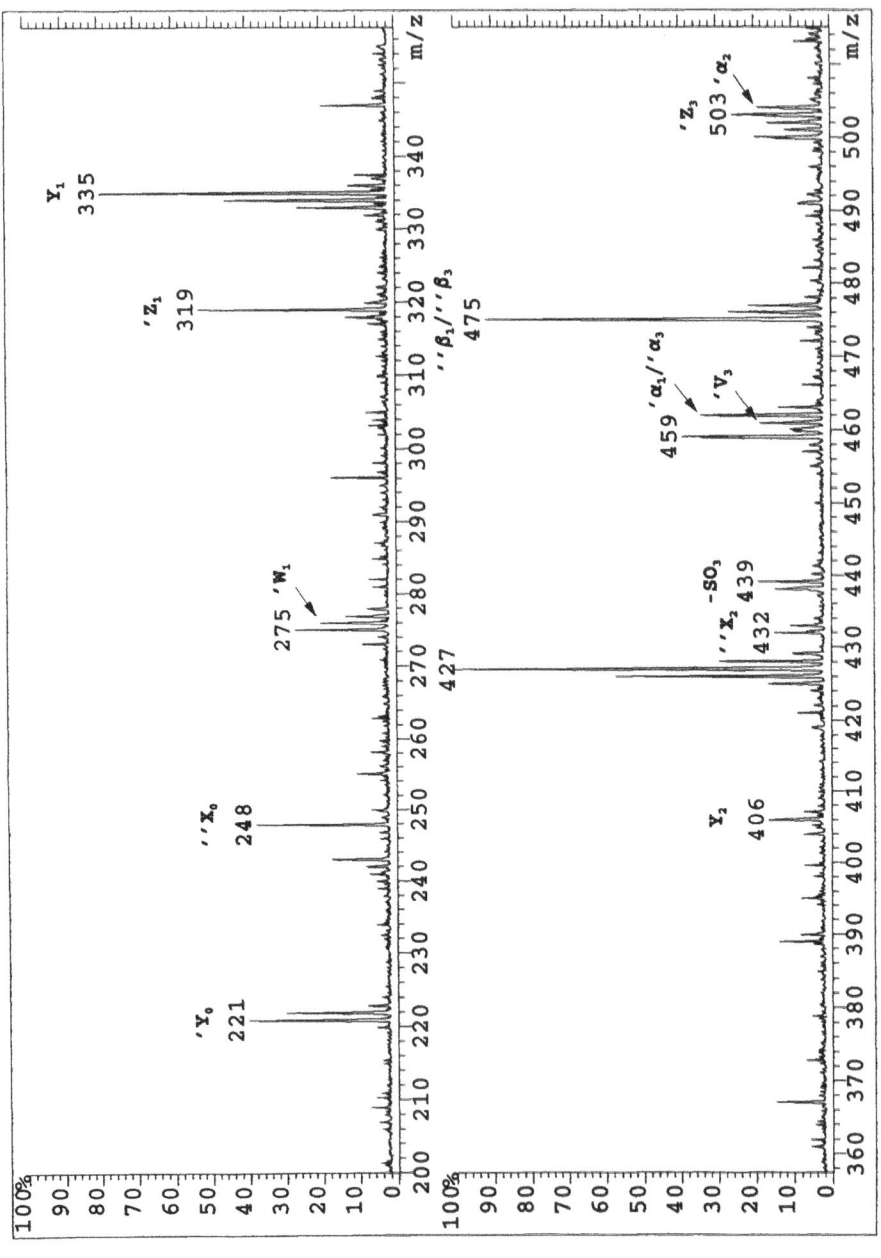

Figure 4. Fast-atom bombardment/collision-induced dissociation spectrum of [M-H]⁻ ions of 500 picomole naphthalenesulphonated peptide 2, m/z = 519. The symbols ' and " before letters indicate the loss of one or two hydrogens, respectively, from the molecule. The symbol * denotes a peak that arises from fragmentation of glycerol adducts. For explanation of letters and indexes, cf. Fig. 5.

factor of 50. The detection limit is further decreased to the low femtomole level with ES ionization and array detection, as revealed by the analysis of peptide 2 (Fig. 3B).

The main objective for incorporation of the naphthalenesulphonic acid group into the peptide, is to localise the charge at a defined position, and thereby prevent it from being directly involved in the fragmentation mechanism. The naphthalenesulphonic acid group is ideal in this respect since the charge is located on the SO_3 which is separated from the peptide by the rigid naphthalene group. This prevents the movement of charge along the peptide backbone. The result is that charge-remote fragmentation, rather than charge-mediated fragmentation, occurs. Negative-ion FAB/CID of 500 picomole naphthalenesulphonated peptide 2 (Fig. 4) shows charge-remote fragmentations "X, Y, 'Z, 'V and 'W (cf. Fig. 5).

These fragmentations are accompanied by side-chain fragmentations 'α and "β. The absence of daughter ions with the charge retained on the N-terminal fragment, makes the assignment of peaks less complicated and facilitates determination of the amino acid sequence.

A peptide model for the structural (non-catalytic) zinc site of horse liver alcohol dehydrogenase has been studied. We have previously shown that a 23-residue synthetic peptide, corresponding to the segment in the enzyme that binds the structural zinc atom, specifically binds zinc and cobalt in a manner that mimics the metal-binding characteristics of the intact protein (Bergman et al., 1992; Bergman et al., 1993). It corresponds to residues 93 - 115 of the parent molecule, and represents a four-residue extension of the protein zinc-containing loop (residues 97 - 111, cf. Eklund et al., 1976) at both ends. Cobalt was used as a spectral probe to closer examine the interaction between peptide and metal. Substitutions in which the paramagnetic coloured cobalt ion replaces the diamagnetic, colourless zinc atom, are effective in spectrally probing the coordination geometry of protein metal-binding sites (Vallee and Holmquist, 1980). Thus, a solution of the reduced peptide was mixed with Co^{2+}, and the absorbance spectrum was monitored (250 - 800 nm). This results in the appearance of a chromophore with maximal absorption at 310 and 675 nm and shoulders at 350, 640 and 720 nm, indicating the formation of a complex between the peptide and cobalt (Bergman et al., 1992; Bergman et al., 1993). The overall absorption pattern of the Co^{2+}/peptide complex in the region 500 - 800 nm is characteristic of tetrahedral ligand geometry (Vallee and Galdes, 1984), and the charge-transfer bands below 400 nm indicate sulphur coordination (Vallee and Galdes, 1984; Green and Berg, 1989). The visible absorption envelope, with a maximum at 675 nm and two shoulders at 640 and 720 nm (Bergman et al., 1992; Bergman et al., 1993), reveals a striking similarity to that of human liver $\beta_1\beta_1$ (Formicka-Kozlowska et al., 1988) and horse liver (Sytkowski and Vallee, 1976; Formicka-Kozlowska and Zeppezauer, 1988) alcohol dehydrogenases, both cobalt-substituted at the structural site only. Four analogues to this peptide have now been synthesized, each with

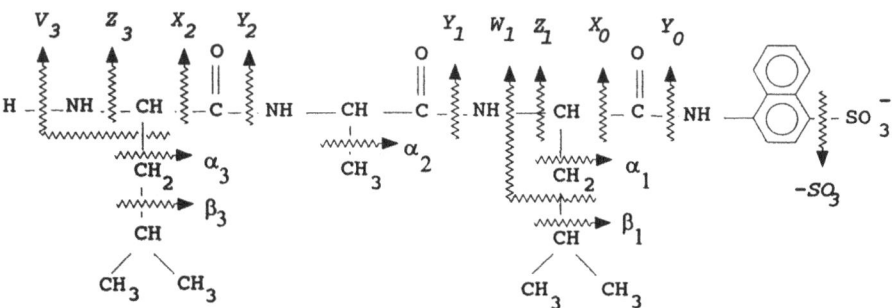

Figure 5. Fragmentation of naphthalenesulphonated peptide 2.

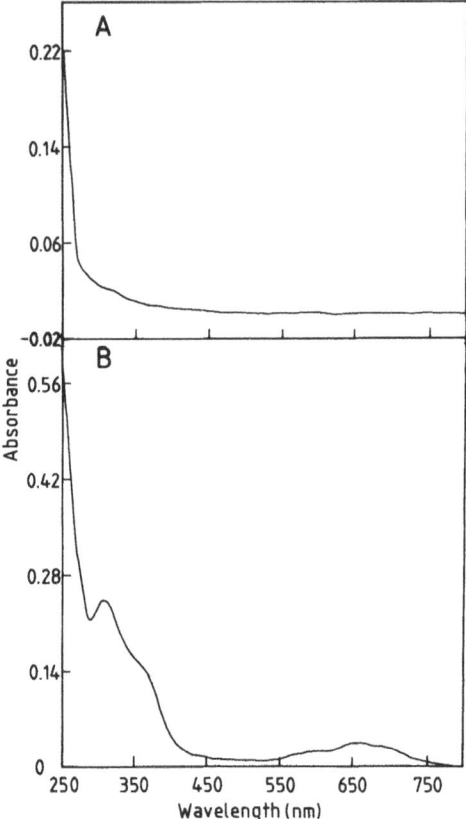

Figure 6. Absorption spectra recorded for ADH-A100: (A) peptide only; (B) after addition of cobalt to the peptide.

one of the four metal-liganding cysteine residues replaced with alanine, in order to determine the importance of individual cysteine residues for metal interaction and coordination. Preliminary tests were carried out with two of these analogues, ADH-A100 and ADH-A111, with Cys exchanged to Ala at positions 100 and 111, respectively (underlined): Phe-Thr-Pro-Gln-**Cys**-Gly-Lys-*Cys*-Arg-Val-**Cys**-Lys-His-Pro-Glu-Gly-Asn-Phe-*Cys*-Leu-Lys-Asn-A sp. Interestingly, when Co^{2+} was added to the reduced analogues, both peptides formed complexes with the metal (Fig. 6). The absorption spectra are similar but not identical to the spectrum obtained with the intact peptide without alanine substitutions. The charge-transfer bands below 400 nm are present, indicating sulphur coordination.

However, a close inspection of the chromophores in the region 500 - 800 nm reveals striking differences. For ADH-A100 (Fig. 7), the absorption envelope contains a maximum and two shoulders, as for the intact peptide, but they are all shifted to a lower wavelength, 675 to 660 nm, 640 to 600 nm and 720 to 690 nm (cf. above).

Also ADH-A111 reveals an absorption envelope significantly different from that of the intact, four cysteine-containing peptide. The maximal absorbance is shifted to 660 nm (as for ADH-A100), the shoulder at 640 nm is now appearing at 620 nm, and the shoulder at 720 nm seems to have disappeared (data not shown). The patterns obtained clearly show that a binding between cobalt and the peptide analogues takes place but with an altered coordination. A potential fourth ligand to cobalt is the glutamic acid residue at position 107.

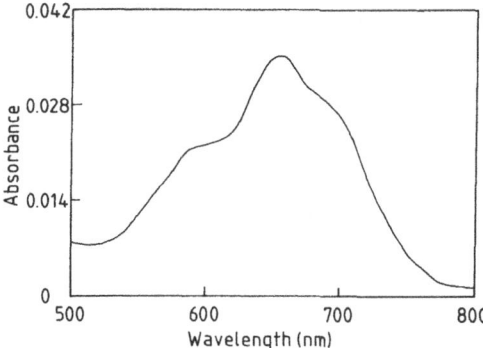

Figure 7. Absorption spectrum recorded for ADH-A100 after incubation with cobalt.

However, the present cobalt-interaction must be further studied employing the other two analogues and with magnetic circular dichroism and nuclear magnetic resonance techniques. An interesting finding was a difference in the resistance to oxidation, where ADH-A111 appears to be more stable than ADH-A100. The chromophore is more or less abolished for ADH-A100 upon a short (2 min) aeration of the solution containing the cobalt/peptide complex, while a significant proportion of the envelope remains when ADH-A111 is treated in the same manner (data not shown). Taken together, the results indicate a strategy for characterization of metal binding and coordination in alcohol dehydrogenases.

In conclusion, N-terminal deacetylation of intact proteins for direct Edman degradation, derivatization of fragments for mass spectrometric sequence determination and synthetic peptides as metal binding models are three efficient approaches to characterization of alcohol dehydrogenase structures.

ACKNOWLEDGMENTS

This work was supported by grants from the Swedish Medical Research Council (projects 13X-3532, 13X-10832, 03X-219 and 03Y-10445), Stiftelsen Lars Hiertas Minne and a travel grant to T.B. from Karolinska Institutet. Carina Palmberg is greatfully acknowledged for the preparation of synthetic peptides and for the drawings. Special thanks to VG Analytical, Manchester.

REFERENCES

Barber, M., Bordoli, R.S., Sedgwick, R.D., and Tyler, A.N., 1981, Fast atom bombardment of solids (F.A.B.): A new ion source for mass spectrometry, *J. Chem. Soc. Chem. Commun.*, 325.

Bergman, T., Agerberth, B., and Jörnvall, H., 1991, Direct analysis of peptides and amino acids from capillary electrophoresis, *FEBS Lett.*, 283:100.

Bergman, T., Jörnvall, H., Holmquist, B., and Vallee, B.L., 1992, A synthetic peptide encompassing the binding site of the second zinc atom (the "structural" zinc) of alcohol dehydrogenase, *Eur. J. Biochem.*, 205:467.

Bergman, T., Jörnvall, H., Härd, T., Holmquist, B., and Vallee, B.L., 1993, A synthetic approach to analysis of the structural zinc site of alcohol dehydrogenase, *in*: Enzymology and Molecular Biology of Carbonyl Metabolism 4, Weiner, H., Crabb, D.W., and Flynn, T.G., eds., Plenum Press, New York, p. 419.

Carr, S.A., Hemling, M.E., Bean, M.F., and Roberts, G.D., 1991, Integration of mass spectrometry in analytical biotechnology, *Anal. Chem.*, 63:2802.

Drum. D.E., Harrison, J.H., IV, Li, T.-K., Bethune, J.L., and Vallee, B.L., 1967, Structural and functional zinc in horse liver alcohol dehydrogenase, *Proc. Natl. Acad. Sci. USA*, 57:1434.

Drum, D.E., and Vallee, B.L., 1970, Differential chemical reactivities of zinc in horse liver alcohol dehydrogenase, *Biochemistry*, 9:4078.

Eklund, H., Nordström, B., Zeppezauer, E., Söderlund, G., Ohlsson, I., Boiwe, T., Söderberg, B.-O., Tapia, O., Brändén, C.-I., and Åkeson, Å., 1976, Three-dimensional structure of horse liver alcohol dehydrogenase at 2.4 Å resolution, *J. Mol. Biol.*, 102:27.

Fenn, J.B., Mann, M., Meng, C.K., Wong, S.F., and Whitehouse, C.M., 1989, Electrospray ionization for mass spectrometry of large biomolecules, *Science*, 246:64.

Formicka-Kozlowska, G., Schneider-Bernlöhr, H., von Wartburg, J.-P., and Zeppezauer, M., 1988, $H_8Zn(c)_2$ and $Zn(c)_2Co(n)_2$ human liver alcohol dehydrogenase, *Eur. J. Biochem.*, 173:281.

Formicka-Kozlowska, G., and Zeppezauer, M., 1988, Horse liver alcohol dehydrogenase derivatives containing nickel(II) and cobalt(II) in the noncatalytic metal binding site, *Inorg. Chim. Acta*, 151:183.

Green, L.M., and Berg, J.M., 1989, A retroviral Cys-Xaa$_2$-Cys-Xaa$_4$-His-Xaa$_4$-Cys peptide binds metal ions: spectroscopic studies and a proposed three-dimensional structure, *Proc. Natl. Acad. Sci. USA*, 86:4047.

Jörnvall, H., 1985, Use of peptides in studies of protein structures and functions, *in*: Synthetic Peptides in Biology and Medicine, Alitalo, K., Partanen, P., and Vaheri, A., eds., Elsevier, Amsterdam, p. 13.

Jörnvall, H., Höög, J.-O., von Bahr-Lindström, H., and Vallee, B.L., 1987a, Mammalian alcohol dehydrogenases of separate classes: intermediates between different enzymes and intraclass isozymes, *Proc. Natl. Acad. Sci. USA*, 84:2580.

Jörnvall, H., Persson, B., and Jeffery, J., 1987b, Characteristics of alcohol/polyol dehydrogenases: the zinc-containing long-chain alcohol dehydrogenases, *Eur. J. Biochem.*, 167:195.

Kaiser, R., Holmquist, B., Hempel, J., Vallee, B.L., and Jörnvall, H., 1988, Class III human liver alcohol dehydrogenase: A novel structural type equidistantly related to the class I and class II enzymes, *Biochemistry*, 27:1132.

Kent, S.B.H., 1988, Chemical synthesis of peptides and proteins, *Annu. Rev. Biochem.*, 57:957.

Lindh, I., Griffiths, W.J., Bergman, T., and Sjövall, J., 1994, Charge remote fragmentation of peptides derivatized with 4-aminonaphthalenesulphonic acid, *Rapid Commun. Mass Spectrom.*, in press.

Persson, B., Krook, M., and Jörnvall, H., 1991, Characteristics of short-chain alcohol dehydrogenases and related enzymes, *Eur. J. Biochem.*, 200:537.

Sytkowski, A.J., and Vallee, B.L., 1976, Chemical reactivities of catalytic and noncatalytic zinc or cobalt atoms of horse liver alcohol dehydrogenase: differentiation by their thermodynamic and kinetic properties, *Proc. Natl. Acad. Sci. USA*, 73:344.

Tsunasawa, S., and Hirano, H., 1993, Deblocking and subsequent microsequence analysis of N-terminally blocked proteins immobilized on PVDF membrane, *in*: Methods in Protein Sequence Analysis, Imahori, K., and Sakiyama, F., eds., Plenum Press, New York, p. 45.

Vallee, B.L., and Auld, D.S., 1990, Active-site zinc ligands and activated H_2O of zinc enzymes, *Proc. Natl. Acad. Sci. USA*, 87:220.

Vallee, B.L., and Auld, D.S., 1991, Zinc chemistry in function and structure of zinc proteins, *in*: Methods in Protein Sequence Analysis, Jörnvall, H., Höög, J.-O., and Gustavsson, A.-M., eds., Birkhäuser, Basel, p. 363.

Vallee, B.L., and Galdes, A., 1984, The metallobiochemistry of zinc enzymes, *Adv. Enzymol.*, 56:283.

Vallee, B.L., and Holmquist, B., 1980, Circular dichroism and magnetic circular dichroism, *Adv. Inorg. Biochem.*, 2:27.

Åkeson, Å., 1964, On the zinc content of horse liver alcohol dehydrogenase, *Biochem. Biophys. Res. Commun.*, 17:211.

50

PURIFICATION AND CHARACTERIZATION OF *S*-FORMYLGLUTATHIONE HYDROLASE FROM HUMAN, RAT AND FISH TISSUES

Martti Koivusalo, Risto Lapatto and Lasse Uotila

Departments of Medical Chemistry and Clinical
Chemistry, University of Helsinki
Helsinki, Finland

INTRODUCTION

Formaldehyde is oxidized in mammalian tissues and in several other sources to formate in two consecutive reactions catalyzed by separate enzymes (Uotila and Koivusalo, 1974a; 1974b) (Scheme 1). Formaldehyde and glutathione react first non-enzymically to form a hemithioacetal adduct (*S*-hydroxymethylglutathione) (Reaction 1). The adduct is the substrate of formaldehyde dehydrogenase (EC 1.2.1.1) and is oxidized in the NAD-dependent reaction to *S*-formylglutathione (Reaction 2). Formaldehyde dehydrogenase has been shown to be identical with the class III alcohol dehydrogenase (Koivusalo et al., 1989). For a comprehensive review on formaldehyde dehydrogenase, see Uotila and Koivusalo (1989). The hydrolysis of *S*-formylglutathione to reduced glutathione and formate (Reaction 3) is catalyzed by a specific enzyme, *S*-formylglutathione hydrolase (EC 3.1.2.12). Formate can be oxidized by the catalase reaction or is transferred to tetrahydrofolate to form 10-formyltetrahydrofolate. Formate is thus incorporated into the reactions of the C-1 metabolism by which it can be oxidized to carbon dioxide in the 10-tetrahydrofolate dehydrogenase reaction (Kutzbach and Stokstad, 1968).

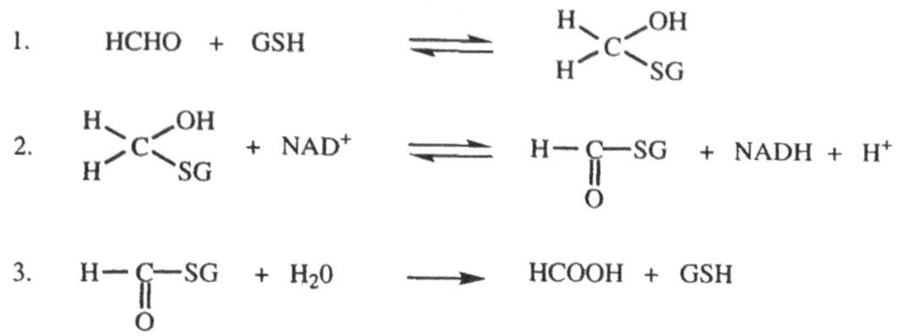

1. $HCHO + GSH \rightleftharpoons$ $\underset{H}{\overset{H}{>}}C\underset{SG}{\overset{OH}{<}}$

2. $\underset{H}{\overset{H}{>}}C\underset{SG}{\overset{OH}{<}} + NAD^+ \rightleftharpoons H-\underset{\underset{O}{\|}}{C}-SG + NADH + H^+$

3. $H-\underset{\underset{O}{\|}}{C}-SG + H_2O \longrightarrow HCOOH + GSH$

Scheme 1.

S-Formylglutathione hydrolase was first described and purified to homogeneity from human liver (Uotila, 1973a; Uotila and Koivusalo, 1974b). The enzyme was found to be highly specific for S-formylglutathione as the substrate. S-Formylglutathione hydrolase is also found in human erythrocytes (Uotila, 1979a; 1984), pea seeds (Uotila and Koivusalo, 1979), Escherichia coli (Uotila and Koivusalo, 1981; 1983), and in the methylotrophic yeasts Kloeckera sp. (Kato et al., 1980) and Candida boidinii (Neben et al., 1980).

Uotila (1984) reported in a Finnish population studied by electrofocusing a polymorphism of red cell S-formylglutathione hydrolase, apparently due to two alleles, FGH^1 and FGH^2 at an autosomal locus. The three phenotypes were characterized by a single cathodic band (FGH 1), a single anodic band (FGH 2) and a three-banded pattern (FGH 2-1), indicating a dimeric enzyme structure which had earlier also been found for the human liver enzyme (Uotila and Koivusalo, 1974b). Board and Coggan (1986) found by electrophoresis a similar type of polymorphism in Australian and some other populations. Later investigators found that the polymorphisms observed for S-formylglutathione hydrolase and esterase D were identical (Apeshiotis and Bender, 1986; Eiberg and Mohr, 1986; Akiyama and Abe, 1986; Tsuge et al., 1987), and it was suggested that S-formylglutathione hydrolase and esterase D are identical enzymes.

Esterase D (EC 3.1.1.56) was first described from human erythrocytes by Hopkinson et al. (1973) as a polymorphic esterase acting on short chain acyl esters of 4-methylumbelliferone but not on naphthyl, indoxyl or thiocholine esters. The enzyme has been investigated extensively by population geneticists, and its gene has been located on human chromosome 13q14.11. The enzyme has also been used as a genetic marker for retinoblastoma susceptibility (Sparkes et al., 1980) and for Wilson disease (Frydman et al., 1985) because of the close association between the genes defective in these diseases and the esterase D gene. The human esterase D gene has been cloned (Lee and Lee, 1986, Squire et al., 1986).

In this work we have purified and characterized S-formylglutathione hydrolase from a number of different sources. The relation of the human red blood cell S-formylglutathione hydrolase to esterase D has been investigated.

METHODS

S-Formylglutathione hydrolase activity was assayed spectrophotometrically at 25 °C with an assay mixture containing 50 mM sodium phosphate buffer pH 7.0, 0.5 mM S-formylglutathione and a suitable dilution of the enzyme preparation. Thiol ester hydrolysis was recorded at 240 nm. Similar assays were made with S-acetylglutathione and with other glutathione thiol esters as the substrates. The glutathione thiol esters were synthesized by the procedures of Uotila (1973a; 1981). Esterase D activity was assayed at 340 nm with an assay mixture containing 50 mM sodium phosphate pH 6.0, 0.1 mM 4-methylumbelliferyl acetate and the enzyme. The activities with p-nitrophenyl acetate and p-nitrophenyl thioacetate as the substrates were assayed at 405 nm with an assay mixture containing 50 mM sodium phosphate pH 7.0, 1 mM of the esters and the enzyme. Nonenzymic blanks were always included. The activity staining on gels for S-formylglutathione hydrolase activity was performed with S-acetylglutathione as the substrate (Uotila, 1979a; 1984). The activity staining for esterase D activity was performed according to Hopkinson et al. (1973).

RESULTS AND DISCUSSION

We purified S-formylglutathione hydrolase from human red blood cells. During the purification we always also assayed the column eluates and enzyme pools for esterase D activity to find out whether or not *S*-formylglutathione hydrolase and esterase D activities are due to a single enzyme. We observed that when the crude soluble fraction of human red cell hemolysate was fractionated on a Sephadex G-100 gel chromatography column, practically all (over 95 %) of the hydrolytic activities for both *S*-formylglutathione and for 4-methyumbelliferyl acetate were coeluted as single, identically located activity peaks slightly after hemoglobin but before ovalbumin used as a standard.

We attempted first the isolation of *S*-formylglutathione hydrolase from human red blood cells by a modification of the procedure of Lee et al. (1986) for esterase D. This did not produce a homogeneous enzyme preparation but electrophoretic homogeneity was achieved after two additional steps, column chromatographies on *p*-chloromercuribenzoate agarose and chromatofocusing (Table 1). The final enzyme preparation catalyzed the hydrolysis of both *S*-formylglutathione and of 4-methylumbelliferyl acetate but the activity with the former substrate was as much as 200-fold higher in comparison to the activity for the latter. A similar ratio of the activities was also found in the crude hemolysate, and the activity ratio for these two substrates remained remarkably constant throughout the purification procedure (Table 1). In the eluates of the various column chomatographies used, the activity peaks for *S*-formylglutathione and for 4-methylumbelliferyl acetate were always identically located. Activity staining of the purified enzyme after gel electrofocusing also produced identically located enzyme forms when the staining was performed with either *S*-acetylglutathione or with 4-methylumbelliferyl acetate as the substrate. The main enzyme form had by this technique a pI of 5.10 which is slightly more acidic than the value (5.4) reported for the human liver enzyme (Uotila and Koivusalo, 1974b). We conclude that *S*-formylglutathione hydrolase and esterase D are indeed identical enzymes in human red blood cells but *S*-formylglutathione is a much more reactive substrate than the artificial substrate 4-methylumbelliferyl acetate used in esterase D assays.

The purified enzyme had approximate M_r values of 53,000 by gel chromatography and 32,000 by dodecyl sulfate gel electrophoresis. These values are compatible with the dimeric structure deduced from genetic studies of the enzyme (Uotila, 1984), and are similar

Table 1. Purification of S-formylglutathione hydrolase from human red blood cells

Step	Activity (μmol x min^{-1})		Protein (mg)	Specific activity (μmol x min^{-1} x mg^{-1})		Yield of activity (%)	
	with formyl-SG	with 4-MUFA		with formyl-SG	with 4-MUFA	with formyl-SG	with 4-MUFA
Dialyzed hemolysate	9 605	46.2	18 560	0.517	0.0025	100	100
CM-Sepharose + amm. sulf. 30 to 90 %	7 037	34.2	875	8.04	0.0391	73.3	74.0
Phenyl-Sepharose	7 339	40.8	171.4	42.8	0.238	76.4	88.3
PBE 94 chromato-focusing I	6 103	33.3	38.8	157.3	0.858	63.5	72.1
Hydroxy-apatite	3 885	25.1	1.97	1972	12.7	40.4	54.3
PCMB-agarose	1 752	11.2	0.689	2543	16.2	18.2	24.2
PBE 94 chromato-focusing II	1 064	5.58	0.314	3390	17.8	11.1	12.1

The activities were assayed in each step with *S*-formylglutathione (formyl-SG) and with 4-methylumbelliferyl acetate (4-MUFA) as the substrates as described in text. PCMB, *p*-chloromercuribenzoate.

Table 2. Partial purification of S-formylglutathione hydrolase from rat liver

Step	Volume (ml)	Activity (U)	Protein (mg)	Specific activity (U/mg)	Yield (%)
Crude supernatant	205	5525	1847	2.99	100
Dialyzed supernatant	250	5538	1200	4.61	100
DEAE-Sephacel (concentrated)	11.3	5139	121.3	42.4	93
Sephadex G-100 (concentrated)	15.5	3416	16.4	208	62

40 g of liver was used for the purification.

to those reported for S-formylglutathione hydrolase from human liver (Uotila and Koivusalo, 1974b). Two groups have in accord with our results reported a dimeric structure with the subunit molecular weight of 34,000-35,000 for esterase D from human erythrocytes (Lee et al., 1986; Okada and Wakabayashi, 1988). Matsuo et al. (1985) in contrast claimed a monomeric structure (M_r approximately 30,000) for the native enzyme. No explanation for this discrepancy is evident.

The purified erythrocyte enzyme was easily inhibited by the thiol group reagents $HgCl_2$ and p-chloromercuribenzoate and was also sensitive to the amino group reagent 2,4,6-trinitrobenzenesulfonate. Phenylmethylsulfonyl fluoride, a serine esterase reagent was not inhibitory up to 1 mM. Similar results have been reported for the human liver enzyme (Uotila and Koivusalo, 1974b).

We have also found and partially purified S-formylglutathione hydrolase from a number of other sources including rat liver, several other rat tissues (kidney, brain, stomach, intestine, heart, spleen, muscle, lung), burbot liver and tissues from mussels, shrimp, crayfish (Astacus fluviatilis) and goldfish (Carassius auratus). All these sources contained quite high activities of S-formylglutathione hydrolase. The partial purifications from rat liver (Table 2) and burbot liver (Table 3) produced enzyme preparations free from other esterases. Therefore the substrate specificities of the preparations could be investigated (Table 4).

Similarly to the enzyme isolated from human erythrocytes, the rat and burbot liver enzymes also could use 4-methylumbelliferyl acetate as the substrate, but only at rates which were 180- to 390-fold lower than the rate with S-formylglutathione. In addition to the activity identified as esterase D, the rat liver extracts contained several other esterase activities which could significantly use 4-methylumbelliferyl acetate as the substrate but were not reactive with S-formylglutathione. Thus, while in the human red cell hemolysates S-formylglutathione hydrolase/esterase D was practically the only enzyme that significantly used 4-methylumbelliferyl acetate as the substrate, the same was not true for the rat liver extracts.

Table 3. Partial purification of S-formylgluthathione hydrolase from burbot liver

Step	Volume (ml)	Activity (U)	Protein (mg)	Specific activity (U/mg)	Yield (%)
Crude supernatant	108	2587	1055	2.45	100
DEAE-Sephacel (concentrated)	7.0	2168	80.5	26.9	84

26 g of liver was used for the purification.

Table 4. Substrate specificity of S-formylglutathione hydrolase purified
from various sources

Substrate	Enzyme isolated from		
	Human erythrocytes	Rat liver	Burbot liver
S-Formylglutathione	100	100	100
S-Acetylglutathione	0.40	0.30	0.26
S-Lactoylglutathione	0	0	0
4-Methylumbelliferyl acetate	0.54	0.56	0.30
p-Nitrophenyl acetate	0.24	0.74	0.20
p-Nitrophenyl thio-acetate	0.95	n.d.	n.d.

The relative activities given by the enzyme preparations with S-formylglutathione as
the substrate have been set to 100. The glutathione thiol esters were used at 0.5 mM
except for the burbot liver enzyme for which 0.3 mM was used (see text), 4-methyl-
umbelliferyl acetate at 0.1 mM, and p-nitrophenyl acetate and p-nitrophenyl thioacetate at 1 mM. n.d., not determined.

The hydrolysis of another oxygen ester, p-nitrophenyl acetate was also slowly catalyzed by
the S-formylglutathione hydrolase from all the sources studied (Table 4). S-Acetylglutathione was hydrolyzed at a rate of only 0.4 % of that for S-formylglutathione by the
erythrocyte enzyme and the rat and burbot liver enzymes gave even lower relative rates. The
burbot liver enzyme showed marked substrate inhibition at S-acetylglutathione concentrations higher than 0.3 mM; therefore the comparison of the thiol esters was made at their 0.3
mM concentrations. S-Lactoylglutathione, the glyoxalase II substrate, was not at all used as
the substrate by any of the enzymes studied (Table 4). It appears that the hydrolases from
the sources studied are highly specific for S-formylglutathione as the substrate similarly to
the enzyme earlier characterized from human liver Uotila and Koivusalo, 1974b). The liver
enzyme was also found to be inactive with coenzyme A thiol esters but could slowly use
formyl thioesters of some other thiols, the best of which was S-formyl-N-acetylcysteine (0.6
% relative activity; Uotila and Koivusalo, 1974b). One additional compound tested in this
work, p-nitrophenyl thioacetate, gave for the human erythrocyte enzyme a relative rate of
approximately 1 % of that for S-formylglutathione (Table 4). The thioacetate ester was thus
far the most reactive substrate after S-formylglutathione found for S-formylglutathione
hydrolase, although still 100 times slower than the principal substrate of the enzyme.

The K_m values determined for the burbot liver enzyme were 0.20 mM for S-formylglutathione, 0.019 mM for 4-methylumbelliferyl acetate and 0.49 mM for p-nitrophenyl
acetate. Although 4-methylumbelliferyl acetate thus gave a lower K_m value than S-formylglutathione, the V_{max}/K_m ratio of the enzyme was 36-fold higher for S-formylglutathione
than for 4-methylumbelliferyl acetate.

Varki et al. (1986) proposed the identity of a sialic acid-specific O-acetylesterase
with esterase D in human erythrocytes. The biochemical and immunological investigations
of Lee et al. (1986) on the purified esterase D, however, strongly suggested that the enzyme
studied by Varki et al. (1986) was not esterase D. The very high substrate specificity of
S-formylglutathione hydrolase from the various sources found in this work indicates that
S-formylglutathione which is produced in the formaldehyde dehydrogenase reaction is the
physiological substrate of S-formylglutathione hydrolase/esterase D.

GLUTATHIONE THIOL ESTERASES

The Enzyme Nomenclature (1992) lists four glutathione thiolesterases: *S*-2-hydroxyacylglutathione hydrolase (glyoxalase II, EC 3.1.2.6), glutathione thiolesterase (EC 3.1.2.7), *S*-formylglutathione hydrolase (EC 3.1.2.12) and *S*-succinylglutathione hydrolase (EC 3.1.2.13). (For a comprehensive review see Uotila, 1989). In addition to *S*-formylglutathione hydrolase, two other enzymes, glyoxalase II and *S*-succinylglutathione hydrolase have been purified to homogeneity and their enzymic properties characterized. All the purified enzymes are highly specific for thiol esters of glutathione. *S*-Succinylglutathione hydrolase is also highly specific for the acyl part of the thiol ester (Uotila, 1979b), whereas glyoxalase II can use several types of thiol esters of glutathione as the substrates although with differing efficiencies (Uotila, 1973b; 1989).

Glyoxalase II, which is the second enzyme of the glyoxalase system, was first purified to homogeneity from human liver by Uotila (1973b). The main isoform is a cytoplasmic monomeric enzyme with a molecular weight of 23,000 and a pI of 8.35. There are also several mitochondrial isoforms; the activity is found both in the intermembrane and the matrix space (Talesa et al., 1989). *S*-Succinylglutathione hydrolase is a cytoplasmic monomeric enzyme, with a molecular weight of 18,000 and a pI of 8.7. It was first described (Uotila, 1973a) and purified (Uotila, 1979b) from human liver. It has a low but easily measurable hydrolytic activity for the oxygen esters *p*-nitrophenyl acetate and 4-methylumbelliferyl acetate, and the enzyme is apparently identical with the activity described separately as esterase B_4 (Coates et al., 1976).

Glutathione thiol esters and thiol esterases have roles at least in the formaldehyde dehydrogenase and glyoxalase systems, and thus in the removal in the cells of the toxic formaldehyde and α-ketoaldehydes. The glutathione thiol esters are high energy compounds and could have additional roles in some acyl group transfer reactions.

REFERENCES

Akiyama, K., and Abe, K., 1986, Gene frequencies of *S*-formylglutathione hydrolase isozyme in a Japanese population, *Jpn. J. Human Genet.* 31: 353-355.

Apeshiotis, F., and Bender, K., 1986, Evidence that *S*-formylglutathione hydrolase and esterase D polymorphisms are identical. *Hum. Genet.* 74: 176 - 177.

Board, P.G., and Coggan, M., 1986, Genetic heterogeneity of *S*-formylglutathione hydrolase, *Ann. Hum. Genet.* 50:3 5-39.

Coates, P.M., Edwards, Y.H., and Hopkinson,D.A., 1976, Purification and properties of an esterase B_4 from human liver, *Eur. J. Biochem.* 61: 331-335.

Eiberg, H., and Mohr, J., 1986, Identity of the polymorphisms for esterase D and S-formylglutathione hydrolase in red cells, *Hum. Genet.* 74:174 -173.

Frydman, M., Bonné-Tamir, B., Farrer, L.A., Conneally, P.M., Magazanik, A. Ashbel, S., and Goldwitch, Z., 1985, Assignment of the gene for Wilson disease to chromosome 13: Linkage to the esterase D locus. *Proc. Natl. Acad. Sci. USA* 82: 1819-1821.

Enzyme Nomenclature 1992, Academic Press Inc., London.

Hopkinson,D.A., Mestriner, M.A., Cortner, J., and Harris, H., 1973, Esterase D: a new human polymorphism, Ann. Hum. Genet., Lond., 37: 119-137.

Kato, N., Sakazawa, C.,Nishizawa, T., Tani, Y.,and Yamada, H., 1980, Purification and characterization of *S*-formylglutatione hydrolase from a methanol-utilizing yeast, Kloeckera sp. No. 2201, *Biochim. Biophys. Acta* 611: 323-332.

Koivusalo, M., Baumann, M., and Uotila, L., 1989, Evidence for the identity of glutathione-dependent formaldehyde dehydrogenase and class III alcohol dehydrogenase, *FEBS Lett.* 257: 105-109.

Kutzbach, C., and Stokstad, E.L.R., 1968, Partial purification of a 10-formyl-tetrahydrofolate:NADP oxidoreductase from mammalian liver, *Biochem.Biophys.Res.Comm.* 30:111-117.

Lee, E.Y.-H., and Lee, W.H., 1986, Molecular cloning of the human esterase D gene, a genetic marker of retinoblastoma, *Proc. Natl. Acad. Sci. USA* 83: 6337-6341.

Lee, W.-H., Wheatley, W., Benedict, W.F., Huang, C.-M., and Lee, E.Y.-H.P., 1986, Purification, biochemical characterization, and biological function of human esterase D, *Proc. Natl. Acad.Sci. USA* 83: 6790-6794.

Matsuo, K., Kobayashi, K., Hagiwara, K., and Kajii, T., 1985, Purification and characterization of esterases D1 and D2 from human erythrocytes. Evidence that they are monomers. *Eur. J. Biochem.* 153:217-222.

Neben, I., Sahm, H., and Kula, M.-R., 1980, Studies on an enzyme, *S*-formylglutathione hydrolase, of the dissimilatory pathway of methanol in Candida boidinii, *Biochim.Biophys.Acta* 614:81-91.

Okada, Y.,and Wakabayashi, K., 1988, Purification and characterization of esterases D-1 and D-2 from human erythrocytes, *Arch. Biochem. Biophys.* 263: 130-136.

Sparkes, R.S., Sparkes, M.C., Wilson, M.G., Towner, J.W., Benedict, W., Murphree, A.L., and Yunis, J.J., 1980, Regional assignment of genes for human esterase D and retinoblastoma to chromosome band 13q14. *Science* 208: 1042-1044.

Squire, J., Dryja, T.P., Dunn, J., Goddard, A., Hofmann, T., Musarella, M., Willard, H.F., Becker, A.J., Gallie, B.L., and Phillips, R.A., 1986, Cloning of the esterase D gene: A polymorphic gene probe closely linked to the retinoblastoma locus on chromosome 13, *Proc. Natl. Acad. Sci. USA* 83: 6573-6577..

Talesa, V., Uotila, L., Koivusalo, M., Principato, G., Giovannini, E., and Rosi, G., 1989, Isolation of glyoxalase II from two different compartments of rat liver mitochondria. Kinetic and immunochemical characterization of the enzymes, *Biochim. Biophys. Acta* 993:7-11.

Tsuge, A., Uchida, H., and Ishimoto, G. , 1987, Erythrocyte *S*-formylglutathione hydrolase polymorphism in Japanese and the relation to esterase D polymorphism. *Jpn. J. Legal Med.* 41: 93-96.

Uotila, L.,1973a, Preparation and assay of glutathione thiol esters. Survey of human liver glutathione thiol esterases, *Biochemistry* 12:3938-3943.

Uotila, L., 1973b, Purification and characterization of *S*-2-hydroxyacylglutathione hydrolase (glyoxalase II) from human liver, *Biochemistry* 12: 3944-3951.

Uotila, L., 1979a, Glutathione thiol esterases of human red blood cells. Fractionation by gel electrophoresis and isoelectric focusing. *Biochim.Biophys.Acta* 580:277-288.

Uotila, L., 1979b, Purification and properties of *S*-succinylglutathione hydrolase from human liver, *J.Biol.Chem.* 254: 7024-7029.

Uotila, L., 1981, Thioesters of glutathione, *Meth.Enzymol.* 77: 424-430.

Uotila, L., 1984, Polymorphism of red cell *S*-formylglutathione hydrolase in a Finnish population, *Hum. Hered.* 34: 273-277.

Uotila, L., 1989, Glutathione thiol esterases, in: "Coenzymes and Cofactors, vol.III, Glutathione. Chemical, Biochemical and Medical Aspects, part A", D. Dolphin, R. Poulson, and O. Avramovic, eds., pp.767-804, John Wiley & Sons, Inc., New York.

Uotila, L., and Koivusalo, M., 1974a, Formaldehyde dehydrogenase from human liver. Purification, properties, and evidence for the formation of glutathione thiol esters by the enzyme, *J. Biol Chem.* 249: 7653-7663.

Uotila,L., and Koivusalo, M., 1974b, Purification and properties of *S*-formylglutathione hydrolase from human liver, *J.Biol.Chem.* 249:7664-7672.

Uotila, L., and Koivusalo, M., 1979, Purification of formaldehyde and formate dehydrogenases from pea seeds by affinity chromatography and *S*-formylglutathione as the intermediate of formaldehyde metabolism, *Arch.Biochem.Biophys.* 196:33-45.

Uotila, L., and Koivusalo, M., 1981, *S*-Formylglutathione hydrolase, *Meth. Enzymol.* 77:320-325.

Uotila, L., and Koivusalo, M., 1983, Formaldehyde dehydrogenase, *in*: "Functions of Glutathione. Biochemical, Physiological, Toxicological and Clinical Aspects", A. Larsson, S. Orrenius, A. Holmgren, B. Mannervik, eds., pp. 175-186, Raven Press, New York.

Uotila, L., and Koivusalo, M., 1989, Glutathione-dependent oxidoreductases: formaldehyde dehydrogenase, in: "Coenzymes and Cofactors, vol. III, Glutathione. Chemical, Biochemical and Medical Aspects, part A", D. Dolphin, R. Poulson and O. Avramovic, eds., pp. 517-551, John Wiley & Sons, Inc., New York.

Varki, A., Muchmore, E., and Diaz, S., 1986, A sialic acid-specific O-acetylesterase in human erythrocytes: Possible identity with esterase D, the genetic marker of retinoblastomas and Wilson disease, *Proc. Natl. Acad. Sci. USA* 83: 882-886.

51

USEFUL MUTANTS OF *ZYMOMONAS MOBILIS* ALCOHOL DEHYDROGENASE-2 OBTAINED BY THE USE OF POLYMERASE CHAIN REACTION RANDOM MUTAGENESIS

Peter Rellos, Bernadette Schwindt and Robert Scopes

School of Biochemistry
La Trobe University
Bundoora, VIC 3083, Australia

INTRODUCTION

The alcohol dehydrogenase-2 isozyme of the ethanologenic bacterium *Zymomonas mobilis* (ZADH-2) is in the class of iron-activated dehydrogenases (Scope, 1983; Neale, *et al.*, 1986; Reid and Fewson, 1994). These enzymes have not yet been characterised structurally except by sequence. No certain sequence similarities exist which allow the identification of active sites or other structural features. ZADH-2 can be purified in large amounts, and the ferrous ion substituted with cobaltous, which stabilizes the product (Neale et *al.*, 1986: Tse, *et al.*, 1989). The gene can be expressed at a high level in *Escherichia coli*, up to 30% of the soluble protein (Neale, *et* al.,1988). We have crystallized the enzyme and anticipate that full structural determination may be possible.

We have developed a simple procedure for random mutagenesis of genes using the polymerase chain reaction, which enables base substitution at a rate of about 1 in 500 (Rellos and Scopes, 1994). By mutating the ZADH-2 gene in this way, a mutant library can be created in which the majority of species contain one or more base changes distributed at random along the gene. Provided that there is a screening method that is able to discriminate the desired mutant from the wild-type and unwanted mutants, then this desired mutant can be picked out of many thousands of clones. Alcohol dehydrogenase activity can be detected by various colour reactions; we have used the tetrazolium salt staining method, after lysing colonies on nitrocellulose filters. The present report outlines some results to date of screening and selection of thermostable mutants of ZADH-2, and selection of mutants that are active on substrates which the wild-type has little activity with.

Enzymology and Molecular Biology of Carbonyl Metabolism 5
Edited by H. Weiner *et al.*, Plenum Press, New York, 1995

CREATION OF MUTANT LIBRARIES

Plasmid pPR6, which is pUC18 containing an insert encoding ZADH-2, was used as template for the polymerase chain reaction. Primers were the forward and reverse sequencing primers. The concentrations of deoxynucleotides in the PCR reaction were varied, with either dCTP or dATP restricted to approximately 10 % of the concentrations of the other three nucleotides (Rellas and Scopes, 1994). 30 cycles of PCR under standard conditions produced DNA, which was trimmed with *Eco*R1 and *Pst*I, and ligated into cut pUC18. The ligation mixture was used to transform *Escherichia coli* strain DH5. Transformed cells were plated out, and nitrocellolose lifts made. Cells were lysed on the filters in conditions that retained the activity of the alcohol dehydrogenase while bound to the nitrocellulose. Activity was detected using ethanol, NAD$^+$, phenazine methosulphate and nitroblue tetrazolium.

SCREENING AND SELECTION OF MUTANTS

The initial objective of this research program was to isolate thermostable mutants of alcohol dehydrogenase. This was done by pre-treating the nitrocellulose filters at a temperature that was just sufficient to denature all the wild-type enzyme. Most such treatments have been carried out at slightly basic pH, in the range 8 to 9, in which the enzyme is somewhat less stable than at pH 6 to 7, but which is in the range that the enzyme must be used for ethanol oxidation. The wild-type enzyme was denatured by incubation for 30 min at 60, but mostly remained active at 55. Screening at 60 of the first mutant library we made led to the isolation of seven thermostable mutants which survived the heat treatment, several of which were studied in detail. Many other libraries have since been made and screened; in some cases thermostable variants occurred at a frequency of as high as 1 in 200.

Later experiments have sought alternative novel properties, in particular the utilization of substrates that the wild-type is virtually inactive on. We have isolated mutants that have increased activity with 1-butanol, and possibly ones with activity on ethanediol and 1,2-propanediol. These alcohols are only acted on very weakly by the wild-type enzyme. We also have NADP$^+$-positive mutants, and mutants that resist inactivation by zinc ions.

SEQUENCING AND CHARACTERISATION OF MUTANTS

Many mutants have been characterised by complete sequencing of the gene, and by kinetic and stability measurements of the extracted and purified enzyme. Sequencing is best carried out by running adjacent A-tracks, etc. of many different mutants; it is then easy to spot where the mutations occur. Oligonucleotides that hybridised at convenient positions along the sequence were used as primers for the cycle sequencing procedure (Promega *fmol*) on double-stranded plasmids.

Thermostability was tested at 10 g/ml ZADH using a continuous sampling method to detect remaining activity. The first order rate constant was measured directly from the exponential decay of activity with time. The addition of 1 mg/ml BSA protected against variable surface denaturation effects at low protein concentration.

Kinetic constants were determined for ethanol oxidation at pH 9.0.

Table 1. Properties of some of the thermostable mutants. Thermostability increase is calculated by measuring the rate constant for denaturation at 60° or 63° in a 50 mM phosphate buffer at pH 7.5. This is then compared to with the wild-type rate, assuming a value of 1.75 per degree (Rellos and Scopes, 1994) for the rate of increase in rate of denaturation wiht temperature for all mutants. Reproducibility of measurement = ± 0.2°. The denaturation rate for the wild-type at 60° is 0.13 min[-1]

Mutant ZADH	Thermostability increase, degrees	Amino acid changes from wild-type
TS-1	3.1	F9S, V295A
TS-2	3.6	M13I,E19K,M192I
TS-5	2.1	F90L
TS-7	4.0	F9S, H355L,A362V
TS-18	3.6	n.d.
TS-25	2.6	n.d.
TS-34	3.4	n.d.
TS-101[a]	3.1	F9S
TS-206[b]	5.5	n.d.

[a] Hybrid of TS-1 with wild-type
[b] 2nd-round mutant from TS-101, with the F9S change plus some other(s) from the second round.

THERMOSTABLE MUTANTS

Some 50 thermostable mutants have been isolated and confirmed by re-screening. In addition, hybrid thermostable mutants have been made by digesting the gene with *Mlu*I, a site nearly half-way down the coding region, and recombining the thermostable half with the wild-type half. Of those investigated, the thermostability predominantly appears to be associated with changes close to the N-terminus. Second-round mutants were made by taking a thermostable variant and re-mutating it with an alternative deoxynucleotide constraint. For instance, A mutant isolated from a dCTP-limiting library was subjected to a second round with limiting dATP.

The method was capable of detecting mutants that had an increased stability of from 1.0 upwards. The strongest-staining clones were studied in detail, and most of these selected first-round mutants had stability increases in the range 2.0 to 4.0. Second-round mutants have reached 5 to 6 greater stablility than the wild-type. Not surprisingly, the second-round mutants, screened at a higher temperature, are less frequent, ocurring at less than 1 in 1000 colonies.

A summary of some of the properties of the thermostable mutants that have been characterised is given in Table 1.

ALTERNATIVE SUBSTRATE MUTANTS

Screenings were made using a variety of alternative alcohols as substrates. Many weakly-positive mutants were found using 1-butanol, ethanediol and 1,2-propanediol as

Table 2. Properties of butanol mutants. Activities are given as percentage compared with activity on ethanol, at 1% substrate concentration, 1 mM NAD$^+$, pH 9.0

ZADH	1-butanol
Wild-type	0.15
BU-1[a]	1.0
BU-2	10
BU-3	10
BU-7	2.7

[a]Mutant BU-1 has two amino acid changes, I106T and A161V. The others have not yet been sequenced.

substrates. Some of the butanol mutants were investigated in more detail, with results as shown in Table 2. Two mutants had butanol activity of around 10% that of ethanol; these mutants are probably identical.

The wild-type enzyme is specific for NAD$^+$, and no activity is found with NADP$^+$. Screening the libraries using NADP$^+$ has found several strong positives, which are capable of utilizing either NAD$^+$ or NADP$^+$. The relative rates and K_m values for the nucleotides for two of these mutants are given in Table 3. In both cases, the K_m for NAD$^+$ was increased compared with the wild-type, and the K_m for NADP$^+$ was approximately 4-fold lower than that for NAD$^+$. The relative rates were such that the k_{cat}/K_m was the same for both nucleotides. The absolute rates have not yet been determined, but the NAD$^+$ rates are estimated to be similar to that of the wild-type.

METAL ION REQUIREMENT

ZADH-2 is naturally an Fe^{2+}-activated enzyme, which loses activity rapidly and irreversibly on exposure to the atmosphere (Scopes, 1983). This loss is prevented by exchange with Co^{2+} ions during extraction, and the Co-substituted enzyme has about 40% of the activity of the Fe-enzyme (Tse, *et al.*, 1989) Unlike other alcohol dehydrogenases, zinc is not only ineffective, but it inactivates the Fe- or Co-enzymes by displacing these metal ions. We have found mutants that are resistant to this inactivation. It is not yet possible to say whether this is due to a much tighter binding of the Fe or Co, so that displacement is slower, or whether the mutants are in fact also active with zinc ions. Preliminary evidence indicates that we have both types of mutants.

LOCATION OF THE MUTATIONS

The positions of the mutations have been determined by total gene sequencing of several mutants. Those thermostable mutants that have been sequenced were found to have alterations close to the N-terminus. Some of the other mutations further along the gene may be incidental. Hybrids of TS-1 and TS-2 with the wild-type were made retaining only the N-terminal mutations; these hybrids had the same thermostablity as the mutant parents (Table 1). The butanol-positive BU-1 had two changes, I106T and A161V, at least one of which must be located close to or at the alcohol binding site.

Table 3. Properties of NADP-positive mutants. Km values were obtained at pH 9.0 using 1 M ethanol as substrate

ZADH	K_m for NAD^+ mM	K_m for $NADP^+$ mM	Ratio $V_{max}s$ $NADP^+/NAD^+$ at 1 mM
Wild-type	0.10	No reaction	0
NP-5	0.18	0.04	0.28
NP-6	0.25	0.06	0.30

GENERAL APPLICABILITY OF RANDOM MUTAGENESIS

The creation of desirable mutants of enzymes can be approached in two ways. First, by designing the properties from full knowledge of the three-dimensional structure, using site-directed mutagenesis, and testing the resultant mutant for the correct property. Second, one can use random mutagenesis coupled with selective screening, in which the only mutants detected are ones that have the desired property. The main limitation of the site-directed method is that considerable detail is needed of the structure before any rational directed mutagenesis can be done; even then it may be difficult to design what is desired. One of the limitations of the random approach is that with most methods only occasional single base changes are achieved, which means that most amino acid changes are not possible, especially when there is considerable bias towards base transitions rather than transversions. To date, most of the methods available for creating random mutant libraries have been laborious; this PCR-based procedure is very simple, and the limiting factor in obtaining mutants is the screening. Although we have not overcome the problem of limited possible mutants due to single base changes, once the general site of useful mutation has been identified, a more rational site-directed cassette mutagenesis can be used to obtain libraries with all possible amino acids at specific positions.

Using more cumbersome methods for obtaining mutant libraries, thermostable mutants of subtilisin (Brayan, *et al.*, 1986) and of glucose dehydrogenase (Makino, *et al.*, 1989), among others, have been produced by random mutagenesis, using screening procedures similar to ours. One great advantage of random mutagenesis is that only successful mutants are isolated; no time is spent investigating unsuccessful mutants.

We are now carrying out random mutagenesis on other enzymes. One aim is to screen for inhibitor resistance, using an enzyme that is the target of an inhibitory drug or pesticide. Unfortunately ZADH-2 is not inhibited by pyrazole, but a pyrazole-sensitive alcohol dehydrogenase could be used with the same technology to test whether inhibitor resistance can be evolved by random mutagenesis.

REFERENCES

Bryan, P.N., Rollence, M.L., Pantolino, M.W., Wood, J., Finzel, B.C., Gillibrand, G.L., Howard, A.J. & Poulos, T.L. (1986) Proteases of enhanced stability: Characterization of a thermostable variant of subtilisin. Proteins Struct. Funct. Genet. **1**, 326-334

Makino, Y., Negoro, S., Urabe, I. & Okada, H. (1989) Stability-increasing mutant of glucose dehydrogenase from *Bacillus megaterium*. J. Biol. Chem. **264**, 6381-6385

Neale, A.D., Scopes, R.K. & Kelly, J.M. (1988) Alcohol production from glucose and xylose using *Escherichia coli* containing *Zymomonas mobilis* genes. Appl.Microbiol. Biotechnol. *29*, 162-167

Neale, A.D., Scopes, R.K., Kelly, J.M. & Wettenhall, R.E.H. (1986) The two alcohol dehydrogenases from *Zymomonas mobilis*: purification by differential dye-ligand chromatography, molecular characterization and physiological roles. Eur. J. Biochem. *154*, 119-124

Reid, M.F. & Fewson, C.A. (1994) Molecular characterization of microbial alcohol dehydrogenases. Crit. Rev. Microbiol. **20**, 13-56

Rellos, P. & Scopes, R.K. (1994) Thermostable mutants of *Zymomonas mobilis* alcohol dehydrogenase-2 created by random mutagenesis using polymerase chain reaction technology. Protein Exp. Purif. **5**, 270-277

Scopes, R.K. (1983) An iron-activated alcohol dehydrogenase. FEBS Letts. *156*, 303-306

Tse, P., Scopes, R.K. & Wedd, A.G. (1989) Iron-activated alcohol dehydrogenase from *Zymomonas mobilis*: isolation of apoenzyme and metal dissociation constants. J. Am. Chem. Soc. *111*, 8703-8706

INDEX

The manufacturer's authorised representative in the EU is Springer
Nature Customer Service Centre GmbH, Europaplatz 3, 69115 Heidelberg,
Germany. If you have any concerns regarding our products, please
contact ProductSafety@springernature.com

Printed and bound by CPI Group (UK) Ltd, Croydon, CR0 4YY
23/04/2026
02095622-0005